Dark Matter and Dark Energy

Astrophysics and Space Science Library

For other titles published in this series, go to
www.springer.com/series/5664

Sabino Matarrese • Monica Colpi
Vittorio Gorini • Ugo Moschella
Editors

Dark Matter and Dark Energy

A Challenge for Modern Cosmology

Editors

Sabino Matarrese
Dipartimento di Fisica G. Galilei
Università degli Studi di Padova
Via Marzolo 8
35131 Padova
Italy

Monica Colpi
Dipartimento di Fisica G. Occhialini
Università de Milano Bicocca
Piazza della Scienza 3
20126 Milano
Italy

Vittorio Gorini
Dipartimento di Fisica e Matematica
Università dell'Insubria
Via Valleggio 11
22100 Como
Italy

Ugo Moschella
Dipartimento di Fisica e Matematica
Università dell'Insubria
Via Valleggio 11
22100 Como
Italy

Published by Springer,
P.O. Box 17, 3300 AA Dordrecht, The Netherlands
In association with
Canopus Academic Publishing Limited,
15 Nelson Parade, Bedminster, Bristol, BS3 4HY, UK

www.springer.com and www.canopusbooks.com

ISSN 0067-0057
ISBN 978-90-481-8684-6 e-ISBN 978-90-481-8685-3
Springer Dordrecht Heidelberg London New York

Library of Congress Control Number: 2011920807

Cover illustration: The planck one-year all-sky survey, printed with kind permission of © ESA, HFI and LFI consortia, July 2010

Printed on acid-free paper

Springer is part of Springer Science+Business Media (www.springer.com)

Contents

Introduction

Sabino Matarrese

The dark side of the Universe

This book aims at presenting a thorough and up-to-date introduction to the fundamental theoretical and observational aspects of the two dark components of the Universe: the dark matter and the dark energy.

During the last decades, many independent observations have provided growing evidence that the Universe is filled with two "dark" ingredients, a collisionless component, able to cluster on sub-horizon scales, called dark matter and an almost uniform component with negative pressure, called dark energy, whose physical nature is still largely unknown. The dark matter component yields almost one-quarter of the total cosmic energy today, while the dark energy is responsible for about 70%. The sum of their contribution to the present-day cosmic energy budget is just impressive: around 96% of the total. The visible material, to which physicists and astronomers paid all of their attention for millennia, appears now as a sort of minor "detail" in the cosmos. Indeed, even the majority of the overall ordinary, baryonic material is invisible to our telescopes: we have indications that at least half of it is concentrated in a network of thin interconnected large-scale filaments/sheet-like structures, the so-called Warm-Hot Inter-Galactic Medium, which we hope to detect with the next generation of X-ray satellites.

The discovery that almost three-quarters of the present cosmic energy density is to be ascribed to an almost uniform dark energy component able to produce, via its negative isotropic pressure, the accelerated expansion of the Universe, represents the most severe crisis of contemporary physics. At the same time, this discovery opens the door to new theoretical speculations and represents the new frontier for observational cosmology, in the joint effort to constrain the dynamical properties of

Sabino Matarrese
Dipartimento di Fisica "G. Galilei", Università degli Studi di Padova and INFN, Sezione di Padova
via Marzolo 8, I-35131 Padova, Italy
e-mail: sabino.matarrese@pd.infn.it

the dark cosmic components and to unveil their physical nature. In this sense, the discovery of the "dark side" of the Universe, represents a formidable challenge for the cosmological research of the 21st Century.

The evidence that the vast majority of the Universe's energy is due to dark components appears as a sort of extreme consequence of the Copernican Principle: not only did we human beings have to accept that we do not live at the center of the Universe, but we are also gradually getting acquainted with the idea that the "ordinary" matter we are made of represents a negligible ingredient in the Universe's composition, playing a minor role in the global dynamics of the Universe.

The very fact that cosmologists had to change so radically their view on the Universe's matter content is also having a strong impact on the more general problem of cosmological model building. Indeed, many cosmologists recently started to question the strict validity of the standard model of the Universe based on the spatially homogeneous and isotropic Friedmann–Lemaître–Robertson–Walker (hereafter FLRW) solution of Einstein's field equations and considered more general/less symmetric background solutions as a viable possibility. The motivation for this new direction of investigation is twofold: observationally based and purely theoretical. From the observational point of view, some large angular scale "anomalies" detected in the Cosmic Microwave Background anisotropy pattern triggered the analysis of alternative background models, such as the homogeneous but anisotropic Bianchi solutions. On the theoretical side, the search for an alternative interpretation of the cosmic acceleration at late times, not requiring dark energy at a fundamental level, stimulated several groups to wonder whether our observable patch of the Universe could be better described by some inhomogeneous and anisotropic background metric. The underlying idea of such an approach is that by averaging over a large (e.g. Hubble radius-size) spatial volume such a non-FLRW metric and fitting it to a FLRW cosmology one unavoidably obtains extra "back-reaction" terms in the effective Friedmann equations (or, alternatively, in cosmological observables, such as the luminosity distance-redshift relation), that would possibly *mimic* a dark energy component. Whether such a back-reaction effect has the right equation of state (with negative pressure) and is large enough to explain the present-day accelerated phase of the cosmic expansion is a matter of controversy: much work has to be done yet before we can have have a realistic inhomogeneous and anisotropic alternative to the FLRW model able to recover the many successes of the standard cosmology and at the same time to explain the cosmic acceleration/dark energy puzzle.

Outline of the book

The book is organized in three, largely complementary, parts.

I Cosmology

The first part of the book starts with a chapter by Norbert Straumann, devoted to the introduction of some fundamental concepts in modern cosmology: cosmic inflation, a phase of accelerated expansion in the early Universe, which plays a fundamental role in providing a causal mechanism for the generation of cosmological perturbations, the seeds which gave rise to all cosmic structures, and to the anisotropies in temperature and polarization of the Cosmic Microwave Background (hereafter CMB) radiation. An introduction to the theory of linear gauge-invariant perturbations and of CMB anisotropies is also provided. This general introduction serves as a background to the analysis of the most important cosmological observables: the CMB and the large-scale structure distribution of matter, as revealed by the spatial clustering of galaxies, which are discussed in the chapter by Licia Verde. Another important cosmological observable, i.e. the gravitational lensing of light by the intervening matter distribution is introduced in the following chapter, written by Alan Heavens. These cosmological observables play a crucial role in constraining the amount and type of dark matter, but they also allow us to place formidable constraints on the physical properties of the dark energy component and of possible alternatives to the latter in the form of modifications of the theory of gravitation with respect to general relativity. This part of the book ends with a chapter, by Lauro Moscardini and Klaus Dolag, which presents an introduction to the techniques and the main results of numerical simulations of the matter distribution in the Universe, the so-called N-body simulations and their extension to include baryons, the hydrodynamical simulations.

II Dark Matter

The second part of this book deals with dark matter both from the astrophysical point of view and from the point of view of particle physics. There are indeed different perspectives under which the many phenomena associated to dark matter can be analyzed. The chapter written by Guido D'Amico, Marc Kamionkowski and Kris Sigurdson presents a review of the astrophysical and cosmological evidence for the existence of dark matter from the galactic to the largest cosmological scales. They also discuss the properties of Weakly Interacting Massive Particles (WIMPS) and of other dark matter candidates, like axions and sterile neutrinos. Antonio Masiero, in his chapter, analyzes the dark matter problem from the particle physics point of view, discussing some aspects of the Standard Model of particle physics and the motivations to go beyond it, and introduces the most likely particle dark matter candidates. The final chapter of this part, by Andrea Giuliani, reviews the status of the direct and indirect searches for the dark matter particles.

III Dark Energy

The third and final part of the book, written by Shinji Tsujikawa, present a very complete and up-to-date introduction to dark energy, both from the phenomenological and from the theoretical point of view. Starting from the observational bounds on dark energy coming from Type Ia Supernovae, CMB and Baryonic Acoustic Oscillations, the chapter reviews the various explanations proposed so far for the dominant component of the Universe today; these include a cosmological constant, several dynamical variants, such as quintessence, k-essence, coupled dark energy and unified models of dark matter and dark energy based on a scalar field component. Possible modifications of gravity, such as $f(R)$ theories are also discussed. The idea that the back-reaction of cosmic inhomogeneities could provide an alternative to dark energy at the fundamental level is also discussed.

Part I
Cosmology

Chapter 1
Relativistic Cosmology

Norbert Straumann

1.1 Introduction

I shall review in two opening sections the Standard Model of cosmology. This includes a brief introduction to *inflation*, a key idea of modern cosmology. More on this can be found at many places, for instance, in the recent textbooks on cosmology [1], [2], [3], [4], [5], [6], [7]. A recent treatise that concentrates mainly on the theoretical aspects of the cosmic microwave background physics is [8].

After this warm up, we shall develop the somewhat involved cosmological perturbation theory. The general formalism will later be applied to two main topics: (1) the generation of primordial fluctuations during an inflationary era and (2) the evolution of these perturbations during the linear regime.

A working knowledge of general relativity (GR) is assumed [9].

At the very beginning, one should presumably reflect a bit about cosmology as a physical discipline. I will not do that, apart from the following few comments.

For some people, cosmology is the science of the "Universe as a whole." I doubt that this is really a scientific concept. By "Universe" I always mean that part of the world which is in principle accessible to us through direct or indirect observations. This restriction is nowadays often abandoned when people talk seriously about parallel universes with other laws of physics and even other spacetime dimensions. Should we then not also include "Heaven" and "Hell" as possible ground states of the string landscape? I have nothing against natural philosophy, but cosmologists should always be aware that there is a real danger to be drifted into some modern form of mythology. I am, however, sure that this danger is absent during this school.

Norbert Straumann
Institute for Theoretical Physics, University of Zurich, Winterthurerstrasse 190, CH–8057 Zurich, Switzerland, e-mail: norbert.straumann@gmail.com

1.2 Essentials of Friedmann–Lemaître Models

For reasons explained in the Introduction I treat in this opening section some standard material that will be needed in the main parts of these notes.

Let me begin with a few historical remarks. It is most remarkable that the simple, highly symmetric cosmological models, which were developed more than 80 years ago by Friedmann and Lemaître, still play such an important role in modern cosmology. After all, they were not put forward on the basis of astronomical observations. When the first paper by Friedmann appeared in 1922 (in *Z.f.Physik*), astronomers had only knowledge of the Milky Way. In particular, the observed velocities of stars were all small. Remember, astronomers only learned later that spiral nebulae are independent star systems outside the Milky Way. This was definitely established when in 1924 Hubble found that there were Cepheid variables in Andromeda and also in other galaxies.

Friedmann's models were based on mathematical simplicity, as he explicitly states. This was already the case with Einstein's static model of 1917, in which space is a metric 3-sphere. About this Einstein wrote to de Sitter that his cosmological model was intended primarily to settle the question "whether the basic idea of relativity can be followed through its completion, or whether it leads to contradictions." And he adds whether the model corresponds to reality was another matter. Friedmann writes in his dynamical generalization of Einstein's model about the metric ansatz that this cannot be justified on the basis of physical or philosophical arguments.

Friedmann's two papers from 1922 to 1924 have a strongly mathematical character. It was too early to apply them to the real Universe. In his second paper, he treated the models with negative spatial curvature. Interestingly, he emphasizes that space can nevertheless be *compact*, an aspect that has only recently come again into the focus of attention. It is really sad that Friedmann died already in 1925, at the age of 37. His papers were largely ignored throughout the 1920s, although Einstein studied them carefully and even wrote a paper about them. He was, however, convinced at the time that Friedmann's models had no physical significance.

The same happened with Lemaître's independent work of 1927. Lemaître was the first person who seriously proposed an expanding universe as a model of the real Universe. He derived the general redshift formula we all know and love, and he showed that it leads for small distances to a linear relation, known as Hubble's law. He also estimated the Hubble constant H_0 based on Slipher's redshift data for about 40 nebulae and Hubble's 1925 distance determinations to Andromeda and some other nearby galaxies, and he found 2 years before Hubble a value only somewhat higher the one of Hubble from 1929. (Actually, Lemaître gave two values for H_0.)

The general attitude is well illustrated by the following remark of Eddington at a Royal Society meeting in January, 1930: *"One puzzling question is why there should be only two solutions. I suppose the trouble is that people look for static solutions."*

Lemaître, who had been for a short time in 1925 a postdoctoral student of Eddington, read this remark in a report to the meeting published in *Observatory* and wrote to Eddington pointing out his 1927 paper. Eddington had seen that paper, but had completely forgotten about it. But now, he was greatly impressed and recommended Lemaître's work in a letter to *Nature*. He also arranged for a translation that appeared in *MNRAS*. It is a curious fact that the crucial paragraph describing how Lemaître estimated H_0 and assessed the evidence for linearity were dropped in the English translation. Because of this omission, Lemaître's role is not sufficiently known among cosmologists who cannot read French.

Hubble, on the other hand, nowhere in his famous 1929 paper even mentions an expanding universe, but interprets his data within the static interpretation of the de Sitter solution (repeating what Eddington wrote in the second edition of his relativity book in 1924). In addition, Hubble never claimed to have discovered the expanding universe, he apparently never believed this interpretation. That Hubble was elevated to the discoverer of the expanding universe belongs to sociology, public relations, and rewriting history.

The following remark is also of some interest. It is true that the instability of Einstein's model is not explicitly stated in Lemaître's 1927 paper, but this was an immediate consequence of his equations. In the words of Eddington: "...it was immediately deducible from his [Lemaître's] formulae that Einstein's world is unstable so that an expanding or a contracting universe is an inevitable result of Einstein's law of gravitation."

Lemaître's successful explanation of Slipher's and Hubble's observations finally changed the viewpoint of the majority of workers in the field. For an excellent, carefully researched book on the early history of cosmology, with a very positive foreword by Allan Sandage, see [10].

1.2.1 Friedmann–Lemaître Spacetimes

There is now good evidence that the (recent as well as the early) Universe[1] is – on large scales – surprisingly homogeneous and isotropic. The most impressive support for this comes from extended redshift surveys of galaxies and from the truly remarkable isotropy of the cosmic microwave background (CMB). In the Two Degree Field (2dF) Galaxy Redshift Survey[2], completed in 2003, the redshifts of about 250'000 galaxies have been measured. The distribution of galaxies out to 4 billion light years shows that there are huge clusters, long filaments, and empty voids measuring over 100 million light years across. But the map also shows that there are *no larger structures*. The more extended Sloan Digital Sky Survey (SDSS) has produced similar results and will in the end have spectra of about a million galaxies[3].

[1] By *Universe*, I always mean that part of the world around us which is in principle accessible to observations. In my opinion, the "Universe as a whole" is not a scientific concept. When talking about *model universes*, we develop on paper or with the help of computers, I tend to use lowercase letters. In this domain we are, of course, free to make extrapolations and venture into speculations.

[2] Consult the Home Page: http://www.mso.anu.edu.au/2dFGRS .

[3] For a description and pictures, see the Home Page: http://www.sdss.org/sdss.html .

One arrives at the Friedmann–Lemaître (–Robertson–Walker) spacetimes by postulating that for each observer, moving along an integral curve of a distinguished 4-velocity field u, the Universe looks spatially isotropic. Mathematically, this means the following: Let $Iso_x(M)$ be the group of local isometries of a Lorentz manifold (M,g), with fixed point $x \in M$, and let $SO_3(u_x)$ be the group of all linear transformations of the tangent space $T_x(M)$ that leave the 4-velocity u_x invariant and induce special orthogonal transformations in the subspace orthogonal to u_x, then

$$\{T_x\phi \,:\, \phi \in Iso_x(M),\ \phi_\star u = u\} \supseteq SO_3(u_x).$$

($\phi_\star$ denotes the push-forward belonging to ϕ; see [9], p. 550). In [11], it is shown that this requirement implies that (M,g) is a Friedmann spacetime, whose structure we now recall. Note that (M,g) is then automatically homogeneous.

A *Friedmann spacetime* (M,g) is a warped product of the form $M = I \times \Sigma$, where I is an interval of $\mathbb{R}$, and the metric g is of the form

$$g = -dt^2 + a^2(t)\gamma, \tag{1.1}$$

such that (Σ,γ) is a Riemannian space of constant curvature $k = 0, \pm 1$. The distinguished time t is the *cosmic time*, and $a(t)$ is the *scale factor* (it plays the role of the warp factor (see Appendix B of [9])). Instead of t, we often use the *conformal time* η, defined by $d\eta = dt/a(t)$. The velocity field is perpendicular to the slices of constant cosmic time, $u = \partial/\partial t$.

1.2.1.1 Spaces of Constant Curvature

For the space (Σ,γ) of constant curvature[4], the curvature is given by

$$R^{(3)}(X,Y)Z = k\,[\gamma(Z,Y)X - \gamma(Z,X)Y]; \tag{1.2}$$

in components:

$$R^{(3)}_{ijkl} = k(\gamma_{ik}\gamma_{jl} - \gamma_{il}\gamma_{jk}). \tag{1.3}$$

Hence, the Ricci tensor and the scalar curvature are

$$R^{(3)}_{jl} = 2k\gamma_{jl}, \quad R^{(3)} = 6k. \tag{1.4}$$

For the curvature two-forms we obtain from (1.3) relative to an orthonormal triad $\{\theta^i\}$

$$\Omega^{(3)}_{ij} = \frac{1}{2}R^{(3)}_{ijkl}\,\theta^k \wedge \theta^l = k\,\theta_i \wedge \theta_j \tag{1.5}$$

($\theta_i = \gamma_{ik}\theta^k$). The simply connected constant curvature spaces are in n dimensions the (n+1)-sphere S^{n+1} ($k = 1$), the Euclidean space ($k = 0$), and the pseudosphere ($k = -1$). Nonsimply connected constant curvature spaces are obtained from these

[4] For a detailed discussion of these spaces I refer – for readers knowing German – to [12] or [14].

by forming quotients with respect to discrete isometry groups. (For detailed derivations, see [12].)

1.2.1.2 Curvature of Friedmann Spacetimes

Let $\{\bar{\theta}^i\}$ be any orthonormal triad on (Σ, γ). On this Riemannian space, the first structure equations read[5] (we use the notation in [9]; quantities referring to this 3-dimensional space are indicated by bars)

$$d\bar{\theta}^i + \bar{\omega}^i{}_j \wedge \bar{\theta}^j = 0. \tag{1.6}$$

On (M, g), we introduce the following orthonormal tetrad:

$$\theta^0 = dt, \ \ \theta^i = a(t)\bar{\theta}^i. \tag{1.7}$$

From this and (1.6), we get

$$d\theta^0 = 0, \ \ d\theta^i = \frac{\dot{a}}{a}\theta^0 \wedge \theta^i - a\,\bar{\omega}^i{}_j \wedge \bar{\theta}^j. \tag{1.8}$$

Comparing this with the first structure equation for the Friedmann manifold implies

$$\omega^0{}_i \wedge \theta^i = 0, \ \ \omega^i{}_0 \wedge \theta^0 + \omega^i{}_j \wedge \theta^j = \frac{\dot{a}}{a}\theta^i \wedge \theta^0 + a\,\bar{\omega}^i{}_j \wedge \bar{\theta}^j, \tag{1.9}$$

whence

$$\boxed{\omega^0{}_i = \frac{\dot{a}}{a}\,\theta^i, \ \ \omega^i{}_j = \bar{\omega}^i{}_j.} \tag{1.10}$$

The worldlines of *comoving observers* are integral curves of the 4-velocity field $u = \partial_t$. We claim that these are geodesics, i.e., that

$$\nabla_u u = 0. \tag{1.11}$$

To show this (and for other purposes), we introduce the basis $\{e_\mu\}$ of vector fields dual to (1.7). Since $u = e_0$ we have, using the connection forms (1.10),

$$\nabla_u u = \nabla_{e_0} e_0 = \omega^\lambda{}_0(e_0) e_\lambda = \omega^i{}_0(e_0) e_i = 0.$$

1.2.1.3 Einstein Equations for Friedmann Spacetimes

Inserting the connection forms (1.10) into the second structure equations, we readily find for the curvature 2-forms $\Omega^\mu{}_\nu$:

[5] Readers who are not familiar with the Cartan calculus should derive the result (1.13) – (1.15) in the traditional way.

$$\Omega^0{}_i = \frac{\ddot{a}}{a}\theta^0 \wedge \theta^i, \quad \Omega^i{}_j = \frac{k+\dot{a}^2}{a^2}\theta^i \wedge \theta^j. \tag{1.12}$$

A routine calculation leads to the following components of the Einstein tensor relative to the basis (1.7)

$$G_{00} = 3\left(\frac{\dot{a}^2}{a^2} + \frac{k}{a^2}\right), \tag{1.13}$$

$$G_{11} = G_{22} = G_{33} = -2\frac{\ddot{a}}{a} - \frac{\dot{a}^2}{a^2} - \frac{k}{a^2}, \tag{1.14}$$

$$G_{\mu\nu} = 0 \ \ (\mu \neq \nu). \tag{1.15}$$

In order to satisfy the field equations, the symmetries of $G_{\mu\nu}$ imply that the energy-momentum tensor *must* have the perfect fluid form (see [9], Sect. 1.4.2):

$$T^{\mu\nu} = (\rho + p)u^\mu u^\nu + p g^{\mu\nu}, \tag{1.16}$$

where u is the comoving velocity field introduced above.

Now, we can write down the field equations (including the cosmological term):

$$3\left(\frac{\dot{a}^2}{a^2} + \frac{k}{a^2}\right) = 8\pi G\rho + \Lambda, \tag{1.17}$$

$$-2\frac{\ddot{a}}{a} - \frac{\dot{a}^2}{a^2} - \frac{k}{a^2} = 8\pi G p - \Lambda. \tag{1.18}$$

Although the "energy-momentum conservation" does not provide an independent equation, it is useful to work this out. As expected, the momentum "conservation" is automatically satisfied. For the "energy conservation," we use the general form (see (1.37) in [9])

$$\nabla_u \rho = -(\rho + p)\nabla \cdot u. \tag{1.19}$$

In our case, we have for the *expansion rate*

$$\nabla \cdot u = \omega^\lambda{}_0(e_\lambda)u^0 = \omega^i{}_0(e_i),$$

thus with (1.10)

$$\nabla \cdot u = 3\frac{\dot{a}}{a}. \tag{1.20}$$

Therefore, Eq. (1.19) becomes

$$\dot{\rho} + 3\frac{\dot{a}}{a}(\rho + p) = 0. \tag{1.21}$$

This should *not* be considered, as it is often done, as an energy conservation law. Because of the equivalence principle, there is in GR no local energy conservation. (For more on this, see Sect. 1.2.3.)

For a given equation of state, $p = p(\rho)$, we can use (1.21) in the form

$$\frac{d}{da}(\rho a^3) = -3pa^2 \tag{1.22}$$

to determine ρ as a function of the scale factor a. Examples: (1) For free massless particles (radiation), we have $p = \rho/3$, thus $\rho \propto a^{-4}$. (2) For dust ($p = 0$), we get $\rho \propto a^{-3}$.

With this knowledge, the *Friedmann equation* (1.17) determines the time evolution of $a(t)$. It is easy to see that (1.18) follows from (1.17) and (1.21).

As an important consequence of (1.17) and (1.18), we obtain for the acceleration of the expansion

$$\ddot{a} = -\frac{4\pi G}{3}(\rho + 3p)a + \frac{1}{3}\Lambda a. \tag{1.23}$$

This shows that as long as $\rho + 3p$ is positive, the first term in (1.23) is decelerating, while a positive cosmological constant is repulsive. This becomes understandable if one writes the field equation as

$$G_{\mu\nu} = \kappa(T_{\mu\nu} + T^{\Lambda}_{\mu\nu}) \qquad (\kappa = 8\pi G), \tag{1.24}$$

with

$$T^{\Lambda}_{\mu\nu} = -\frac{\Lambda}{8\pi G} g_{\mu\nu}. \tag{1.25}$$

This vacuum contribution has the form of the energy-momentum tensor of an ideal fluid, with energy density $\rho_\Lambda = \Lambda/8\pi G$ and pressure $p_\Lambda = -\rho_\Lambda$. Hence, the combination $\rho_\Lambda + 3p_\Lambda$ is equal to $-2\rho_\Lambda$ and is thus negative. In what follows we shall often include in ρ and p the vacuum pieces.

1.2.1.4 Redshift

As a result of the expansion of the Universe, the light of distant sources appears redshifted. The amount of redshift can be simply expressed in terms of the scale factor $a(t)$.

Consider two integral curves of the average velocity field u. We imagine that one describes the worldline of a distant comoving source and the other that of an observer at a telescope (see Fig. 1.1). Since light is propagating along null geodesics, we conclude from (1.1) that along the worldline of a light ray $dt = a(t)d\sigma$, where $d\sigma$ is the line element on the 3-dimensional space (Σ, γ) of constant curvature $k = 0, \pm 1$. Hence, the integral on the left of

$$\int_{t_e}^{t_o} \frac{dt}{a(t)} = \int_{source}^{obs.} d\sigma, \tag{1.26}$$

between the time of emission (t_e) and the arrival time at the observer (t_o), is independent of t_e and t_o. Therefore, if we consider a second light ray that is emitted at

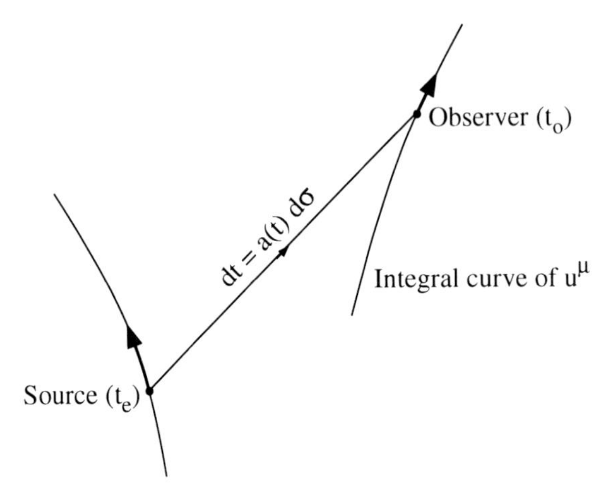

Fig. 1.1 Redshift for Friedmann models.

the time $t_e + \Delta t_e$ and is received at the time $t_o + \Delta t_o$, we obtain from the last equation

$$\int_{t_e+\Delta t_e}^{t_o+\Delta t_o} \frac{dt}{a(t)} = \int_{t_e}^{t_o} \frac{dt}{a(t)}. \tag{1.27}$$

For a small Δt_e, this gives

$$\frac{\Delta t_o}{a(t_o)} = \frac{\Delta t_e}{a(t_e)}.$$

The observed and the emitted frequencies ν_o and ν_e, respectively, are thus related according to

$$\frac{\nu_o}{\nu_e} = \frac{\Delta t_e}{\Delta t_o} = \frac{a(t_e)}{a(t_o)}. \tag{1.28}$$

The redshift parameter z is defined by

$$z := \frac{\nu_e - \nu_o}{\nu_o}, \tag{1.29}$$

and is given by the key equation

$$\boxed{1 + z = \frac{a(t_o)}{a(t_e)}.} \tag{1.30}$$

One can also express this by the equation $\nu \cdot a = const$ along a null geodesic.

1.2.1.5 Cosmic Distance Measures

We now introduce a further important tool, namely operational definitions of three different distance measures, and show that they are related by simple redshift factors.

If D is the physical (proper) extension of a distant object and δ is its angle subtended, then the *angular diameter distance* D_A is defined by

$$D_A := D/\delta. \tag{1.31}$$

If the object is moving with the proper transversal velocity $V_\perp$ and with an apparent angular motion $d\delta/dt_0$, then the *proper-motion distance* is by definition

$$D_M := \frac{V_\perp}{d\delta/dt_0}. \tag{1.32}$$

Finally, if the object has the intrinsic luminosity $\mathscr{L}$ and $\mathscr{F}$ is the received energy flux, then the *luminosity distance* is naturally defined as

$$D_L := (\mathscr{L}/4\pi\mathscr{F})^{1/2}. \tag{1.33}$$

Here, we show that these three distances are related as follows:

$$\boxed{D_L = (1+z)D_M = (1+z)^2 D_A.} \tag{1.34}$$

It will be useful to introduce on (Σ, γ) "polar" coordinates (r, ϑ, φ) (obtained by stereographic projection), such that

$$\gamma = \frac{dr^2}{1-kr^2} + r^2 d\Omega^2, \quad d\Omega^2 = d\vartheta^2 + \sin^2\vartheta d\varphi^2. \tag{1.35}$$

One easily verifies that the curvature forms of this metric satisfy (1.5). (This follows without doing any work by using in [9] the curvature forms (3.9) in the ansatz (3.3) for the Schwarzschild metric.)

To prove (1.34), we show that the three distances can be expressed as follows, if r_e denotes the comoving radial coordinate (in (1.35)) of the distant object and the observer is (without loss of generality) at $r = 0$:

$$D_A = r_e a(t_e), \quad D_M = r_e a(t_0), \quad D_L = r_e a(t_0)\frac{a(t_0)}{a(t_e)}. \tag{1.36}$$

Once this is established, (1.34) follows from (1.30).

From Fig. 1.2 and (1.35), we see that

$$D = a(t_e) r_e \delta, \tag{1.37}$$

hence the first equation in (1.36) holds.

To prove the second one, we note that the source moves in a time dt_0 a proper transversal distance

$$dD = V_\perp dt_e = V_\perp dt_0 \frac{a(t_e)}{a(t_0)}.$$

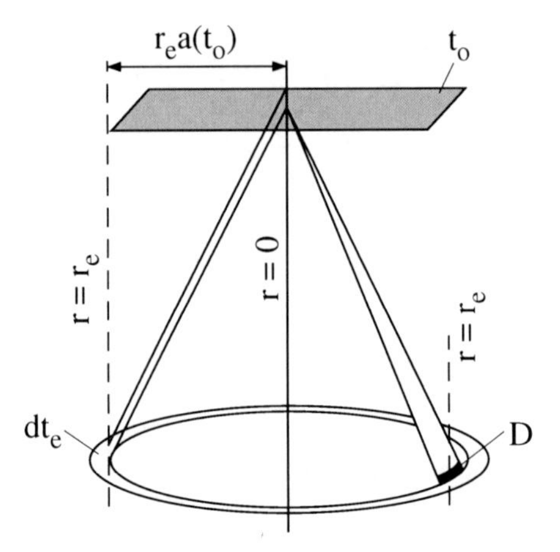

Fig. 1.2 Spacetime diagram for cosmic distance measures.

Using again the metric (1.35), we see that the apparent angular motion is

$$d\delta = \frac{dD}{a(t_e)r_e} = \frac{V_\perp dt_0}{a(t_0)r_e}.$$

Inserting this into the definition (1.32) shows that the second equation in (1.36) holds. For the third equation, we have to consider the observed energy flux. In a time dt_e, the source emits an energy $\mathscr{L}dt_e$. This energy is redshifted to the present by a factor $a(t_e)/a(t_0)$ and is now distributed by (1.35) over a sphere with proper area $4\pi(r_e a(t_0))^2$ (see Fig. 1.2). Hence, the received flux (*apparent luminosity*) is

$$\mathscr{F} = \mathscr{L}dt_e \frac{a(t_e)}{a(t_0)} \frac{1}{4\pi(r_e a(t_0))^2} \frac{1}{dt_0},$$

thus

$$\mathscr{F} = \frac{\mathscr{L}a^2(t_e)}{4\pi a^4(t_0) r_e^2}.$$

Inserting this into the definition (1.33) establishes the third equation in (1.36). For later applications, we write the last equation in the more transparent form

$$\boxed{\mathscr{F} = \frac{\mathscr{L}}{4\pi(r_e a(t_0))^2} \frac{1}{(1+z)^2}.} \tag{1.38}$$

The last factor is due to redshift effects.

Two of the discussed distances as a function of z are shown in Fig. 1.3 for two Friedmann models with different cosmological parameters. The other two distance measures will be introduced in Sect. 1.3.2.

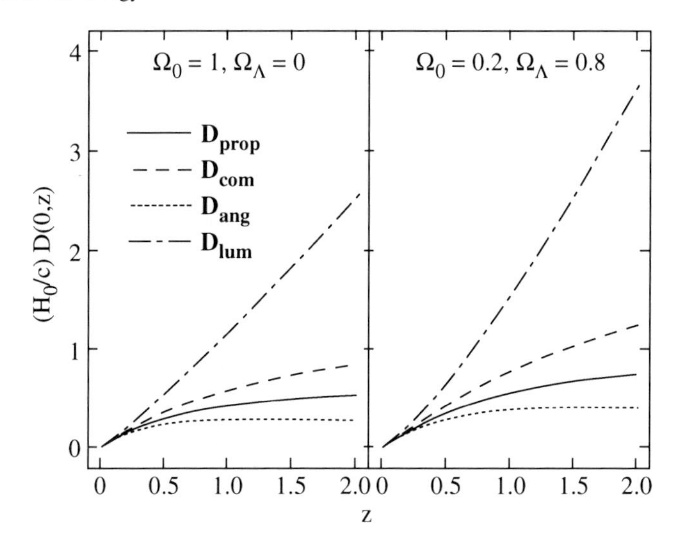

Fig. 1.3 Cosmological distance measures as a function of source redshift for two cosmological models. The angular diameter distance $D_{ang} \equiv D_A$ and the luminosity distance $D_{lum} \equiv D_L$ have been introduced in this Section. The other two will be introduced in Sect. 1.3.2.

1.2.2 Thermal History Below 100 MeV

1.2.2.1 Overview

Below the transition at about 200 *MeV* from a quark-gluon plasma to the confinement phase, the Universe was initially dominated by a complicated dense hadron soup. The abundance of pions, for example, was so high that they nearly overlapped. The pions, kaons, and other hadrons soon began to decay and most of the nucleons and antinucleons annihilated, leaving only a tiny baryon asymmetry. The energy density is then almost completely dominated by radiation and the stable leptons ($e^{\pm}$, the three neutrino flavors and their antiparticles). For some time, all these particles are in thermodynamic equilibrium. For this reason, only a few initial conditions have to be imposed. The Universe was never as simple as in this lepton era. (At this stage, it is almost inconceivable that the complex world around us would eventually emerge.)

The first particles that freeze out of this equilibrium are the weakly interacting neutrinos. Let us estimate when this happened. The coupling of the neutrinos in the lepton era is dominated by the reactions:

$$e^- + e^+ \leftrightarrow \nu + \bar{\nu}, \;\; e^{\pm} + \nu \rightarrow e^{\pm} + \nu, \;\; e^{\pm} + \bar{\nu} \rightarrow e^{\pm} + \bar{\nu}.$$

For dimensional reasons, the cross-sections are all of magnitude

$$\sigma \simeq G_F^2 T^2, \tag{1.39}$$

where G_F is the Fermi coupling constant ($\hbar = c = k_B = 1$). Numerically, $G_F m_p^2 \simeq 10^{-5}$. On the other hand, the electron and neutrino densities n_e, n_ν are about T^3. For this reason, the reaction rates Γ for ν-scattering and ν-production per electron are of magnitude $c \cdot \nu \cdot n_e \simeq G_F^2 T^5$. This has to be compared with the expansion rate of the Universe

$$H = \frac{\dot{a}}{a} \simeq (G\rho)^{1/2}.$$

Since $\rho \simeq T^4$, we get

$$H \simeq G^{1/2} T^2, \tag{1.40}$$

and thus

$$\frac{\Gamma}{H} \simeq G^{-1/2} G_F^2 T^3 \simeq (T/10^{10}\ K)^3. \tag{1.41}$$

This ration is larger than 1 for $T > 10^{10}\ K \simeq 1\ MeV$, and the neutrinos thus remain in thermodynamic equilibrium until the temperature has decreased to about 1 MeV. But even below this temperature, the neutrinos remain Fermi distributed,

$$n_\nu(p)dp = \frac{1}{2\pi^2} \frac{1}{e^{p/T_\nu} + 1} p^2 dp\,, \tag{1.42}$$

as long as they can be treated as massless. The reason is that the number density decreases as a^{-3} and the momenta with a^{-1}. Because of this, we also see that the neutrino temperature T_ν decreases after decoupling as a^{-1}. The same is, of course, true for photons. The reader will easily find out how the distribution evolves when neutrino masses are taken into account. (Since neutrino masses are so small, this is only relevant at very late times.)

1.2.2.2 Chemical Potentials of the Leptons

The equilibrium reactions below 100 MeV, say, conserve several additive quantum numbers[6], namely the electric charge Q, the baryon number B, and the three lepton numbers L_e, L_μ, L_τ. Correspondingly, there are five independent chemical potentials. Since particles and antiparticles can annihilate to photons, their chemical potentials are oppositely equal: $\mu_{e^-} = -\mu_{e^+}$, etc. From the following reactions

$$e^- + \mu^+ \to \nu_e + \bar{\nu}_\mu, \quad e^- + p \to \nu_e + n, \quad \mu^- + p \to \nu_\mu + n,$$

we infer the equilibrium conditions

$$\mu_{e^-} - \mu_{\nu_e} = \mu_{\mu^-} - \mu_{\nu_\mu} = \mu_n - \mu_p. \tag{1.43}$$

[6] Even if B, L_e, L_μ, L_τ should not be strictly conserved, this is not relevant within a Hubble time H_0^{-1}.

As independent chemical potentials, we can thus choose

$$\boxed{\mu_p,\ \mu_{e^-},\ \mu_{\nu_e},\ \mu_{\nu_\mu},\ \mu_{\nu_\tau}.} \tag{1.44}$$

Because of local electric charge neutrality, the charge number density n_Q vanishes. From observations (see Sect. 1.2.2.3), we also know that the baryon number density n_B is much smaller than the photon number density ($\sim$ entropy density s_γ). The ratio n_B/s_γ remains constant for adiabatic expansion (both decrease with a^{-3}; see the next section). Moreover, the lepton number densities are

$$n_{L_e} = n_{e^-} + n_{\nu_e} - n_{e^+} - n_{\bar{\nu}_e},\quad n_{L_\mu} = n_{\mu^-} + n_{\nu_\mu} - n_{\mu^+} - n_{\bar{\nu}_\mu},\quad etc. \tag{1.45}$$

Since in the present Universe, the number density of electrons is equal to that of the protons (bound or free), we know that after the disappearance of the muons $n_{e^-} \simeq n_{e^+}$ (recall $n_B \ll n_\gamma$), thus $\mu_{e^-}\ (= -\mu_{e^+}) \simeq 0$. It is conceivable that the chemical potentials of the neutrinos and antineutrinos cannot be neglected, i.e., n_{L_e} is not much smaller than the photon number density. In analogy to what we know about the baryon density, we make the reasonable *assumption* that the lepton number densities are also much smaller than s_γ. Then we can take the chemical potentials of the neutrinos equal to zero ($|\mu_\nu|/kT \ll 1$). With what we said before, we can then put the five chemical potentials (1.44) equal to zero because the charge number densities are all odd in them. Of course, n_B does not really vanish (otherwise we would not be here), but for the thermal history in the era we are considering they can be ignored.

1.2.2.3 Constancy of Entropy

Let ρ_{eq}, p_{eq} denote (in this subsection only) the total energy density and pressure of all particles in thermodynamic equilibrium. Since the chemical potentials of the leptons vanish, these quantities are only functions of the temperature T. According to the second law, the differential of the entropy $S(V,T)$ is given by

$$dS(V,T) = \frac{1}{T}[d(\rho_{eq}(T)V) + p_{eq}(T)dV]. \tag{1.46}$$

This implies

$$\begin{aligned} d(dS) = 0 &= d\left(\frac{1}{T}\right) \wedge d(\rho_{eq}(T)V) + d\left(\frac{p_{eq}(T)}{T}\right) \wedge dV \\ &= -\frac{\rho_{eq}}{T^2} dT \wedge dV + \frac{d}{dT}\left(\frac{p_{eq}(T)}{T}\right) dT \wedge dV, \end{aligned}$$

i.e., the Maxwell relation

$$\boxed{\frac{dp_{eq}(T)}{dT} = \frac{1}{T}[\rho_{eq}(T) + p_{eq}(T)].} \tag{1.47}$$

If we use this in (1.46), we get

$$dS = d\left[\frac{V}{T}(\rho_{eq} + p_{eq})\right],$$

so the entropy density of the particles in equilibrium is

$$\boxed{s = \frac{1}{T}[\rho_{eq}(T) + p_{eq}(T)].} \tag{1.48}$$

For an adiabatic expansion, the entropy in a comoving volume remains constant:

$$S = a^3 s = const. \tag{1.49}$$

This constancy is equivalent to the energy equation (1.21) for the equilibrium part. Indeed, the latter can be written as

$$a^3 \frac{dp_{eq}}{dt} = \frac{d}{dt}[a^3(\rho_{eq} + p_{eq})],$$

and by (1.48), this is equivalent to $dS/dt = 0$.

In particular, we obtain for massless particles ($p = \rho/3$) from (1.47) again $\rho \propto T^4$ and from (1.48) that $S =$ constant implies $T \propto a^{-1}$.

It is sometimes said that for a Friedmann model the expansion always proceeds adiabatically because the symmetries forbid a heat current to flow into a comoving volume. Although there is indeed no heat current, entropy can be generated if the cosmic fluid has a nonvanishing bulk viscosity. This follows formally from general relativistic thermodynamics. Eq. (B.36) in Appendix B of [13] shows that the divergence of the entropy current contains the term $(\zeta/T)\theta^2$, where ζ is the bulk viscosity and θ the expansion rate (=$3(\dot{a}/a)$ for a Friedmann spacetime).

Once the electrons and positrons have annihilated below $T \sim m_e$, the equilibrium components consist of photons, electrons, protons and – after the big bang nucleosynthesis – of some light nuclei (mostly He^4). Since the charged particle number densities are much smaller than the photon number density, the photon temperature T_γ still decreases as a^{-1}. Let us show this formally. For this we consider beside the photons an ideal gas in thermodynamic equilibrium with the black body radiation. The total pressure and energy density are then (we use units with $\hbar = c = k_B = 1$; n is the number density of the nonrelativistic gas particles with mass m):

$$p = nT + \frac{\pi^2}{45}T^4, \quad \rho = nm + \frac{nT}{\gamma - 1} + \frac{\pi^2}{15}T^4 \tag{1.50}$$

($\gamma = 5/3$ for a monoatomic gas). The conservation of the gas particles, $na^3 = const.$, together with the energy equation (1.22) implies, if $\sigma := s_\gamma/n$,

$$\frac{d\ln T}{d\ln a} = -\left[\frac{\sigma + 1}{\sigma + 1/[3(\gamma - 1)]}\right].$$

For $\sigma \ll 1$, this gives the well-known relation $T \propto a^{3(\gamma-1)}$ for an adiabatic expansion of an ideal gas.

We are, however, dealing with the opposite situation $\sigma \gg 1$, and then we obtain, as expected, $a \cdot T = const.$

Let us look more closely at the famous ratio n_B/s_γ. We need

$$s_\gamma = \frac{4}{3T}\rho_\gamma = \frac{4\pi^2}{45}T^3 = 3.60 n_\gamma, \quad n_B = \rho_B/m_p = \Omega_B \rho_{crit}/m_p. \tag{1.51}$$

From the present value of $T_\gamma \simeq 2.7\ K$ and (1.89), $\rho_{crit} = 1.12 \times 10^{-5}\ h_0^2(m_p/cm^3)$, we obtain as a measure for the baryon asymmetry of the Universe

$$\boxed{\frac{n_B}{s_\gamma} = 0.75 \times 10^{-8}(\Omega_B h_0^2).} \tag{1.52}$$

It is one of the great challenges to explain this tiny number. So far, this has been achieved at best qualitatively in the framework of grand unified theories (GUTs).

1.2.2.4 Neutrino Temperature

During the electron-positron annihilation below $T = m_e$, the a-dependence is complicated since the electrons can no more be treated as massless. We want to know at this point what the ratio T_γ/T_ν is after the annihilation. This can easily be obtained by using the constancy of comoving entropy for the photon-electron-positron system, which is sufficiently strongly coupled to maintain thermodynamic equilibrium.

We need the entropy for the electrons and positrons at $T \gg m_e$, long before annihilation begins. To compute this, note the identity

$$\int_0^\infty \frac{x^n}{e^x - 1}dx - \int_0^\infty \frac{x^n}{e^x + 1}dx = 2\int_0^\infty \frac{x^n}{e^{2x} - 1}dx = \frac{1}{2^n}\int_0^\infty \frac{x^n}{e^x - 1}dx,$$

whence

$$\int_0^\infty \frac{x^n}{e^x + 1}dx = (1 - 2^{-n})\int_0^\infty \frac{x^n}{e^x - 1}dx. \tag{1.53}$$

In particular, we obtain for the entropies s_e, s_γ the following relation

$$s_e = \frac{7}{8}s_\gamma \quad (T \gg m_e). \tag{1.54}$$

Equating the entropies for $T_\gamma \gg m_e$ and $T_\gamma \ll m_e$ gives

$$(T_\gamma a)^3\big|_{before}\left[1 + 2 \times \frac{7}{8}\right] = (T_\gamma a)^3\big|_{after} \times 1$$

because the neutrino entropy is conserved. Therefore, we obtain

$$(aT_\gamma)\big|_{after} = \left(\frac{11}{4}\right)^{1/3} (aT_\gamma)\big|_{before}. \tag{1.55}$$

But $(aT_\nu)|_{after} = (aT_\nu)|_{before} = (aT_\gamma)\big|_{before}$, hence we obtain the important relation

$$\boxed{\left(\frac{T_\gamma}{T_\nu}\right)\Big|_{after} = \left(\frac{11}{4}\right)^{1/3} = 1.401.} \tag{1.56}$$

1.2.2.5 Epoch of Matter-Radiation Equality

In the main sections of this book, the epoch when radiation (photons and neutrinos) has about the same energy density as nonrelativistic matter (dark matter and baryons) plays a very important role. Let us determine the redshift, z_{eq}, when there is equality.

For the three neutrino and antineutrino flavors, the energy density is according to (1.53)

$$\rho_\nu = 3 \times \frac{7}{8} \times \left(\frac{4}{11}\right)^{4/3} \rho_\gamma. \tag{1.57}$$

Using

$$\frac{\rho_\gamma}{\rho_{crit}} = 2.47 \times 10^{-5} h_0^{-2} (1+z)^4, \tag{1.58}$$

we obtain for the total radiation energy density, ρ_r,

$$\frac{\rho_r}{\rho_{crit}} = 4.15 \times 10^{-5} h_0^{-2} (1+z)^4, \tag{1.59}$$

Equating this to

$$\frac{\rho_M}{\rho_{crit}} = \Omega_M (1+z)^3, \tag{1.60}$$

we obtain

$$\boxed{1 + z_{eq} = 2.4 \times 10^4 \Omega_M h_0^2.} \tag{1.61}$$

Only a small fraction of Ω_M is baryonic. There are several methods to determine the fraction Ω_B in baryons. A traditional one comes from the abundances of the light elements. This is treated in most texts on cosmology. (German-speaking readers find a detailed discussion in my lecture notes [14], which are available in the Internet.) The comparison of the straightforward theory with observation gives a value in the range $\Omega_B h_0^2 = 0.021 \pm 0.002$. Other determinations are all compatible with this value[7]. For instance, it will be shown in other chapters that Ω_B can be obtained from the CMB anisotropies. The striking agreement of different methods, sensitive to different physics, strongly supports our standard big bang picture of the Universe.

1.2.2.6 Recombination and Decoupling

The plasma era ends when electrons combine with protons and helium ions to form neutral atoms. The details of the physics of recombination are a bit complicated, but for a rough estimate of the recombination time one can assume thermodynamic equilibrium conditions. (When the ionization fraction becomes low, a kinetic treatment is needed.) For simplicity, we ignore helium and study the thermodynamic equilibrium of $e^- + p \rightleftharpoons H + \gamma$. The condition for chemical equilibrium is

$$\mu_{e^-} + \mu_p = \mu_H, \tag{1.62}$$

where μ_i $(i = e^-, p, H)$ are the chemical potentials of e^-, p, and neutral hydrogen H. These are related to the particle number densities as follows: for electrons,

$$n_e = \int \frac{2d^3p}{(2\pi)^3} \frac{1}{e^{(E_e(p)-\mu_e)/T} + 1} \simeq \int \frac{2d^3p}{(2\pi)^3} e^{-(\mu_e - m_e)/T} e^{-p^2/2mT},$$

in the nonrelativistic and nondegenerate case. In our problem, we can thus use

$$n_e = 2e^{(\mu_e - m_e)/T} \left(\frac{m_e T}{2\pi} \right)^{3/2}, \tag{1.63}$$

and similarly for the proton component

$$n_p = 2e^{(\mu_p - m_p)/T} \left(\frac{m_p T}{2\pi} \right)^{3/2}. \tag{1.64}$$

For a composite system like H statistical mechanics gives

$$n_H = 2e^{(\mu_H - m_H)/T} Q \left(\frac{m_H T}{2\pi} \right)^{3/2}, \tag{1.65}$$

where Q is the partition sum of the internal degrees of freedom

$$Q = \sum_n g_n e^{-\varepsilon_n/T}$$

[7] For a critical discussion, see, e.g., [15].

(ε_n is measured from the ground state). Usually only the ground state is taken into account, $Q \simeq 4$.

For hydrogen, the partition sum of an isolated atom is obviously infinite, as a result of the long-range of the Coulomb potential. However, in a plasma, the latter is screened, and for our temperature and density range, the ground state approximation is very good (estimate the Debye length). Then, we obtain the *Saha equation*:

$$\frac{n_e n_p}{n_H} = e^{-\Delta/T}\left(\frac{m_e T}{2\pi}\right)^{3/2}, \tag{1.66}$$

where Δ is the ionization energy $\Delta = \frac{1}{2}\alpha^2 m_e \simeq 13.6$ eV. (In the last factor, we have replaced m_p/m_H by unity.)

Let us rewrite this in terms of the ionization fraction $x_e := n_e/n_B$, $n_B = n_p + n_H = n_e + n_H$:

$$\frac{x_e^2}{1-x_e} = \frac{1}{n_B}\left(\frac{m_e T}{2\pi}\right)^{3/2} e^{-\Delta/T}. \tag{1.67}$$

It is important to see the role of the large ratio $\sigma := s_\gamma/n_B = \frac{4\pi^2}{45}T^3/n_B$ given in (1.56). In terms of this, we have

$$\frac{x_e^2}{1-x_e} = \frac{45}{4\pi^2}\sigma\left(\frac{m_e T}{2\pi}\right)^{3/2} e^{-\Delta/T}. \tag{1.68}$$

So, when the temperature is of order Δ, the right-hand side is of order $10^9(m_e/T)^{3/2} \sim 10^{15}$. Hence x_e is very close to 1. Recombination only occurs when T drops far below Δ. Using (1.56), we see that $x_e = 1/2$ for

$$\left(\frac{T_{rec}}{1\ eV}\right)^{-3/2} \exp(-13.6\ eV/T_{rec}) = 1.3\cdot 10^{-6}\Omega_B h_0^2.$$

For $\Omega_B h_0^2 \simeq 0.02$, this gives

$$T_{rec} \simeq 3760\ K = 0.32\ eV,\ \ z_{rec} \simeq 1380.$$

Decoupling occurs roughly when the Thomson scattering rate is comparable to the expansion rate. The first is $n_e\sigma_T = x_e m_p n_B \sigma_T/m_p = x_e \sigma_T \Omega_B \rho_{crit}/m_p$. For H, we use Eqs. (1.90) and (1.91) below: $H(z) = H_0 E(z)$, where for large redshifts $E(z) \simeq \Omega_M^{1/2}(1+z)^{3/2}[1+(1+z)/(1+z_{eq})]^{1/2}$. So, we get

$$\frac{n_e\sigma_T}{H} = \frac{x_e\sigma_T\Omega_B}{H_0\Omega_M^{1/2}}\frac{\rho_{crit}}{m_p}(1+z)^{3/2}[1+(1+z)/(1+z_{eq})]^{1/2}. \tag{1.69}$$

For best-fit values of the cosmological parameters, the right-hand side is for $z \simeq 1000$ about $10^2 x_e$. Hence, photons decouple when x_e drops below $\sim 10^{-2}$.

Kinetic Treatment

For an accurate kinetic treatment, one has to take into account some complications connected with the population of the $1s$ state and the Ly-α background. We shall add later some remarks on this, but for the moment, we are satisfied with a simplified treatment.

We replace the photon number density n_γ by the equilibrium distribution of temperature T. If σ_{rec} denotes the recombination cross section of $e^- + p \to H + \gamma$, the electron number density satisfies the rate equation

$$a^{-3}(t)\frac{d}{dt}(n_e a^3) = -n_e n_p \langle \sigma_{rec} \cdot v_e \rangle + n_\gamma^{eq} n_H \langle \sigma_{ion} \cdot c \rangle. \tag{1.70}$$

The last term represents the contribution of the inverse reaction $\gamma + H \to p + e^-$. This can be obtained from *detailed balance*: For equilibrium, the right-hand side must vanish, thus

$$n_e^{eq} n_p^{eq} \langle \sigma_{rec} \cdot v_e \rangle = n_\gamma^{eq} n_H^{eq} \langle \sigma_{ion} \cdot c \rangle. \tag{1.71}$$

Hence,

$$\frac{dx_e}{dt} = \langle \sigma_{rec} \cdot v_e \rangle \left[-x_e^2 n_B + (1 - x_e)\frac{n_e^{eq} n_p^{eq}}{n_H^{eq}} \right] \tag{1.72}$$

or with the Saha-equation

$$\frac{dx_e}{dt} = \langle \sigma_{rec} \cdot v_e \rangle \left[-n_B x_e^2 + (1 - x_e)\left(\frac{m_e T}{2\pi}\right)^{3/2} e^{-\Delta/T} \right]; \tag{1.73}$$

The recombination rate $\langle \sigma_{rec} \cdot v_e \rangle$ for a transition to the nth excited state of H is usually denoted by α_n. In Eq. (1.74) we have to take the sum

$$\alpha^{(2)} := \sum_{n=2}^{\infty} \alpha_n, \tag{1.74}$$

ignoring $n = 1$ because transitions to the ground state level $n = 1$ produce photons that are sufficiently energetic to ionize other hydrogen atoms.

With this, the rate equation (1.74) takes the form

$$\frac{dx_e}{dt} = -n_B \alpha^{(2)} x_e^2 + \beta(1 - x_e), \tag{1.75}$$

where

$$\beta := \alpha^{(2)} \left(\frac{m_e T}{2\pi}\right)^{3/2} e^{-\Delta/T}. \tag{1.76}$$

In the relevant range one finds with Dirac's radiation theory the approximate formula

$$\alpha^{(2)} \simeq 10.9 \frac{\alpha^2}{m_e^2} \left(\frac{\Delta}{T}\right)^{1/2} \ln\left(\frac{\Delta}{T}\right). \tag{1.77}$$

Our kinetic equation is too simple. Especially, the relative population of the $1s$ and $2s$ states requires some detailed study in which the two-photon transition $2s \to 1s + 2\gamma$ enters. The interested reader finds the details in [1], Sect. 6 or [7], Sect. 2.3.

A tiny residual ionization played an important role in the formation of the first stars.

1.2.3 Luminosity-Redshift Relation

In 1998 the Hubble diagram for Type Ia supernovae gave, as a big surprize, the first serious evidence for a currently accelerating Universe. This will be discussed in detail in a later chapter of this book. Here, we develop some theoretical background.

If the comoving radial coordinate r is chosen such that the Friedmann–Lemaître metric takes the form

$$g = -dt^2 + a^2(t)\left[\frac{dr^2}{1-kr^2} + r^2 d\Omega^2\right], \quad k = 0, \pm 1, \tag{1.78}$$

then the luminosity distance of a source at redshift z is according to (1.36)

$$D_L(z) = a_0(1+z)r(z) \quad (a_0 \equiv a(t_0)). \tag{1.79}$$

We need the function $r(z)$. From

$$dz = -\frac{a_0}{a}\frac{\dot{a}}{a}dt, \quad dt = -a(t)\frac{dr}{\sqrt{1-kr^2}}$$

for light rays, we obtain the two differential relations

$$\frac{dr}{\sqrt{1-kr^2}} = \frac{1}{a_0}\frac{dz}{H(z)} = -\frac{dt}{a(t)} \quad \left(H(z) = \frac{\dot{a}}{a}\right). \tag{1.80}$$

Now, we make use of the Friedmann equation

$$H^2 + \frac{k}{a^2} = \frac{8\pi G}{3}\rho. \tag{1.81}$$

Let us decompose the total energy-mass density ρ into nonrelativistic (NR), relativistic (R), Λ, quintessence (Q), and possibly other contributions

$$\rho = \rho_{NR} + \rho_R + \rho_\Lambda + \rho_Q + \cdots. \tag{1.82}$$

For the relevant cosmic period, we can assume that the energy equation

$$\frac{d}{da}(\rho a^3) = -3pa^2 \tag{1.83}$$

also holds for the individual components $X = NR, R, \Lambda, Q, \cdots$. If $w_X \equiv p_X/\rho_X$ is constant, this implies that

$$\rho_X a^{3(1+w_X)} = const. \tag{1.84}$$

Therefore,

$$\rho = \sum_X \left(\rho_X a^{3(1+w_X)}\right)_0 \frac{1}{a^{3(1+w_X)}} = \sum_X (\rho_X)_0 (1+z)^{3(1+w_X)}. \tag{1.85}$$

Hence the Friedmann equation (1.81) can be written as

$$\frac{H^2(z)}{H_0^2} + \frac{k}{H_0^2 a_0^2}(1+z)^2 = \sum_X \Omega_X (1+z)^{3(1+w_X)}, \tag{1.86}$$

where Ω_X is the dimensionless density parameter for the species X,

$$\Omega_X = \frac{(\rho_X)_0}{\rho_{crit}}, \tag{1.87}$$

where ρ_{crit} is the critical density:

$$\begin{aligned} \rho_{crit} &= \frac{3H_0^2}{8\pi G} \\ &= 1.88 \times 10^{-29}\, h_0^2\, g\ cm^{-3} \\ &= 8 \times 10^{-47} h_0^2\, GeV^4. \end{aligned} \tag{1.88}$$

Here, h_0 denotes the *reduced Hubble parameter*

$$h_0 = H_0/(100\ km\ s^{-1}\ Mpc^{-1}) \simeq 0.7. \tag{1.89}$$

Using also the curvature parameter $\Omega_K \equiv -k/H_0^2 a_0^2$, we obtain the useful form

$$\boxed{H^2(z) = H_0^2 E^2(z; \Omega_K, \Omega_X),} \tag{1.90}$$

with

$$E^2(z; \Omega_K, \Omega_X) = \Omega_K (1+z)^2 + \sum_X \Omega_X (1+z)^{3(1+w_X)}. \tag{1.91}$$

Especially for $z = 0$ this gives

$$\Omega_K + \Omega_0 = 1, \quad \Omega_0 \equiv \sum_X \Omega_X. \tag{1.92}$$

If we use (1.90) in (1.80), we get

$$\int_0^{r(z)} \frac{dr}{\sqrt{1-kr^2}} = \frac{1}{H_0 a_0} \int_0^z \frac{dz'}{E(z')} \tag{1.93}$$

and thus

$$r(z) = \mathscr{S}(\chi(z)), \tag{1.94}$$

where

$$\chi(z) = \frac{1}{H_0 a_0} \int_0^z \frac{dz'}{E(z')} \tag{1.95}$$

and

$$\mathscr{S}(\chi) = \begin{cases} \sin\chi & : \quad k = 1 \\ \chi & : \quad k = 0 \\ \sinh\chi & : \quad k = 1. \end{cases} \tag{1.96}$$

Inserting this in (1.79) gives finally the relation we were looking for

$$D_L(z) = \frac{1}{H_0} \mathscr{D}_L(z; \Omega_K, \Omega_X), \tag{1.97}$$

with

$$\mathscr{D}_L(z; \Omega_K, \Omega_X) = (1+z) \frac{1}{|\Omega_K|^{1/2}} \mathscr{S}\left(|\Omega_K|^{1/2} \int_0^z \frac{dz'}{E(z')} \right) \tag{1.98}$$

for $k = \pm 1$. For a flat universe, $\Omega_K = 0$ or equivalently $\Omega_0 = 1$, the "Hubble-constant-free" luminosity distance is

$$\mathscr{D}_L(z) = (1+z) \int_0^z \frac{dz'}{E(z')}. \tag{1.99}$$

Astronomers use as logarithmic measures of $\mathscr{L}$ and $\mathscr{F}$ the *absolute and apparent magnitudes*[8], denoted by M and m, respectively. The conventions are chosen such that the *distance modulus* $\mu := m - M$ is related to D_L as follows

$$m - M = 5 \log \left(\frac{D_L}{1\, Mpc} \right) + 25. \tag{1.100}$$

Inserting the representation (1.97), we obtain the following relation between the apparent magnitude m and the redshift z:

$$m = \mathscr{M} + 5 \log \mathscr{D}_L(z; \Omega_K, \Omega_X), \tag{1.101}$$

where, for our purpose, $\mathscr{M} = M - 5 \log H_0 + 25$ is an uninteresting fit parameter. The comparison of this theoretical *magnitude redshift relation* with data will lead to interesting restrictions for the cosmological Ω-parameters. In practice often only Ω_M and Ω_Λ are kept as independent parameters, where from now on the subscript M denotes (as in most papers) nonrelativistic matter.

The following remark about *degeneracy curves* in the Ω-plane is important in this context. For a fixed z in the presently explored interval, the contours defined by the equations $\mathscr{D}_L(z; \Omega_M, \Omega_\Lambda) = const$ have little curvature, and thus we can associate an

[8] Beside the (bolometric) magnitudes m, M, astronomers also use magnitudes m_B, m_V, ... referring to certain wavelength bands B (blue), V (visual), and so on.

approximate slope to them. For $z = 0.4$, the slope is about 1 and increases to 1.5-2 by $z = 0.8$ over the interesting range of Ω_M and Ω_Λ. Hence even quite accurate data can at best select a strip in the Ω-plane, with a slope in the range just discussed.

In this context, it is also interesting to determine the dependence of the *deceleration parameter*

$$q_0 = -\left(\frac{a\ddot{a}}{\dot{a}^2}\right)_0 \tag{1.102}$$

on Ω_M and Ω_Λ. At an any cosmic time we obtain from (1.23) and (1.85) for the deceleration function

$$q(z) \equiv -\frac{\ddot{a}a}{\dot{a}^2} = \frac{1}{2}\frac{1}{E^2(z)}\sum_X \Omega_X(1+z)^{3(1+w_X)}(1+3w_X). \tag{1.103}$$

For $z = 0$ this gives

$$q_0 = \frac{1}{2}\sum_X \Omega_X(1+3w_X) = \frac{1}{2}(\Omega_M - 2\Omega_\Lambda + \cdots). \tag{1.104}$$

The line $q_0 = 0$ ($\Omega_\Lambda = \Omega_M/2$) separates decelerating from accelerating universes at the present time. For given values of Ω_M, Ω_Λ, etc, (1.103) vanishes for z determined by

$$\Omega_M(1+z)^3 - 2\Omega_\Lambda + \cdots = 0. \tag{1.105}$$

This equation gives the redshift at which the deceleration period ends (coasting redshift).

Remark

Without using the Friedmann equation one can express the luminosity distance $D_L(z)$ purely kinematically in terms of the deceleration variable $q(z)$. With the help of the previous tools, the reader may derive the following relations for a spatially flat Friedmann spacetime:

$$H^{-1}(z) = H_0^{-1}\exp\left\{-\int_0^z \frac{1+q(z')}{1+z'}dz'\right\}, \tag{1.106}$$

$$D_L(z) = (1+z)H_0^{-1}\int_0^z dz'\exp\left\{-\int_0^{z'}[1+q(z'')]\,d\ln(1+z'')\right\}. \tag{1.107}$$

It has been claimed that the existing supernova data imply an accelerating phase at late times [16].

Generalization for Dynamical Models of Dark Energy

If the vacuum energy constitutes the missing two-thirds of the average energy density of the *present* Universe, we would be confronted with the following *cosmic coincidence* problem: Since the vacuum energy density is constant in time – at least after the QCD phase transition – while the matter energy density decreases as the Universe expands, it would be more than surprising if the two are comparable just at about the present time, while their ratio was tiny in the early Universe and would become very large in the distant future. The goal of dynamical models of dark energy is to avoid such an extreme fine-tuning. The ratio p/ρ of this component then becomes a function of redshift, which we denote by $w_Q(z)$ (because so-called quintessence models are particular examples). Then the function $E(z)$ in (1.91) gets modified.

To see how, we start from the energy equation (1.83) and write this as

$$\frac{d\ln(\rho_Q a^3)}{d\ln(1+z)} = 3w_Q.$$

This gives

$$\rho_Q(z) = \rho_{Q0}(1+z)^3 \exp\left(\int_0^{\ln(1+z)} 3w_Q(z')d\ln(1+z')\right)$$

or

$$\rho_Q(z) = \rho_{Q0} \exp\left(3\int_0^{\ln(1+z)} (1+w_Q(z'))d\ln(1+z')\right). \tag{1.108}$$

Hence, we have to perform on the right of (1.91) the following substitution:

$$\Omega_Q(1+z)^{3(1+w_Q)} \to \Omega_Q \exp\left(3\int_0^{\ln(1+z)} (1+w_Q(z'))d\ln(1+z')\right). \tag{1.109}$$

As indicated above, a much discussed class of dynamical models for Dark Energy are *quintessence models*. In many ways, people thereby repeat what has been done in inflationary cosmology. The main motivation there was (see Sect. 1.3) to avoid excessive fine tunings of standard big bang cosmology (horizon and flatness problems). It has to be emphasized, however, that quintessence models do *not* solve the vacuum energy problem, so far also not the coincidence puzzle.

Finally, I mention another theoretical complication. In the analysis of the data, the luminosity distance for an ideal Friedmann universe was always used. But the data are taken in the real inhomogeneous Universe. This may perhaps not be good enough, especially for high-redshift standard candles. The magnitude-redshift relation for a perturbed Friedmann model has been derived in [19] and was later used to determine the angular power spectrum of the luminosity distance [20]. One of the numerical results was that the uncertainties in determining cosmological parameters via the magnitude-redshift relation caused by fluctuations are small compared with the intrinsic dispersion in the absolute magnitude of Type Ia supernovae.

This subject was recently taken up in [21] as part of a program to develop the tools for extracting cosmological parameters, when much extended supernovae data become available.

1.3 Inflationary Scenario

1.3.1 Introduction

The horizon and flatness problems of standard big bang cosmology are so serious that the proposal of a very early accelerated expansion, preceding the hot era dominated by relativistic fluids, appears quite plausible. This general qualitative aspect of "inflation" is now widely accepted. However, when it comes to concrete model building the situation is not satisfactory. Since we do not know the fundamental physics at superhigh energies not too far from the Planck scale, models of inflation are usually of a phenomenological nature. Most models consist of a number of scalar fields, including a suitable form for their potential. Usually there is no direct link to fundamental theories, like supergravity, however, there have been many attempts in this direction. For the time being, inflationary cosmology should be regarded as an attractive scenario, and not yet as a theory.

The most important aspect of inflationary cosmology is that *the generation of perturbations on large scales from initial quantum fluctuations is unavoidable and predictable*. For a given model, these fluctuations can be calculated accurately because they are tiny and cosmological perturbation theory can be applied. And, most importantly, these predictions can be *confronted with the cosmic microwave anisotropy measurements*. We are in the fortunate position to witness rapid progress in this field. The results from various experiments, most recently from WMAP, give already strong support of the basic predictions of inflation. Further experimental progress can be expected in the coming years.

1.3.2 The Horizon Problem and the General Idea of Inflation

I begin by describing the famous horizon puzzle, which is a very serious causality problem of standard big bang cosmology.

Past and Future Light Cone Distances

Consider our past light cone for a Friedmann spacetime model (Fig. 1.4). For a radial light ray, the differential relation $dt = a(t)dr/(1-kr^2)^{1/2}$ holds for the coordinates (t,r) of the metric (1.78). The proper radius of the past light sphere at time t (cross section of the light cone with the hypersurface $\{t = const\}$) is

$$l_p(t) = a(t)\int_0^{r(t)} \frac{dr}{\sqrt{1-kr^2}}, \tag{1.110}$$

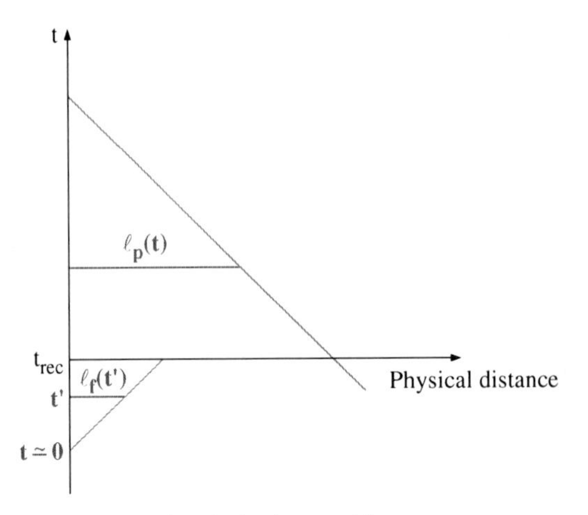

Fig. 1.4 Spacetime diagram illustrating the horizon problem.

where the coordinate radius is determined by

$$\int_0^{r(t)} \frac{dr}{\sqrt{1-kr^2}} = \int_t^{t_0} \frac{dt'}{a(t')}. \tag{1.111}$$

Hence,

$$l_p(t) = a(t) \int_t^{t_0} \frac{dt'}{a(t')}. \tag{1.112}$$

We rewrite this in terms of the redhift variable. From $1+z = a_0/a$, we get $dz = -(1+z)Hdt$, so

$$\frac{dt}{dz} = -\frac{1}{H_0(1+z)E(z)}, \quad H(z) = H_0E(z).$$

Therefore,

$$l_p(z) = \frac{1}{H_0(1+z)} \int_0^z \frac{dz'}{E(z')}. \tag{1.113}$$

Similarly, the extension $l_f(t)$ of the forward light cone at time t of a very early event ($t \simeq 0$, $z \simeq \infty$) is

$$l_f(t) = a(t) \int_0^t \frac{dt'}{a(t')} = \frac{1}{H_0(1+z)} \int_z^\infty \frac{dz'}{E(z')}. \tag{1.114}$$

For the present Universe (t_0), this becomes what is called the *particle horizon distance*

$$D_{hor} = H_0^{-1} \int_0^\infty \frac{dz'}{E(z')}, \tag{1.115}$$

and gives the size of the *observable Universe* .

Analytical expressions for these distances are only available in special cases. For orientation, we consider first the Einstein-de Sitter model ($K=0,\ \Omega_\Lambda=0,\ \Omega_M=1$), for which $a(t)=a_0(t/t_0)^{2/3}$ and thus

$$D_{hor}=3t_0=2H_0^{-1},\ \ l_f(t)=3t,\ \ \frac{l_p}{l_f}=\left(\frac{t_0}{t}\right)^{1/3}-1=\sqrt{1+z}-1. \tag{1.116}$$

For a flat universe, a good fitting formula for cases of interest is (Hu and White)

$$D_{hor}\simeq 2H_0^{-1}\frac{1+0.084\ln\Omega_M}{\sqrt{\Omega_M}}. \tag{1.117}$$

It is often convenient to work with "comoving distances," by rescaling distances referring to time t (like $l_p(t),l_f(t)$) with the factor $a(t_0)/a(t)=1+z$ to the present. We indicate this by the superscript c. For instance,

$$l_p^c(z)=\frac{1}{H_0}\int_0^z\frac{dz'}{E(z')}. \tag{1.118}$$

This distance is plotted in Fig. 1.3 as $D_{com}(z)$. Note that for $a_0=1:\ l_f^c(\eta)=\eta,\ l_p^c(\eta)=\eta_0-\eta$. Hence, (1.114) gives the following relation between η and z:

$$\eta=\frac{1}{H_0}\int_z^\infty\frac{dz'}{E(z')}.$$

The Number of Causality Distances on the Cosmic Photosphere

The number of causality distances at redshift z between two antipodal emission points is equal to $l_p(z)/l_f(z)$, and thus the ratio of the two integrals on the right of (1.113) and (1.114). We are particularly interested in this ratio at the time of last scattering with $z_{rec}\simeq 1100$. Then, we can use for the numerator a flat universe with nonrelativistic matter, while for the denominator we can neglect in the standard hot big bang model Ω_K and Ω_Λ. A reasonable estimate is already obtained by using the simple expression in (1.116), i.e., $z_{rec}^{1/2}\approx 30$. A more accurate evaluation would increase this number to about 40. The length $l_f(z_{rec})$ subtends an angle of about 1 degree (Exercise). How can it be that there is such a large number of causally disconnected regions we see on the microwave sky all having the same temperature? This is what is meant by the *horizon problem* and was a troublesome mystery before the invention of inflation.

Vacuum-Like Energy and Exponential Expansion

This causality problem is potentially avoided, if $l_f(t)$ would be increased in the very early Universe as a result of different physics. If, for instance, a vacuum-like energy density would dominate, the Universe would undergo an *exponential expansion*. Indeed, in this case the Friedmann equation is

$$\left(\frac{\dot{a}}{a}\right)^2 + \frac{k}{a^2} = \frac{8\pi G}{3}\rho_{vac}, \quad \rho_{vac} \simeq const, \tag{1.119}$$

and has the solutions

$$a(t) \propto \begin{cases} \cosh H_{vac}t & : \quad k = 1 \\ e^{H_{vac}t} & : \quad k = 0 \\ \sinh H_{vac}t & : \quad k = 1, \end{cases} \tag{1.120}$$

with

$$H_{vac} = \sqrt{\frac{8\pi G}{3}\rho_{vac}}\,. \tag{1.121}$$

Assume that such an exponential expansion starts for some reason at time t_i and ends at the *reheating time* t_e, after which standard expansion takes over. From

$$a(t) = a(t_i)e^{H_{vac}(t-t_i)} \quad (t_i < t < t_e), \tag{1.122}$$

for $k = 0$ we get

$$l_f^c(t_e) \simeq a_0 \int_{t_i}^{t_e} \frac{dt}{a(t)} = \frac{a_0}{H_{vac}a(t_i)}\left(1 - e^{-H_{vac}\Delta t}\right) \simeq \frac{a_0}{H_{vac}a(t_i)},$$

where $\Delta t := t_e - t_i$. We want to satisfy the condition $l_f^c(t_e) \gg l_p^c(t_e) \simeq H_0^{-1}$ (see (1.117)), i.e.,

$$a_i H_{vac} \ll a_0 H_0 \quad \Leftrightarrow \frac{a_i}{a_e} \ll \frac{a_0 H_0}{a_e H_{vac}} \tag{1.123}$$

or

$$e^{H_{vac}\Delta t} \gg \frac{a_e H_{vac}}{a_0 H_0} = \frac{H_{eq}a_{eq}}{H_0 a_0}\frac{H_{vac}a_e}{H_{eq}a_{eq}}.$$

Here, eq indicates the values at the time t_{eq} when the energy densities of nonrelativistic and relativistic matter were equal. We now use the Friedmann equation for $k = 0$ and $w := p/\rho = const$. From (1.84), it follows that in this case

$$Ha \propto a^{-(1+3w)/2},$$

and hence we arrive at

$$e^{H_{vac}\Delta t} \gg \left(\frac{a_0}{a_{eq}}\right)^{1/2}\left(\frac{a_{eq}}{a_e}\right) = (1+z_{eq})^{1/2}\left(\frac{T_e}{T_{eq}}\right) = (1+z_{eq})^{-1/2}\frac{T_{Pl}}{T_0}\frac{T_e}{T_{Pl}}, \tag{1.124}$$

where we used $aT = const$. So the number of e-folding periods during the inflationary period, $\mathcal{N} = H_{vac}\Delta t$, should satisfy

$$\mathcal{N} \gg \ln\left(\frac{T_{Pl}}{T_0}\right) - \frac{1}{2}\ln z_{eq} + \ln\left(\frac{T_e}{T_{Pl}}\right) \simeq 70 + \ln\left(\frac{T_e}{T_{Pl}}\right). \tag{1.125}$$

For a typical GUT scale, $T_e \sim 10^{14}\ GeV$, we arrive at the condition $\mathcal{N} \gg 60$.

Such an exponential expansion would also solve the *flatness problem*, another worry of standard big bang cosmology. Let me recall how this problem arises.

The Friedmann equation (1.17) can be written as

$$(\Omega^{-1} - 1)\rho a^2 = -\frac{3k}{8\pi G} = const.,$$

where

$$\Omega(t) := \frac{\rho(t)}{3H^2/8\pi G} \tag{1.126}$$

(ρ includes vacuum energy contributions). Thus,

$$\Omega^{-1} - 1 = (\Omega_0^{-1} - 1)\frac{\rho_0 a_0^2}{\rho a^2}. \tag{1.127}$$

Without inflation, we have

$$\rho = \rho_{eq}\left(\frac{a_{eq}}{a}\right)^4 \quad (z > z_{eq}), \tag{1.128}$$

$$\rho = \rho_0\left(\frac{a_0}{a}\right)^3 \quad (z < z_{eq}). \tag{1.129}$$

According to (1.85), z_{eq} is given by

$$1 + z_{eq} = \frac{\Omega_M}{\Omega_R} \simeq 10^4\ \Omega_0 h_0^2. \tag{1.130}$$

For $z > z_{eq}$, we obtain from (1.127) and (1.128)

$$\Omega^{-1} - 1 = (\Omega_0^{-1} - 1)\frac{\rho_0 a_0^2}{\rho_{eq} a_{eq}^2}\frac{\rho_{eq} a_{eq}^2}{\rho a^2} = (\Omega_0^{-1} - 1)(1 + z_{eq})^{-1}\left(\frac{a}{a_{eq}}\right)^2 \tag{1.131}$$

or

$$\Omega^{-1} - 1 = (\Omega_0^{-1} - 1)(1 + z_{eq})^{-1}\left(\frac{T_{eq}}{T}\right)^2 \simeq 10^{-60}(\Omega_0^{-1} - 1)\left(\frac{T_{Pl}}{T}\right)^2. \tag{1.132}$$

Let us apply this equation for $T = 1 MeV$, $\Omega_0 \simeq 0.2 - 0.3$. Then $|\ \Omega - 1\ | \leq 10^{-15}$, thus the Universe was already incredibly flat at modest temperatures, not much higher than at the time of nucleosynthesis.

Such a fine tuning must have a physical reason. This is naturally provided by inflation because our observable Universe could originate from a small patch at t_e. (A tiny part of the Earth surface is also practically flat.)

Beside the horizon scale $l_f(t)$, the *Hubble length* $H^{-1}(t) = a(t)/\dot{a}(t)$ also plays an important role. One might call this the "microphysics horizon" because this is the maximal distance microphysics that can operate coherently in one expansion time. It is this length scale that enters in basic evolution equations, such as the equation of motion for a scalar field (see Eq. (1.139) below).

We sketch in Figs. 1.5 – 1.7 the various length scales in inflationary models that are for models with a period of accelerated (e.g., exponential) expansion. From these, it is obvious that there can be – at least in principle – a *causal generation mechanism for perturbations*. This topic will be discussed in great detail in later parts of this chapter.

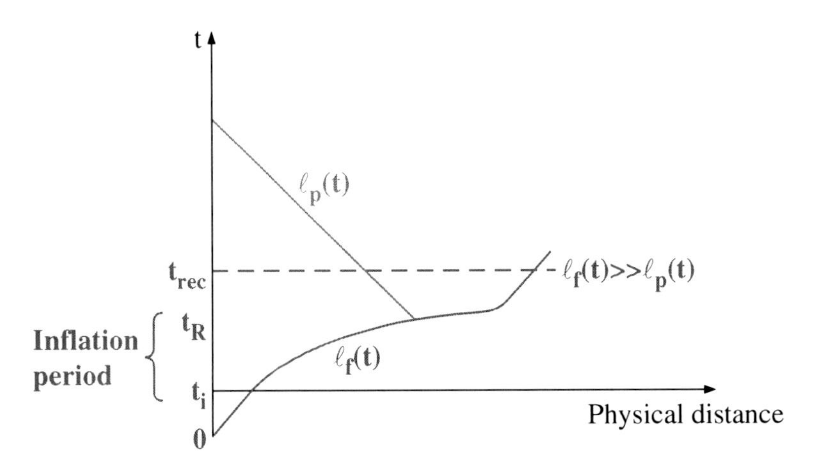

Fig. 1.5 Past and future light cones in models with an inflationary period.

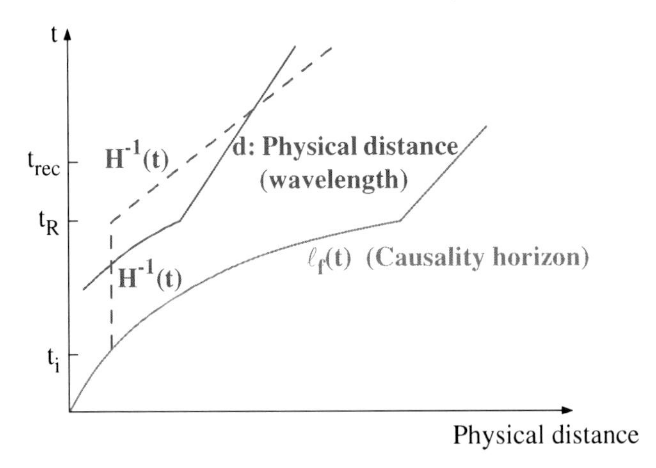

Fig. 1.6 Physical distance (e.g., between clusters of galaxies), Hubble distance, and causality horizon in inflationary models.

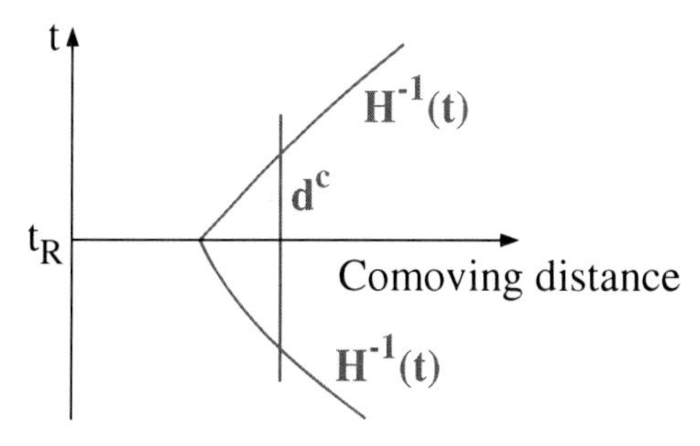

Fig. 1.7 Part of Fig. 3.3 expressed in terms of comoving distances.

Exponential inflation is just an example. What we really need is an early phase during which the *comoving Hubble length decreases* (Fig. 1.7). This means that (for Friedmann spacetimes)

$$\boxed{\left(H^{-1}(t)/a\right)^{\cdot} < 0.} \tag{1.133}$$

This is the *general definition of inflation*; equivalently, $\ddot{a} > 0$ (accelerated expansion). For a Friedmann model, Eq. (1.23) tells us that

$$\ddot{a} > 0 \Leftrightarrow p < -\rho/3. \tag{1.134}$$

This is, of course, not satisfied for "ordinary" fluids.

Assume, as another example, *power-law inflation*: $a \propto t^p$. Then $\ddot{a} > 0 \Leftrightarrow p > 1$.

1.3.3 Scalar Field Models

Models with $p < -\rho/3$ are naturally obtained in scalar field theories. Most of the time, we shall consider the simplest case of *one* neutral scalar field φ minimally coupled to gravity. Thus, the Lagrangian density is assumed to be

$$\mathscr{L} = \frac{M_{pl}^2}{16\pi} R[g] - \frac{1}{2}\nabla_\mu \varphi \nabla^\mu \varphi - V(\varphi), \tag{1.135}$$

where $R[g]$ is the Ricci scalar for the metric g. The scalar field equation is

$$\Box \varphi = V_{,\varphi}, \tag{1.136}$$

and the energy-momentum tensor in the Einstein equation

$$G_{\mu\nu} = \frac{8\pi}{M_{Pl}^2} T_{\mu\nu} \tag{1.137}$$

is

$$T_{\mu\nu} = \nabla_\mu \varphi \nabla_\nu \varphi + g_{\mu\nu} \mathscr{L}_\varphi \tag{1.138}$$

($\mathscr{L}_\varphi$ is the scalar field part of (3.26)).

We consider first Friedmann spacetimes. Using previous notation, we obtain from (1.1)

$$\sqrt{-g} = a^3 \sqrt{\gamma}, \quad \Box\varphi = \frac{1}{\sqrt{-g}} \partial_\mu (\sqrt{-g} g^{\mu\nu} \partial_\nu \varphi) = -\frac{1}{a^3}(a^3\dot{\varphi})^{\cdot} + \frac{1}{a^2} \triangle_\gamma \varphi.$$

The field equation (1.136) becomes

$$\boxed{\ddot{\varphi} + 3H\dot{\varphi} - \frac{1}{a^2}\triangle_\gamma \varphi = -V_{,\varphi}(\varphi).} \tag{1.139}$$

Note that the expansion of the Universe induces a "friction" term. In this basic equation one also sees the appearance of the Hubble length. From (1.138), we obtain the energy density and the pressure of the scalar field

$$\rho_\varphi = T_{00} = \frac{1}{2}\dot{\varphi}^2 + V + \frac{1}{2a^2}(\nabla\varphi)^2, \tag{1.140}$$

$$p_\varphi = \frac{1}{3} T^i{}_i = \frac{1}{2}\dot{\varphi}^2 - V - \frac{1}{6a^2}(\nabla\varphi)^2. \tag{1.141}$$

(Here, $(\nabla\varphi)^2$ denotes the squared gradient on the 3-space (Σ,γ).)

Suppose the gradient terms can be neglected, and that φ is during a certain phase slowly varying in time, then we get

$$\rho_\varphi \approx V, \quad p_\varphi \approx -V. \tag{1.142}$$

Thus $p_\varphi \approx -\rho_\varphi$, as for a cosmological term.

Let us ignore for the time being the spatial inhomogeneities in the previous equations. Then, these reduce to

$$\ddot{\varphi} + 3H\dot{\varphi} + V_{,\varphi}(\varphi) = 0; \tag{1.143}$$

$$\rho_\varphi = \frac{1}{2}\dot{\varphi}^2 + V, \quad p_\varphi = \frac{1}{2}\dot{\varphi}^2 - V. \tag{1.144}$$

Beside (1.143), the other dynamical equation is the Friedmann equation

$$\boxed{H^2 + \frac{K}{a^2} = \frac{8\pi}{3M_{Pl}^2}\left[\frac{1}{2}\dot{\varphi}^2 + V(\varphi)\right].} \tag{1.145}$$

Equations (1.143) and (1.145) define a nonlinear dynamical system for the dynamical variables $a(t), \varphi(t)$, which can be studied in detail (see, e.g., [22]).

Let us ignore the curvature term K/a^2 in (1.145). Differentiating this equation and using (1.143) shows that

$$\dot{H} = -\frac{4\pi}{M_{Pl}^2}\dot{\varphi}^2. \tag{1.146}$$

Regard H as a function of φ, then

$$\frac{dH}{d\varphi} = -\frac{4\pi}{M_{Pl}^2}\dot{\varphi}. \tag{1.147}$$

This allows us to write the Friedmann equation as

$$\left(\frac{dH}{d\varphi}\right)^2 - \frac{12\pi}{M_{Pl}^2}H^2(\varphi) = -\frac{32\pi^2}{M_{Pl}^4}V(\varphi). \tag{1.148}$$

For a given potential $V(\varphi)$, this is a differential equation for $H(\varphi)$. Once this function is known, we obtain $\varphi(t)$ from (1.147) and $a(t)$ from (1.146).

1.3.3.1 Power-Law Inflation

We now proceed in the reverse order, assuming that $a(t)$ follows a power law

$$a(t) = const.\, t^p. \tag{1.149}$$

Then $H = p/t$, so by (1.146)

$$\dot{\varphi} = \sqrt{\frac{p}{4\pi}}M_{Pl}\frac{1}{t}, \quad \varphi(t) = \sqrt{\frac{p}{4\pi}}M_{Pl}\ln(t) + const.,$$

hence

$$H(\varphi) \propto \exp\left(-\sqrt{\frac{4\pi}{p}}\frac{\varphi}{M_{Pl}}\right). \tag{1.150}$$

Using this in (1.148) leads to an exponential potential

$$V(\varphi) = V_0 \exp\left(-4\sqrt{\frac{\pi}{p}}\frac{\varphi}{M_{Pl}}\right). \tag{1.151}$$

1.3.3.2 Slow-Roll Approximation

An important class of solutions is obtained in the slow-roll approximation (SLA), in which the basic Eqs. (1.143) and (1.145) can be replaced by

$$H^2 = \frac{8\pi}{3M_{Pl}^2}V(\varphi), \tag{1.152}$$

$$3H\dot{\varphi} = -V_{,\varphi}. \tag{1.153}$$

A necessary condition for their validity is that the *slow-roll parameters*

$$\varepsilon_V(\varphi) := \frac{M_{Pl}^2}{16\pi}\left(\frac{V_{,\varphi}}{V}\right)^2, \tag{1.154}$$

$$\eta_V(\varphi) := \frac{M_{Pl}^2}{8\pi}\frac{V_{,\varphi\varphi}}{V} \tag{1.155}$$

are small:

$$\varepsilon_V \ll 1, \quad |\,\eta_V\,| \ll 1. \tag{1.156}$$

These conditions, which guarantee that the potential is flat, are, however, not sufficient.

The simplified system (1.152) and (1.153) implies

$$\boxed{\dot{\varphi}^2 = \frac{M_{Pl}^2}{24\pi}\frac{1}{V}\left(V_{,\varphi}\right)^2.} \tag{1.157}$$

This is a differential equation for $\varphi(t)$.

Let us consider potentials of the form

$$V(\varphi) = \frac{\lambda}{n}\varphi^n. \tag{1.158}$$

Then, Eq. (1.157) becomes

$$\boxed{\dot{\varphi}^2 = \frac{n^2 M_{Pl}^2}{24\pi}\frac{1}{\varphi^2}V.} \tag{1.159}$$

Hence, (1.152) implies

$$\frac{\dot{a}}{a} = -\frac{4\pi}{nM_{Pl}^2}(\varphi^2)^{\cdot},$$

and so

$$\boxed{a(t) = a_0 \exp\left[\frac{4\pi}{nM_{Pl}^2}(\varphi_0^2 - \varphi^2(t))\right].} \tag{1.160}$$

We see from (1.159) that $\frac{1}{2}\dot{\varphi}^2 \ll V(\varphi)$ for

$$\varphi \gg \frac{n}{4\sqrt{3\pi}}M_{Pl}. \tag{1.161}$$

Consider first the example $n = 4$. Then, (1.159) implies

$$\frac{\dot{\varphi}}{\varphi} = \sqrt{\frac{\lambda}{6\pi}}M_{Pl} \Rightarrow \varphi(t) = \varphi_0 \exp\left(-\sqrt{\frac{\lambda}{6\pi}}M_{Pl}\,t\right). \tag{1.162}$$

For $n \neq 4$:

$$\varphi(t)^{2-n/2} = \varphi_0^{2-n/2} + t\left(2-\frac{n}{2}\right)\sqrt{\frac{n\lambda}{24\pi}}M_{Pl}^{3-n/2}. \tag{1.163}$$

For the special case $n = 2$ this gives, using the notation $V = \frac{1}{2}m^2\varphi^2$, the simple result

$$\varphi(t) = \varphi_0 - \frac{mM_{Pl}}{2\sqrt{3\pi}}t. \tag{1.164}$$

Inserting this into (1.160) provides the time dependence of $a(t)$.

1.3.3.3 Why did Inflation Start?

Attempts to answer this and related questions are *very speculative* indeed. A reasonable direction is to imagine random initial conditions and try to understand how inflation can emerge, perhaps generically, from such a state of matter. A.Linde first discussed a scenario along these lines, which he called *chaotic inflation*. In the context of a single scalar field model, he argued that typical initial conditions correspond to $\frac{1}{2}\dot{\varphi}^2 \sim \frac{1}{2}(\partial_i\varphi)^2 \sim V(\varphi) \sim 1$ (in Planckian units). The chance that the potential energy dominates in some domain of size $> \mathcal{O}(1)$ is presumably not very small. In this situation, inflation could begin and $V(\varphi)$ would rapidly become even more dominant, which ensures continuation of inflation. Linde concluded from such considerations that chaotic inflation occurs under rather natural initial conditions. For this to happen, the form of the potential $V(\varphi)$ can even be a simple power law of the form (1.158). Many questions remain, however, open.

The chaotic inflationary Universe will look on very large scales – much larger than the present Hubble radius – extremely inhomogeneous. For a review of this scenario, I refer to [23]. A much more extended discussion of inflationary models, including references, can be found in [3].

1.3.3.4 The Trans-Planckian Problem

Another serious worry is this: If the period of inflation lasted sufficiently long (see the inequality (1.125)), then the scales inside today's Hubble radius started out at the beginning of inflation with physical wavelengths *smaller* than the Planck scale. In this domain, classical GR can most probably no more be trusted.

Optimistically, one can hope that observations of primordial spectra may turn out to be a window to unknown physics not far from the Planck scale.

1.4 Cosmological Perturbation Theory

The astonishing isotropy of the CMB radiation provides direct evidence that the early Universe can be described in a good first approximation by a Friedmann model. At the time of recombination, deviations from homogeneity and isotropy have been very small indeed ($\sim 10^{-5}$). Thus, there was a long period during which deviations from Friedmann models can be studied perturbatively, i.e., by linearizing the Einstein and matter equations about solutions of the idealized Friedmann–Lemaître models.

Cosmological perturbation (CPT) theory is a very important tool that is by now well developed. (Some reviews are [24], [25], and [26].) Here, I give an abbreviated version of my Combo-Lectures [27] where the detailed derivations (and sometimes rather lengthy calculations) can be found.

The formalism, developed in this part, will later be applied to two main problems: (1) The generation of primordial fluctuations during an inflationary era. (2) The evolution of these perturbations during the linear regime. A main goal will be to determine the CMB power spectrum as a function of certain cosmological parameters. Among these, the fractions of *Dark Matter* and *Dark Energy* are particularly interesting.

In this section, we develop the model independent parts of CPT. This forms the basis of much that follows. The development is in principle quite straightforward. Unfortunately, a lot of symbols have to be introduced, to a large extent because of the gauge freedom implied by the diffeomorphism invariance of GR.

1.4.1 Generalities

For the unperturbed Friedmann models, the metric is denoted by $g^{(0)}$ and has the form

$$g^{(0)} = -dt^2 + a^2(t)\gamma = a^2(\eta)\left[-d\eta^2 + \gamma\right]; \tag{1.165}$$

γ is the metric of a space with constant curvature K. In addition, we have matter variables for the various components (radiation, neutrinos, baryons, cold dark matter [CDM], etc). We shall linearize all basic equations about the unperturbed solutions.

1.4.1.1 Decomposition into Scalar, Vector, and Tensor Contributions

We may regard the various perturbation amplitudes as time-dependent functions on a three-dimensional Riemannian space (Σ, γ) of constant curvature K. Since such a space is highly symmetric, we can perform two types of decompositions.

Consider first the set $\mathscr{X}(\Sigma)$ of smooth vector fields on Σ. This module can be decomposed into an orthogonal sum of "scalar" and "vector" contributions

$$\mathscr{X}(\Sigma) = \mathscr{X}^S \bigoplus \mathscr{X}^V, \tag{1.166}$$

where $\mathscr{X}^S$ consists of all gradients and $\mathscr{X}^V$ of all vector fields with vanishing divergence. The scalar product of two vector fields ξ^i and η^j is defined by

$$(\xi,\eta)=\int_\Sigma \gamma_{ij}\xi^i\eta^j dv_\gamma\,. \tag{1.167}$$

(Prove the claimed orthonormality.)

Similarly, we can decompose a symmetric tensor $t \in \mathscr{S}(\Sigma)$ (= set of all symmetric tensor fields on Σ) into "scalar," "vector," and "tensor" contributions:

$$t_{ij}=t_{ij}^{(S)}+t_{ij}^{(V)}+t_{ij}^{(T)}\,, \tag{1.168}$$

where

$$t_{ij}^{(S)} = \frac{1}{3}t^k{}_k\gamma_{ij}+(\nabla_i\nabla_j-\frac{1}{3}\gamma_{ij}\nabla^2)f\,, \tag{1.169}$$

$$t_{ij}^{(V)} = \nabla_i\xi_j+\nabla_j\xi_i, \tag{1.170}$$

$$t_{ij}^{(T)} \;:\; t^{(T)i}{}_i=0;\;\; \nabla_j t^{(T)ij}=0. \tag{1.171}$$

In these equations, f is a function on Σ and ξ^i a vector field with vanishing divergence. In what follows ∇^2 always denotes $\gamma^{ij}\nabla_i\nabla_j$ on (Σ,γ). (Note that this does not agree with the Laplace–Beltrami operator for differential forms, except for functions. But for tensor fields, this is the natural extension of the Laplace operator on functions.) Show that the three components are orthogonal to each other with respect to the obvious generalization of the scalar product (1.167). This fact implies that the decomposition of t_{ij} is unique. A rigorous existence proof is given in [28].

In addition, these decompositions are respected by the covariant derivatives. For example, if $\xi\in\mathscr{X}(\Sigma)$, $\xi=\xi_*+\nabla f$, $\nabla\cdot\xi_*=0$, then

$$\nabla^2\xi=\nabla^2\xi_*+\nabla\left[\nabla^2 f+2Kf\right] \tag{1.172}$$

(prove this as an exercise). Here, the first term on the right has a vanishing divergence (show this), and the second term (the gradient) involves only f. For other cases, see Appendix B of [24]. Is there a conceptual proof based on the symmetrics of (Σ,γ)?

1.4.1.2 Decomposition into Spherical Harmonics

In a second step, we perform a harmonic decomposition. For $K=0$, this is just Fourier analysis. In this case, the spherical harmonics $\{Y\}$ of (Σ,γ) are the functions $Y(\mathbf{x};\mathbf{k})=\exp(i\mathbf{k}\cdot\mathbf{x})$ (for $\gamma=\delta_{ij}dx^i dx^j$). The *scalar* parts of vector and symmetric tensor fields can be expanded in terms of

$$Y_i := -k^{-1}\nabla_i Y, \tag{1.173}$$

$$Y_{ij} := k^{-2}\nabla_i\nabla_j Y+\frac{1}{3}\gamma_{ij}Y, \tag{1.174}$$

and $\gamma_{ij}Y$.

There are corresponding complete sets of spherical harmonics for $K \neq 0$. They are eigenfunctions of the Laplace operator on (Σ, γ):

$$(\nabla^2 + k^2)Y = 0. \tag{1.175}$$

Indices referring to the various modes are usually suppressed. By making use of the Riemann tensor of (Σ, γ), one can easily derive the following identities (using repeatedly the Ricci identity):

$$\begin{aligned}
\nabla_i Y^i &= kY, \\
\nabla^2 Y_i &= -(k^2 - 2K)Y_i, \\
\nabla_j Y_i &= -k(Y_{ij} - \frac{1}{3}\gamma_{ij}Y), \\
\nabla^j Y_{ij} &= \frac{2}{3}k^{-1}(k^2 - 3K)Y_i, \\
\nabla_j \nabla^m Y_{im} &= \frac{2}{3}(3K - k^2)(Y_{ij} - \frac{1}{3}\gamma_{ij}Y), \\
\nabla^2 Y_{ij} &= -(k^2 - 6K)Y_{ij}, \\
\nabla_m Y_{ij} - \nabla_j Y_{im} &= \frac{k}{3}\left(1 - \frac{3K}{k^2}\right)(\gamma_{im}Y_j - \gamma_{ij}Y_m).
\end{aligned} \tag{1.176}$$

The main point of the harmonic decomposition is, of course, that different modes in the linearized approximation do not couple. Hence, it suffices to consider a generic mode.

For the time being, we consider only scalar perturbations. Tensor perturbations (gravity modes) will be studied later. For the harmonic analysis of vector and tensor perturbations, I refer again to [24].

Exercise

Verify some of the relations in (1.176). ◇

1.4.1.3 Gauge Transformations, Gauge Invariant Amplitudes

In GR, the diffeomorphism group of spacetime is an invariance group. This means that we can replace the metric g and the matter fields by their pull-backs $\phi^\star(g)$, etc., for any diffeomorphism ϕ, without changing the physics. Consider, in particular, the flow ϕ_λ of a vector field ξ. By definition of the Lie derivative L_ξ, we have for the pull-back of a physical variable Q (metric g, etc)

$$\phi_\lambda^* Q = Q + \lambda L_\xi Q + \mathcal{O}(\lambda^2).$$

If

$$Q = Q^{(0)} + \lambda Q^{(1)} + \mathcal{O}(\lambda^2)$$

is the expansion of Q into background plus perturbations, we have

$$\phi_\lambda^* Q = Q^{(0)} + \lambda Q^{(1)} + \lambda L_\xi Q^{(0)} + \mathcal{O}(\lambda^2).$$

So, the first-order perturbation of $\phi_\lambda^* Q$ is $\lambda(Q^{(1)} + L_\xi Q^{(0)})$. In other words, $Q^{(1)}$ transforms as

$$Q^{(1)} \to Q^{(1)} + L_\xi Q^{(0)}.$$

This shows that for small-amplitude departures in

$$g = g^{(0)} + \delta g, \text{ etc.,} \tag{1.177}$$

we have the *gauge freedom*

$$\boxed{\delta g \to \delta g + L_\xi g^{(0)}}, \text{ etc.,} \tag{1.178}$$

where ξ is any vector field and L_ξ denotes its Lie derivative. (For further explanations, see [9], Sect. 4.1). These transformations will induce changes in the various perturbation amplitudes. It is clearly desirable to write all independent perturbation equations in a manifestly *gauge invariant* manner. In this way one can, for instance, avoid misinterpretations of the growth of density fluctuations, especially on super-horizon scales. Moreover, one gets rid of uninteresting gauge modes.

I find it astonishing that it took so long until the gauge invariant formalism was widely used.

1.4.1.4 Parametrization of the Metric Perturbations

The most general *scalar* perturbation of the metric can be parametrized as follows:

$$\delta g = a^2(\eta)\left[-2A d\eta^2 - 2B_{,i}dx^i d\eta + (2D\gamma_{ij} + 2E_{|ij})dx^i dx^j\right]. \tag{1.179}$$

The functions $A(\eta, x^i)$, B, D, E are the scalar perturbation amplitudes; $E_{|ij}$ denotes $\nabla_i \nabla_j E$ on (Σ, γ). Thus, the true metric is

$$\boxed{g = a^2(\eta)\left\{-(1+2A)d\eta^2 - 2B_{,i}dx^i d\eta + [(1+2D)\gamma_{ij} + 2E_{|ij}]dx^i dx^j\right\}.} \tag{1.180}$$

Let us work out how A, B, D, E change under a gauge transformation (1.178), provided the vector field is of the "scalar" type[9]:

$$\xi = \xi^0 \partial_0 + \xi^i \partial_i, \quad \xi^i = \gamma^{ij}\xi_{|j}. \tag{1.181}$$

[9] It suffices to consider this type of vector fields, since vector fields from $\mathscr{X}^V$ do not affect the scalar amplitudes; check this.

(The index 0 refers to the conformal time η.) For this, we need (' $\equiv d/d\eta$)

$$L_\xi a^2(\eta) = 2aa'\xi^0 = 2a^2\mathcal{H}\xi^0, \quad \mathcal{H} := a'/a,$$

$$L_\xi d\eta = dL_\xi \eta = (\xi^0)'d\eta + \xi^0{}_{|i}dx^i,$$

$$L_\xi dx^i = dL_\xi x^i = d\xi^i = \xi^i{}_{,j}dx^j + (\xi^i)'d\eta = \xi^i{}_{,j}dx^j + \xi'^{|i}d\eta,$$

implying

$$\begin{aligned} L_\xi \left(a^2(\eta)d\eta^2\right) &= 2a^2\left\{(\mathcal{H}\xi^0 + (\xi^0)')d\eta^2 + \xi^0{}_{|i}dx^i d\eta\right\}, \\ L_\xi \left(\gamma_{ij}dx^i dx^j\right) &= 2\xi_{|ij}dx^i dx^j + 2\xi'_{|i}dx^i d\eta. \end{aligned}$$

This gives the transformation laws:

$$A \to A + \mathcal{H}\xi^0 + (\xi^0)', \quad B \to B + \xi^0 - \xi', \quad D \to D + \mathcal{H}\xi^0, \quad E \to E + \xi. \tag{1.182}$$

From this, one concludes that the following *Bardeen potentials*

$$\Psi = A - \frac{1}{a}\left[a(B+E')\right]', \tag{1.183}$$

$$\Phi = D - \mathcal{H}(B+E'), \tag{1.184}$$

are gauge invariant.

Note that the transformations of A and D involve *only* ξ^0. This is also the case for the combinations

$$\chi := a(B+E') \to \chi + a\xi^0 \tag{1.185}$$

and

$$\kappa := \frac{3}{a}(\mathcal{H}A - D') - \frac{1}{a^2}\nabla^2\chi \tag{1.186}$$

$$\longrightarrow \kappa + \frac{3}{a}\left[\mathcal{H}(\mathcal{H}\xi^0 + (\xi^0)') - (\mathcal{H}\xi^0)'\right] - \frac{1}{a^2}\nabla^2\xi^0. \tag{1.187}$$

Therefore, it is good to work with A, D, χ, κ. This was emphasized in [29]. Below we will show that χ and κ have a simple geometrical meaning. Moreover, it will turn out that the perturbation of the Einstein tensor can be expressed completely in terms of the amplitudes A, D, χ, κ.

Exercise

The most general vector perturbation of the metric is obviously of the form

$$(\delta g_{\mu\nu}) = a^2(\eta)\begin{pmatrix} 0 & \beta_i \\ \beta_i & H_{i|j} + H_{j|i} \end{pmatrix},$$

with $B_i{}^{|i} = H_i{}^{|i} = 0$. Derive the gauge transformations for β_i and H_i. Show that H_i can be gauged away. Compute $R^0{}_j$ in this gauge. Result:

$$R^0{}_j = \frac{1}{2}\left(\nabla^2\beta_j + 2K\beta_j\right). \diamondsuit$$

1.4.1.5 Geometrical Interpretation

Let us first compute the scalar curvature $R^{(3)}$ of the slices with constant time η with the induced metric

$$g^{(3)} = a^2(\eta)\left[(1+2D)\gamma_{ij} + 2E_{|ij}\right]dx^i dx^j. \tag{1.188}$$

If we drop the factor a^2, then the Ricci tensor does not change, but $R^{(3)}$ has to be multiplied afterwards with a^{-2}.

For the metric $\gamma_{ij} + h_{ij}$, the *Palatini identity* (Eq. (4.20) in [9])

$$\delta R_{ij} = \frac{1}{2}\left[h^k{}_{i|jk} - h^k{}_{k|ij} + h^k{}_{j|ik} - \nabla^2 h_{ij}\right] \tag{1.189}$$

gives

$$\delta R^i{}_i = h^{ij}{}_{|ij} - \nabla^2 h \quad (h := h^i_i), \;\; h_{ij} = 2D\gamma_{ij} + 2E_{|ij}.$$

We also use

$$\begin{aligned} h = 6D + 2\nabla^2 E, \;\; E^{|ij}{}_{|ij} &= \nabla_j(\nabla^2\nabla^j E) = \nabla_j(\nabla^j\nabla^2 E - 2K\nabla^j E) \\ &= (\nabla^2)^2 E - 2K\nabla^2 E \end{aligned}$$

(we used $(\nabla_i\nabla^2 - \nabla^2\nabla_i)f = -R^{(0)}_{ij}\nabla^j f$, for a function f). This implies

$$\begin{aligned} h^{ij}{}_{|ij} &= 2\nabla^2 D + 2((\nabla^2)^2 E - 2K\nabla^2 E), \\ \delta R^i{}_i &= -4D - 4K\nabla^2 E), \end{aligned}$$

whence

$$\delta R = \delta R^i{}_i + h^{ij} R^{(0)}_{ij} = -4\nabla^2 D + 12KD.$$

This shows that D determines the scalar curvature perturbation

$$\boxed{\delta R^{(3)} = \frac{1}{a^2}(-4\nabla^2 D + 12KD).} \tag{1.190}$$

Next, we compute the second fundamental form[10] K_{ij} for the time slices. We shall show that

$$\boxed{\kappa = \delta K^i{}_i,} \tag{1.191}$$

[10] This geometrical concept is introduced in Appendix A of [9].

and

$$K_{ij} - \frac{1}{3} g_{ij} K^l{}_l = -(\chi_{|ij} - \frac{1}{3}\gamma_{ij}\nabla^2\chi). \qquad (1.192)$$

Derivation

In the following derivation, we make use of Sect. 2.9 of [9] on the $3+1$ formalism. According to Eq. (2.287) of this reference, the second fundamental form is determined in terms of the lapse α, the shift $\beta = \beta^i \partial_i$, and the induced metric $\bar{g}$ as follows (dropping indices)

$$K = -\frac{1}{2\alpha}(\partial_t - L_\beta)\bar{g}. \qquad (1.193)$$

To first order, this gives in our case

$$K_{ij} = -\frac{1}{2a(1+A)}\left[a^2(1+2D)\gamma_{ij} + 2a^2 E_{|ij}\right]' - aB_{|ij}. \qquad (1.194)$$

(Note that $\beta_i = -a^2 B_{,i}$, $\beta^i = -\gamma^{ij} B_{,j}$.)

In zeroth order, this gives

$$K^{(0)}_{ij} = -\frac{1}{a}\mathcal{H} g^{(0)}_{ij}. \qquad (1.195)$$

Collecting the first-order terms gives the claimed Eqs. (1.191) and (1.192). (Note that the trace-free part must be of first order because the zeroth order vanishes according to (1.195).)

Conformal Gauge

According to (1.182) and (1.185), we can always chose the gauge such that $B = E = 0$. This so-called *conformal Newtonian (or longitudinal) gauge* is often particularly convenient to work with. Note that in this gauge

$$\chi = 0, \; A = \Psi, \; D = \Phi, \; \kappa = \frac{3}{a}(\mathcal{H}\Psi - \Phi').$$

1.4.1.6 Scalar Perturbations of the Energy-Momentum Tensor

At this point, we do not want to specify the matter model. For a convenient parametrization of the scalar perturbations of the energy-momentum tensor $T_{\mu\nu} = T^{(0)}_{\mu\nu} + \delta T_{\mu\nu}$, we define the four-velocity u^μ as a normalized time-like eigenvector of $T^{\mu\nu}$:

$$T^\mu{}_\nu u^\nu = -\rho u^\mu, \tag{1.196}$$
$$g_{\mu\nu} u^\mu u^\nu = -1. \tag{1.197}$$

The eigenvalue ρ is the *proper energy-mass density*.

For the unperturbed situation, we have

$$u^{(0)0} = \frac{1}{a}, \; u_0^{(0)} = -a, \; u^{(0)i} = 0, \; T^{(0)0}{}_0 = -\rho^{(0)}, \; T^{(0)i}{}_j = p^{(0)} \delta^i{}_j, \; T^{(0)0}{}_i = 0. \tag{1.198}$$

Remark. There may be additional contributions to the unperturbed $T^{(0)}_{\mu\nu}$, for instance from the Λ term or unclustered dark energy. These change only the background evolution, but not the perturbation equations (as long as only the general form of the background equations is used).

Setting $\rho = \rho^{(0)} + \delta\rho, \; u^\mu = u^{(0)\mu} + \delta u^\mu$, etc, we obtain from (1.197)

$$\delta u^0 = -\frac{1}{a} A, \quad \delta u_0 = -aA. \tag{1.199}$$

The first-order terms of (1.196) give, using (1.198),

$$\delta T^\mu{}_0 u^{(0)0} + \delta^\mu{}_0 u^{(0)0} \delta\rho + \left(T^{(0)\mu}{}_\nu + \rho^{(0)} \delta^\mu{}_\nu \right) \delta u^\nu = 0.$$

For $\mu = 0$ and $\mu = i$, this leads to

$$\delta T^0{}_0 = -\delta\rho, \tag{1.200}$$
$$\delta T^i{}_0 = -a(\rho^{(0)} + p^{(0)}) \delta u^i. \tag{1.201}$$

From this, we can determine the components of $\delta T^0{}_j$:

$$\begin{aligned}
\delta T^0{}_j &= \delta \left[g^{0\mu} g_{j\nu} T^\nu{}_\mu \right] \\
&= \delta g^{0k} g^{(0)}_{ij} T^{(0)i}{}_k + g^{(0)00} \delta g_{0j} T^{(0)0}{}_0 + g^{(0)00} g^{(0)}_{ij} \delta T^i{}_0 \\
&= \left(-\frac{1}{a^2} \gamma^{ki} B_{|i} \right) (a^2 \gamma_{ij}) p^{(0)} \delta^i{}_k + \left(-\frac{1}{a^2} \right) (-a^2 B_{|j}) (-\rho^{(0)}) - \gamma_{ij} \delta T^i{}_0 .
\end{aligned}$$

Collecting terms gives

$$\delta T^0{}_j = a(\rho^{(0)} + p^{(0)}) \underbrace{\left[\gamma_{ij} \delta u^i - \frac{1}{a} B_{|j} \right]}_{a^{-2} \delta u_j}. \tag{1.202}$$

Scalar perturbations of δu^i can be represented as

$$\delta u^i = \frac{1}{a} \gamma^{ij} v_{|j}. \tag{1.203}$$

Inserting this above gives

$$\delta T^0{}_j = (\rho^{(0)} + p^{(0)})(v - B)_{|j}. \tag{1.204}$$

The scalar perturbations of the spatial components $\delta T^i{}_j$ can be represented as follows

$$\delta T^i{}_j = \delta^i{}_j\, \delta p + p^{(0)} \left(\Pi^{|i}{}_{|j} - \frac{1}{3}\delta^i{}_j\, \nabla^2 \Pi \right). \tag{1.205}$$

Let us collect these formulae (dropping (0) for the unperturbed quantities $\rho^{(0)}$, etc):

$$\begin{aligned} \delta u^0 &= -\frac{1}{a}A, \;\; \delta u_0 = -aA, \;\; \delta u^i = \frac{1}{a}\gamma^{ij} v_{|j} \;\; \Rightarrow \delta u_i = a(v - B)_{|i}; \\ \delta T^0{}_0 &= -\delta\rho, \\ \delta T^0{}_i &= (\rho + p)(v - B)_{|i}, \;\; \delta T^i{}_0 = -(\rho + p)\gamma^{ij} v_{|j}, \\ \delta T^i{}_j &= \delta p\, \delta^i{}_j + p \left(\Pi^{|i}{}_{|j} - \frac{1}{3}\delta^i{}_j\, \nabla^2 \Pi \right). \end{aligned} \tag{1.206}$$

Sometimes, we shall also use the quantity

$$\mathcal{Q} := a(\rho + p)(v - B),$$

in terms of which the energy flux density can be written as

$$\delta T^0{}_i = \frac{1}{a}\mathcal{Q}_{,i}\,, \;\; (\Rightarrow T^t{}_i = \mathcal{Q}_{,i}). \tag{1.207}$$

For fluids, one often decomposes δp as

$$p\pi_L := \delta p = c_s^2 \delta\rho + p\Gamma, \tag{1.208}$$

where c_s is the sound velocity

$$c_s^2 = \dot{p}/\dot{\rho}. \tag{1.209}$$

Γ measures the deviation between $\delta p/\delta\rho$ and $\dot{p}/\dot{\rho}$. One can show [30] that the divergence of the entropy current is proportional to Γ.

As for the metric, we have four perturbation amplitudes:

$$\boxed{\delta := \delta\rho/\rho\,,\;\; v\,,\;\; \Gamma\,,\;\; \Pi\,.} \tag{1.210}$$

Let us see how they change under gauge transformations:

$$\delta T^\mu{}_\nu \to \delta T^\mu{}_\nu + (L_\xi T^{(0)})^\mu{}_\nu, \;\; (L_\xi T^{(0)})^\mu{}_\nu = \xi^\lambda T^{(0)\mu}{}_{\nu,\lambda} - T^{(0)\lambda}{}_\nu \xi^\mu{}_{,\lambda} + T^{(0)\mu}{}_\lambda \xi^\lambda{}_{,\nu}. \tag{1.211}$$

Now,

$$(L_\xi T^{(0)})^0{}_0 = \xi^0 T^{(0)0}{}_{0,0} = \xi^0(-\rho)',$$

hence

$$\delta\rho \to \delta\rho + \rho'\xi^0\,; \quad \delta \to \delta + \frac{\rho'}{\rho}\xi^0 = \delta - 3(1+w)\mathscr{H}\xi^0 \tag{1.212}$$

($w := p/\rho$). Similarly ($\xi^i = \gamma^{ij}\xi_{|j}$):

$$(L_\xi T^{(0)})^0{}_i = 0 - T^{(0)j}{}_i\xi^0{}_{|j} + T^{(0)0}{}_0\xi^0{}_{,i} = -\rho\xi^0{}_{|i} - p\xi^0{}_{|i};$$

so

$$v - B \to (v - B) - \xi^0. \tag{1.213}$$

Finally,

$$(L_\xi T^{(0)})^i{}_j = p'\delta^i{}_j\xi^0,$$

hence

$$\delta p \to \delta p + p'\xi^0, \tag{1.214}$$

$$\Pi \to \Pi. \tag{1.215}$$

From (1.208), (1.212), and (1.214), we also obtain

$$\Gamma \to \Gamma. \tag{1.216}$$

We see that Γ, Π are gauge invariant. Note that the transformation of δ and $v - B$ involves only ξ^0, while v transforms as

$$v \to v - \xi'.$$

For $\mathscr{Q}$, we get

$$\mathscr{Q} \to \mathscr{Q} - a(\rho + p)\xi^0. \tag{1.217}$$

We can introduce various gauge invariant quantities. It is useful to adopt the following notation: For example, we use the symbol $\delta_{\mathscr{Q}}$ for that gauge invariant quantity, which is equal to δ in the gauge where $\mathscr{Q} = 0$, often called the *comoving gauge*. Thus,

$$\delta_{\mathscr{Q}} = \delta - \frac{3}{a\rho}\mathscr{H}\mathscr{Q} = \delta - 3(1+w)\mathscr{H}(v - B). \tag{1.218}$$

Similarly, gauge invariant perturbations related to the *zero-shear gauge* $\chi = 0$ are

$$\delta_\chi = \delta + 3\frac{(1+w)\mathscr{H}}{a}\chi = \delta + 3\mathscr{H}(1+w)(B + E'); \tag{1.219}$$

$$V := (v - B)_\chi = v - B + a^{-1}\chi = v + E' = \frac{1}{a}\left(\chi + \frac{1}{\rho + p}\mathscr{Q}\right); \tag{1.220}$$

$$\mathscr{Q}_\chi = \mathscr{Q} + (\rho + p)\chi = a(\rho + p)V. \tag{1.221}$$

Another important gauge invariant amplitude, often called the *curvature perturbation* (see (1.190)), is

$$\mathscr{R} := D_{\mathscr{Q}} = D + \mathscr{H}(v - B) = D_\chi + \mathscr{H}(v - B)_\chi = D_\chi + \mathscr{H}V. \tag{1.222}$$

1.4.2 Explicit form of the Energy-Momentum Conservation

After these preparations, we work out the consequences $\nabla \cdot T = 0$ of Einstein's field equations for the metric (1.180) and $T^\mu{}_\nu$ as given by (1.198) and (1.206). The details of the calculations are presented in Sect. 3.5 of [27].

The energy equation reads:

$$(\rho\delta)' + 3\mathscr{H}\rho\delta + 3\mathscr{H}p\pi_L + (\rho + p)\left[\nabla^2(v + E') + 3D'\right] = 0 \tag{1.223}$$

or, with $(\rho\delta)'/\rho = \delta' - 3\mathscr{H}(1 + w)\delta$ and (1.220),

$$\delta' + 3\mathscr{H}(c_s^2 - w)\delta + 3\mathscr{H}w\Gamma = -(1 + w)(\nabla^2 V + 3D'). \tag{1.224}$$

This gives, putting an index χ, the gauge invariant equation

$$\delta_\chi' + 3\mathscr{H}(c_s^2 - w)\delta_\chi + 3\mathscr{H}w\Gamma = -(1 + w)(\nabla^2 V + 3D_\chi'). \tag{1.225}$$

Conversely, Eq. (1.224) follows from (1.225): the χ-terms cancel, as is easily verified by using the zeroth-order equation

$$w' = -3(c_s^2 - w)(1 + w)\mathscr{H}, \tag{1.226}$$

that is easily derived from the Friedman equations in Sect. 1.1.3. From the definitions, it follows readily that the last factor in (1.224) is equal to $-(a\kappa - 3\mathscr{H}A - \nabla^2(v - B))$.

The momentum equation becomes

$$[(\rho + p)(v - B)]' + 4\mathscr{H}(\rho + p)(v - B) + (\rho + p)A + p\pi_L + \frac{2}{3}(\nabla^2 + 3K)p\Pi = 0. \tag{1.227}$$

Using (1.208) in the form

$$p\pi_L = \rho(c_s^2\delta + w\Gamma), \tag{1.228}$$

and putting the index χ at the perturbation amplitudes gives the gauge invariant equation

$$[(\rho + p)V]' + 4\mathscr{H}(\rho + p)V + (\rho + p)A_\chi + \rho c_s^2\delta_\chi + \rho w\Gamma + \frac{2}{3}(\nabla^2 + 3K)p\Pi = 0 \tag{1.229}$$

or[11]

$$V' + (1 - 3c_s^2)\mathscr{H}V + A_\chi + \frac{c_s^2}{1+w}\delta_\chi + \frac{w}{1+w}\Gamma + \frac{2}{3}(\nabla^2 + 3K)\frac{w}{1+w}\Pi = 0. \quad (1.230)$$

For later use, we write (1.227) also as

$$(v - B)' + (1 - 3c_s^2)\mathscr{H}(v - B) + A + \frac{c_s^2}{1+w}\delta + \frac{w}{1+w}\Gamma + \frac{2}{3}(\nabla^2 + 3K)\frac{w}{1+w}\Pi = 0 \quad (1.231)$$

(from which (1.230) follows immediately).

1.4.3 Einstein Equations

A direct computation of the first order changes $\delta G^\mu{}_\nu$ of the Einstein tensor for (1.179) is complicated. It is much simpler to proceed as follows: Compute first $\delta G^\mu{}_\nu$ in the *longitudinal gauge* $B = E = 0$. (That these gauge conditions can be imposed follows from (1.182).) Then, one can write the perturbed Einstein equations in a gauge invariant form. It is then easy to rewrite these equations without imposing any gauge conditions, thus obtaining the equations one would get for the general form (1.179). For the details, we refer again to [27].

Below, we collect the first-order Einstein equations, valid in any gauge (indicating also their origin). As perturbation amplitudes, we use A, D, χ, κ (metric functions) and $\delta, \mathscr{Q}, \Pi, \Gamma$ (matter functions) because these are either gauge invariant or their gauge transformations involve only the component ξ^0 of the vector field ξ^μ.

- definition of κ:

$$\kappa = 3(HA - \dot{D}) - \frac{1}{a^2}\nabla^2\chi; \quad (1.232)$$

- $\delta G^0{}_0$:

$$\frac{1}{a^2}(\nabla^2 + 3K)D + H\kappa = -4\pi G\rho\delta; \quad (1.233)$$

- $\delta G^0{}_j$:

$$\kappa + \frac{1}{a^2}(\nabla^2 + 3K)\chi = -12\pi G\mathscr{Q}; \quad (1.234)$$

- $\delta G^i{}_j - \frac{1}{3}\delta^i{}_j\,\delta G^k{}_k$:

$$\dot{\chi} + H\chi - A - D = 8\pi G a^2 p\Pi; \quad (1.235)$$

- $\delta G^i{}_i - \delta G^0{}_0$:

$$\dot{\kappa} + 2H\kappa = -\left(\frac{1}{a^2}\nabla^2 + 3\dot{H}\right)A + \underbrace{4\pi G(1 + 3c_s^2)\rho\delta + 12\pi G p\Gamma}_{4\pi G\rho(\delta + 3w\pi_L)}; \quad (1.236)$$

[11] Note that $h := \rho + p$ satisfies $h' = -3\mathscr{H}(1 + c_s^2)h$.

- $T^{0\nu}{}_{;\nu}$ (Eq. (1.224)):

$$\dot{\delta} + 3H(c_s^2 - w)\delta + 3Hw\Gamma = (1+w)(\kappa - 3HA) - \frac{1}{\rho a^2}\nabla^2 \mathcal{Q} \tag{1.237}$$

or

$$(\rho\delta)^{\cdot} + 3H\rho(\delta + \underbrace{w\pi_L}_{c_s^2\delta + w\Gamma}) = (\rho + p)(\kappa - 3HA) - \frac{1}{a^2}\nabla^2 \mathcal{Q}; \tag{1.238}$$

- $T^{i\nu}{}_{;\nu} = 0$ (Eq. (1.227)):

$$\dot{\mathcal{Q}} + 3H\mathcal{Q} = -(\rho + p)A - p\pi_L - \frac{2}{3}(\nabla^2 + 3K)p\Pi. \tag{1.239}$$

These equations are, of course, not all independent. Putting an index χ or $\mathcal{Q}$, etc., at the perturbation amplitudes in any of them gives a gauge invariant equation. We write these down for $A_\chi, D_\chi, \cdots$ (instead of $\mathcal{Q}_\chi$ we use V; see also (1.225) and (1.230)):

$$\kappa_\chi = 3(HA_\chi - \dot{D}_\chi); \tag{1.240}$$

$$\frac{1}{a^2}(\nabla^2 + 3K)D_\chi + H\kappa_\chi = -4\pi G\rho\delta_\chi; \tag{1.241}$$

$$\kappa_\chi = -12\pi G\mathcal{Q}_\chi; \tag{1.242}$$

$$A_\chi + D_\chi = -8\pi G a^2 p\Pi; \tag{1.243}$$

$$\dot{\kappa}_\chi + 2H\kappa_\chi = -\left(\frac{1}{a^2}\nabla^2 + 3\dot{H}\right)A_\chi + \underbrace{4\pi G(1 + 3c_s^2)\rho\delta_\chi + 12\pi G p\Gamma}_{4\pi G\rho(\delta_\chi + 3w\pi_L)}; \tag{1.244}$$

$$\dot{\delta}_\chi + 3H(c_s^2 - w)\delta_\chi + 3Hw\Gamma = -3(1+w)\dot{D}_\chi - \frac{1+w}{a}\nabla^2 V; \tag{1.245}$$

$$\dot{V} + (1 - 3c_s^2)HV = -\frac{1}{a}A_\chi - \frac{1}{a}\left[\frac{c_s^2}{1+w}\delta_\chi + \frac{w}{1+w}\Gamma + \frac{2}{3}(\nabla^2 + 3K)\frac{w}{1+w}\Pi\right]. \tag{1.246}$$

Harmonic Decomposition

We write these equations once more for the amplitudes of harmonic decompositions, adopting the following conventions. For those amplitudes which enter in $g_{\mu\nu}$ and $T_{\mu\nu}$ without spatial derivatives (i.e., A, D, δ, Γ), we set

$$A = A_{(k)}Y_{(k)}\ , etc\ ; \tag{1.247}$$

those which appear only through their gradients (B, v) are decomposed as

$$B = -\frac{1}{k} B_{(k)} Y_{(k)} \text{ ,} etc \text{ ,} \tag{1.248}$$

and, finally, we set for E and Π, entering only through second derivatives,

$$E = \frac{1}{k^2} E_{(k)} Y_{(k)} \quad (\Rightarrow \nabla^2 E = -E_{(k)} Y_{(k)}). \tag{1.249}$$

The reason for this is that we then have, using the definitions (1.173) and (1.174),

$$B_{|i} = B_{(k)} Y_{(k)i}, \quad \Pi_{|ij} - \frac{1}{3}\gamma_{ij}\nabla^2 \Pi = \Pi_{(k)} Y_{(k)ij}. \tag{1.250}$$

The spatial part of the metric in (1.180) then becomes

$$g_{ij} dx^i dx^j = a^2(\eta) \left[\gamma_{ij} + 2(D - \frac{1}{3}E)\gamma_{ij} Y + 2E Y_{ij} \right] dx^i dx^j. \tag{1.251}$$

The basic equations (1.232) – (1.239) imply for $A_{(k)}, B_{(k)}$, etc[12], dropping the index (k),

$$\kappa = 3(HA - \dot{D}) + \frac{k^2}{a^2}\chi, \tag{1.252}$$

$$-\frac{k^2 - 3K}{a^2} D + H\kappa = -4\pi G \rho \delta, \tag{1.253}$$

$$\kappa - \frac{k^2 - 3K}{a^2} \chi = -12\pi G \mathscr{Q}, \tag{1.254}$$

$$\dot{\chi} + H\chi - A - D = 8\pi G a^2 p \Pi / k^2, \tag{1.255}$$

$$\dot{\kappa} + 2H\kappa = \left(\frac{k^2}{a^2} - 3\dot{H} \right) A + \underbrace{4\pi G (1 + 3c_s^2)\rho\delta + 12\pi G p \Gamma}_{4\pi G \rho (\delta + 3w\pi_L)}, \tag{1.256}$$

$$(\rho\delta)^{\cdot} + 3H\rho(\delta + \underbrace{w\pi_L}_{c_s^2 \delta + w\Gamma}) = (\rho + p)(\kappa - 3HA) + \frac{k^2}{a^2}\mathscr{Q}, \tag{1.257}$$

$$\dot{\mathscr{Q}} + 3H\mathscr{Q} = -(\rho + p)A - p\pi_L + \frac{2}{3}\frac{k^2 - 3K}{k^2} p\Pi. \tag{1.258}$$

For later use, we also collect the gauge invariant Eqs. (1.240) – (1.246) for the Fourier amplitudes:

$$\kappa_\chi = 3(HA_\chi - \dot{D}_\chi), \tag{1.259}$$

$$-\frac{k^2 - 3K}{a^2} D_\chi + H\kappa_\chi = -4\pi G \rho \delta_\chi, \tag{1.260}$$

[12] We replace χ by $\chi_{(k)} Y_{(k)}$, where according to (1.185) $\chi_{(k)} = -(a/k)(B - k^{-1}E')$; Eq. (1.252) is then just the translation of (1.186) to the Fourier amplitudes, with $\kappa \to \kappa_{(k)} Y_{(k)}$. Similarly, $\mathscr{Q} \to \mathscr{Q}_{(k)} Y_{(k)}$, $\mathscr{Q}_{(k)} = -(1/k)a(\rho + p)(v - B)_{(k)}$.

$$\kappa_\chi = -12\pi G \mathcal{Q}_\chi \quad \left(\mathcal{Q}_\chi = -\frac{a}{k}(\rho+p)V\right), \tag{1.261}$$

$$k^2(A_\chi + D_\chi) = -8\pi G a^2 p\Pi, \tag{1.262}$$

$$\dot{\kappa}_\chi + 2H\kappa_\chi = \left(\frac{k^2}{a^2} - 3\dot{H}\right)A_\chi + \underbrace{4\pi G(1+3c_s^2)\rho\delta_\chi + 12\pi G p\Gamma}_{4\pi G\rho(\delta_\chi + 3w\pi_L)}, \tag{1.263}$$

$$\dot{\delta}_\chi + 3H(c_s^2 - w)\delta_\chi + 3Hw\Gamma = -3(1+w)\dot{D}_\chi - (1+w)\frac{k}{a}V, \tag{1.264}$$

$$\dot{V} + (1-3c_s^2)HV = \frac{k}{a}A_\chi + \frac{c_s^2}{1+w}\frac{k}{a}\delta_\chi + \frac{w}{1+w}\frac{k}{a}\Gamma - \frac{2}{3}\frac{w}{1+w}\frac{k^2-3K}{k^2}\frac{k}{a}\Pi. \tag{1.265}$$

1.4.3.1 Alternative Basic Systems of Equations

From the basic equations (1.232) – (1.246), we now derive another set which is sometimes useful, as we shall see. We want to work with[13] $\delta_\mathcal{Q}, V$ and D_χ.

The energy equation (1.237) with index $\mathcal{Q}$ gives

$$\dot{\delta}_\mathcal{Q} + 3H(c_s^2 - w)\delta_\mathcal{Q} + 3Hw\Gamma = (1+w)(\kappa_\mathcal{Q} - 3HA_\mathcal{Q}). \tag{1.266}$$

Similarly, the momentum equation (1.239) implies

$$A_\mathcal{Q} = -\frac{1}{1+w}\left[c_s^2\delta_\mathcal{Q} + w\Gamma + \frac{2}{3}(\nabla^2 + 3K)w\Pi\right]. \tag{1.267}$$

From (1.234), we obtain

$$\kappa_\mathcal{Q} + \frac{1}{a^2}(\nabla^2 + 3K)\chi_\mathcal{Q} = 0. \tag{1.268}$$

But from (1.220), we see that

$$\chi_\mathcal{Q} = aV, \tag{1.269}$$

hence,

$$\kappa_\mathcal{Q} = -\frac{1}{a}(\nabla^2 + 3K)V. \tag{1.270}$$

Now, we insert (1.267) and (1.270) in (1.266) and obtain

$$\boxed{\dot{\delta}_\mathcal{Q} - 3Hw\delta_\mathcal{Q} = -(1+w)\frac{1}{a}(\nabla^2 + 3K)V + 2H(\nabla^2 + 3K)w\Pi.} \tag{1.271}$$

[13] A detailed analysis in [31] shows that the equations for $\delta_\mathcal{Q}, V$ and D_χ are for pressureless fluids, but general scales, of the same form as the corresponding Newtonian equations.

Next, we use (1.246) and the relation

$$\delta_\chi = \delta_\mathscr{Q} + 3(1+w)HV, \tag{1.272}$$

which follows from (1.218), to obtain

$$\dot{V} + HV = -\frac{1}{a}A_\chi - \frac{1}{a(1+w)}\left[c_s^2\delta_\mathscr{Q} + w\Gamma + \frac{2}{3}(\nabla^2 + 3K)w\Pi\right]. \tag{1.273}$$

Here we make use of (1.243), with the result

$$\boxed{\dot{V} + HV = \tfrac{1}{a}D_\chi - \tfrac{1}{a(1+w)}\left[c_s^2\delta_\mathscr{Q} + w\Gamma - 8\pi Ga^2(1+w)p\Pi + \tfrac{2}{3}(\nabla^2 + 3K)w\Pi\right].} \tag{1.274}$$

From (1.240), (1.242), (1.243), and (1.221), we find

$$\boxed{\dot{D}_\chi + HD_\chi = 4\pi Ga(\rho + p)V - 8\pi Ga^2 Hp\Pi.} \tag{1.275}$$

Finally, we replace in (1.241) δ_χ by $\delta_\mathscr{Q}$ (making use of (1.272)) and κ_χ by V according to (1.242), giving the Poisson-like equation

$$\boxed{\frac{1}{a^2}(\nabla^2 + 3K)D_\chi = -4\pi G\rho\delta_\mathscr{Q}.} \tag{1.276}$$

The system we were looking for consists of (1.271), (1.274), (1.275), or (1.276).

From these equations, we now derive an interesting expression for $\dot{\mathscr{R}}$. Recall (1.222):

$$\mathscr{R} = D_\mathscr{Q} = D_\chi + aHV = D_\chi + \dot{a}V. \tag{1.277}$$

Thus,

$$\dot{\mathscr{R}} = \dot{D}_\chi + \ddot{a}V + \dot{a}\dot{V}.$$

On the right of this equation, we use for the first term, Eq. (1.275); for the second term, the following consequence of the Friedmann equations (1.17) and (1.23)

$$\ddot{a} = -\frac{1}{2}(1+3w)a\left(H^2 + \frac{K}{a^2}\right); \tag{1.278}$$

and for the last term, we use (1.274). The result becomes relatively simple for $K = 0$ (the V-terms cancel):

$$\dot{\mathscr{R}} = -\frac{H}{1+w}\left[c_s^2\delta_\mathscr{Q} + w\Gamma + \frac{2}{3}w\nabla^2\Pi\right].$$

Using also (1.276) and the Friedmann equation (1.17) (for $K = 0$) leads to

$$\dot{\mathscr{R}} = \frac{H}{1+w}\left[\frac{2}{3}c_s^2\frac{1}{(Ha)^2}\nabla^2 D_\chi - w\Gamma - \frac{2}{3}w\nabla^2\Pi\right]. \tag{1.279}$$

This is an important equation that will show, for instance, that $\mathscr{R}$ remains constant on superhorizon scales, provided Γ and Π can be neglected.

As another important application, we can derive through elimination a second-order equation for $\delta_{\mathscr{Q}}$. For this, we perform again a harmonic decomposition and rewrite the basic equations (1.271), (1.274), (1.275), and (1.276) for the Fourier amplitudes:

$$\dot{\delta}_{\mathscr{Q}} - 3Hw\delta_{\mathscr{Q}} = -(1+w)\frac{k}{a}\frac{k^2-3K}{k^2}V - 2H\frac{k^2-3K}{k^2}w\Pi, \tag{1.280}$$

$$\dot{V} + HV = -\frac{k}{a}D_\chi + \frac{1}{1+w}\frac{k}{a}\left[c_s^2\delta_{\mathscr{Q}} + w\Gamma - 8\pi G(1+w)\frac{a^2}{k^2}p\Pi - \frac{2}{3}\frac{k^2-3K}{k^2}w\Pi\right] \tag{1.281}$$

$$\frac{k^2-3K}{a^2}D_\chi = 4\pi G\rho\delta_{\mathscr{Q}}, \tag{1.282}$$

$$\dot{D}_\chi + HD_\chi = -4\pi G(\rho+p)\frac{a}{k}V - 8\pi GH\frac{a^2}{k^2}p\Pi. \tag{1.283}$$

Through elimination, one can derive the following important second-order equation for $\delta_{\mathscr{Q}}$ (including the Λ term)

$$\ddot{\delta}_{\mathscr{Q}} + (2+3c_s^2-6w)H\dot{\delta}_{\mathscr{Q}} + \left[c_s^2\frac{k^2}{a^2} - 4\pi G\rho(1+w)(1+3c_s^2) \right.$$
$$\left. -3\dot{H}(w+c_s^2) + 3H^2(3c_s^2-5w)\right]\delta_{\mathscr{Q}} = \mathscr{S}, \tag{1.284}$$

where

$$\mathscr{S} = -\frac{k^2-3K}{a^2}w\Gamma - 2\left(1-\frac{3K}{k^2}\right)Hw\dot{\Pi}$$
$$-\left(1-\frac{3K}{k^2}\right)\left[-\frac{1}{3}\frac{k^2}{a^2} + 2\dot{H} + (5-3c_s^2/w)H^2\right]2w\Pi. \tag{1.285}$$

This is obtained by differentiating (1.280) and eliminating V and $\dot{V}$ with the help of (1.280) and (1.281). In addition, one has to use several zeroth-order equations. We leave the details to the reader. Note that $\mathscr{S} = 0$ for $\Gamma = \Pi = 0$.

For the special case of dust ($c_s^2 = w = \Pi = \Gamma = 0$) and $K = 0$, we get for (1.280) – (1.282) and (1.284) the same equations as in Newtonian theory:

$$\dot{\delta}_{\mathscr{Q}} = -\frac{k}{a}V, \quad \dot{V} + HV = -\frac{k}{a}\Phi, \quad \frac{k^2}{a^2}\Phi = 4\pi G\rho\delta_{\mathscr{Q}},$$

$$\ddot{\delta}_{\mathscr{Q}} + 2H\dot{\delta}_{\mathscr{Q}} - 4\pi G\rho\delta_{\mathscr{Q}} = 0.$$

1.5 Some Applications of CPT

In this section, we discuss some applications of the general formalism. More relevant applications will follow in later chapters.

Before studying realistic multicomponent fluids, we consider first the simplest case when one component, for instance CDM, dominates. First, we study, however, a general problem.

Let us write down the basic equations (1.280) – (1.283) in the notation adopted later ($A_\chi = \Psi, D_\chi = \Phi, \delta_{\mathcal{Q}} = \Delta$):

$$\dot{\Delta} - 3Hw\Delta = -(1+w)\frac{k}{a}\frac{k^2-3K}{k^2}V - 2H\frac{k^2-3K}{k^2}w\Pi, \tag{1.286}$$

$$\dot{V} + HV = -\frac{k}{a}\Phi + \frac{1}{1+w}\frac{k}{a}\Big[c_s^2\Delta + w\Gamma \\ -8\pi G(1+w)\frac{a^2}{k^2}p\Pi - \frac{2}{3}\frac{k^2-3K}{k^2}w\Pi\Big], \tag{1.287}$$

$$\frac{k^2-3K}{a^2}\Phi = 4\pi G\rho\Delta, \tag{1.288}$$

$$\dot{\Phi} + H\Phi = -4\pi G(\rho+p)\frac{a}{k}V - 8\pi GH\frac{a^2}{k^2}p\Pi. \tag{1.289}$$

Recall also (1.262):

$$\Phi + \Psi = -8\pi G\frac{a^2}{k^2}p\Pi. \tag{1.290}$$

Note that $\Phi = -\Psi$ for $\Pi = 0$.

From these perturbation equations, we derived through elimination the second-order equation (1.284) for Δ, which we repeat for $\Pi = 0$ (vanishing anisotropic stresses) and $\Gamma = 0$ (vanishing entropy production):

$$\ddot{\Delta} + (2+3c_s^2-6w)H\dot{\Delta} + \Big[c_s^2\frac{k^2}{a^2} - 4\pi G\rho(1+w)(1+3c_s^2) \\ -3\dot{H}(w+c_s^2) + 3H^2(3c_s^2-5w)\Big]\Delta = 0. \tag{1.291}$$

Remarkably, this can be written as [31]

$$\frac{1+w}{a^2H}\left[\frac{H^2}{a(\rho+p)}\left(\frac{a^3\rho}{H}\Delta\right)^{\cdot}\right]^{\cdot} + c_s^2\frac{k^2}{a^2}\Delta = 0 \tag{1.292}$$

(Exercise).

Sometimes it is convenient to write this in terms of the conformal time for the quantity $\rho a^3\Delta$. Making use of $(\rho a^3)^{\cdot} = -3Hw(\rho a^3)$ (see (1.22)), one finds

$$(\rho a^3\Delta)'' + (1+3c_s^2)\mathscr{H}(\rho a^3\Delta)' + \left[(k^2-3K)c_s^2 - 4\pi G(\rho+p)a^2\right](\rho a^3\Delta) = 0. \tag{1.293}$$

Using (1.288) we obtain from (1.293) the following compact second-order equation for Φ:

$$\frac{\rho+p}{H}\left[\frac{H^2}{a(\rho+p)}\left(\frac{a}{H}\Phi\right)^{\cdot}\right]^{\cdot} + c_s^2\frac{k^2}{a^2}\Phi = 0. \tag{1.294}$$

With (1.222) and (1.289), it is easy to see that for $K=0$ and $\Pi=0$ the curvature perturbation can be written as

$$\mathscr{R} = \frac{H^2}{4\pi G a(\rho+p)}\left(\frac{a}{H}\Phi\right)^{\cdot}. \tag{1.295}$$

Hence (1.294) again implies that $\mathscr{R}$ remains constant on large scales ($c_s k/(aH) \ll 1$).

1.5.1 Nonrelativistic Limit

It is instructive to first consider a one-component nonrelativistic fluid. The nonrelativistic limit of the second-order equation (1.291) is

$$\ddot{\Delta} + 2H\dot{\Delta} = 4\pi G\rho\Delta - c_s^2\left(\frac{k}{a}\right)^2\Delta. \tag{1.296}$$

From this basic equation, one can draw various conclusions.

The Jeans Criterion

One sees from (1.296) that gravity wins over the pressure term $\propto c_s^2$ for $k < k_J$, where

$$k_J^2\left(\frac{c_s}{a}\right)^2 = 4\pi G\rho \tag{1.297}$$

defines the *comoving Jeans wave number*. The corresponding *Jeans length* (wave length) is

$$\lambda_J = \frac{2\pi}{k_J}a = \left(\frac{\pi c_s^2}{G\rho}\right)^{1/2}, \quad \frac{\lambda_J}{2\pi} \simeq \frac{c_s}{H}. \tag{1.298}$$

For $\lambda < \lambda_J$, we expect that the fluid oscillates, while for $\lambda \gg \lambda_J$ an over-density will increase.

Let us illustrate this for a polytropic equation of state $p = const\ \rho^\gamma$. We consider, as a simple example, a matter dominated Einstein-de Sitter model ($K=0$), for which $a(t) \propto t^{2/3}$, $H = 2/(3t)$. Eq. (1.296) then becomes (taking ρ from the Friedmann equation, $\rho = 1/(6\pi G t^2)$)

$$\ddot{\Delta} + \frac{4}{3t}\dot{\Delta} + \left(\frac{L^2}{t^{2\gamma-2/3}} - \frac{2}{3t^2}\right)\Delta = 0, \tag{1.299}$$

where L^2 is the constant

$$L^2 = \frac{t^{2\gamma-2/3}c_s^2k^2}{a^2}. \tag{1.300}$$

The solutions of (1.299)are

$$\Delta_\pm(t) \propto t^{-1/6}J_{\mp 5/6\nu}\left(\frac{Lt^{-\nu}}{\nu}\right), \quad \nu := \gamma - \frac{4}{3} > 0. \tag{1.301}$$

The Bessel functions J oscillate for $t \ll L^{1/\nu}$, whereas for $t \gg L^{1/\nu}$ the solutions behave like

$$\Delta_\pm(t) \propto t^{-\frac{1}{6}\pm\frac{5}{6}}. \tag{1.302}$$

Now, $t > L^{1/\nu}$ signifies $c_s^2k^2/a^2 < 6\pi G\rho$. This is essentially again the Jeans criterion $k < k_J$. At the same time, we see that

$$\Delta_+ \propto t^{2/3} \propto a, \tag{1.303}$$

$$\Delta_- \propto t^{-1}. \tag{1.304}$$

Thus, the *growing mode increases like the scale factor*. This means that the growth factor in linear theory from recombination to redshifts of a few is only about 10^3. So, initial fluctuations of $\sim 10^{-5}$ cannot become of order unity until the present. Since long, this is considered as strong evidence for the existence of a dominant dark matter component, whose fluctuations could grow already long before recombination.

1.5.2 Large-Scale Solutions

Consider, as an important application, wavelengths *larger than the Jeans length*, i.e., $c_s(k/aH) \ll 1$. Then, we can drop the last term in Eq. (1.294) and solve for Φ in terms of quadratures:

$$\Phi(t,\mathbf{k}) = C(\mathbf{k})\frac{H}{a}\int_0^t \frac{a(\rho+p)}{H^2}dt + \frac{H}{a}d(\mathbf{k}). \tag{1.305}$$

We write this differently by using in the integrand the following background equation (implied by (1.17)) and (1.18))

$$\frac{a(\rho+p)}{H^2} = \left(\frac{a}{H}\right)^{\cdot} - a\left(1 - \frac{K}{\dot{a}^2}\right).$$

With this, we obtain

$$\Phi(t,\mathbf{k}) = C(\mathbf{k})\left[1 - \frac{H}{a}\int_0^t a\left(1 - \frac{K}{\dot{a}^2}\right)dt\right] + \frac{H}{a}d(\mathbf{k}). \tag{1.306}$$

Let us work this out for a mixture of dust ($p = 0$) and radiation ($p = \frac{1}{3}\rho$). We use the "normalized" scale factor $\zeta := a/a_{eq}$, where a_{eq} is the value of a when the energy densities of dust (CDM) and radiation are equal. Then (see Sect. 1.1.3),

$$\rho = \frac{1}{2}\zeta^{-4} + \frac{1}{2}\zeta^{-3}, \quad p = \frac{1}{6}\zeta^{-4}. \tag{1.307}$$

Note that

$$\zeta' = kx\zeta, \quad x := \frac{Ha}{k}. \tag{1.308}$$

From now on, we assume $K = 0$, $\Lambda = 0$. Then, the Friedmann equation gives

$$H^2 = H_{eq}^2 \frac{\zeta+1}{2}\zeta^{-4}, \tag{1.309}$$

thus

$$x^2 = \frac{\zeta+1}{2\zeta^2}\frac{1}{\omega^2}, \quad \omega := \frac{1}{x_{eq}} = \frac{k}{(aH)_{eq}}. \tag{1.310}$$

In (1.306), we need the integral

$$\frac{H}{a}\int_0^t a dt = Ha_{eq}\frac{1}{\zeta}\int_0^\eta \zeta^2 d\eta = \frac{\sqrt{\zeta+1}}{\zeta^3}\int_0^\zeta \frac{\zeta^2}{\sqrt{\zeta+1}} d\zeta.$$

As a result, we get for the growing mode

$$\Phi(\zeta, \mathbf{k}) = C(\mathbf{k})\left[1 - \frac{\sqrt{\zeta+1}}{\zeta^3}\int_0^\zeta \frac{\zeta^2}{\sqrt{\zeta+1}} d\zeta\right]. \tag{1.311}$$

From (1.288) and the definition of x, we obtain

$$\Phi = \frac{3}{2}x^2\Delta, \tag{1.312}$$

hence with (1.310)

$$\Delta = \frac{4}{3}\omega^2 C(\mathbf{k})\frac{\zeta^2}{\zeta+1}\left[1 - \frac{\sqrt{\zeta+1}}{\zeta^3}\int_0^\zeta \frac{\zeta^2}{\sqrt{\zeta+1}} d\zeta\right]. \tag{1.313}$$

The integral is elementary. One finds that Δ is proportional to

$$U_g = \frac{1}{\zeta(\zeta+1)}\left[\zeta^3 + \frac{2}{9}\zeta^2 - \frac{8}{9}\zeta - \frac{16}{9} + \frac{16}{9}\sqrt{\zeta+1}\right]. \tag{1.314}$$

This is a well-known result.

The decaying mode corresponds to the second term in (1.306) and is thus proportional to

$$U_d = \frac{1}{\zeta\sqrt{\zeta+1}}. \tag{1.315}$$

Limiting approximations of (1.314) are

$$U_g = \begin{cases} \frac{10}{9}\zeta^2 & : \quad \zeta \ll 1 \\ \zeta & : \quad \zeta \gg 1. \end{cases} \tag{1.316}$$

For the potential $\Phi \propto x^2\Delta$, the growing mode is given by

$$\Phi(\zeta) = \Phi(0)\frac{9}{10}\frac{\zeta+1}{\zeta^2}U_g. \tag{1.317}$$

Thus,

$$\Phi(\zeta) = \Phi(0)\begin{cases} 1 & : \quad \zeta \ll 1 \\ \frac{9}{10} & : \quad \zeta \gg 1. \end{cases} \tag{1.318}$$

In particular, Φ stays *constant both in the radiation and in the matter dominated eras*. Recall that this holds only for $c_s(k/aH) \ll 1$. We shall later study Eq. (1.294) for arbitrary scales.

1.5.3 Solution for Dust

Using the Poisson equation (1.288), we can write (1.294) in terms of Δ

$$\frac{1+w}{a^2H}\left[\frac{H^2}{a(\rho+p)}\left(\frac{a^3\rho}{H}\Delta\right)^{\cdot}\right]^{\cdot} + c_s^2\frac{k^2}{a^2}\Delta = 0. \tag{1.319}$$

For dust, this reduces to (using $\rho a^3 = const$)

$$\left[a^2H^2\left(\frac{\Delta}{H}\right)^{\cdot}\right]^{\cdot} = 0. \tag{1.320}$$

The general solution of this equation is

$$\Delta(t,\mathbf{k}) = C(\mathbf{k})H(t)\int_0^t \frac{dt'}{a^2(t')H^2(t')} + d(k)H(t). \tag{1.321}$$

This result can also be obtained in Newtonian perturbation theory. The first term gives the growing mode and the second term the decaying mode.

Let us rewrite (1.321) in terms of the redshift z. From $1+z = a_0/a$, we get $dz = -(1+z)Hdt$, so by (1.90)

$$\frac{dt}{dz} = -\frac{1}{H_0(1+z)E(z)}, \quad H(z) = H_0E(z). \tag{1.322}$$

Therefore, the growing mode $D_g(z)$ can be written in the form

$$D_g(z) = \frac{5}{2}\Omega_M E(z) \int_z^\infty \frac{1+z'}{E^3(z')} dz'. \tag{1.323}$$

Here, the normalization is chosen such that $D_g(z) = (1+z)^{-1} = a/a_0$ for $\Omega_M = 1$, $\Omega_\Lambda = 0$. This growth function is plotted in Fig. 7.12 of [4].

1.5.4 A Simple Relativistic Example

As an additional illustration, we now solve (1.293) for a single perfect fluid with $w = c_s^2 = const$, $K = \Lambda = 0$. For a flat universe, the background equations are then

$$\rho' + 3\frac{a'}{a}(1+w)\rho = 0, \quad \left(\frac{a'}{a}\right)^2 = \frac{8\pi G}{3}a^2\rho.$$

Inserting the ansatz

$$\rho a^2 = A\eta^{-\nu}, \quad a = a_0(\eta/\eta_0)^\beta,$$

we get

$$\frac{\beta^2}{\eta^2} = \frac{8\pi G}{3}A\eta^{-\nu} \quad \Rightarrow \nu = 2,\ A = \frac{3}{8\pi G}\beta^2.$$

The energy equation then gives $\beta = 2/(1+3w)$ ($= 1$ if radiation dominates). Let $x := k\eta$ and

$$f := x^{\beta-2}\Delta \propto \rho a^3 \Delta.$$

Also note that $k/(aH) = x/\beta$. With all this, we obtain from (1.293) for f

$$\left[\frac{d^2}{dx^2} + \frac{2}{x}\frac{d}{dx} + c_s^2 - \frac{\beta(\beta+1)}{x^2}\right] f = 0. \tag{1.324}$$

The solutions are given in terms of Bessel functions:

$$f(x) = C_0 j_\beta(c_s x) + D_0 n_\beta(c_s x). \tag{1.325}$$

This implies acoustic oscillations for $c_s x \gg 1$, i.e., for $\beta(k/aH) \gg 1$ (subhorizon scales). In particular, if the radiation dominates ($\beta = 1$)

$$\Delta \propto x[C_0 j_1(c_s x) + D_0 n_1(c_s x)], \tag{1.326}$$

and the growing mode is soon proportional to $x\cos(c_s x)$, whereas the term going with $\sin(c_s x)$ dies out.

On the other hand, on superhorizon scales ($c_s x \ll 1$), one obtains

$$f \simeq Cx^\beta + Dx^{-(\beta+1)},$$

and thus

$$\begin{aligned}
\Delta &\simeq Cx^2 + Dx^{-(2\beta-1)}, \\
\Phi &\simeq \frac{3}{2}\beta^2(C + Dx^{-(2\beta+1)}, \\
V &\simeq \frac{3}{2}\beta\left(-\frac{1}{\beta+1}Cx + Dx^{-2\beta}\right).
\end{aligned} \tag{1.327}$$

We see that the growing mode behaves as $\Delta \propto a^2$ in the radiation dominated phase and $\Delta \propto a$ in the matter dominated era.

The characteristic Jeans wave number is obtained when the square bracket in (1.293) vanishes. This gives

$$\lambda_J = \left(\frac{\pi c_s^2}{Gh}\right)^{1/2}, \quad h = \rho + p. \tag{1.328}$$

Exercises

1. Derive the exact expression for V. 2. Specialize the differential equation (1.292) for Φ to the model of this section and solve the resulting equation for $w = c_s^2 = 1/3$ (radiation). Discuss the result. ♢

Remark

Equation (1.291) for radiation domination ($w = c_s^2 = 1/3$) and $K = 0 = \Lambda$ becomes

$$\ddot{\Delta} + H\dot{\Delta} + \frac{1}{3}\frac{k^2}{a^2}\Delta = -\frac{16\pi}{3}G\rho\Delta.$$

As was pointed out in [31], several textbooks arrive instead at an incorrect equation.

Later, we shall study more complicated coupled fluid models that are important for the evolution of perturbations before recombination. In the next part, the general theory will be applied in attempts to understand the generation of primordial perturbations from original quantum fluctuations.

1.6 CPT for Scalar Field Models

We begin by repeating the set up of Sect. 1.3.3.

We consider a minimally coupled scalar field φ, with Lagrangian density

$$\mathscr{L} = -\frac{1}{2}g^{\mu\nu}\partial_\mu\varphi\partial_\nu\varphi - U(\varphi) \tag{1.329}$$

and corresponding field equation

$$\Box\varphi = U_{,\varphi}. \tag{1.330}$$

As a result of this, the energy-momentum tensor

$$T^{\mu}{}_{\nu} = \partial^{\mu}\varphi\partial_{\nu}\varphi - \delta^{\mu}{}_{\nu}\left(\frac{1}{2}\partial^{\lambda}\varphi\partial_{\lambda}\varphi + U(\varphi)\right) \tag{1.331}$$

is covariantly conserved. The unperturbed quantities ρ_{φ}, etc., are

$$\rho_{\varphi} = -T^{0}{}_{0} = \frac{1}{2a^2}(\varphi')^2 + U(\varphi), \tag{1.332}$$

$$p_{\varphi} = \frac{1}{3}T^{i}{}_{i} = \frac{1}{2a^2}(\varphi')^2 - U(\varphi), \tag{1.333}$$

$$h_{\varphi} = \rho_{\varphi} + p_{\varphi} = \frac{1}{a^2}(\varphi')^2. \tag{1.334}$$

Furthermore,

$$\rho'_{\varphi} = -3\frac{a'}{a}h_{\varphi}. \tag{1.335}$$

It is not very sensible to introduce a "velocity of sound" c_{φ}.

1.6.1 Basic Perturbation Equations

Now, we consider small deviations from the ideal Friedmann behavior:

$$\varphi \to \varphi_0 + \delta\varphi, \;\; \rho_{\varphi} \to \rho_{\varphi} + \delta\rho, \;\; \text{etc.} \tag{1.336}$$

(The index 0 is only used for the unperturbed field φ.) Since $L_{\xi}\varphi_0 = \xi^0\varphi_0'$, the gauge transformation of $\delta\varphi$ is

$$\delta\varphi \to \delta\varphi + \xi^0\varphi_0'. \tag{1.337}$$

Therefore,

$$\delta\varphi_{\chi} = \delta\varphi - \frac{1}{a}\varphi_0'\chi = \delta\varphi - \varphi_0'(B+E') \tag{1.338}$$

is gauge invariant (see (1.185)). Further perturbations are

$$\delta T^{0}{}_{0} = -\frac{1}{a^2}\left[-\varphi_0'^2 A + \varphi_0'\delta\varphi' + U_{,\varphi}a^2\delta\varphi\right], \tag{1.339}$$

$$\delta T^{0}{}_{i} = -\frac{1}{a^2}\varphi_0'\delta\varphi_{,i}, \tag{1.340}$$

$$\delta T^{i}{}_{j} = -\frac{1}{a^2}[\varphi_0'^2 A - \varphi_0'\delta\varphi' + U_{,\varphi}a^2\delta\varphi]\delta^{i}{}_{j}. \tag{1.341}$$

This gives (recall (1.207)).

$$\delta\rho = \frac{1}{a^2}[-\varphi_0'^2 A + \varphi_0'\delta\varphi' + a^2 U_{,\varphi}\delta\varphi], \tag{1.342}$$

$$\delta p = p\pi_L = \frac{1}{a^2}[\varphi_0'\delta\varphi' - \varphi_0'^2 A - a^2 U_{,\varphi}\delta\varphi], \tag{1.343}$$

$$\Pi = 0, \quad \mathcal{Q} = -\dot{\varphi}_0\delta\varphi. \tag{1.344}$$

Einstein Equations

We insert these expressions into the general perturbation equations (1.232) – (1.239) and obtain

$$\kappa = 3(HA - \dot{D}) - \frac{1}{a^2}\nabla^2\chi, \tag{1.345}$$

$$\frac{1}{a^2}(\nabla^2 + 3K)D + H\kappa = -4\pi G[\dot{\varphi}_0\delta\dot{\varphi} - \dot{\varphi}_0^2 A + U_{,\varphi}\delta\varphi], \tag{1.346}$$

$$\kappa + \frac{1}{a^2}(\nabla^2 + 3K)\chi = 12\pi G\dot{\varphi}_0\delta\varphi, \tag{1.347}$$

$$A + D = \dot{\chi} + H\chi. \tag{1.348}$$

Equation (1.236) is in the present notation

$$\dot{\kappa} + 2H\kappa = -\left(\frac{1}{a^2}\nabla^2 + 3\dot{H}\right)A + 4\pi G[\delta\rho + 3\delta p],$$

with

$$\delta\rho + 3\delta p = 2(-2\dot{\varphi}_0^2 A + 2\dot{\varphi}_0\delta\dot{\varphi} - U_{,\varphi}\delta\varphi).$$

If we also use (recall (1.278))

$$\dot{H} = -4\pi G\dot{\varphi}_0^2 + \frac{K}{a^2},$$

we obtain

$$\dot{\kappa} + 2H\kappa = -\left(\frac{\nabla^2 + 3K}{a^2} + 4\pi G\dot{\varphi}_0^2\right)A + 8\pi G(2\dot{\varphi}_0\delta\dot{\varphi} - U_{,\varphi}\delta\varphi). \tag{1.349}$$

The two remaining equations (1.238) and (1.239) are

$$(\delta\rho)^{\cdot} + 3H(\delta\rho + \delta p) = (\rho + p)(\kappa - 3HA) - \frac{1}{a^2}\nabla^2\mathcal{Q}, \tag{1.350}$$

and

$$\dot{\mathcal{Q}} + 3H\mathcal{Q} = -(\rho + p)A - \delta p, \tag{1.351}$$

with the expressions (1.342) – (1.344). Since these last two equations express energy-momentum "conservation," they are not independent of the others if we add the field equation for φ; we shall not make use of them below.

Equations (1.345) – (1.349) can immediately be written in a gauge invariant form:

$$\kappa_\chi = 3(HA_\chi - \dot{D}_\chi), \tag{1.352}$$

$$\frac{1}{a^2}(\nabla^2 + 3K)D_\chi + H\kappa_\chi = -4\pi G[\dot{\varphi}_0\delta\dot{\varphi}_\chi - \dot{\varphi}_0^2 A_\chi + U_{,\varphi}\delta\varphi_\chi], \tag{1.353}$$

$$\kappa_\chi = 12\pi G\dot{\varphi}_0\delta\varphi_\chi, \tag{1.354}$$

$$A_\chi + D_\chi = 0 \tag{1.355}$$

$$\dot{\kappa}_\chi + 2H\kappa_\chi = -\left(\frac{\nabla^2 + 3K}{a^2} + 4\pi G\dot{\varphi}_0^2\right)A_\chi + 8\pi G(2\dot{\varphi}_0\delta\dot{\varphi}_\chi - U_{,\varphi}\delta\varphi_\chi). \tag{1.356}$$

From now on, we set $\mathbf{K} = \mathbf{0}$. Use of (1.355) then gives us the following four basic equations:

$$\kappa_\chi = 3(\dot{A}_\chi + HA_\chi), \tag{1.357}$$

$$\frac{1}{a^2}\nabla^2 A_\chi - H\kappa_\chi = 4\pi G[\dot{\varphi}_0\delta\dot{\varphi}_\chi - \dot{\varphi}_0^2 A_\chi + U_{,\varphi}\delta\varphi_\chi], \tag{1.358}$$

$$\kappa_\chi = 12\pi G\dot{\varphi}_0\delta\varphi_\chi, \tag{1.359}$$

$$\dot{\kappa}_\chi + 2H\kappa_\chi = -\frac{1}{a^2}\nabla^2 A_\chi - 4\pi G\dot{\varphi}_0^2 A_\chi + 8\pi G(2\dot{\varphi}_0\delta\dot{\varphi}_\chi - U_{,\varphi}\delta\varphi_\chi). \tag{1.360}$$

Recall also

$$4\pi G\dot{\varphi}_0^2 = -\dot{H}. \tag{1.361}$$

From (1.357) and (1.359), we get

$$\boxed{\dot{A}_\chi + HA_\chi = 4\pi G\dot{\varphi}_0\delta\varphi_\chi.} \tag{1.362}$$

The difference of (1.360) and (1.358) gives (using also (1.357))

$$(\dot{A}_\chi + HA_\chi)^{\cdot} + 3H(\dot{A}_\chi + HA_\chi) = 4\pi G(\dot{\varphi}_0\delta\dot{\varphi}_\chi - U_{,\varphi}\delta\varphi_\chi)$$

i.e.,

$$\boxed{\ddot{A}_\chi + 4H\dot{A}_\chi + (\dot{H} + 3H^2)A_\chi = 4\pi G(\dot{\varphi}_0\delta\dot{\varphi}_\chi - U_{,\varphi}\delta\varphi_\chi).} \tag{1.363}$$

Beside (1.362) and (1.363), we keep (1.358) in the form (making use of (1.361))

$$\boxed{\frac{1}{a^2}\nabla^2 A_\chi - 3H\dot{A}_\chi - (\dot{H} + 3H^2)A_\chi = 4\pi G(\dot{\varphi}_0\delta\dot{\varphi}_\chi + U_{,\varphi}\delta\varphi_\chi).} \tag{1.364}$$

Scalar Field Equation

We now turn to the φ equation (1.330). Recall (the index 0 denotes in this subsection the t-coordinate)

$$\begin{aligned} g_{00} &= -(1+2A), \;\; g_{0j} = -aB_{,j}, \;\; g_{ij} = a^2[\gamma_{ij} + 2D\gamma_{ij} + 2E_{|ij}]; \\ g^{00} &= -(1-2A), \;\; g^{0j} = -\frac{1}{a}B^{,j}, \;\; g^{ij} = \frac{1}{a^2}[\gamma^{ij} - 2D\gamma^{ij} - 2E^{|ij}]; \\ \sqrt{-g} &= a^3\sqrt{\gamma}(1+A+3D+\nabla^2 E. \end{aligned}$$

Up to first order, we have (note that $\partial_j\varphi$ and g^{0j} are of first order)

$$\Box\varphi = \frac{1}{\sqrt{-g}}\partial_\mu(\sqrt{-g}g^{\mu\nu}\partial_\nu\varphi) = \frac{1}{\sqrt{-g}}(\sqrt{-g}g^{00}\dot{\varphi})^{\cdot} + \frac{1}{a^2}\nabla^2\delta\varphi - \frac{1}{a}\dot{\varphi}_0\nabla^2 B.$$

Using the zeroth-order field equation (1.143), we readily find

$$\begin{aligned} &\delta\ddot{\varphi} + 3H\delta\dot{\varphi} + \left(-\frac{1}{a^2}\nabla^2 + U_{,\varphi\varphi}\right)\delta\varphi = \\ &(\dot{A} - 3\dot{D} - \nabla^2\dot{E} + 3HA - \frac{1}{a}\nabla^2 B)\dot{\varphi}_0 - (3H\dot{\varphi}_0 + 2U_{,\varphi})A. \end{aligned}$$

Recalling the definition of κ,

$$\kappa = 3(HA - \dot{D}) - \frac{1}{a}\nabla^2(B + a\dot{E}),$$

we finally obtain the perturbed field equation in the form

$$\delta\ddot{\varphi} + 3H\delta\dot{\varphi} + \left(-\frac{1}{a^2}\nabla^2 + U_{,\varphi\varphi}\right)\delta\varphi = (\kappa + \dot{A})\dot{\varphi}_0 - (3H\dot{\varphi}_0 + 2U_{,\varphi})A. \quad (1.365)$$

By putting the index χ at all perturbation amplitudes, one obtains a gauge invariant equation. Using also (1.357), one arrives at

$$\boxed{\delta\ddot{\varphi}_\chi + 3H\delta\dot{\varphi}_\chi + \left(-\frac{1}{a^2}\nabla^2 + U_{,\varphi\varphi}\right)\delta\varphi_\chi = 4\dot{\varphi}_0\dot{A}_\chi - 2U_{,\varphi}A_\chi.} \quad (1.366)$$

Our basic – but not independent – equations are (1.362), (1.363), (1.364), and (1.366).

1.6.2 Consequences and Reformulations

In (1.222), we have introduced the curvature perturbation (recall also (1.344))

$$\mathscr{R} := D_{\mathscr{Q}} = D_\chi - \frac{H}{\dot{\varphi}_0}\delta\varphi_\chi = D - \frac{H}{\dot{\varphi}_0}\delta\varphi. \quad (1.367)$$

It will turn out to be convenient to work also with

$$u = -z\mathcal{R}, \quad z := \frac{a\dot{\varphi}_0}{H}, \tag{1.368}$$

thus

$$u = a\left[\delta\varphi_\chi - \frac{\dot{\varphi}_0}{H}D_\chi\right] = a\left[\delta\varphi - \frac{\dot{\varphi}_0}{H}D\right]. \tag{1.369}$$

This amplitude will play an important role because we shall obtain from the previous formulae the simple equation

$$\boxed{u'' - \nabla^2 u - \frac{z''}{z}u = 0.} \tag{1.370}$$

This is a Klein–Gordon equation with a time-dependent mass.

We next rewrite the basic equations in terms of the conformal time:

$$\nabla^2 A_\chi - 3\mathcal{H}A'_\chi - (\mathcal{H}' + 3\mathcal{H}^2)A_\chi = 4\pi G(\varphi'_0\delta\varphi'_\chi + U_{,\varphi}a^2\delta\varphi_\chi), \tag{1.371}$$

$$A'_\chi + \mathcal{H}A_\chi = 4\pi G\varphi'_0\delta\varphi_\chi, \tag{1.372}$$

$$A''_\chi + 3\mathcal{H}A'_\chi + (\mathcal{H}' + 2\mathcal{H}^2)A_\chi = 4\pi G(\varphi'_0\delta\varphi'_\chi - U_{,\varphi}a^2\delta\varphi_\chi), \tag{1.373}$$

$$\delta\varphi''_\chi + 2\mathcal{H}\delta\varphi'_\chi - \nabla^2\delta\varphi_\chi + U_{,\varphi\varphi}a^2\delta\varphi_\chi = 4\varphi'_0A'_\chi - 2U_{,\varphi}a^2A_\chi. \tag{1.374}$$

Let us first express u (or $\mathcal{R}$) in terms of A_χ. From (4.40), (4.39) we obtain in a first step

$$4\pi Gzu = 4\pi Gz^2A_\chi + 4\pi G\frac{z^2\mathcal{H}}{\varphi'_0}\delta\varphi_\chi.$$

For the first term on the right, we use the unperturbed equation (see (1.361))

$$4\pi G\varphi_0'^2 = \mathcal{H}^2 - \mathcal{H}', \tag{1.375}$$

and in the second term, we make use of (1.372). Collecting terms gives

$$\boxed{4\pi Gzu = \left(\frac{a^2A_\chi}{\mathcal{H}}\right)'.} \tag{1.376}$$

Next, we derive an equation for A_χ alone. For this, we subtract (1.371) from (1.373) and use (1.372) to express $\delta\varphi_\chi$ in terms of A_χ and A'_χ. Moreover, we make use of (1.375) and the unperturbed equation (1.143),

$$\varphi''_0 + 2\mathcal{H}\varphi'_0 + U_{,\varphi}(\varphi_0)a^2 = 0. \tag{1.377}$$

Detailed Derivation

The quoted equations give

$$A''_\chi + 6\mathcal{H}A'_\chi - \nabla^2 A_\chi + 2(\mathcal{H}' + 2\mathcal{H}^2)A_\chi =$$
$$-8\pi G U_{,\varphi} a^2 \delta\varphi_\chi = \frac{2}{\varphi'_0}(\varphi''_0 + 2\mathcal{H}\varphi'_0)(A'_\chi + \mathcal{H}A_\chi),$$

thus

$$A''_\chi + 2(\mathcal{H} - \varphi''_0/\varphi'_0)A'_\chi - \nabla^2 A_\chi + 2(\mathcal{H}' - \mathcal{H}\varphi''_0/\varphi'_0)A_\chi = 0.$$

Rewriting the coefficients of A_χ, A'_χ slightly, we obtain the important equation:

$$\boxed{A''_\chi + 2\frac{(a/\varphi'_0)'}{a/\varphi'_0}A'_\chi - \nabla^2 A_\chi + 2\varphi'_0(\mathcal{H}/\varphi'_0)'A_\chi = 0.} \tag{1.378}$$

Now, we return to (1.376) and write this, using (1.375), as follows:

$$\frac{u}{z} = A_\chi + \frac{A'_\chi + \mathcal{H}A_\chi}{L}, \tag{1.379}$$

where

$$L = 4\pi G\frac{z^2\mathcal{H}}{a^2} = 4\pi G(\varphi'_0)^2/\mathcal{H} = \mathcal{H} - \mathcal{H}'/\mathcal{H}. \tag{1.380}$$

Differentiating (1.379) implies

$$\left(\frac{u}{z}\right)' = A'_\chi + \frac{A''_\chi + (\mathcal{H}A_\chi)'}{L} - \frac{A'_\chi + \mathcal{H}A_\chi}{L^2}L'$$

or, making use of (4.52) and (4.50),

$$L\left(\frac{u}{z}\right)' = (\mathcal{H} - \mathcal{H}'/\mathcal{H})A'_\chi - 2\frac{(a/\varphi'_0)'}{a/\varphi'_0}A'_\chi + \nabla^2 A_\chi$$
$$-2\varphi'_0(\mathcal{H}/\varphi'_0)'A_\chi + (\mathcal{H}A_\chi)' - (A'_\chi + \mathcal{H}A_\chi)\frac{(\varphi_0'^2/\mathcal{H})'}{\varphi_0'^2/\mathcal{H}}.$$

From this, one easily finds the simple equation

$$\boxed{4\pi G\frac{\mathcal{H}z^2}{a^2}\left(\frac{u}{z}\right)' = \nabla^2 A_\chi.} \tag{1.381}$$

Finally, we derive the announced Eq. (1.370). To this end, we rewrite the last equation as

$$\nabla^2 A_\chi = 4\pi G\frac{\mathcal{H}}{a^2}(zu' - z'u),$$

from which we get

$$\nabla^2 A'_\chi = 4\pi G\left(\frac{\mathcal{H}}{a^2}\right)'(zu' - z'u) + 4\pi G\frac{\mathcal{H}}{a^2}(zu'' - z''u).$$

Taking the Laplacian of (4.51) gives

$$4\pi G\frac{\mathcal{H}}{a^2}z\nabla^2 u = L\nabla^2 A_\chi + \nabla^2 A'_\chi + \mathcal{H}\nabla^2 A_\chi.$$

Combining the last two equations and making use of (1.380) shows that indeed (1.370) holds.

Summarizing, we have the basic equations

$$u'' - \nabla^2 u - \frac{z''}{z}u = 0, \tag{1.382}$$

$$\nabla^2 A_\chi = 4\pi G\frac{\mathcal{H}}{a^2}(zu' - z'u), \tag{1.383}$$

$$\left(\frac{a^2 A_\chi}{\mathcal{H}}\right)' = 4\pi G z u. \tag{1.384}$$

We now discuss some important consequences of these equations. The first concerns the curvature perturbation $\mathcal{R} = -u/z$ (original definition in (1.367)). In terms of this quantity, Eq. (1.383) can be written as

$$\frac{\dot{\mathcal{R}}}{H} = \frac{1}{1 - \mathcal{H}'/\mathcal{H}^2}\frac{1}{(aH)^2}(-\nabla^2 A_\chi). \tag{1.385}$$

The right-hand side is of order $(k/aH)^2$, hence very small on scales much larger than the Hubble radius. It is common practice to use the terms "Hubble length" and "horizon" interchangeably and to call length scales satisfying $k/aH \ll 1$ to be *super-horizon*. (This can cause confusion; "super-Hubble" might be a better term, but the jargon can probably not be changed anymore.)

We have studied $\dot{\mathcal{R}}$ already at the end of Sect. 1.4.3. I recall (1.279):

$$\dot{\mathcal{R}} = \frac{H}{1+w}\left[\frac{2}{3}c_s^2\frac{1}{(Ha)^2}\nabla^2 D_\chi - w\Gamma - \frac{2}{3}w\nabla^2\Pi\right]. \tag{1.386}$$

This general equation also holds for our scalar field model, for which $\Pi = 0$, $D_\chi = -A_\chi$. The first term on the right in (1.386) is again small on super-horizon scales. So the nonadiabatic piece $p\Gamma = \delta p - c_s^2\delta\rho$ must also be small on large scales. This means that the perturbations are **adiabatic**. We shall show this more directly further below, by deriving the following expression for Γ:

$$\boxed{p\Gamma = -\frac{U_{,\varphi}}{6\pi G H\dot{\varphi}}\frac{1}{a^2}\nabla^2 A_\chi.} \tag{1.387}$$

After inflation, when relativistic fluids dominate the matter content, Eq. (1.386) still holds. The first term on the right is small on scales larger than the *sound horizon*. Since Γ and Π are then not important, we see that for super-horizon scales $\mathscr{R}$ *remains constant also after inflation*. This will become important in the study of CMB anisotropies.

Later, it will be useful to have a handy expression of A_χ in terms of $\mathscr{R}$. According to (1.222) and (1.221), we have

$$\mathscr{R} = D_\chi + \frac{\mathscr{H}}{a(\rho + p)} \mathscr{Q}_\chi. \tag{1.388}$$

We rewrite this by combining (1.240) and (1.242)

$$\mathscr{R} = D_\chi - \frac{\mathscr{H}}{4\pi G a^2(\rho + p)} (\mathscr{H} A_\chi - D'_\chi). \tag{1.389}$$

At this point, we specialize again to $K = 0$ and use the background equation

$$4\pi G a^2(\rho + p) = \mathscr{H}^2(1 - \mathscr{H}'/\mathscr{H}^2)$$

to obtain

$$\boxed{\mathscr{R} = D_\chi - \frac{1}{\varepsilon \mathscr{H}} (\mathscr{H} A_\chi - D'_\chi),} \tag{1.390}$$

where

$$\varepsilon := 1 - \mathscr{H}'/\mathscr{H}^2. \tag{1.391}$$

If $\Pi = 0$, then $D_\chi = -A_\chi$, so

$$\boxed{-\mathscr{R} = A_\chi + \frac{1}{\varepsilon \mathscr{H}} (\mathscr{H} A_\chi + A'_\chi),} \tag{1.392}$$

I claim that for a constant $\mathscr{R}$

$$A_\chi = -\left(1 - \frac{\mathscr{H}}{a^2} \int a^2 d\eta \right) \mathscr{R}. \tag{1.393}$$

We prove this by showing that (1.393) satisfies (1.392). Differentiating the last equation gives by the same equation and (1.391) our claim.

As a special case, we consider (always for $K = 0$) $w = const$. Then, as shown in Sect. 1.5.4,

$$a = a_0(\eta/\eta_0)^\beta, \quad \beta = \frac{2}{3w + 1}. \tag{1.394}$$

Thus,

$$\frac{\mathscr{H}}{a^2} \int a^2 d\eta = \frac{\beta}{2\beta + 1},$$

hence,

$$\boxed{A_\chi = -\frac{3(w+1)}{3w+5}\mathscr{R}.} \tag{1.395}$$

This will be important later.

Derivation of (1.387): By definition

$$p\Gamma = \delta p - c_s^2\delta\rho, \ c_s^2 = \dot{p}/\dot{\rho} \Rightarrow p\Gamma = \frac{\dot{\rho}\delta p - \dot{p}\delta\rho}{\dot{\rho}}. \tag{1.396}$$

Now, by (1.335) and (1.333)

$$\dot{\rho} = -3H\dot{\varphi}^2, \ \dot{p} = \dot{\varphi}(\ddot{\varphi} - U_{,\varphi}) = -\dot{\varphi}(3H\dot{\varphi} + 2U_{,\varphi}),$$

and by (1.342) and (1.343)

$$\delta\rho = -\dot{\varphi}^2 A + \dot{\varphi}\delta\dot{\varphi} + U_{,\varphi}\delta\varphi\ , \ \delta p = \dot{\varphi}\delta\dot{\varphi} - \dot{\varphi}^2 A - U_{,\varphi}\delta\varphi.$$

With these expressions, one readily finds

$$p\Gamma = -\frac{2}{3}\frac{U_{,\varphi}}{H\dot{\varphi}}[-\ddot{\varphi}\delta\varphi + \dot{\varphi}(\delta\dot{\varphi} - \dot{\varphi}A)]. \tag{1.397}$$

Till now, we have not used the perturbed field equations. The square bracket on the right of the last equation appears in the combination (1.346)-$H\cdot$ (1.347) for the right-hand sides. Since the right-hand side of (1.397) must be gauge invariant, we can work in the gauge $\chi = 0$, and obtain (for $K = 0$) from (1.346), (1.347)

$$\frac{1}{a^2}\nabla^2 A = 4\pi G[-\ddot{\varphi}\delta\varphi + \dot{\varphi}(\delta\dot{\varphi} - \dot{\varphi}A)],$$

thus (1.387) since in the longitudinal gauge $A = A_\chi$.

Application. We return to Eq. (1.385) and use there (1.387) to obtain

$$\boxed{\dot{\mathscr{R}} = 4\pi G\frac{\rho p}{\dot{U}}\Gamma.} \tag{1.398}$$

As a result of (1.387), Γ is small on super-horizon scales and hence (1.398) tells us that $\mathscr{R}$ is almost constant (as we knew before).

The crucial conclusion is that the perturbations are **adiabatic**, which is not obvious (I think). For multifield inflation this is, in general, not the case (see, e.g., [32]).

1.7 Quantization, Primordial Power Spectra

The main goal of this section is to derive the primordial power spectra that are generated as a result of quantum fluctuations during an inflationary period.

1.7.1 Power Spectrum of the Inflaton Field

For the quantization of the scalar field that drives the inflation, we note that the equation of motion (1.370) for the scalar perturbation (1.369),

$$u = a\left[\delta\varphi_\chi - \frac{\dot{\varphi}_0}{H} D_\chi\right] = a\left[\delta\varphi_\chi + \frac{\varphi_0'}{\mathscr{H}} A_\chi\right], \tag{1.399}$$

is the Euler–Lagrange equation for the effective action

$$S_{eff} = \frac{1}{2}\int d^3x d\eta \left[(u')^2 - (\nabla u)^2 + \frac{z''}{z}u^2\right]. \tag{1.400}$$

The normalization is chosen such that S_{eff} reduces to the correct action when gravity is switched off. (In [25], this action is obtained by considering the quadratic piece of the full action with Lagrange density (1.135), but this calculation is extremely tedious.)

The effective Lagrangian of (1.399) is

$$\mathscr{L} = \frac{1}{2}\left[(u')^2 - (\nabla u)^2 + \frac{z''}{z}u^2\right]. \tag{1.401}$$

This is just a free theory with a time-dependent mass $m^2 = -z''/z$. Therefore, the quantization is straightforward. Once u is quantized, the quantization of $\Psi = A_\chi$ is then fixed (see Eq. (1.383)).

The canonical momentum is

$$\pi = \frac{\partial \mathscr{L}}{\partial u'} = u', \tag{1.402}$$

and the canonical commutation relations are the usual ones:

$$\left[\hat{u}(\eta,\mathbf{x}),\hat{u}(\eta,\mathbf{x}')\right] = \left[\hat{\pi}(\eta,\mathbf{x}),\hat{\pi}(\eta,\mathbf{x}')\right] = 0, \quad \left[\hat{u}(\eta,\mathbf{x}),\hat{\pi}(\eta,\mathbf{x}')\right] = i\delta^{(3)}(\mathbf{x}-\mathbf{x}'). \tag{1.403}$$

Let us expand the field operator $\hat{u}(\eta,\mathbf{x})$ in terms of eigenmodes $u_k(\eta)e^{i\mathbf{k}\cdot\mathbf{x}}$ of Eq. (1.370), for which

$$u_k'' + \left(k^2 - \frac{z''}{z}\right)u_k = 0. \tag{1.404}$$

The time-independent normalization is chosen to be

$$u_k^* u_k' - u_k u_k'^* = -i. \tag{1.405}$$

In the decomposition

$$\hat{u}(\eta,\mathbf{x}) = (2\pi)^{-3/2}\int d^3k \left[u_k(\eta)\hat{a}_{\mathbf{k}}e^{i\mathbf{k}\cdot\mathbf{x}} + u_k^*(\eta)\hat{a}_{\mathbf{k}}^\dagger e^{-i\mathbf{k}\cdot\mathbf{x}}\right], \tag{1.406}$$

the coefficients $\hat{a}_{\mathbf{k}}, \hat{a}_{\mathbf{k}}^{\dagger}$ are annihilation and creation operators with the usual commutation relations:

$$[\hat{a}_{\mathbf{k}}, \hat{a}_{\mathbf{k}'}] = [\hat{a}_{\mathbf{k}}^{\dagger}, \hat{a}_{\mathbf{k}'}^{\dagger}] = 0, \quad [\hat{a}_{\mathbf{k}}, \hat{a}_{\mathbf{k}'}^{\dagger}] = \delta^{(3)}(\mathbf{k} - \mathbf{k}'). \tag{1.407}$$

With the normalization (1.405), these imply indeed the commutation relations (1.403). (Translate (1.406) with the help of (1.383) into a similar expansion of Ψ, whose mode functions are determined by $u_k(\eta)$.)

The modes $u_k(\eta)$ are chosen such that at very short distances ($k/aH \to \infty$), they approach the plane waves of the gravity free case with positive frequencies

$$u_k(\eta) \sim \frac{1}{\sqrt{2k}} e^{-ik\eta} \qquad (k/aH \gg 1). \tag{1.408}$$

In the opposite long-wave regime, where k can be neglected in (1.404), we see that the *growing mode* solution is

$$u_k \propto z \qquad (k/aH \ll 1), \tag{1.409}$$

i.e., u_k/z and thus $\mathscr{R}$ is constant on super-horizon scales. This has to be so on the basis of what we saw in Sect. 1.6.2. The power spectrum is conveniently defined in terms of $\mathscr{R}$. We have (we do not put a hat on $\mathscr{R}$)

$$\mathscr{R}(\eta, \mathbf{x}) = (2\pi)^{-3/2} \int \mathscr{R}_{\mathbf{k}}(\eta) e^{i\mathbf{k}\cdot\mathbf{x}} d^3k, \tag{1.410}$$

with

$$\mathscr{R}_{\mathbf{k}}(\eta) = \left[\frac{u_k(\eta)}{z} \hat{a}_{\mathbf{k}} + \frac{u_k^*(\eta)}{z} \hat{a}_{-\mathbf{k}}^{\dagger} \right]. \tag{1.411}$$

The *power spectrum* is defined by (see also Appendix A)

$$\langle 0 | \mathscr{R}_{\mathbf{k}} \mathscr{R}_{\mathbf{k}'}^{\dagger} | 0 \rangle =: \frac{2\pi^2}{k^3} P_{\mathscr{R}}(k) \delta^{(3)}(\mathbf{k} - \mathbf{k}'). \tag{1.412}$$

From (1.411), we obtain

$$\boxed{P_{\mathscr{R}}(k) = \frac{k^3}{2\pi^2} \frac{|u_k(\eta)|^2}{z^2}.} \tag{1.413}$$

Below we shall work this out for the inflationary models considered in Chap. 1.6. Before, we should address the question why we considered the two-point correlation for the Fock vacuum relative to our choice of modes $u_k(\eta)$ (often called the *Bunch–Davies vacuum*). A priori, the initial state could contain all kinds of excitations, for instance a thermal distribution. These would, however, be redshifted away by the enormous inflationary expansion, and the final power spectrum on interesting scales, much larger than the Hubble length, should be largely independent of possible initial excitations. Plausibility arguments for the choice of the Bunch–Davies vacuum state are discussed in [33].

There is also the important question of how the quantum fields and (vacuum) expectations of products of them can be reinterpreted on large scales at the end of inflation in terms of *classical* random fields. There must be some kind of decoherence at work, but it is not obvious how this happens. A necessary condition is that the commutator $[\hat{u}(\mathbf{x},\eta),\ \hat{u}(\mathbf{x}',\eta')]$ can be neglected. It is easy to express this as a Fourier integral of products of the mode functions $u_k(\eta)$ for different times η. Using expressions for these valid well outside the horizon, e.g. (1.423) below, one can see explicitly that such modes do not contribute to the commutator. Unfortunately, I cannot say more about this issue.

1.7.1.1 Power Spectrum for Power-Law Inflation

For power-law inflation, one can derive an exact expression for (1.413). For the mode equation (1.404), we need z''/z. To compute this, we insert in the definition (1.368) of z the results of Sect. 1.3.3.1, giving immediately $z \propto a(t) \propto t^p$. In addition, (1.149) implies $t \propto \eta^{1/1-p}$, so $a(\eta) \propto \eta^{p/1-p}$. Hence,

$$\boxed{\frac{z''}{z} = \left(\nu^2 - \frac{1}{4}\right)\frac{1}{\eta^2}}, \tag{1.414}$$

where

$$\nu^2 - \frac{1}{4} = \frac{p(2p-1)}{(p-1)^2}. \tag{1.415}$$

Using this in (1.404) gives the mode equation

$$\boxed{u_k'' + \left(k^2 - \frac{\nu^2 - 1/4}{\eta^2}\right) u_k = 0.} \tag{1.416}$$

This can be solved in terms of Bessel functions. Before proceeding with this, we note two further relations that will be needed later. First, from $H = p/t$ and $a(t) = a_0 t^p$, we get

$$\eta = -\frac{1}{aH}\frac{1}{1-1/p}. \tag{1.417}$$

In addition,

$$\frac{z}{a} = \frac{\dot{\varphi}}{H} = \sqrt{\frac{p}{4\pi}\frac{M_{Pl}/t}{(p/t)}} = \frac{1}{\sqrt{4\pi p}} M_{Pl},$$

so

$$\boxed{\varepsilon := -\frac{\dot{H}}{H^2} = \frac{1}{p} = \frac{4\pi}{M_{Pl}^2}\frac{z^2}{a^2}.} \tag{1.418}$$

Let us now turn to the mode equation (5.18). According to [34], 9.1.49, the functions $w(z) = z^{1/2}\mathscr{C}_\nu(\lambda z)$, $\mathscr{C}_\nu \propto H_\nu^{(1)}, H_\nu^{(2)}, \ldots$ satisfy the differential equation

$$w'' + \left(\lambda^2 - \frac{\nu^2 - 1/4}{z^2}\right) w = 0. \tag{1.419}$$

From the asymptotic formula for large z ([34], 9.2.3),

$$H_\nu^{(1)} \sim \sqrt{\frac{2}{\pi z}} e^{i(z - \frac{1}{2}\nu\pi - \frac{1}{4}\pi)} \quad (-\pi < \arg z < \pi), \tag{1.420}$$

we see that the correct solutions are

$$u_k(\eta) = \frac{\sqrt{\pi}}{2} e^{i(\nu + \frac{1}{2})\frac{\pi}{2}} (-\eta)^{1/2} H_\nu^{(1)}(-k\eta). \tag{1.421}$$

Indeed, since $-k\eta = (k/aH)(1 - 1/p)^{-1}$, $k/aH \gg 1$ means large $-k\eta$, hence (1.421) satisfies (1.408). Moreover, the Wronskian is normalized according to (1.405) (use 9.1.9 in [34]).

In what follows, we are interested in modes that are well outside the horizon: $(k/aH) \ll 1$. In this limit, we can use (9.1.9 in [34])

$$iH_\nu^{(1)}(z) \sim \frac{1}{\pi}\Gamma(\nu)\left(\frac{1}{2}z\right)^{-\nu} \quad (z \to 0) \tag{1.422}$$

to find

$$u_k(\eta) \simeq 2^{\nu - 3/2} e^{i(\nu - 1/2)\pi/2} \frac{\Gamma(\nu)}{\Gamma(3/2)} \frac{1}{\sqrt{2k}} (-k\eta)^{-\nu + 1/2}. \tag{1.423}$$

Therefore, by (1.417) and (1.418)

$$|u_k| = 2^{\nu - 3/2} \frac{\Gamma(\nu)}{\Gamma(3/2)} (1 - \varepsilon)^{\nu - 1/2} \frac{1}{\sqrt{2k}} \left(\frac{k}{aH}\right)^{-\nu + 1/2}. \tag{1.424}$$

The form (1.424) will turn out to hold also in more general situations studied below, however, with a different ε. We write (1.424) as

$$|u_k| = C(\nu) \frac{1}{\sqrt{2k}} \left(\frac{k}{aH}\right)^{-\nu + 1/2}, \tag{1.425}$$

with

$$C(\nu) = 2^{\nu - 3/2} \frac{\Gamma(\nu)}{\Gamma(3/2)} (1 - \varepsilon)^{\nu - 1/2} \tag{1.426}$$

(recall $\nu = \frac{3}{2} + \frac{1}{p-1}$).

The power spectrum is thus

$$P_{\mathscr{R}}(k) = \frac{k^3}{2\pi^2}\left|\frac{u_k(\eta)}{z^2}\right|^2 = \frac{k^3}{2\pi^2}\frac{1}{z^2}C^2(\nu)\frac{1}{2k}\left(\frac{k}{aH}\right)^{1-2\nu}. \tag{1.427}$$

For z, we could use (1.418). There is, however, a formula which holds more generally: From the definition (1.149) of z and (1.147), we get

$$\boxed{z = -\frac{M_{Pl}^2}{4\pi}\frac{a}{H}\frac{dH}{d\varphi}.} \tag{1.428}$$

Inserting this in the previous equation, we obtain for the power spectrum on superhorizon scales

$$P_{\mathscr{R}}(k) = C^2(\nu)\frac{4}{M_{Pl}^4}\frac{H^4}{(dH/d\varphi)^2}\left(\frac{k}{aH}\right)^{3-2\nu}. \tag{1.429}$$

For power-law inflation, a comparison of (1.418) and (1.428) shows that

$$\frac{M_{Pl}^2}{4\pi}\frac{(dH/d\varphi)^2}{H^2} = \frac{1}{p} = \varepsilon. \tag{1.430}$$

The asymptotic expression (1.429), valid for $k/aH \ll 1$, remains, as we know, constant in time[14]. Therefore, we can evaluate it at *horizon crossing* $k = aH$:

$$\boxed{P_{\mathscr{R}}(k) = C^2(\nu)\frac{4}{M_{Pl}^4}\frac{H^4}{(dH/d\varphi)^2}\bigg|_{k=aH}.} \tag{1.431}$$

We emphasize that this is *not* the value of the spectrum at the moment when the scale crosses outside the Hubble radius. We have just rewritten the asymptotic value for $k/aH \ll 1$ in terms of quantities at horizon crossing.

Note also that $C(\nu) \simeq 1$. The result (1.431) holds, as we shall see below, also in the slow-roll approximation.

1.7.1.2 Power Spectrum in the Slow-Roll Approximation

We now define two slow-roll parameters and rewrite them with the help of (1.146) and (1.147):

---

[14] Let us check this explicitly. Using (1.430), we can write (1.429) as

$$P_{\mathscr{R}}(k) = C^2(\nu)\frac{1}{\pi M_{Pl}^2}\frac{H^2}{\varepsilon}\left(\frac{k}{aH}\right)^{3-2\nu},$$

and we thus have to show that $H^2(aH)^{2\nu-3}$ is time independent. This is indeed the case since $aH \propto 1/\eta$, $H = p/t$, $t \propto \eta^{1/(1-p)} \Rightarrow H \propto \eta^{-1/(1-p)}$.

$$\varepsilon = -\frac{\dot{H}}{H^2} = \frac{4\pi}{M_{Pl}^2}\frac{\dot{\varphi}^2}{H^2} = \frac{M_{Pl}^2}{4\pi}\left(\frac{dH/d\varphi}{H(\varphi)}\right)^2, \quad (1.432)$$

$$\delta = -\frac{\ddot{\varphi}}{H\dot{\varphi}} = \frac{M_{Pl}^2}{4\pi}\frac{d^2H/d\varphi^2}{H} \quad (1.433)$$

($|\varepsilon|, |\delta| \ll 1$ in the slow-roll approximation). These parameters are approximately related to ε_U, η_U introduced in (1.154) and (1.155), as we now show. From (1.145) for $K = 0$ and (1.146), we obtain

$$H^2(1-\frac{\varepsilon}{3}) = \frac{8\pi}{3M_{Pl}^2}U(\varphi). \quad (1.434)$$

For small $|\varepsilon|$, we obtain from this the following approximate expressions for the slow-roll parameters:

$$\varepsilon \simeq \frac{M_{Pl}^2}{16\pi}\left(\frac{U_{,\varphi}}{U}\right)^2 = \varepsilon_U, \quad (1.435)$$

$$\delta \simeq \frac{M_{Pl}^2}{8\pi}\frac{U_{,\varphi\varphi}}{U} - \frac{M_{Pl}^2}{16\pi}\left(\frac{U_{,\varphi}}{U}\right)^2 = \eta_U - \varepsilon_U. \quad (1.436)$$

(In the literature, the letter η is often used instead of δ, but η is already occupied for the conformal time.)

We use these small parameters to approximate various quantities, such as the effective mass z''/z.

First, we note that (1.432) and (1.428) imply the relations[15]

$$\varepsilon = 1 - \frac{\mathcal{H}'}{\mathcal{H}^2} = \frac{4\pi}{M_{Pl}^2}\frac{z^2}{a^2}. \quad (1.437)$$

According to (1.433), we have $\delta = 1 - \varphi''/\varphi'\mathcal{H}$. For the last term, we obtain from the definition $z = a\varphi'/\mathcal{H}$

$$\frac{\varphi''}{\varphi'\mathcal{H}} = \frac{z'}{z\mathcal{H}} - (1 - \mathcal{H}'/\mathcal{H}^2).$$

Hence,

$$\delta = 1 + \varepsilon - \frac{z'}{z\mathcal{H}}. \quad (1.438)$$

[15] Note also that

$$\frac{\ddot{a}}{a} \equiv \dot{H} + H^2 = (1-\varepsilon)H^2,$$

so $\ddot{a} > 0$ for $\varepsilon < 1$.

Next, we look for a convenient expression for the conformal time. From (1.437), we get

$$\frac{\varepsilon}{a\mathscr{H}}da = \varepsilon d\eta = d\eta - (\mathscr{H}'/\mathscr{H}^2)d\eta = d\eta + d\left(\frac{1}{\mathscr{H}}\right),$$

so

$$\eta = -\frac{1}{\mathscr{H}} + \int \frac{\varepsilon}{a\mathscr{H}}da. \tag{1.439}$$

Now, we determine z''/z to first order in ε and δ. From (1.438), i.e., $z'/z = \mathscr{H}(1+\varepsilon-\delta)$, we get

$$\frac{z''}{z} - \left(\frac{z'}{z}\right)^2 = (\varepsilon' - \delta')\mathscr{H} + (1+\varepsilon-\delta)\mathscr{H}',$$

hence,

$$z''/z = \mathscr{H}^2\left[\frac{\varepsilon'-\delta'}{\mathscr{H}} + (1+\varepsilon-\delta)(2-\delta)\right]. \tag{1.440}$$

We can consider ε', δ' as of second order: For instance, by (1.437)

$$\varepsilon' = \frac{4\pi}{M_{Pl}^2}\frac{2zz'}{a^2} - 2\varepsilon\mathscr{H}$$

or

$$\varepsilon' = 2\mathscr{H}\varepsilon(\varepsilon-\delta). \tag{1.441}$$

Treating ε, δ as constant, Eq. (1.439) gives $\eta = -(1/\mathscr{H}) + \varepsilon\eta$, thus

$$\eta = -\frac{1}{\mathscr{H}}\frac{1}{1-\varepsilon}. \tag{1.442}$$

This generalizes (1.417), in which $\varepsilon = 1/p$ (see (1.418)). Using this in (1.440), we obtain first order

$$\frac{z''}{z} = \frac{1}{\eta^2}(2+2\varepsilon-3\delta).$$

We write this as (1.414), but with a different ν:

$$\frac{z''}{z} = \left(\nu^2 - \frac{1}{4}\right)\frac{1}{\eta^2}, \quad \nu := \frac{1+\varepsilon-\delta}{1-\varepsilon} + \frac{1}{2}. \tag{1.443}$$

As a result of all this, we can immediately write down the power spectrum in the slow-roll approximation. From the derivation it is clear that the formula (1.431) still holds, and the same is true for (1.426). Since ν is close to 3/2, we have $C(\nu) \simeq 1$. In sufficient approximation, we thus finally obtain the important result:

$$\boxed{P_{\mathscr{R}}(k) = \frac{4}{M_{Pl}^4}\frac{H^4}{(dH/d\varphi)^2}\bigg|_{k=aH} = \frac{1}{\pi M_{Pl}^2}\frac{H^2}{\varepsilon}\left(\frac{k}{aH}\right)^{3-2\nu}.} \tag{1.444}$$

This spectrum is *nearly scale-free*. This is evident if we use the formula (1.429), from which we get

$$\boxed{n-1 := \frac{d\ln P_{\mathscr{R}(k)}}{d\ln k} = 3-2\nu = 2\delta - 4\varepsilon,} \tag{1.445}$$

so n is *close to unity*.

Exercise

Show that (1.445) follows also from the first equation in (1.444).

Solution: In a first step, we get

$$n-1 = \frac{d}{d\varphi}\ln\left[\left.\frac{H^4}{(dH/d\varphi)^2}\right|_{k=aH}\right]\frac{d\varphi}{d\ln k}.$$

For the last factor, we note that $k = aH$ implies

$$d\ln k = \frac{da}{a} + \frac{dH}{H} \Rightarrow \frac{d\ln k}{d\varphi} = \frac{H}{\dot{\varphi}} + \frac{dH/d\varphi}{H}$$

or, with (1.146),

$$\frac{d\ln k}{d\varphi} = \frac{4\pi}{M_{Pl}^2}\frac{H}{dH/d\varphi}\left[\frac{M_{Pl}^2}{4\pi}\left(\frac{dH/d\varphi}{H}\right)^2 - 1\right].$$

Hence, using (1.432),

$$\frac{d\varphi}{d\ln k} = \frac{M_{Pl}^2}{4\pi}\frac{dH/d\varphi}{H}\frac{1}{\varepsilon-1}.$$

Therefore,

$$n-1 = \frac{M_{Pl}^2}{4\pi}\frac{dH/d\varphi}{H}\frac{1}{\varepsilon-1}\left[4\frac{dH/d\varphi}{H} - 2\frac{d^2H/d\varphi^2}{dH/d\varphi}\right] = \frac{1}{\varepsilon-1}(4\varepsilon - 2\delta)$$

by (1.432) and (1.433). ◇

1.7.1.3 Power Spectrum for Density Fluctuations

Let $P_{\Phi}(k)$ be the power spectrum for the Bardeen potential $\Phi = D_{\chi}$. The latter is related to the density fluctuation Δ by the Poisson equation (1.167),

$$k^2\Phi = 4\pi G\rho a^2\Delta. \tag{1.446}$$

Recall also that for $\Pi = 0$, we have $\Phi = -\Psi$ $(= -A_{\chi})$, and according to (1.395), the following relation for a period with $w = const$.

$$\boxed{\Phi = \frac{3(w+1)}{3w+5}\mathscr{R},} \tag{1.447}$$

and thus,

$$P_{\Phi}^{1/2}(k) = \frac{3(w+1)}{3w+5} P_{\mathscr{R}}^{1/2}(k). \tag{1.448}$$

Inserting (1.444) gives for the *primordial* spectrum on super-horizon scales

$$P_{\Phi}(k) = \left[\frac{3(w+1)}{3w+5}\right]^2 \frac{4}{M_{Pl}^4} \frac{H^4}{(dH/d\varphi)^2}\bigg|_{k=aH}. \tag{1.449}$$

From (1.446), we obtain

$$\Delta(k) = \frac{2(w+1)}{3w+5}\left(\frac{k}{aH}\right)^2 \mathscr{R}(k), \tag{1.450}$$

and thus for the power spectrum of Δ:

$$P_{\Delta}(k) = \frac{4}{9}\left(\frac{k}{aH}\right)^4 P_{\Phi}(k) = \frac{4}{9}\left[\frac{3(w+1)}{3w+5}\right]^2 \left(\frac{k}{aH}\right)^4 P_{\mathscr{R}}(k). \tag{1.451}$$

During the plasma era until recombination, the primordial spectra (1.444) and (1.449) are modified in a way that will be studied in Part III of this book. The modification is described by the so-called *transfer function*[16] $T(k,z)$, normalized such that $T(k) \simeq 1$ for $(k/aH) \ll 1$. Including this, we have in the (dark) matter dominated era (in particular at the time of recombination)

$$P_{\Delta}(k) = \frac{4}{25}\left(\frac{k}{aH}\right)^4 P_{\mathscr{R}}^{prim}(k) T^2(k), \tag{1.452}$$

where $P_{\mathscr{R}}^{prim}(k)$ denotes the primordial spectrum ((1.444) for our simple model of inflation).

Remark

The fact that $\mathscr{R}$ is constant on super-horizon scales allows us to establish the relation between $\Delta_H(k) := \Delta(k,\eta)\,|_{k=aH}$ and $\Delta(k,\eta)$ on these scales. From (1.450), we see that

$$\Delta(k,\eta) = \left(\frac{k}{aH}\right)^2 \Delta_H(k). \tag{1.453}$$

[16] For more on this, see Sect. 1.8.2.4, where the z-dependence of $T(k,z)$ is explicitly split off.

In particular, if $|\mathscr{R}(k)| \propto k^{n-1}$, thus $|\Delta(k,\eta)|^2 = Ak^{n+3}$, then

$$\boxed{|\Delta_H(k)|^2 = Ak^{n-1},} \tag{1.454}$$

and this is *independent* of k for $n = 1$. In this case, the density fluctuation for each mode at horizon crossing has the same magnitude. This explains why the case $n = 1$ – also called the *Harrison-Zel'dovich spectrum* – is called *scale free*.

1.7.2 Generation of Gravitational Waves

In this section, we determine the power spectrum of gravitational waves by quantizing tensor perturbations of the metric.

These are parametrized as follows:

$$g_{\mu\nu} = a^2(\eta)[\gamma_{\mu\nu} + 2H_{\mu\nu}], \tag{1.455}$$

where $a^2(\eta)\gamma_{\mu\nu}$ is the Friedmann metric ($\gamma_{\mu 0} = 0,\ \gamma_{ij}$: metric of (Σ,γ)), and $H_{\mu\nu}$ satisfies the *transverse traceless* (TT) gauge conditions

$$H_{00} = H_{0i} = H^i{}_i = H_i{}^j{}_{|j} = 0. \tag{1.456}$$

The tensor perturbation amplitudes H_{ij} remain invariant under gauge transformations (1.178). Indeed, as in Sect. 1.4.1.4, one readily finds

$$\begin{aligned} L_\xi g^{(0)} = 2a^2(\eta)\{&-(\mathscr{H}\xi^0 + (\xi^0)')d\eta^2 + (\xi_i' - \xi^0{}_{|i})dx^i d\eta \\ &+ (\mathscr{H}\gamma_{ij}\xi^0 + \xi_{i|j})dx^i dx^j\}. \end{aligned}$$

Decomposing ξ^μ into scalar and vector parts gives the scalar and vector contributions of $L_\xi g^{(0)}$, but there are obviously *no* tensor contributions.

The perturbations of the Einstein tensor belonging to $H_{\mu\nu}$ are derived in the Appendix to this chapter. The result is

$$\begin{aligned} \delta G^0{}_0 &= \delta G^0{}_j = \delta G^i{}_0 = 0, \\ \delta G^i{}_j &= \frac{1}{a^2}\left[(H^i{}_j)'' + 2\frac{a'}{a}(H^i{}_j)' + (-\nabla^2 + 2K)H^i{}_j\right]. \end{aligned} \tag{1.457}$$

We claim that the quadratic part of the Einstein–Hilbert action is

$$S^{(2)} = \frac{M_{Pl}^2}{16\pi}\int\left[(H^i{}_k)'(H^k{}_i)' - H^i{}_{k|l}H^k{}_i{}^{|l} - 2KH^i{}_k H^k{}_i\right]a^2(\eta)d\eta\sqrt{\gamma}d^3x. \tag{1.458}$$

(Remember that the indices are raised and lowered with γ_{ij}.) Note first that $\sqrt{-g}d^4x = \sqrt{\gamma}a^4(\eta)d\eta d^3x+$ quadratic terms in H_{ij} because H_{ij} is traceless. A direct derivation of (1.458) from the Einstein–Hilbert action would be extremely tedious (see [25]).

It suffices, however, to show that the variation of (1.458) is just the linearization of the general variation formula (see Sect. 2.3 of [9])

$$\delta S = -\frac{M_{Pl}^2}{16\pi}\int G^{\mu\nu}\delta g_{\mu\nu}\sqrt{-g}d^4x \tag{1.459}$$

for the Einstein–Hilbert action

$$S = \frac{M_{Pl}^2}{16\pi}\int R\sqrt{-g}d^4x. \tag{1.460}$$

Now, we have after the usual partial integrations,

$$\delta S^{(2)} = -\frac{M_{Pl}^2}{8\pi}\int\left[\frac{(a^2H^i{}_k)')'}{a^2} + (-\nabla^2 + 2K)H^i{}_k\right]\delta H^k{}_i a^2(\eta)d\eta\sqrt{\gamma}d^3x.$$

Since $\delta H^k{}_i = \frac{1}{2}\delta g^k{}_i$ this is, with the expression (1.457), indeed the linearization of (1.459).

We absorb in (1.458), the factor $a^2(\eta)$ by introducing the rescaled perturbation

$$P^i{}_j(x) := \left(\frac{M_{Pl}^2}{8\pi}\right)^{1/2} a(\eta)H^i{}_j(x). \tag{1.461}$$

Then $S^{(2)}$ becomes, after another partial integration,

$$\boxed{S^{(2)} = \frac{1}{2}\int\left[(P^i{}_k)'(P^k{}_i)' - P^i{}_{k|l}P^k{}_i{}^{|l} + \left(\frac{a''}{a} - 2K\right)P^i{}_kP^k{}_i\right]d\eta\sqrt{\gamma}d^3x.} \tag{1.462}$$

In what follows, we take again $K = 0$. Then, we have the following Fourier decomposition: Let $\varepsilon_{ij}(\mathbf{k},\lambda)$ be the two polarization tensors, satisfying

$$\begin{aligned}&\varepsilon_{ij} = \varepsilon_{ji},\ \varepsilon^i{}_i = 0,\ k^i\varepsilon_{ij}(\mathbf{k},\lambda) = 0,\ \varepsilon_i{}^j(\mathbf{k},\lambda)\varepsilon_j{}^i(\mathbf{k},\lambda')^* = \delta_{\lambda\lambda'},\\ &\varepsilon_{ij}(-\mathbf{k},\lambda) = \varepsilon_{ij}^*(\mathbf{k},\lambda),\end{aligned} \tag{1.463}$$

then

$$P^i{}_j(\eta,\mathbf{x}) = (2\pi)^{-3/2}\int d^3k\sum_\lambda v_{\mathbf{k},\lambda}(\eta)\varepsilon^i{}_j(\mathbf{k},\lambda)e^{i\mathbf{k}\cdot\mathbf{x}}. \tag{1.464}$$

In terms of $v_{\mathbf{k},\lambda}(\eta)$, the action becomes

$$S^{(2)} = \frac{1}{2}\int d\eta\sum_\lambda\int d^3k\left[|v'_{\mathbf{k},\lambda}|^2 - \left(k^2 - \frac{a''}{a}\right)|v_{\mathbf{k},\lambda}|^2\right]$$

as for two scalar fields in Minkowski spacetime, each with an effective mass a''/a. The field is now quantized by interpreting $v_{\mathbf{k},\lambda}(\eta)$ as the operator

$$\hat{v}_{\mathbf{k},\lambda}(\eta) = v_k(\eta)\hat{a}_{\mathbf{k},\lambda} + v_k^*(\eta)\hat{a}^\dagger_{-\mathbf{k},\lambda}, \tag{1.465}$$

where $v_k(\eta)\varepsilon_{ij}(\mathbf{k},\lambda)e^{i\mathbf{k}\cdot\mathbf{x}}$ satisfies the field equation[17] corresponding to the action (1.462), that is (for $K=0$)

$$v_k'' + \left(k^2 - \frac{a''}{a}\right) v_k = 0. \tag{1.466}$$

(Instead of z''/z in (1.404), we now have the "mass" a''/a.)

In the long-wavelength regime, the growing mode now behaves as $v_k \propto a$, hence v_k/a remains constant.

Again, we have to impose the normalization (1.405):

$$v_k^* v_k' - v_k v_k'^* = -i, \tag{1.467}$$

and the asymptotic behavior

$$v_k(\eta) \sim \frac{1}{\sqrt{2k}} e^{-ik\eta} \qquad (k/aH \gg 1). \tag{1.468}$$

The decomposition (1.464) translates to

$$H^i{}_j(\eta,\mathbf{x}) = (2\pi)^{-3/2} \int d^3k \sum_\lambda \hat{h}_{\mathbf{k},\lambda}(\eta)\varepsilon^i{}_j(\mathbf{k},\lambda)e^{i\mathbf{k}\cdot\mathbf{x}}, \tag{1.469}$$

where

$$\hat{h}_{\mathbf{k},\lambda}(\eta) = \left(\frac{8\pi}{M_{Pl}^2}\right)^{1/2} \frac{1}{a}\hat{v}_{\mathbf{k},\lambda}(\eta). \tag{1.470}$$

We define the *power spectrum of gravitational waves* by

$$\frac{2\pi^2}{k^3} P_g(k)\delta^{(3)}(\mathbf{k}-\mathbf{k}') = \sum_\lambda \langle 0|\hat{h}_{\mathbf{k},\lambda}\hat{h}^\dagger_{\mathbf{k}',\lambda}|0\rangle, \tag{1.471}$$

thus

$$\sum_\lambda \langle 0|\hat{v}_{\mathbf{k},\lambda}\hat{v}^\dagger_{\mathbf{k}',\lambda}|0\rangle = \frac{M_{Pl}^2 a^2}{8\pi}\frac{2\pi^2}{k^3} P_g(k)\delta^{(3)}(\mathbf{k}-\mathbf{k}'). \tag{1.472}$$

Using (1.465) for the left-hand side, we obtain instead of (1.413)[18]

$$\boxed{P_g(k) = 2\frac{8\pi}{M_{Pl}^2 a^2}\frac{k^3}{2\pi^2}|v_k(\eta)|^2.} \tag{1.473}$$

[17] We ignore possible tensor contributions to the energy-momentum tensor.

[18] In the literature, one often finds an expression for $P_g(k)$ which is four times larger because the power spectrum is defined in terms of $h_{ij} = 2H_{ij}$.

The factor 2 on the right is due to the two polarizations. Note that

$$\langle H_{ij}(\eta,\mathbf{x})H^{ij}(\eta,\mathbf{x})\rangle = \int \frac{dk}{k} P_g(\eta,k). \tag{1.474}$$

1.7.2.1 Power Spectrum for Power-Law Inflation

For the modes $v_k(\eta)$, we need a''/a. From

$$\frac{a''}{a} = (a\mathscr{H})'/a = \mathscr{H}^2 + \mathscr{H}' = 2\mathscr{H}^2\left[1 - \frac{1}{2}(1 - \mathscr{H}'/\mathscr{H}^2)\right]$$

and (1.437), we obtain the generally valid formula

$$\frac{a''}{a} = 2\mathscr{H}^2(1 - \varepsilon/2). \tag{1.475}$$

For power-law inflation, we had $\varepsilon = 1/p$, $a(\eta) \propto \eta^{p/(1-p)}$, thus

$$\mathscr{H} = \frac{p}{p-1}\frac{1}{\eta}$$

and hence,

$$\frac{a''}{a} = \left(\mu^2 - \frac{1}{4}\right)\frac{1}{\eta^2}, \quad \mu := \frac{3}{2} + \frac{1}{p-1}. \tag{1.476}$$

This shows that for power-law inflation, $v_k(\eta)$ is identical to $u_k(\eta)$. Therefore, we have by Eq. (1.425)

$$|v_k| = C(\mu)\frac{1}{\sqrt{2k}}\left(\frac{k}{aH}\right)^{-\mu+1/2}, \tag{1.477}$$

with

$$C(\mu) = 2^{\mu-3/2}\frac{\Gamma(\mu)}{\Gamma(3/2)}(1-\varepsilon)^{\mu-1/2}. \tag{1.478}$$

Inserting this in (1.473) gives

$$P_g(k) = \frac{16\pi}{M_{Pl}^2}\frac{k^3}{2\pi^2}\frac{1}{a^2}C^2(\mu)\frac{1}{2k}\left(\frac{k}{aH}\right)^{1-2\mu} \tag{1.479}$$

or

$$P_g(k) = C^2(\mu)\frac{4}{\pi}\left(\frac{H}{M_{Pl}}\right)^2\left(\frac{k}{aH}\right)^{3-2\mu}. \tag{1.480}$$

Alternatively, we have

$$\boxed{P_g(k) = C^2(\mu)\frac{4}{\pi}\frac{H^2}{M_{Pl}^2}\bigg|_{k=aH}.} \tag{1.481}$$

1.7.2.2 Slow-Roll Approximation

From (1.475) and (1.442), we obtain again the first equation in (1.476), but with a different μ:

$$\mu = \frac{1}{1-\varepsilon} + \frac{1}{2}. \tag{1.482}$$

Hence, $v_k(\eta)$ is equal to $u_k(\eta)$ if ν is replaced by μ. The formula (1.481), with $C(\mu)$ given by (1.478), remains therefore valid, but now μ is given by (1.482), where ε is the slow-roll parameter in (1.432) or (1.437). Again $C(\mu) \simeq 1$.

The power index for tensor perturbations,

$$n_T(k) := \frac{d\ln P_g(k)}{d\ln k}, \tag{1.483}$$

can be read off from (1.480):

$$\boxed{n_T \simeq -2\varepsilon,} \tag{1.484}$$

showing that the *power spectrum is almost flat*[19].

Consistency Equation

Let us collect some of the important formulas:

$$P_{\mathscr{R}}^{1/2}(k) = 2\,\frac{H^2}{M_{Pl}^2|dH/d\varphi|}\bigg|_{k=aH}, \tag{1.485}$$

$$P_g^{1/2}(k) = \frac{2}{\sqrt{\pi}}\frac{H}{M_{Pl}}\bigg|_{k=aH}, \tag{1.486}$$

$$n - 1 = 2\delta - 4\varepsilon, \tag{1.487}$$

$$n_T = -2\varepsilon. \tag{1.488}$$

[19] The result (1.485) can also be obtained from (1.481). Making use of an intermediate result in the solution of the Exercise at the end of Sect. 1.7.1.2 and (1.432), we get

$$n_T = \frac{d\ln H^2}{d\varphi}\frac{d\varphi}{d\ln k} = \frac{2\varepsilon}{\varepsilon - 1} \simeq -2\varepsilon.$$

The relative amplitude of the two spectra (scalar and tensor) is thus given by

$$\boxed{r := \frac{4P_g}{P_{\mathscr{R}}} = 16\varepsilon.} \tag{1.489}$$

More importantly, we obtain the *consistency condition*

$$\boxed{n_T = -8r,} \tag{1.490}$$

which is characteristic for inflationary models. In principle, this can be tested with CMB measurements, but there is a long way before this can be done in practice.

For attempts to discriminate among various single-field inflationary models on the basis of WMAP and SDSS data, see for example [35].

1.7.2.3 Stochastic Gravitational Background Radiation

The spectrum of gravitational waves, generated during the inflationary era and stretched to astronomical scales by the expansion of the Universe, contributes to the background energy density. Using the results of the previous section, we can compute this.

I first recall a general formula for the effective energy-momentum tensor of gravitational waves. (For detailed derivations, see Sect. 4.4 of [9].)

By "gravitational waves," we mean propagating ripples in curvature on scales much smaller than the characteristic scales of the background spacetime (the Hubble radius for the situation under study). For sufficiently high-frequency waves, it is meaningful to associate them – in an *averaged* sense – an energy-momentum tensor. Decomposing the full metric $g_{\mu\nu}$ into a background $\bar{g}_{\mu\nu}$ plus fluctuation $h_{\mu\nu}$, the effective energy-momentum tensor is given by the following expression

$$T^{(GW)}_{\alpha\beta} = \frac{1}{32\pi G}\left\langle h_{\mu\nu|\alpha} h^{\mu\nu}{}_{|\beta}\right\rangle, \tag{1.491}$$

if the gauge is chosen such that $h^{\mu\nu}{}_{|\nu} = 0$, $h^{\mu}{}_{\mu} = 0$. Here, a vertical stroke indicates covariant derivatives with respect to the background metric, and $\langle \cdots \rangle$ denotes a four-dimensional average over regions of several wavelengths.

For a Friedmann background, we have in the TT gauge for $h_{\mu\nu} = 2H_{\mu\nu}$: $h_{\mu 0} = 0$, $h_{ij|0} = h_{ij,0}$, thus

$$T^{(GW)}_{00} = \frac{1}{8\pi G}\left\langle \dot{H}_{ij}\dot{H}^{ij}\right\rangle. \tag{1.492}$$

As in (1.469), we perform (for $K = 0$) a Fourier decomposition

$$H_{ij}(\eta, \mathbf{x}) = (2\pi)^{-3/2}\int d^3k \sum_{\lambda} h_{\lambda}(\eta, \mathbf{k})\varepsilon_{ij}(\mathbf{k}, \lambda)e^{i\mathbf{k}\cdot\mathbf{x}}. \tag{1.493}$$

The gravitational background energy density, ρ_g, is obtained by taking the space-time average in (1.492). At this point, we regard $h_\lambda(\eta,\mathbf{k})$ as a random field, indicated by a hat (since it is on macroscopic scales equivalent to the original quantum field $\hat{h}_\lambda(\eta,\mathbf{k})$), and replace the spatial average by the *stochastic average* (for which we use the same notation). Clearly, this is only justified if some *ergodicity* property holds. This issue is discussed in Appendix C of [27].

The power spectrum at time η is defined by

$$\langle \hat{h}_\lambda(\eta,\mathbf{k})\hat{h}^*_{\lambda'}(\eta,\mathbf{k}')\rangle = \delta_{\lambda\lambda'}\delta^{(3)}(\mathbf{k}-\mathbf{k}')\frac{\pi^2}{k^3}P_g(k,\eta). \tag{1.494}$$

The normalization is chosen such that (1.474) holds. The time evolution of the stochastic variable $\hat{h}_\lambda(\eta,\mathbf{k})$ is determined by that of the mode functions $h_k(\eta)$:

$$\hat{h}_\lambda(\eta,\mathbf{k}) = \frac{h_k(\eta)}{h_k(\eta_i)}\hat{h}_\lambda(\eta_i,\mathbf{k}),$$

where η_i is some early time. Therefore, we obtain for ρ_g

$$\rho_g = \frac{2}{8\pi G a^2(2\pi)^3}\int d^3k \left\langle \left|\frac{h'_k(\eta)}{h_k(\eta_i)}\right|^2\right\rangle \sum_\lambda \frac{\pi^2}{k^3}P_g(k,\eta_i), \tag{1.495}$$

where from now on $\langle\cdots\rangle$ denotes the *average over several periods*. For the spectral density, this gives

$$\boxed{k\frac{d\rho_g(k)}{dk} = \frac{M_{Pl}^2}{8\pi a^2}\left\langle \left|\frac{h'_k(\eta)}{h_k(\eta_i)}\right|^2\right\rangle P_g(k,\eta_i).} \tag{1.496}$$

When the radiation is well inside the horizon, we can replace h'_k by kh_k.

The differential equation (1.466) reads in terms of $h_k(\eta)$

$$h'' + 2\frac{a'}{a}h' + k^2 h = 0. \tag{1.497}$$

For the matter dominated era ($a(\eta) \propto \eta^2$), this becomes

$$h'' + \frac{4}{\eta}h' + k^2 h = 0.$$

Using 9.1.53 of [34], one sees that this is satisfied by $j_1(k\eta)/k\eta$. Furthermore, by 10.1.4 of the same reference, we have $3j_1(x)/x \to 1$ for $x \to 0$ and

$$\left(\frac{j_1(x)}{x}\right)' = -\frac{1}{x}j_2(x) \ \to 0 \quad (x \to 0).$$

So the correct solution is

$$\frac{h_k(\eta)}{h_k(0)} = 3\frac{j_1(k\eta)}{k\eta} \tag{1.498}$$

if the modes cross inside the horizon during the matter dominated era. Note also that

$$j_1(x) = \frac{\sin x}{x^2} - \frac{\cos x}{x}. \tag{1.499}$$

For modes which enter the horizon earlier, we introduce a *transfer function* $T_g(k)$ by

$$\frac{h_k(\eta)}{h_k(0)} =: 3\frac{j_1(k\eta)}{k\eta}T_g(k), \tag{1.500}$$

that has to be determined numerically from the differential equation (1.497)[20]. We can then write the result (1.496) as

$$k\frac{d\rho_g(k)}{dk} = \frac{M_{Pl}^2}{8\pi}\frac{k^2}{a^2}P_g^{prim}(k)|T_g(k)|^2\left\langle\left[\frac{3j_1(k\eta)}{k\eta}\right]^2\right\rangle, \tag{1.501}$$

where $P_g^{prim}(k)$ denotes the primordial power spectrum. This holds in particular at the present time η_0 ($a_0 = 1$). Since the time average $\langle\cos^2 k\eta\rangle = \frac{1}{2}$, we finally obtain for $\Omega_g(k) := \rho_g(k)/\rho_{crit}$

$$\boxed{\frac{d\Omega_g(k)}{d\ln k} = \frac{3}{2}P_g^{prim}(k)|T_g(k)|^2\frac{1}{(k\eta_0)^2(H_0\eta_0)^2}.} \tag{1.502}$$

Here, one may insert the inflationary result (1.481), giving

$$\frac{d\Omega_g(k)}{d\ln k} = \frac{6}{\pi}\left.\frac{H^2}{M_{Pl}^2}\right|_{k=aH}|T_g(k)|^2\frac{1}{(k\eta_0)^2(H_0\eta_0)^2}. \tag{1.503}$$

Numerical Results

Since the normalization in (1.481) cannot be predicted, it is reasonable to choose it, for illustration, to be equal to the observed CMB normalization at large scales. (In reality, the tensor contribution is presumably only a small fraction of this; see (1.489).) Then, one obtains the result shown in Fig. 1.8, taken from [36]. This shows that the spectrum of the stochastic gravitational background radiation is predicted to be flat in the interesting region, with $d\Omega_g/d\ln(k\eta_0) \sim 10^{-14}$. Unfortunately, this is too small to be detectable by the future LISA interferometer in space.

---

[20] After neutrino decoupling, an accurate treatment should include tensor contributions to the energy-momentum tensor due to neutrino free-streaming. This would lead to an integro-differential equation. (This has been solved numerically for instance in [37].)

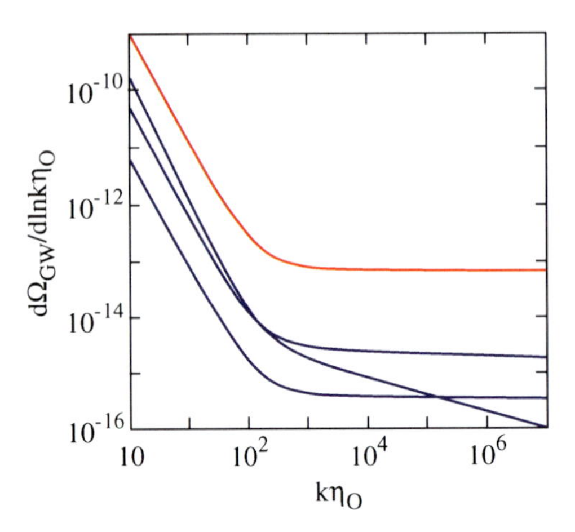

Fig. 1.8 Differential energy density (5.108) of the stochastic background of inflation-produced gravitational waves. The normalization of the upper curve, representing the scale-invariant limit, is arbitrary. The blue curves are normalized to the COBE quadrupole and show the result for $n_T = -0.003, \ -0.03$, and -0.3. (Adapted from [36].)

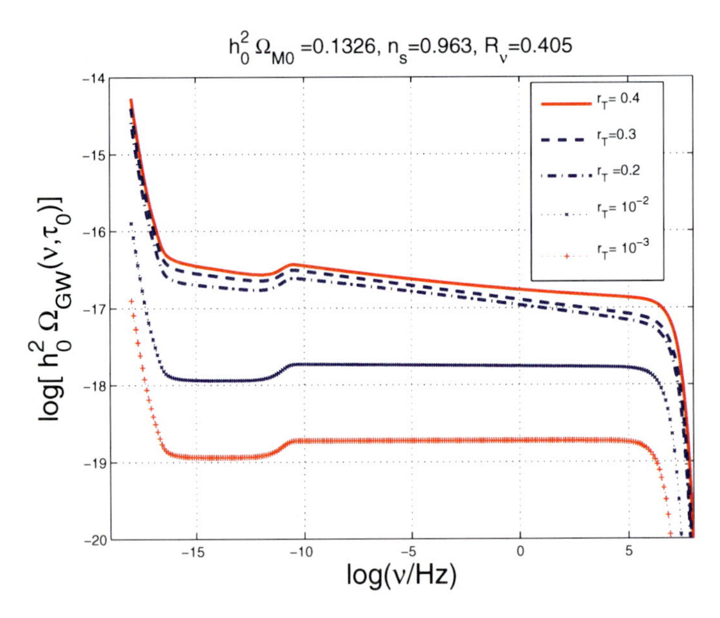

Fig. 1.9 GW spectra for ΛCDM models. Obviously, the background is too small to be within reach by wide-band detectors. From [38].

It would be of great importance if one day the stochastic gravitational wave background could be detected because it has been formed in the very early Universe. In the high-frequency region, accessible to wide-band interferometers, the spectrum depends on the expansion rate after inflation and thus on poorly known physics.

For a recent review, we refer to [38]. Fig. 1.9, taken from this reference, shows the spectrum of relict gravitational radiation for a minimal ΛCDM scenario for various values of r. For orientation, recall that $\nu_{eq} := k_{eq}/2\pi \simeq 10^{-17}$ Hz and $\nu_p = k_p/2\pi \simeq 10^{-18}$ Hz, where k_p is the "pivot" wave-number used by WMAP (corresponding to $l \approx 30$). The Ligo/Virgo frequency band is $\sim$ 10–100 Hz.

This relic spectrum was obtained from a numerical integration of the evolutionary equations for the transfer function and the background geometry across the matter-radiation transition. The coupling to the anisotropic neutrino stress (see Appendix E) is included.

Exercise

Consider a massive free scalar field ϕ (mass m) and discuss the quantum fluctuations for a de Sitter background (neglecting gravitational back reaction). Compute the power spectrum as a function of conformal time for $m/H < 3/2$.

Hint: Work with the field $a\phi$ as a function of conformal time.

Remark: This exercise was solved at an astonishingly early time ($\sim$ 1940) by E. Schrödinger. $\diamondsuit$

1.7.3 Appendix to Section 1.7: Einstein Tensor for Tensor Perturbations

In this appendix, we derive the expressions (1.457) for the tensor perturbations of the Einstein tensor.

The metric (1.455) is conformal to $\tilde{g}_{\mu\nu} = \gamma_{\mu\nu} + 2H_{\mu\nu}$. We first compute the Ricci tensor $\tilde{R}_{\mu\nu}$ of this metric and then use the general transformation law of Ricci tensors for conformally related metrics (see Eq. (2.264) of [9]).

Let us first consider the simple case $K = 0$ that we considered in Sect. 1.7.2. Then, $\gamma_{\mu\nu}$ is the Minkowski metric. In the following computation of $\tilde{R}_{\mu\nu}$, we drop temporarily the tildes.

The Christoffel symbols are immediately found (to first order in $H_{\mu\nu}$)

$$\Gamma^{\mu}{}_{00} = \Gamma^{0}{}_{0i} = 0, \;\; \Gamma^{0}{}_{ij} = H'_{ij}, \;\; \Gamma^{i}{}_{0j} = (H^{i}{}_{j})',$$
$$\Gamma^{i}{}_{jk} = H^{i}{}_{j,k} + H^{i}{}_{k,j} - H_{jk}{}^{,i}. \tag{1.504}$$

So these vanish or are of first-order small. Hence, up to higher orders,

$$R_{\mu\nu} = \partial_{\lambda}\Gamma^{\lambda}{}_{\nu\mu} - \partial_{\nu}\Gamma^{\lambda}{}_{\lambda\mu}. \tag{1.505}$$

Inserting (1.504) and using the TT conditions, (1.456) readily gives

$$R_{00} = 0, \quad R_{0i} = 0, \tag{1.506}$$

$$R_{ij} = \partial_\lambda \Gamma^\lambda{}_{ij} - \partial_j \Gamma^\lambda{}_{\lambda i} = \partial_0 \Gamma^0{}_{ij} + \partial_k \Gamma^k{}_{ij} - \partial_j \Gamma^0{}_{0i} - \partial_j \Gamma^k{}_{ki}$$
$$= H''_{ij} + (H^k{}_{i,j} + H^k{}_{j,i} - H_{ij}{}^{,k})_{,k}.$$

Thus,

$$R_{ij} = H''_{ij} - \nabla^2 H_{ij}. \tag{1.507}$$

Now, we use the quoted general relation between the Ricci tensors for two metrics related as $g_{\mu\nu} = e^f \tilde{g}_{\mu\nu}$. In our case $e^f = a^2(\eta)$, hence

$$\tilde{\nabla}_\mu f = 2\mathcal{H}\delta_{\mu 0}, \quad \tilde{\nabla}_\mu \tilde{\nabla}_\nu f = \partial_\mu (2\mathcal{H}\delta_{\nu 0}) - \Gamma^\lambda{}_{\mu\nu} 2\mathcal{H}\delta_{\lambda 0}$$
$$= 2\mathcal{H}'\delta_{\mu 0}\delta_{\nu 0} - 2\mathcal{H}H'_{\mu\nu}, \quad \tilde{\nabla}^2 f = \tilde{g}^{\mu\nu}\tilde{\nabla}_\mu \tilde{\nabla}_\nu f = 2\mathcal{H}'.$$

As a result, we find

$$R_{\mu\nu} = \tilde{R}_{\mu\nu} + (-2\mathcal{H}' + 2\mathcal{H}^2)\delta_{\mu 0}\delta_{\nu 0} + (\mathcal{H}' + 2\mathcal{H}^2)\tilde{g}_{\mu\nu} + 2\mathcal{H}H'_{\mu\nu}, \tag{1.508}$$

thus

$$\delta R_{00} = \delta R_{0i} = 0,$$
$$\delta R_{ij} = H''_{ij} - \nabla^2 H_{ij} + 2(\mathcal{H}' + 2\mathcal{H}^2)H_{ij} + 2\mathcal{H}H'_{ij}. \tag{1.509}$$

From this, it follows that

$$\delta R = g^{(0)\mu\nu}\delta R_{\mu\nu} + \delta g^{\mu\nu} R^{(0)}_{\mu\nu} = 0. \tag{1.510}$$

The result (1.457) for the Einstein tensor is now easily obtained. For the generalization to $K \neq 0$, see [27]

1.8 Tight Coupling Phase

Long before recombination (at temperatures $T > 6000K$, say), photons, electrons, and baryons were so strongly coupled that these components may be treated together as a single fluid. In addition to this, there is also a dark matter component. For all practical purposes, the two interact only gravitationally. The investigation of such a two-component fluid for small deviations from an idealized Friedmann behavior is a well-studied application of CPT and will be treated in this section.

At a later stage, when decoupling is approached, this approximate treatment breaks down because the mean free path of the photons becomes longer (and finally "infinite" after recombination). Although the electrons and baryons can still be treated as a single fluid, the neutrinos, photons and their coupling to the electrons have to be described by the general relativistic Boltzmann equation. The latter is, of course, again linearized about the idealized Friedmann solution. Together with the linearized fluid equations (for baryons and cold dark matter, say) and the linearized Einstein equations, one arrives at a complete system of equations for the various perturbation amplitudes of the metric and matter variables. (For detailed derivations, we refer again to [27].)

We now discuss in detail the strong coupling phase. As already emphasized, photons, electrons, and baryons are so strongly coupled that these components may be treated as a single fluid, indexed by r in what follows. Beside this, we have to include a CDM component for which we we use the index d (for "dust" or dark). Since these two fluids interact only gravitationally, the extension of the perturbation theory in Sect. 1.5 is easy. (Note that we neglect fluctuations of the neutrinos.)

1.8.1 Basic Equations

Let $T^{\mu}_{(\alpha)\nu}$ denote the energy-momentum tensor of species (α). The total $T^{\mu}{}_{\nu}$ is assumed to be just the sum

$$T^{\mu}{}_{\nu} = \sum_{(\alpha)} T^{\mu}_{(\alpha)\nu}, \tag{1.511}$$

and is, of course, "conserved". For the unperturbed background, we have, as in (1.198),

$$T^{(0)}_{(\alpha)\mu}{}^{\nu} = (\rho_\alpha^{(0)} + p_\alpha^{(0)}) u_\mu^{(0)} u^{(0)\nu} + p_\alpha^{(0)} \delta_\mu{}^\nu \tag{1.512}$$

with

$$\left(u^{(0)\mu}\right) = \left(\frac{1}{a}, \mathbf{0}\right). \tag{1.513}$$

The divergence of $T^{\mu}_{(\alpha)\nu}$ does, in general, not vanish. This is, however, the case for our two-component system $\alpha = (r, d)$ during the tight coupling phase. (The *phenomenological* description of general multicomponent systems is, for instance, described in [24] and [27].) From

$$T^{\nu}_{(\alpha)\mu;\nu} = 0. \tag{1.514}$$

we obtain for the background

$$\dot{\rho}_\alpha^{(0)} = -3H(\rho_\alpha^{(0)} + p_\alpha^{(0)}) = -3Hh_\alpha, \tag{1.515}$$

where

$$h_\alpha = \rho_\alpha^{(0)} + p_\alpha^{(0)}. \tag{1.516}$$

Clearly,

$$\rho^{(0)} = \sum_\alpha \rho_\alpha^{(0)}, \quad p^{(0)} = \sum_\alpha p_\alpha^{(0)}, \quad h := \rho^{(0)} + p^{(0)} = \sum_\alpha h_\alpha. \tag{1.517}$$

We again consider only *scalar perturbations* and proceed with each component as in Sect. 1.4.1.6. In particular, Eqs. (1.196), (1.197), (1.206), and (1.208) become

$$T^{\mu}_{(\alpha)\nu}u^{\nu}_{(\alpha)} = -\rho_{(\alpha)}u^{\mu}_{(\alpha)}, \tag{1.518}$$

$$g_{\mu\nu}u^{\mu}_{(\alpha)}u^{\nu}_{(\alpha)} = -1, \tag{1.519}$$

$$\begin{aligned}
\delta u^0_{(\alpha)} &= -\frac{1}{a}A, \quad \delta u^i_{(\alpha)} = \frac{1}{a}\gamma^{ij}v_{\alpha|j} \quad \Rightarrow \delta u_{(\alpha)i} = a(v_\alpha - B)_{|i}, \\
\delta T^0_{(\alpha)0} &= -\delta\rho_\alpha, \\
\delta T^0_{(\alpha)j} &= h_\alpha(v_\alpha - B)_{|j}, \quad T^i_{(\alpha)0} = -h_\alpha\gamma^{ij}v_{\alpha|j}, \\
\delta T^i_{(\alpha)j} &= \delta p_\alpha \delta^i{}_j + p_\alpha\left(\Pi^{|i}_{\alpha|j} - \frac{1}{3}\delta^i{}_j\nabla^2\Pi_\alpha\right), \\
\delta p_\alpha &= c^2_\alpha\delta\rho_\alpha + p_\alpha\Gamma_\alpha \equiv p_\alpha\pi_{L\alpha}, \quad c^2_\alpha := \dot{p}_\alpha/\dot{\rho}_\alpha.
\end{aligned} \tag{1.520}$$

In (1.520) and in what follows, the index (0) is dropped.

Summation of these equations give ($\delta_\alpha := \delta\rho_\alpha/\rho_\alpha$):

$$\rho\delta = \sum_\alpha \rho_\alpha\delta_\alpha, \tag{1.521}$$

$$hv = \sum_\alpha h_\alpha v_\alpha, \tag{1.522}$$

$$p\pi_L = \sum_\alpha p_\alpha \pi_{L\alpha}, \tag{1.523}$$

$$p\Pi = \sum_\alpha p_\alpha \Pi_\alpha. \tag{1.524}$$

We turn to the gauge transformation properties. As long as we do not use the zeroth-order energy equation (1.515), the transformation laws for $\delta_\alpha, v_\alpha, \pi_{L\alpha}, \Pi_\alpha$ remain the same as those in Sect. 1.4.1.6 for δ, v, π_L, and Π. Thus, using (1.515) and the notation $w_\alpha = p_\alpha/\rho_\alpha$, we have

$$\begin{aligned}
\delta_\alpha &\to \delta_\alpha + \frac{\rho'_\alpha}{\rho_\alpha}\xi^0 = \delta_\alpha - 3(1+w_\alpha)\mathscr{H}\xi^0, \\
v_\alpha - B &\to (v_\alpha - B) - \xi^0, \\
\delta p_\alpha &\to \delta p_\alpha + p'_\alpha\xi^0, \\
\Pi_\alpha &\to \Pi_\alpha, \\
\Gamma_\alpha &\to \Gamma_\alpha.
\end{aligned} \tag{1.525}$$

The quantity $\mathscr{Q}$, introduced below (1.206), will also be used for each component:

$$\delta T^0_{(\alpha)i} =: \frac{1}{a}\mathscr{Q}_{\alpha|i}, \quad \Rightarrow \quad \mathscr{Q} = \sum_\alpha \mathscr{Q}_{\alpha|i}. \tag{1.526}$$

The transformation law of $\mathscr{Q}_\alpha$ is

$$\mathscr{Q}_\alpha \to \mathscr{Q}_\alpha - ah_\alpha\xi^0. \tag{1.527}$$

For each α, we define gauge invariant density perturbations $(\delta_\alpha)_{\mathscr{Q}_\alpha}, (\delta_\alpha)_\chi$ and velocities $V_\alpha = (v_\alpha - B)_\chi$. Because of the modification in the first of Eq. (1.525), we have instead of (1.218)

$$\Delta_\alpha := (\delta_\alpha)_{\mathscr{Q}_\alpha} = \delta_\alpha - 3\mathscr{H}(1+w_\alpha)(v_\alpha - B). \tag{1.528}$$

Similarly, adopting the notation of [24], Eq. (1.219) generalizes to

$$\Delta_{s\alpha} := (\delta_\alpha)_\chi = \delta_\alpha + 3(1+w_\alpha)H\chi. \tag{1.529}$$

If we replace in (1.528) $v_\alpha - B$ by $v - B$, we obtain another gauge invariant density perturbation

$$\Delta_{c\alpha} := (\delta_\alpha)_{\mathscr{Q}} = \delta_\alpha - 3\mathscr{H}(1+w_\alpha)(v - B), \tag{1.530}$$

which reduces to δ_α for the *comoving gauge*: $v = B$.

The following relations between the three gauge invariant density perturbations are useful. Putting an index χ on the right of (1.528) gives

$$\Delta_\alpha = \Delta_{s\alpha} - 3\mathscr{H}(1+w_\alpha)V_\alpha. \tag{1.531}$$

Similarly, putting χ as an index on the right of (1.530) implies

$$\Delta_{cs} = \Delta_{s\alpha} - 3\mathscr{H}(1+w_\alpha)V. \tag{1.532}$$

For V_α, we have, as in (1.220),

$$V_\alpha = v_\alpha + E'. \tag{1.533}$$

From now on, we use similar notations for the total density perturbations:

$$\Delta := \delta_{\mathscr{Q}}, \quad \Delta_s := \delta_\chi \quad (\Delta \equiv \Delta_c). \tag{1.534}$$

Let us translate the identities (1.521) – (1.524). For instance,

$$\sum_\alpha \rho_\alpha \Delta_{c\alpha} = \sum \alpha \rho_\alpha \delta_\alpha + 3\mathscr{H}(v-B)\sum_\alpha h_\alpha = \rho\delta + 3\mathscr{H}(v-B)h = \rho\Delta.$$

We collect this and related identities:

$$\rho\Delta = \sum_\alpha \rho_\alpha \Delta_{c\alpha} \tag{1.535}$$

$$= \sum_\alpha \rho_\alpha \Delta_\alpha, \tag{1.536}$$

$$\rho\Delta_s = \sum_\alpha \rho_\alpha \Delta_{s\alpha}, \tag{1.537}$$

$$hV = \sum_\alpha h_\alpha V_\alpha, \tag{1.538}$$

$$p\Pi = \sum_\alpha p_\alpha \Pi_\alpha. \tag{1.539}$$

We would like to write also $p\Gamma$ in a manifestly gauge invariant form. From (using (1.521), (1.523), and (1.520))

$$p\Gamma = p\pi_L - c_s^2\rho\delta = \sum_\alpha \underbrace{p_\alpha\pi_{L\alpha}}_{c_\alpha^2\rho_\alpha\delta_\alpha + p_\alpha\Gamma_\alpha} - c_s^2\sum_\alpha \rho_\alpha\delta_\alpha = \sum_\alpha p_\alpha\Gamma_\alpha + \sum_\alpha (c_\alpha^2 - c_s^2)\rho_\alpha\delta_\alpha,$$

we get

$$p\Gamma = p\Gamma_{int} + p\Gamma_{rel}, \tag{1.540}$$

with

$$p\Gamma_{int} = \sum_\alpha p_\alpha\Gamma_\alpha \tag{1.541}$$

and

$$p\Gamma_{rel} = \sum_\alpha (c_\alpha^2 - c_s^2)\rho_\alpha\delta_\alpha. \tag{1.542}$$

Since $p\Gamma_{int}$ is obviously gauge invariant, this must also be the case for $p\Gamma_{rel}$. We want to exhibit this explicitly. First note, using (1.517) and (1.515) that

$$c_s^2 = \frac{p'}{\rho'} = \sum_\alpha \frac{p'_\alpha}{\rho'} = \sum_\alpha c_\alpha^2 \frac{\rho'_\alpha}{\rho'} = \sum_\alpha c_\alpha^2 \frac{h_\alpha}{h}, \tag{1.543}$$

i.e.,

$$c_s^2 = \sum_\alpha \frac{h_\alpha}{h} c_\alpha^2. \tag{1.544}$$

Now, we replace δ_α in (1.542) with the help of (1.530) and use (1.543), with the result

$$p\Gamma_{rel} = \sum_\alpha (c_\alpha^2 - c_s^2)\rho_\alpha\Delta_{c\alpha}. \tag{1.545}$$

One can write this in a physically more transparent fashion by using (1.543) again, as well as (1.517),

$$p\Gamma_{rel} = \sum_{\alpha,\beta} (c_\alpha^2 - c_\beta^2)\frac{h_\beta}{h}\rho_\alpha\Delta_{c\alpha},$$

or

$$p\Gamma_{rel} = \frac{1}{2}\sum_{\alpha,\beta} (c_\alpha^2 - c_\beta^2)\frac{h_\alpha h_\beta}{h} S_{\alpha\beta}\,; \tag{1.546}$$

$$S_{\alpha\beta} := \frac{\Delta_{c\alpha}}{1+w_\alpha} - \frac{\Delta_{c\beta}}{1+w_\beta}. \tag{1.547}$$

Note that $\delta_\alpha/(1+w_\alpha) - \delta_\beta/(1+w_\beta)$ is gauge invariant, and thus agrees with $S_{\alpha\beta}$.

Dynamical Equations

We now turn to the dynamical equations that follow from

$$\delta T^{\nu}_{(\alpha)\mu;\nu} = 0, \tag{1.548}$$

and the expressions for $\delta T^{\nu}_{(\alpha)\mu;\nu}$ given in (1.520). Obviously, we obtain equations (1.223) and (1.227) for each component. We write these in a harmonic decomposition. From (1.223) and the last line in (1.520), we get

$$(\rho_\alpha \delta_\alpha)' + 3\frac{a'}{a}\rho_\alpha \delta_\alpha + 3\frac{a'}{a} p_\alpha \pi_{L\alpha} + h_\alpha (k v_\alpha + 3D' - E') = 0. \tag{1.549}$$

In the longitudinal gauge, we have $\Delta_{s\alpha} = \delta_\alpha, V_\alpha = v_\alpha, E = 0$, and $A = A_\chi, D = D_\chi$. We also note that, according to the definitions (1.183) and (1.184), the Bardeen potentials can be expressed as

$$\boxed{A_\chi = \Psi, \quad D_\chi = \Phi.} \tag{1.550}$$

Equation (1.549) can thus be written in the following gauge invariant form

$$(\rho_\alpha \Delta_{s\alpha})' + 3\frac{a'}{a}\rho_\alpha \Delta_{s\alpha} + 3\frac{a'}{a} p_\alpha \left(\frac{c_\alpha^2}{w_\alpha}\Delta_{s\alpha} + \Gamma_\alpha\right) + h_\alpha (k V_\alpha + 3\Phi') = 0. \tag{1.551}$$

Similarly, we obtain from (1.227) the momentum equation

$$\begin{aligned} &[h_\alpha (v_\alpha - B)]' + 4\frac{a'}{a} h_\alpha (v_\alpha - B) - k h_\alpha A - k p_\alpha \pi_{L\alpha} \\ &+ \frac{2}{3}\frac{k^2 - 3K}{k} p_\alpha \Pi_\alpha = 0. \end{aligned} \tag{1.552}$$

The gauge invariant form of this is

$$\begin{aligned} &(h_\alpha V_\alpha)' + 4\frac{a'}{a} h_\alpha V_\alpha - k p_\alpha \left(\frac{c_\alpha^2}{w_\alpha}\Delta_{s\alpha} + \Gamma_\alpha\right) \\ &- k h_\alpha \Psi + \frac{2}{3}\frac{k^2 - 3K}{k} p_\alpha \Pi_\alpha = 0. \end{aligned} \tag{1.553}$$

Equations (1.551) and (1.553) constitute our basic system describing the dynamics of matter. It will be useful to rewrite the momentum equation by using

$$(h_\alpha V_\alpha)' = h_\alpha V_\alpha' + V_\alpha h_\alpha', \quad h_\alpha' = \rho_\alpha'(1 + c_s^2) = -3\frac{a'}{a}(1 + c_\alpha^2) h_\alpha.$$

Using also (1.516), we obtain

$$V_\alpha' - 3\frac{a'}{a}(1+c_\alpha^2)V_\alpha + 4\frac{a'}{a}V_\alpha - k\frac{p_\alpha}{h_\alpha}\left(\frac{c_\alpha^2}{w_\alpha}\Delta_{s\alpha} + \Gamma_\alpha\right) - k\Psi + \frac{2}{3}\frac{k^2-3K}{k}\frac{p_\alpha}{h_\alpha}\Pi_\alpha = 0$$

or

$$\begin{aligned} V_\alpha' + \frac{a'}{a}V_\alpha &= k\Psi + 3\frac{a'}{a}c_\alpha^2 V_\alpha \\ &+k\left[\frac{c_\alpha^2}{1+w_\alpha}\Delta_{s\alpha} + \frac{w_\alpha}{1+w_\alpha}\Gamma_\alpha\right] - \frac{2}{3}\frac{k^2-3K}{k}\frac{w_\alpha}{1+w_\alpha}\Pi_\alpha. \end{aligned} \tag{1.554}$$

Here, we use (1.531) in the harmonic decomposition, i.e.,

$$\Delta_\alpha = \Delta_{s\alpha} + 3(1+w_\alpha)\frac{a'}{a}\frac{1}{k}V_\alpha, \tag{1.555}$$

and finally get

$$V_\alpha' + \frac{a'}{a}V_\alpha = k\Psi + k\left[\frac{c_\alpha^2}{1+w_\alpha}\Delta_\alpha + \frac{w_\alpha}{1+w_\alpha}\Gamma_\alpha\right] - \frac{2}{3}\frac{k^2-3K}{k}\frac{w_\alpha}{1+w_\alpha}\Pi_\alpha. \tag{1.556}$$

Below it will be useful to have an equation for $V_{\alpha\beta} := V_\alpha - V_\beta$. We derive this for $\Gamma_\alpha = 0$ ($\Rightarrow \Gamma_{int} = 0$), since this is a good approximation. From (1.556), we get

$$V_{\alpha\beta}' + \frac{a'}{a}V_{\alpha\beta} = +k\left[\frac{c_\alpha^2}{1+w_\alpha}\Delta_\alpha - \frac{c_\beta^2}{1+w_\beta}\Delta_\beta\right] - \frac{2}{3}\frac{k^2-3K}{k}\Pi_{\alpha\beta}, \tag{1.557}$$

where

$$\Pi_{\alpha\beta} = \frac{w_\alpha}{1+w_\alpha}\Pi_\alpha - \frac{w_\beta}{1+w_\beta}\Pi_\beta. \tag{1.558}$$

Beside (1.555), we also use (1.532) in the harmonic decomposition,

$$\Delta_{c\alpha} = \Delta_{s\alpha} + 3(1+w_\alpha)\frac{a'}{a}\frac{1}{k}V, \tag{1.559}$$

to get

$$\Delta_\alpha = \Delta_{c\alpha} + 3(1+w_\alpha)\frac{a'}{a}\frac{1}{k}(V_\alpha - V). \tag{1.560}$$

From now on, we consider only a *two-component system* α, β. (The generalization is easy; see [24].) Then, $V_\alpha - V = (h_\beta/h)V_{\alpha\beta}$, and therefore the second term on the right of (1.557) is

$$k\left[\frac{c_\alpha^2}{1+w_\alpha}\Delta_\alpha - \frac{c_\beta^2}{1+w_\beta}\Delta_\beta\right] =$$
$$k\left[\frac{c_\alpha^2}{1+w_\alpha}\Delta_{c\alpha} - \frac{c_\beta^2}{1+w_\beta}\Delta_{c\beta}\right] + 3\frac{a'}{a}\left(c_\alpha^2 V_{\alpha\beta}\frac{h_\beta}{h} + c_\beta^2 V_{\alpha\beta}\frac{h_\alpha}{h}\right). \quad (1.561)$$

At this point, we use the identity[21]

$$\frac{\Delta_{c\alpha}}{1+w_\alpha} = \frac{\Delta}{1+w} + \frac{h_\beta}{h} S_{\alpha\beta}. \quad (1.562)$$

Introducing also the abbreviation

$$c_z^2 := c_\alpha^2 \frac{h_\beta}{h} + c_\beta^2 \frac{h_\alpha}{h} \quad (1.563)$$

the right-hand side of (1.561) becomes $k(c_\alpha^2 - c_\beta^2)\frac{\Delta}{1+w} + kc_z^2 S_{\alpha\beta} + 3\frac{a'}{a}c_z^2 V_{\alpha\beta}$. So finally, we arrive at

$$V'_{\alpha\beta} + \frac{a'}{a}(1-3c_z^2)V_{\alpha\beta}$$
$$= k(c_\alpha^2 - c_\beta^2)\frac{\Delta}{1+w} + kc_z^2 S_{\alpha\beta} - \frac{2}{3}\frac{k^2-3K}{k}\Pi_{\alpha\beta}. \quad (1.564)$$

For the generalization of this equation, without the simplifying assumptions, see (II.5.27) in [24].

Under the same assumptions, we can simplify the energy equation (1.551). Using

$$\left(\frac{\rho_\alpha \Delta_{s\alpha}}{h_\alpha}\right)' = \frac{1}{h_\alpha}(\rho_\alpha \Delta_{s\alpha})' - \frac{h'_\alpha}{h_\alpha}\frac{\rho_\alpha}{h_\alpha}\Delta_{s\alpha}, \quad \frac{h'_\alpha}{h_\alpha}\frac{\rho_\alpha}{h_\alpha} = -3\frac{a'}{a}(1+c_\alpha^2)\frac{1}{1+w_\alpha}$$

in (1.551) yields

$$\boxed{\left(\frac{\Delta_{s\alpha}}{1+w_\alpha}\right)' = -kV_\alpha - 3\Phi'.} \quad (1.565)$$

From this, (1.559) and the defining equation (1.547) of $S_{\alpha\beta}$, we obtain the useful equation

$$\boxed{S'_{\alpha\beta} = -kV_{\alpha\beta}.} \quad (1.566)$$

[21] From (1.547), we obtain for an arbitrary number of components (making use of (1.535))

$$\sum_\beta \frac{h_\beta}{h} S_{\alpha\beta} = \frac{\Delta_{c\alpha}}{1+w_\alpha} - \sum_\beta \underbrace{\frac{h_\beta}{h}\frac{1}{1+w_\beta}}_{\rho_\beta/h}\Delta_{c\beta} = \frac{\Delta_{c\alpha}}{1+w_\alpha} - \frac{\rho}{h}\Delta = \frac{\Delta_{c\alpha}}{1+w_\alpha} - \frac{\Delta}{1+w}.$$

It is sometimes useful to have an equation for $(\Delta_{c\alpha}/(1+w_\alpha))'$. From (1.559) and (1.565), we get

$$\left(\frac{\Delta_{c\alpha}}{1+w_\alpha}\right)' = -kV_\alpha - 3\Phi' + 3\left(\frac{a'}{a}\frac{1}{k}V\right)'.$$

For the last term make use of (1.278), (1.281), and (1.262). If one also uses the following consequence of (1.259) and (1.261)

$$\frac{a'}{a}\Psi - \Phi' = 4\pi G\rho a^2(1+w)k^{-1}V = \frac{3}{2}\left[\left(\frac{a'}{a}\right)^2 + K\right](1+w)k^{-1}V, \qquad (1.567)$$

one obtains after some manipulations

$$\left(\frac{\Delta_{c\alpha}}{1+w_\alpha}\right)' = -kV_\alpha + 3\frac{K}{k}V + 3\frac{a'}{a}c_s^2\frac{\Delta}{1+w} + 3\frac{a'}{a}\frac{w}{1+w}\Gamma$$
$$- 3\frac{a'}{a}\frac{w}{1+w}\frac{2}{3}\left(1 - \frac{3K}{k^2}\right)\Pi. \qquad (1.568)$$

Let us summarize the basic equations for the two-component fluid for $K=0$ and $\Gamma_\alpha = 0$ (no intrinsic entropy production of each component r and d). In addition, it is certainly a good approximation to neglect in the tight coupling era the anisotropic stresses Π_α. Then, $\Psi = -\Phi$ and since $\Gamma_{int} = 0$ the amplitude Γ for entropy production is proportional to

$$S := S_{dr} = \frac{\Delta_{cd}}{1+w_d} - \frac{\Delta_{cr}}{1+w_r}, \qquad \frac{w}{1+w}\Gamma = \frac{h_d h_r}{h^2}(c_d^2 - c_r^2)S. \qquad (1.569)$$

We also recall the definition (1.563)

$$c_z^2 = \frac{h_r}{h}c_d^2 + \frac{h_d}{h}c_r^2. \qquad (1.570)$$

The energy and momentum equations are

$$\Delta' - 3\frac{a'}{a}w\Delta = -k(1+w)V, \qquad (1.571)$$

$$V' + \frac{a'}{a}V = k\Psi + k\frac{c_s^2}{1+w}\Delta + k\frac{w}{1+w}\Gamma. \qquad (1.572)$$

By (1.566), the derivative of S is given by

$$S' = -kV_{dr}, \qquad (1.573)$$

and that of V_{dr} follows from (1.564):

$$V_{dr}' + \frac{a'}{a}(1-3c_z^2)V_{dr} = k(c_d^2 - c_r^2)\frac{\Delta}{1+w} + kc_z^2 S. \qquad (1.574)$$

In the constraint equation (1.282), we use the Friedmann equation for $K = 0$,

$$\frac{8\pi G\rho}{3H^2} = 1, \tag{1.575}$$

and obtain

$$\Phi = -\Psi = \frac{3}{2}\left(\frac{Ha}{k}\right)^2 \Delta. \tag{1.576}$$

It will be convenient to introduce the comoving wave number in units of the Hubble length $x := Ha/k$ and the renormalized scale factor $\zeta := a/a_{eq}$, where a_{eq} is the scale factor at the "equality time" (see Sect. 1.2.2.5). Then, the last equation becomes

$$\boxed{\Phi = -\Psi = \frac{3}{2}x^2\Delta.} \tag{1.577}$$

Using $\zeta' = kx\zeta$ and introducing the operator $D := \zeta d/d\zeta$, we can write (1.571) as

$$\boxed{(D - 3w)\Delta = -\frac{1}{x}(1+w)V.} \tag{1.578}$$

Similarly, (1.572) (together with (1.569)) gives

$$\boxed{(D+1)V = \frac{\Psi}{x} + \frac{c_s^2}{x}\frac{\Delta}{1+w} + \frac{1}{x}\frac{h_d h_r}{h^2}(c_d^2 - c_r^2)S.} \tag{1.579}$$

We also rewrite (1.573) and (1.574)

$$\boxed{DS = -\frac{1}{x}V_{dr},} \tag{1.580}$$

$$\boxed{(D+1-3c_z^2)V_{dr} = \frac{1}{x}(c_d^2 - c_r)\frac{\Delta}{1+w} + \frac{1}{x}c_z^2 S.} \tag{1.581}$$

It will turn out to be useful to work alternatively with the equations of motion for V_α and

$$X_\alpha := \frac{\Delta_{c\alpha}}{1+w_\alpha} \quad (\alpha = r, d). \tag{1.582}$$

From (1.556), we obtain

$$V_\alpha' + \frac{a'}{a}V_\alpha = k\Psi + k\frac{c_\alpha^2}{1+w_\alpha}\Delta_\alpha, \tag{1.583}$$

Here, we replace Δ_α by $\Delta_{c\alpha}$ with the help of (1.531) and (1.532), implying (in the harmonic decomposition)

$$\Delta_\alpha = \Delta_{c\alpha} + 3(1+w_\alpha)\frac{a'}{a}\frac{1}{k}(V_\alpha - V). \tag{1.584}$$

We then get

$$V'_\alpha + \frac{a'}{a}(1-3c_\alpha^2)V_\alpha = k\Psi + kc_\alpha^2 X_\alpha - 3\frac{a'}{a}c_\alpha^2 V. \tag{1.585}$$

From (1.568), we find, using (1.569),

$$X'_\alpha = -kV_\alpha + 3\frac{a'}{a}c_s^2\frac{\Delta}{1+w} + 3\frac{a'}{a}\frac{h_d h_r}{h^2}(c_d^2 - c_r^2)S. \tag{1.586}$$

Rewriting the last two equations as above, we arrive at the system

$$(D+1-3c_\alpha^2)V_\alpha = \frac{\Psi}{x} + \frac{c_\alpha^2}{x}X_\alpha - 3c_\alpha^2 V, \tag{1.587}$$

$$DX_\alpha = -\frac{V_\alpha}{x} + 3c_s^2\frac{\Delta}{1+w} + 3\frac{h_d h_r}{h^2}(c_d^2 - c_r^2)S. \tag{1.588}$$

This system is closed since by (1.569), (1.191), and (1.538)

$$S = X_d - X_r, \quad \frac{\Delta}{1+w} = \sum_\alpha \frac{h_\alpha}{h}X_\alpha, \quad V = \sum_\alpha \frac{h_\alpha}{h}V_\alpha. \tag{1.589}$$

Also, note that according to (1.562)

$$\frac{\Delta}{1+w} = X_r + \frac{h_d}{h}S = X_d - \frac{h_r}{h}S. \tag{1.590}$$

From these basic equations, we now deduce second-order equations for the pair (Δ, S), respectively, for X_α $(\alpha = r, d)$. For doing this, we note that for any function $f, f' = (a'/a)Df$, in particular (using $\dot{H} = -4\pi G(\rho + p)$ and (1.226))

$$Dx = -\frac{1}{2}(3w+1)x, \quad Dw = -3(1+w)(c_s^2 - w). \tag{1.591}$$

The result of the somewhat tedious but straightforward calculation is [39]:

$$D^2\Delta + \left[\frac{1-3w}{2} + 3c_s^2 - 6w\right]D\Delta + \left[\frac{c_s^2}{x^2} - 3w + 9(c_s^2 - w) + \frac{3}{2}(3w^2 - 1)\right]\Delta = \frac{1}{x^2}\frac{h_r h_d}{\rho h}(c_r^2 - c_d^2)S, \tag{1.592}$$

$$D^2 S + \left[\frac{1-3w}{2} - 3c_z^2\right]DS + \frac{c_z^2}{x^2}S = \frac{c_r^2 - c_d^2}{x^2(1+w)}\Delta \tag{1.593}$$

for the pair Δ, S, and

$$
\begin{aligned}
&D^2 X_\alpha + \left[\frac{1-3w}{2} - 3c_\alpha^2\right] D X_\alpha \\
&+ \left\{\frac{c_\alpha^2}{x^2} - \frac{h_\alpha}{h}\left[\frac{3}{2}(1+w) + \frac{3}{2}(1-3w)c_\alpha^2 + 9c_\alpha^2(c_s^2 - c_\alpha^2) + 3Dc_\alpha^2\right]\right\} X_\alpha \\
&= 3\frac{h_\beta}{h}\left[(c_\beta^2 - c_\alpha^2)D + \frac{1+w}{2} + \frac{1-3w}{2}c_\beta^2 + 3c_\beta^2(c_s^2 - c_\beta^2) + Dc_\beta^2\right] X_\beta
\end{aligned}
\tag{1.594}
$$

for the pair X_α.

Alternative System for Tight Coupling Limit

Instead of the first-order system (1.585) and (1.586), one may work with similar equations for the amplitudes $\Delta_{s\alpha}$ and V_α. From (1.554), we obtain instead of (1.585) for $\Pi_\alpha = 0$

$$
V_\alpha' + \frac{a'}{a}(1-3c_\alpha^2)V_\alpha = k\Psi + k\frac{c_\alpha^2}{1+w_\alpha}\Delta_{s\alpha}. \tag{1.595}
$$

Beside this, we have Eq. (1.272)

$$
\left(\frac{\Delta_{s\alpha}}{1+w_\alpha}\right)' = -kV_\alpha - 3\Phi'. \tag{1.596}
$$

To this, we add the following consequence of the constraint equations (1.288), (1.289), and the relations (1.272), (1.537), and (1.538):

$$
k^2\Psi = -4\pi G a^2 \sum_\alpha \left[\rho_\alpha \Delta_{s\alpha} + 3\frac{aH}{k}\rho_\alpha(1+w_\alpha)V_\alpha\right]. \tag{1.597}
$$

Instead one can also use, for instance for generating numerical solutions, the following first-order differential equation that is obtained similarly

$$
k^2\Psi + 3\frac{a'}{a}(\Psi' + \frac{a'}{a}\Psi) = -4\pi G a^2 \sum_\alpha \rho_\alpha \Delta_{s\alpha}. \tag{1.598}
$$

Adiabatic and Isocurvature Perturbations

These differential equations have to be supplemented with initial conditions. Two linearly independent types are considered for some very early stage, for instance, at the end of the inflationary era:

- **adiabatic** perturbations: all $S_{\alpha\beta} = 0$, but $\mathscr{R} \neq 0$;
- **isocurvature** perturbations: some $S_{\alpha\beta} \neq 0$, but $\mathscr{R} = 0$.

Recall that $\mathscr{R}$ measures the spatial curvature for the slicing $\mathscr{Q}=0$. According to the initial definition (1.222) of $\mathscr{R}$ and the Eqs. (1.577) and (1.578), we have

$$\mathscr{R}=\Phi-xV=\frac{x^2}{1+w}\left[D+\frac{3}{2}(1-w)\right]\Delta. \tag{1.599}$$

Explicit forms of the Two-Component Differential Equations

At this point, we make use of the equation of state for the two-component model under consideration. It is convenient to introduce a parameter c by

$$R:=\frac{3\rho_b}{4\rho_\gamma}=\frac{\zeta}{c}\ \Rightarrow\ \frac{\Omega_d}{\Omega_b}=\frac{3c}{4}-1. \tag{1.600}$$

We then have for various background quantities

$$\begin{aligned}
\frac{\rho_d}{\rho_{eq}}&=\frac{1}{2}\left(1-\frac{4}{3c}\right)\frac{1}{\zeta^3},\ p_d=0,\\
\frac{\rho_r}{\rho_{eq}}&=\frac{2}{3}\frac{\zeta+3c/4}{c}\frac{1}{\zeta^4},\ \frac{p_r}{\rho_{eq}}=\frac{1}{6}\frac{1}{\zeta^4},\\
\frac{\rho}{\rho_{eq}}&=\frac{1}{2}(\zeta+1)\frac{1}{\zeta^4},\ \frac{p}{\rho_{eq}}=\frac{1}{6}\frac{1}{\zeta^4},\\
\frac{h_r}{h}&=\frac{4}{3}\frac{\zeta+c}{c(\zeta+4/3)},\ \frac{h_d}{h}=\left(1-\frac{4}{3c}\right)\frac{\zeta}{\zeta+4/3},\\
w&=\frac{1}{3(\zeta+1)},\ w_r=\frac{c}{4\zeta+3c},\ w_d=0,\\
c_d^2&=0,\ c_r^2=\frac{1}{3}\frac{c}{\zeta+c},\ c_s^2=\frac{4}{9}\frac{1}{\zeta+4/3},\ c_z^2=\frac{1}{3}\frac{(c-4/3)\zeta}{(\zeta+c)(\zeta+4/3)},\\
H^2&=H_{eq}^2\frac{\zeta+1}{2}\frac{1}{\zeta^4},\ x^2=\frac{\zeta+1}{2\zeta^2}\frac{1}{\omega^2},\ \omega:=\frac{1}{x_{eq}}=\left(\frac{k}{aH}\right)_{eq}.
\end{aligned} \tag{1.601}$$

Since we now know that the dark matter fraction is much larger than the baryon fraction, we write the basic equations only in the limit $c\to\infty$. (For finite c, these are given in [39].) Equation (1.594) leads to the pair

$$\begin{aligned}
&D^2X_r+\left(\frac{1}{2}\frac{\zeta}{1+\zeta}-1\right)DX_r\\
&\quad+\left\{\frac{2}{3}\frac{\omega^2\zeta^2}{1+\zeta}+\frac{4}{3}\frac{1}{\zeta+4/3}\left[\frac{\zeta}{\zeta+4/3}-2\right]\right\}X_r=\left[\frac{3}{2}\frac{\zeta}{\zeta+1}-\frac{\zeta}{\zeta+4/3}D\right]X_d,
\end{aligned} \tag{1.602}$$

$$\left\{D^2+\frac{1}{2}\frac{\zeta}{1+\zeta}D-\frac{3}{2}\frac{\zeta}{1+\zeta}\right\}X_d=\frac{4}{3}\frac{1}{\zeta+4/3}\left[D+2-\frac{\zeta}{\zeta+4/3}\right]X_r. \tag{1.603}$$

From (6.24) and (6.25), we obtain on the other hand

$$
\begin{aligned}
&D^2\Delta + \left(-1 + \frac{5}{2}\frac{\zeta}{\zeta+1} - \frac{\zeta}{\zeta+4/3}\right)D\Delta \\
&+\left\{-2 + \frac{3}{4}\zeta + \frac{1}{2}\left(\frac{\zeta}{\zeta+1}\right)^2 - \frac{3\zeta^2}{\zeta+1} + \frac{9\zeta^2}{4(\zeta+4/3)}\right\}\Delta \\
&= \frac{8}{9}\omega^2\frac{\zeta^2}{(\zeta+1)^2(\zeta+4/3)}\left[\zeta S - (\zeta+1)\Delta\right],
\end{aligned}
\tag{1.604}
$$

$$
\begin{aligned}
&D^2 S + \left(\frac{1}{2}\frac{1}{\zeta+1} - \frac{1}{\zeta+4/3}\right)\zeta DS \\
&+\frac{2}{3}\omega^2\frac{\zeta^3}{(\zeta+1)(\zeta+4/3)}S = \frac{2}{3}\omega^2\frac{\zeta^2}{\zeta+4/3}\Delta.
\end{aligned}
\tag{1.605}
$$

We also note that (1.599) becomes

$$
\mathscr{R} = \frac{1}{2\omega^2}\frac{\zeta+1}{\zeta^2(\zeta+4/3)}\left[(\zeta+1)D + \frac{3}{2}\zeta + 1\right]\Delta. \tag{1.606}
$$

We can now define more precisely what we mean by the two types of primordial initial perturbations by considering solutions of our perturbation equations for $\zeta \ll 1$.

- *adiabatic* (or *curvature*) perturbations: growing mode behaves as

$$
\begin{aligned}
\Delta &= \zeta^2\left[1 - \frac{17}{16}\zeta + \cdots\right] - \frac{\omega^2}{15}\zeta^4[1-\cdots], \\
S &= \frac{\omega^2}{32}\zeta^4\left[1 - \frac{28}{25}\zeta + \cdots\right]; \quad \Rightarrow \mathscr{R} = \frac{9}{8\omega^2}(1 + \mathscr{O}(\zeta)).
\end{aligned}
\tag{1.607}
$$

- *isocurvature* perturbations: growing mode behaves as

$$
\begin{aligned}
\Delta &= \frac{\omega^2}{6}\zeta^3\left[1 - \frac{17}{10}\zeta + \cdots\right], \\
S &= 1 - \frac{\omega^2}{18}\zeta^3[1-\cdots]; \quad \Rightarrow \mathscr{R} = \frac{1}{4}\zeta(1 + \mathscr{O}(\zeta)).
\end{aligned}
\tag{1.608}
$$

From (1.589) and (1.590), we obtain the relation between the two sets of perturbation amplitudes:

$$
X_r = \frac{\zeta+1}{\zeta+4/3}\Delta - \frac{\zeta}{\zeta+4/3}S, \quad X_d = \frac{\zeta+1}{\zeta+4/3}\Delta + \frac{4}{3}\frac{1}{\zeta+4/3}S, \tag{1.609}
$$

$$
\Delta = \frac{1}{\zeta+1}\left(\frac{4}{3}X_r + \zeta X_d\right), \quad S = X_d - X_r. \tag{1.610}
$$

Let us also write the alternative system (1.595) – (1.598) explicitly in terms of the independent variable ζ. As before one finds

$$D\left(\frac{\Delta_{s\alpha}}{1+w_\alpha}\right) = -\frac{V_\alpha}{x} - 3D\Phi, \tag{1.611}$$

$$(D+1-3c_\alpha^2)V_\alpha = -\frac{\Phi}{x} + \frac{1}{x}\frac{c_\alpha^2}{1+w_\alpha}\Delta_{s\alpha}, \tag{1.612}$$

and for Φ:

$$\Phi + 3x^2(D\Phi+\Phi) = \frac{3}{2}x^2\sum_\alpha \frac{\rho_\alpha}{\rho}\Delta_{s\alpha}, \tag{1.613}$$

$$\Phi = \frac{3}{2}x^2\sum_\alpha \frac{\rho_\alpha}{\rho}[\Delta_{s\alpha} + 3x(1+w_\alpha)V_\alpha]. \tag{1.614}$$

With (1.601), i.e.,

$$x^2 = \frac{\zeta+1}{2\zeta^2}\frac{1}{\omega^2}, \quad \frac{\rho_d}{\rho} = \frac{1}{2}\frac{\zeta}{\zeta+1}, \quad \frac{\rho_r}{\rho} = \frac{1}{2}\frac{1}{\zeta+1},$$

everything is explicit. The initial conditions for the growing modes follow from the expansions (1.607) and (1.608), once we have expressed the five amplitudes $\Delta_{s\alpha}(\zeta)$, $V_\alpha(\zeta)$, $\Phi(\zeta)$ in terms of Δ and S.

Φ is related to Δ by (1.577). From (1.559), we obtain

$$\frac{\Delta_{s\alpha}}{1+w_\alpha} = X_\alpha - 3xV.$$

For the last term, we use (1.578), which implies

$$3xV = -3\frac{x^2}{1+w}(D-3w)\Delta. \tag{1.615}$$

The amplitudes X_α are given in terms of Δ, S by (1.609).

From these equations, it is now easy to determine the initial conditions for our first-order differential equations. For *adiabatic* perturbations, one finds for the growing modes

$$\Phi(0) = \frac{2}{3}\mathscr{R}, \; \Delta_{sd}(0) = \mathscr{R}, \; \Delta_{sr}(0) = \frac{4}{3}\mathscr{R}, \; V_d(0) = V_r(0) = 0. \tag{1.616}$$

Note that, as a result of (1.580), the difference $V_d - V_r$ must vanish for small ζ as $\mathscr{O}(\zeta^3)$.

1.8.2 Analytical and Numerical Analysis

The system of linear differential equations (1.602)–(1.605) has been discussed analytically in great detail in [39]. One learns, however, more about the physics of the gravitationally coupled fluids in a mixed analytical-numerical approach.

1.8.2.1 Solutions for Super-Horizon Scales

For super-horizon scales ($x \gg 1$), Eq. (1.580) implies that S is constant. If the mode enters the horizon in the matter dominated era, then the parameter ω in (1.601) is small. For $\omega \ll 1$, Eq. (1.604) reduces to

$$\begin{aligned} &D^2\Delta + \left(-1 + \frac{5}{2}\frac{\zeta}{\zeta+1} - \frac{\zeta}{\zeta+4/3}\right) D\Delta \\ &+ \left\{-2 + \frac{3}{4}\zeta + \frac{1}{2}\left(\frac{\zeta}{\zeta+1}\right)^2 - \frac{3\zeta^2}{\zeta+1} + \frac{9\zeta^2}{4(\zeta+4/3)}\right\}\Delta \\ &= \frac{8}{9}\omega^2 \frac{\zeta^3}{(\zeta+1)^2(\zeta+4/3)} S. \end{aligned} \tag{1.617}$$

For *adiabatic* modes, we are led to the homogeneous equation already studied in Sect. 1.5.1, with the two independent solutions U_g and U_d given in (1.314) and (1.315). Recall that the Bardeen potentials remain constant both in the radiation and in the matter dominated eras. According to (1.318), Φ decreases to 9/10 of the primordial value Φ^{prim}.

For *isocurvature* modes, we can solve (1.609) with the Wronskian method and obtain for the growing mode [39]

$$\Delta_{iso} = \frac{4}{15}\omega^2 S \zeta^3 \frac{3\zeta^2 + 22\zeta + 24 + 4(3\zeta+4)\sqrt{1+\zeta}}{(\zeta+1)(3\zeta+4)[1+(1+\zeta)^{1/2}]^4}. \tag{1.618}$$

Thus,

$$\Delta_{iso} \simeq \begin{cases} \frac{1}{6}\omega^2 S\zeta^3 & : \ \zeta \ll 1 \\ \frac{4}{15}\omega^2 S\zeta & : \ \zeta \gg 1. \end{cases} \tag{1.619}$$

1.8.2.2 Horizon Crossing

We will now study the behavior of adiabatic modes more closely, in particular, what happens in horizon crossing.

Crossing in Radiation Dominated Era

When the mode enters the horizon in the radiation dominated phase, we can neglect in (1.604) the term proportional to S for $\zeta < 1$. As long as the radiation dominates ζ is small, whence (1.604) gives in leading order

$$(D^2 - D - 2)\Delta = -\frac{2}{3}\omega^2\zeta^2\Delta. \tag{1.620}$$

(This could also be directly obtained from (1.592), setting $c_s^2 \simeq 1/3$, $w \simeq 1/3$.) Since $D^2 - D = \zeta^2 d^2/d\zeta^2$, this perturbation equation can be written as

$$\left[\zeta^2 \frac{d^2}{d\zeta^2} + \left(\frac{2}{3}\omega^2\zeta^2 - 2\right)\right]\Delta = 0. \tag{1.621}$$

Instead of ζ, we choose independent variable the comoving sound horizon r_s times k. We have

$$r_s = \int c_s d\eta = \int c_s \frac{d\eta}{d\zeta} d\zeta,$$

with $c_s \simeq 1/\sqrt{3}$, $d\zeta/d\eta = kx\zeta = aH\zeta = (aH)/(aH)_{eq})(k/\omega)\zeta \simeq (k/\omega\sqrt{2})$, thus $\zeta \simeq (k/\sqrt{2}\omega)\eta$ and

$$u := kr_s \simeq \sqrt{\frac{2}{3}}\omega\zeta \simeq k\eta/\sqrt{3}. \tag{1.622}$$

Therefore, (1.621) is equivalent to

$$\boxed{\left[\frac{d^2}{du^2} + \left(1 - \frac{2}{u^2}\right)\right]\Delta = 0.} \tag{1.623}$$

This differential equation is well known. According to 9.1.49 of [34], the functions $w(x) \propto x^{1/2}\mathscr{C}_\nu(\lambda x)$, $\mathscr{C}_\nu \propto H_\nu^{(1)}, H_\nu^{(2)}$ satisfy

$$w'' + \left(\lambda - \frac{\nu^2 - \frac{1}{4}}{x^2}\right) w = 0. \tag{1.624}$$

Since $j_\nu(x) = \sqrt{\pi/2x}J_{\nu+1/2}(x)$ and $n_\nu(x) = \sqrt{\pi/2x}Y_{\nu+1/2}(x)$, we see that Δ is a linear combination of $uj_1(u)$ and $un_1(u)$:

$$\Delta(\zeta) = Cuj_1(u) + Dun_1(u); \quad u = \sqrt{\frac{2}{3}}\omega\zeta \quad (u = kr_s = \frac{k\eta}{\sqrt{3}}). \tag{1.625}$$

Now,

$$xj_1(x) = \frac{1}{x}\sin x - \cos x, \quad xn_1(x) = -\frac{1}{x}\cos x - \sin x. \tag{1.626}$$

On super-horizon scales $u = kr_s \ll 1$ and $uj_1(u) \approx u \propto a$, while $un_1(u) \approx -1/u \propto 1/a$. Thus, the first term in (1.625) corresponds to the growing mode. If we only keep this, we have

$$\Delta(\zeta) \approx C\left(\frac{1}{u}\sin u - \cos u\right). \tag{1.627}$$

Once the mode is deep within the Hubble horizon only the cos-term survives. This is an important result because if this happens long before recombination, we can use for adiabatic modes the *initial condition*

$$\boxed{\Delta(\eta) \propto cos[kr_s(\eta)].} \tag{1.628}$$

We conclude that all adiabatic modes are temporally correlated (**synchronized**), while they are spatially uncorrelated (random phases). This is one of the basic reasons for the appearance of acoustic peaks in the CMB anisotropies. Note also that, as a result of (1.577) and (1.601), $\Phi \propto \Delta/\zeta^2 \propto \Delta/u^2$, i.e.,

$$\Psi = 3\Psi^{(prim)}\left[\frac{\sin u - u\cos u}{u^3}\right]. \tag{1.629}$$

Thus, If the mode enters the horizon during the radiation dominated era, its *potential begins to decay*.

As an exercise show that for isocurvature perturbations, the cos in (1.628) has to be replaced by the sin (out of phase).

We could have used in the discussion above the system (1.602) and (1.603). In the same limit, it reduces to

$$\left(D^2 - D - 2 + \frac{2}{3}\omega^2\zeta^2\right)X_r \simeq 0, \quad D^2X_d \simeq (D+2)X_r. \tag{1.630}$$

As expected, the equation for X_r is the same as for Δ. One also sees that X_d is driven by X_r and is growing logarithmically for $\omega \gg 1$.

The previous analysis can be improved by not assuming radiation domination and also including baryons (see [39]). It turns out that for $\omega \gg 1$, the result (1.628) is not much modified: The cos-dependence remains, but with the exact sound horizon; only the amplitude is slowly varying in time $\propto (1+R)^{-1/4}$.

Since the matter perturbation is driven by the radiation, we may use the potential (1.629) and work out its influence on the matter evolution. It is more convenient to do this for the amplitude Δ_{sd} (instead of Δ_{cd}) making use of the Eqs. (1.595) and (1.596) for $\alpha = d$:

$$\Delta'_{sd} = -kV_d - 3\Phi', \quad V'_d = -\frac{a'}{a}V_d - k\Phi. \tag{1.631}$$

Let us eliminate V_d:

$$\Delta''_{sd} = -V'_d - 3\Phi'' = \frac{a'}{a}kV_d + k^2\Phi - 3\Phi'' = \frac{a'}{a}(-\Delta'_{sd} - 3\Phi') + k^2\Phi - 3\Phi''.$$

The resulting equation

$$\Delta_{sd}'' + \frac{a'}{a}\Delta_{sd}' = k^2\Phi - 3\Phi'' - 3\frac{a'}{a}\Phi' \tag{1.632}$$

can be solved with the Wronskian method. Two independent solutions of the homogeneous equation are $\Delta_{sd} = const.$ and $\Delta_{sd} = \ln(a)$. These determine the Green's function in the standard manner. One then finds in the radiation dominated regime (for details, see [4], p.198)

$$\Delta_{sd}(\eta) = A\Phi^{prim}\ln(Bk\eta), \tag{1.633}$$

with $A \simeq 9.0,\ B \simeq 0.62$.

Matter Dominated Approximation

As a further illustration, we now discuss the matter dominated approximation. For this ($\zeta \gg 1$), the system (1.602) and (1.603) becomes

$$\left(D^2 - \frac{1}{2}D + \frac{2}{3}\omega^2\zeta\right)X_r = \left(-D + \frac{3}{2}\right)X_d, \tag{1.634}$$

$$\left(D^2 + \frac{1}{2}D - \frac{3}{2}\right)X_d = 0. \tag{1.635}$$

As expected, the equation for X_d is independent of X_r, while the radiation perturbation is driven by the dark matter. The solution for X_d is

$$X_d = A\zeta + B\zeta^{-3/2}. \tag{1.636}$$

Keeping only the growing mode, (1.634) becomes

$$\frac{d}{d\zeta}\left(\zeta\frac{dX_r}{d\zeta}\right) - \frac{1}{2}\frac{dX_r}{d\zeta} + \frac{2}{3}\omega^2\left(X_r - \frac{3A}{4\omega^2}\right) = 0. \tag{1.637}$$

Substituting

$$X_r =: \frac{3A}{4\omega^2} + \zeta^{-3/4}f(\zeta),$$

we get for $f(\zeta)$ the following differential equation

$$f'' = -\left(\frac{3}{16}\frac{1}{\zeta^2} + \frac{2}{3}\frac{\omega^2}{\zeta}\right)f. \tag{1.638}$$

For $\omega \gg 1$, we can use the WKB approximation

$$f = \frac{\zeta^{1/4}}{\sqrt{\omega}}\exp\left(\pm i\sqrt{\frac{8}{3}}\omega\zeta^{1/2}\right),$$

implying the following oscillatory behavior of the radiation

$$X_r = \frac{3A}{4\omega^2} + B\frac{1}{\sqrt{\omega\zeta}} \exp\left(\pm i\sqrt{\frac{8}{3}}\omega\zeta^{1/2}\right). \tag{1.639}$$

A look at (1.610) shows that this result for X_d, X_r implies the constancy of the Bardeen potentials in the matter dominated era.

1.8.2.3 Sub-Horizon Evolution

For $\omega \gg 1$, one may expect on physical grounds that the dark matter perturbation X_d eventually evolves independently of the radiation. Unfortunately, I cannot see this from the basic equations (1.602) and (1.603). Therefore, we choose a different approach, starting from the alternative system (1.595) – (1.597). This implies

$$\Delta'_{sd} = -kV_d - 3\Phi', \tag{1.640}$$

$$V'_d = -\frac{a'}{a}V_d - k\Phi, \tag{1.641}$$

$$k^2\Phi = 4\pi G a^2[\rho_d\Delta_{sd} + \cdots]. \tag{1.642}$$

As an approximation, we drop in the last equation the radiative[22] and velocity contributions that have not been written out. Then, we get a closed system which we again write in terms of the variable ζ:

$$D\Delta_{sd} = -\frac{1}{x}V_d - 3D\Phi, \tag{1.643}$$

$$DV_d = -V_d - \frac{1}{x}\Phi, \tag{1.644}$$

$$\Phi \simeq \frac{3}{4}\frac{1}{\omega^2}\frac{1}{\zeta}\Delta_{sd}. \tag{1.645}$$

In the last equation, we used $\rho_d = (\zeta/\zeta+1)\rho$, (1.575) and the expression (1.601) for x^2.

For large ω, we can easily deduce a second-order equation for Δ_{sd}: Applying D to (1.643) and using (1.644) gives

$$\begin{aligned} D^2\Delta_{sd} &= -\frac{1}{x}DV_d + \frac{1}{x^2}(Dx)V_d - 3D^2\Phi \\ &= \frac{1}{x^2}\Phi + \frac{1}{2}(1-3w)\frac{1}{x}V_d - 3D^2\Phi \\ &= \frac{1}{x^2}\Phi - \frac{1}{2}(1-3w)D\Delta_{sd} - \frac{3}{2}(1-3w)D\Phi - 3D^2\Phi. \end{aligned}$$

[22] The growth in the matter perturbations implies that eventually $\rho_d\Delta_{sd} > \rho_r\Delta_{sr}$ even if $\Delta_{sd} < \Delta_{sr}$.

Because of (1.645), the last two terms are small, and we end up (using again (1.601)) with

$$\left\{D^2 + \frac{1}{2}\frac{\zeta}{1+\zeta}D - \frac{3}{2}\frac{\zeta}{1+\zeta}\right\}\Delta_{sd} = 0 \tag{1.646}$$

known in the literature as the *Meszaros equation*. Note that this agrees, as was to be expected, with the homogeneous equation belonging to (1.603).

The Meszaros equation can be solved analytically. On the basis of (1.636), one may guess that one solution is linear in ζ. Indeed, one finds that

$$X_d(\zeta) = D_1(\zeta) = \zeta + 2/3 \tag{1.647}$$

is a solution. A linearly independent solution can then be found by quadratures. It is a general fact that $f(\zeta) := \Delta_{sd}/D_1(\zeta)$ must satisfy a differential equation, which is first order for f'. One readily finds that this equation is

$$(1+\frac{3\zeta}{2})f'' + \frac{1}{4\zeta(\zeta+1)}[21\zeta^2 + 24\zeta + 4]f' = 0.$$

The solution for f' is

$$f' \propto (\zeta + 2/3)^{-2}\zeta^{-1}(\zeta+1)^{-1/2}.$$

Integrating once more provides the second solution of (1.646)

$$D_2(\zeta) = D_1(\zeta)\ln\left[\frac{\sqrt{1+\zeta}+1}{\sqrt{1+\zeta}-1}\right] - 2\sqrt{1+\zeta}. \tag{1.648}$$

For late times, the two solutions approach to those found in (1.636).

The growing and the decaying solutions D_1, D_2 have to be superposed such that a match to (1.633) is obtained.

1.8.2.4 Transfer Function and Numerical Results

According to (1.317) and (1.318), the early evolution of Φ on super-horizon scales is given by[23]

$$\Phi(\zeta) = \Phi^{(prim)}\frac{9}{10}\frac{\zeta+1}{\zeta^2}U_g \simeq \frac{9}{10}\Phi^{(prim)}, \; for\ \zeta \gg 1. \tag{1.649}$$

At sufficiently late times in the matter dominated regime, all modes evolve identically with the *growth function* $D_g(\zeta)$ given in (1.323). I recall that this function is normalized such that it is equal to a/a_0 when we can still ignore the dark energy (at $z > 10$, say). The growth function describes the evolution of Δ, thus by the Poisson

[23] The origin of the factor 9/10 is best seen from the constancy of $\mathcal{R}$ for super-horizon perturbations and Eq. (1.395).

equation (1.288) Φ grows with $D_g(a)/a$. We, therefore, define the *transfer function* $T(k)$ by (we choose the normalization $a_0 = 1$)

$$\Phi(k,a) = \Phi^{(prim)} \frac{9}{10} \frac{D_g(a)}{a} T(k) \tag{1.650}$$

for late times. This definition is chosen such that $T(k) \to 1$ for $k \to 0$ and does not depend on time.

At these late times $\rho_M = \Omega_M a^{-3} \rho_{crit}$, hence the Poisson equation gives the following relation between Φ and Δ

$$\Phi = \left(\frac{a}{k}\right)^2 4\pi G \rho_M \Delta = \frac{3}{2} \frac{1}{ak^2} H_0^2 \Omega_M \Delta.$$

Therefore, (1.650) translates to

$$\boxed{\Delta(a) = \frac{3}{5} \frac{k^2}{\Omega_M H_0^2} \Phi^{(prim)} D_g(a) T(k).} \tag{1.651}$$

The transfer function can be determined by solving numerically the pair (1.592) and (1.593) of basic perturbation equations. One can derive even a reasonably good analytic approximation by putting our previous results together (for details, see again [4], Sect. 7.4). For a CDM model, the following accurate fitting formula to the numerical solution in terms of the variable $\tilde{q} = k/k_{eq}$, where k_{eq} is defined such that the corresponding value of the parameter ω in (1.601) is equal to 1 (i.e., $k_{eq} = a_{eq} H_{eq} = \sqrt{2\Omega_M} H_0 / \sqrt{a_{eq}}$, using (1.90)) was given in [40]:

$$T_{BBKS}(\tilde{q}) = \frac{\ln(1+0.171\tilde{q})}{0.171\tilde{q}} [1 + 0.284\tilde{q} + (1.18\tilde{q})^2 + (0.399\tilde{q})^3 + (0.490\tilde{q})^4]^{-1/4}. \tag{1.652}$$

Note that $\tilde{q}$ depends on the cosmological parameters through the combination[24] $\Omega_M h_0$, usually called the *shape parameter* Γ. In terms of the variable, $q = k/(\Gamma h_0 Mpc^{-1})$ (1.652) can be written as

$$T_{BBKS}(q) = \frac{\ln(1+2.34q)}{2.34q} [1 + 3.89q + (16.1q)^2 + (5.46q)^3 + (6.71q)^4]^{-1/4}. \tag{1.653}$$

This result for the transfer function is based on a simplified analysis. The tight coupling approximation is no more valid when the decoupling temperature is approached. Moreover, anisotropic stresses and baryons have been ignored. It will, of course, be very interesting to compare the theory with available observational data. For this, one has to keep in mind that the linear theory only applies to sufficiently large scales. For late times and small scales, it has to be corrected by numerical simulations for nonlinear effects.

[24] Since k is measured in units of $h_0\ Mpc^{-1}$ and $a_{eq} = 4.15 \times 10^{-5}/(\Omega_M h_0^2)$.

For a given primordial power spectrum, the transfer function determines the power spectrum after the "transfer regime" (when all modes evolve with the growth function D_g). From (1.651), we obtain for the power spectrum of Δ

$$P_\Delta(z) = \frac{9}{25}\frac{k^4}{\Omega_M^2 H_0^4} P_\Phi^{(prim)} D_g^2(z) T^2(k). \tag{1.654}$$

We choose $P_\Phi^{(prim)} \propto k^{n-1}$ and the amplitude such that

$$\boxed{P_\Delta(z) = \delta_H^2 \left(\frac{k}{H_0}\right)^{3+n} T^2(k) \left(\frac{D_g(z)}{D_g(0)}\right)^2.} \tag{1.655}$$

Note that $P_\Delta(0) = \delta_H^2$ for $k = H_0$. The normalization factor δ_H has to be determined from observations (e.g., from CMB anisotropies at large scales). Comparison of (1.654) and (1.655) and use of (1.448) implies

$$P_{\mathscr{R}}^{(prim)}(k) = \frac{9}{4} P_\Phi^{(prim)}(k) = \frac{25}{4}\delta_H^2 \left(\frac{\Omega_M}{D_g(0)}\right)^2 \left(\frac{k}{H_0}\right)^{n-1}. \tag{1.656}$$

Exercise

Write the Eqs. (1.595) – (1.598) in explicit form, using (1.601) in the limit when baryons are neglected ($c \to \infty$). (For a truncated subsystem, this was done in (1.643) – (1.645)). Solve the five first-order differential equations (1.595), (1.596) for $\alpha = d, r$ and (1.598) numerically. Determine, in particular, the transfer function defined in (1.650). (A standard code gives this in less than a second.) ♢

1.9 General Relativistic Boltzmann Equation

For the description of photons and neutrinos before recombination, we need the general relativistic version of the Boltzmann equation.

1.9.1 One-Particle Phase Space, Liouville Operator

For what follows, we first have to develop some kinematic and differential geometric tools. Our goal is to generalize the standard description of Boltzmann in terms of one-particle distribution functions.

Let g be the metric of the spacetime manifold M. On the cotangent bundle $T^*M = \bigcup_{p\in M} T_p^*M$, we have the natural symplectic 2-form ω, which is given in natural bundle coordinates[25] (x^μ, p_ν) by

$$\omega = dx^\mu \wedge dp_\mu. \tag{1.657}$$

(For an intrinsic description, see Chap. 6 of [41].) So far, no metric is needed. The pair (T^*M, ω) is always a symplectic manifold.

The metric g defines a natural diffeomorphism between the tangent bundle TM and T^*M, which can be used to pull ω back to a symplectic form ω_g on TM. In natural bundle coordinates, the diffeomorphism is given by $(x^\mu, p^\alpha) \mapsto (x^\mu, p_\alpha = g_{\alpha\beta} p^\beta)$, hence

$$\omega_g = dx^\mu \wedge d(g_{\mu\nu} p^\nu). \tag{1.658}$$

On TM, we can consider the "Hamiltonian function"

$$L = \frac{1}{2} g_{\mu\nu} p^\mu p^\nu \tag{1.659}$$

and its associated Hamiltonian vector field X_g, determined by the equation

$$i_{X_g} \omega_g = dL. \tag{1.660}$$

It is not difficult to show that in bundle coordinates

$$X_g = p^\mu \frac{\partial}{\partial x^\mu} - \Gamma^\mu{}_{\alpha\beta} p^\alpha p^\beta \frac{\partial}{\partial p^\mu} \tag{1.661}$$

(Exercise). The Hamiltonian vector field X_g on the symplectic manifold (TM, ω_g) is the *geodesic spray*. Its integral curves satisfy the canonical equations:

$$\frac{dx^\mu}{d\lambda} = p^\mu, \tag{1.662}$$

$$\frac{dp^\mu}{d\lambda} = -\Gamma^\mu{}_{\alpha\beta} p^\alpha p^\beta. \tag{1.663}$$

The *geodesic flow* is the flow of the vector field X_g.

Let Ω_{ω_g} be the volume form belonging to ω_g, i.e., the Liouville volume

$$\Omega_{\omega_g} = const\ \omega_g \wedge \cdots \wedge \omega_g,$$

or $(g = \det(g_{\alpha\beta}))$

$$\begin{aligned} \Omega_{\omega_g} &= (-g)(dx^0 \wedge dx^1 \wedge dx^2 \wedge dx^3) \wedge (dp^0 \wedge dp^1 \wedge dp^2 \wedge dp^3) \\ &\equiv (-g) dx^{0123} \wedge dp^{0123}. \end{aligned} \tag{1.664}$$

[25] If x^μ are coordinates of M, then the dx^μ form in each point $p \in M$ is a basis of the cotangent space $T^*_p M$. The *bundle coordinates* of $\beta \in T^*_p M$ are then (x^μ, β_ν) if $\beta = \beta_\nu dx^\nu$ and x^μ are the coordinates of p. With such bundle coordinates, one can define an atlas, by which T^*M becomes a differentiable manifold.

The *one-particle phase space* for particles of mass m is the following submanifold of TM:

$$\Phi_m = \{v \in TM \mid v \text{ future directed}, \ g(v,v) = -m^2\}. \tag{1.665}$$

This is invariant under the geodesic flow. The restriction of X_g to Φ_m will also be denoted by X_g. Ω_{ω_g} induces a volume form Ω_m (see below) on Φ_m, which is also invariant under X_g:

$$L_{X_g}\Omega_m = 0. \tag{1.666}$$

Ω_m is determined as follows (known from Hamiltonian mechanics): Write Ω_{ω_g} in the form

$$\Omega_{\omega_g} = -dL \wedge \sigma,$$

(this is always possible, but σ is not unique), then Ω_m is the pull-back of Ω_{ω_g} by the injection $i: \ \Phi_m \to TM$,

$$\Omega_m = i^*\sigma. \tag{1.667}$$

Although σ is not unique (one can, for instance, add a multiple of dL), the form Ω_m is independent of the choice of σ (show this). In natural bundle coordinates, a possible choice is

$$\sigma = (-g)dx^{0123} \wedge \frac{dp^{123}}{(-p_0)}$$

because

$$-dL \wedge \sigma = [-g_{\mu\nu}p^\mu dp^\nu + \cdots] \wedge \sigma = (-g)dx^{0123} \wedge g_{\mu 0}p^\mu dp^0 \wedge \frac{dp^{123}}{p_0} = \Omega_{\omega_g}.$$

Hence,

$$\Omega_m = \eta \wedge \Pi_m, \tag{1.668}$$

where η is the volume form of (M,g),

$$\eta = \sqrt{-g}dx^{0123}, \tag{1.669}$$

and

$$\Pi_m = \sqrt{-g}\frac{dp^{123}}{|p_0|}, \tag{1.670}$$

with $p^0 > 0$ and $g_{\mu\nu}p^\mu p^\nu = -m^2$.

We shall need some additional tools. Let Σ be a hypersurface of Φ_m transversal to X_g. On Σ, we can use the volume form

$$vol_\Sigma = i_{X_g}\Omega_m \mid \Sigma. \tag{1.671}$$

Now, we note that the 6-form

$$\omega_m := i_{X_g}\Omega_m \tag{1.672}$$

on Φ_m is closed,

$$d\omega_m = 0, \tag{1.673}$$

because

$$d\omega_m = di_{X_g}\Omega_m = L_{X_g}\Omega_m = 0$$

(we used $d\Omega_m = 0$ and (1.666)). From (1.668), we obtain

$$\omega_m = (i_{X_g}\eta) \wedge \Pi_m + \eta \wedge i_{X_g}\Pi_m. \tag{1.674}$$

In the special case, when Σ is a "time section," i.e., in the inverse image of a space-like submanifold of M under the natural projection $\Phi_m \to M$, then the second term in (1.674) vanishes on Σ, while the first term is on Σ according to (1.661) equal to $i_p\eta \wedge \Pi_m,\ p = p^\mu\partial/\partial x^\mu$. Thus, we have on a time section[26] Σ

$$\boxed{vol_\Sigma = \omega_m \mid \Sigma = i_p\eta \wedge \Pi_m.} \tag{1.675}$$

Let f be a one-particle distribution function on Φ_m, defined such that the number of particles in a time section Σ is

$$N(\Sigma) = \int_\Sigma f\omega_m. \tag{1.676}$$

The particle number current density is

$$n^\mu(x) = \int_{P_m(x)} f p^\mu \Pi_m, \tag{1.677}$$

where $P_m(x)$ is the fiber over x in Φ_m (all momenta with $\langle p, p\rangle = -m^2$). Similarly,, one defines the energy-momentum tensor, etc.

Let us show that

$$n^\mu{}_{;\mu} = \int_{P_m} (L_{X_g}f)\,\Pi_m. \tag{1.678}$$

We first note that (always in Φ_m)

$$d(f\omega_m) = (L_{X_g}f)\,\Omega_m. \tag{1.679}$$

Indeed, because of (1.673), the left-hand side of this equation is

$$df \wedge \omega_m = df \wedge i_{X_g}\Omega_m = (i_{X_g}df) \wedge \Omega_m = (L_{X_g}f)\,\Omega_m.$$

Now, let D be a domain in Φ_m, which is the inverse of a domain $\bar{D} \subset M$ under the projection $\Phi_m \to M$. Then, we have on the one hand by (1.674), setting $i_X\eta \equiv X^\mu\sigma_\mu$,

$$\int_{\partial D} f\omega_m = \int_{\partial\bar{D}} \sigma_\mu \int_{P_m(x)} p^\mu f \Pi_m = \int_{\partial\bar{D}} \sigma_\mu n^\mu = \int_{\partial\bar{D}} i_n\eta = \int_{\bar{D}} (\nabla\cdot n)\eta.$$

[26] Note that in Minkowski spacetime, we get for a constant time section $vol_\Sigma = dx^{123} \wedge dp^{123}$.

On the other hand, by (1.679) and (1.668)

$$\int_{\partial D} f\omega_m = \int_D d(f\omega_m) = \int_D (L_{X_g} f)\,\Omega_m = \int_{\bar{D}} \eta \int_{P_m(x)} (L_{X_g} f)\,\Pi_m.$$

Since $\bar{D}$ is arbitrary, we indeed obtain (1.678).

The divergence of the energy-momentum tensor

$$T^{\mu\nu} = \int_{P_m} p^\mu p^\nu f \Pi_m \tag{1.680}$$

is given by

$$T^{\mu\nu}{}_{;\nu} = \int_{P_m} p^\mu \left(L_{X_g} f\right) \Pi_m. \tag{1.681}$$

This follows from the previous proof by considering instead of n^ν the vector field $N^\nu := v_\mu T^{\mu\nu}$, where v_μ is geodesic in x.

1.9.2 The General Relativistic Boltzmann Equation

Let us first consider particles for which collisions can be neglected (e.g., neutrinos at temperatures much below 1 MeV). Then, the conservation of the particle number in a domain that is comoving with the flow ϕ_s of X_g means that the integrals

$$\int_{\phi_s(\Sigma)} f\omega_m,$$

Σ as before a hypersurface of Φ_m transversal to X_g, are independent of s. We now show that this implies the *collisionless Boltzmann equation*

$$\boxed{L_{X_g} f = 0.} \tag{1.682}$$

The proof of this expected result proceeds as follows. Consider a "cylinder" $\mathscr{G}$, sweping by Σ under the flow ϕ_s in the interval $[0,s]$ (see Fig. 1.10), and the integral

$$\int_{\mathscr{G}} L_{X_g} f\Omega_m = \int_{\partial\mathscr{G}} f\omega_m$$

(we used Eq. (1.679)). Since $i_{X_g}\omega_m = i_{X_g}(i_{X_g}\Omega_m) = 0$, the integral over the mantle of the cylinder vanishes while those over Σ and $\phi_s(\Sigma)$ cancel (conservation of particles). Because Σ and s are arbitrary, we conclude that (1.682) must hold.

From (1.678) and (1.679) we obtain, as expected, the conservation of the particle number current density: $n^\mu{}_{;\mu} = 0$.

With collisions, the Boltzmann equation has the symbolic form

$$\boxed{L_{X_g} f = C[f]}, \tag{1.683}$$

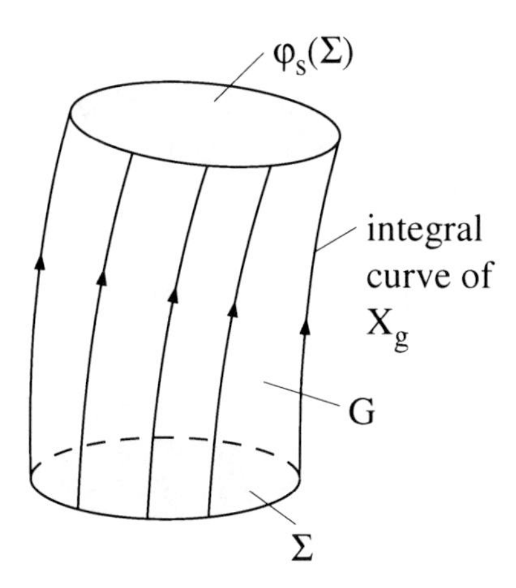

Fig. 1.10 Picture for the proof of (1.682).

where $C[f]$ is the "collision term." For the general form of this in terms of the invariant transition matrix element for a two-body collision, see (B.9). In the appendix, we also work this out explicitly for photon-electron scattering.

By (1.681) and (1.683), we have

$$T^{\mu\nu}{}_{;\nu} = Q^{\mu}, \tag{1.684}$$

with

$$Q^{\mu} = \int_{P_m} p^{\mu} C[f] \Pi_m. \tag{1.685}$$

1.9.3 Gauge Transformations

We consider again small deviations from Friedmann models and set correspondingly

$$f = f^{(0)} + \delta f. \tag{1.686}$$

How does δf change under a gauge transformation? This is derived in detail in [27] and allows us to introduce the gauge invariant perturbations $(\delta f)_{\chi}$, $(\delta f)_{\mathscr{Q}}$, etc.

1.9.4 Liouville Operator in the Longitudinal Gauge

We want to determine the action of the Liouville operator $\mathscr{L} := L_{X_g}$ on $(\delta f)_{\chi}$. The simplest way to do this is to work in the longitudinal gauge $B = E = 0$. In what follows, we always assume $K = 0$. The metric is then

$$g = a^2(\eta)\{-(1+2A)d\eta^2 + (1+2D)\delta_{ij}dx^i dx^j\}. \tag{1.687}$$

It is convenient to introduce an adapted orthonormal (to first order) tetrad

$$e_{\hat{0}} = \frac{1}{a(1+A)}\partial_\eta, \quad e_{\hat{k}} = \frac{1}{a(1+D)}\partial_k. \tag{1.688}$$

From $p^{\hat{\mu}} e_{\hat{\mu}} = p^\mu \partial_\mu$, we get

$$p^0 = \frac{p^{\hat{0}}}{a(1+A)}, \quad p^k = \frac{p^{\hat{k}}}{a(1+D)}.$$

Let $p := \sqrt{\sum_k (p^{\hat{k}})^2}$. In what follows, we consider the case of rest mass zero[27] and leave the generalization $m \neq 0$ to the reader. Then $p^{\hat{0}} = p$, and in terms of the comoving momentum $q = ap$, we have

$$p^0 = \frac{q}{a^2}(1-A), \quad p^i = \frac{q}{a^2}(1-D)\gamma^i, \tag{1.689}$$

where γ^i denotes the unit vector $p^{\hat{i}}/p$.

We consider the distribution function as a function of the independent variables η, x^i, q, γ^i. To determine the action of the Liouville operator, we compute the total derivative of f along a geodesic motion[28]:

$$\frac{df}{d\eta} = \frac{\partial f}{\partial \eta} + \frac{\partial f}{\partial x^i}\frac{dx^i}{d\eta} + \frac{\partial f}{\partial q}\frac{dq}{d\eta} + \frac{\partial f}{\partial \gamma^i}\frac{d\gamma^i}{d\eta}.$$

The last term in this equation is obviously of second order. For the further evaluation, we need $dx^i/d\eta$ and $dq/d\eta$. Let λ be the affine parameter in the geodesic equations (1.662) and (1.663). We have

$$\frac{dx^i}{d\eta} = \frac{dx^i}{d\lambda}\frac{d\lambda}{d\eta}, \quad \frac{dx^0}{d\eta} = 1 = \frac{dx^0}{d\lambda}\frac{d\lambda}{d\eta} = p^0\frac{d\lambda}{d\eta},$$

[27] For this case, the following calculations become a bit simpler if one makes use of the general fact that null geodesics remain null geodesics under conformal changes of the metric.

[28] Recall that the Lie derivative (directional derivative) L_X of a function f on a manifold with respect to a vector field X can be obtained from the total derivative along an integral curve $x(\lambda)$ of X from the relation

$$\frac{d}{d\lambda}f(x(\lambda)) = (L_X f)(x(\lambda)).$$

so $\frac{d\lambda}{d\eta} = 1/p^0$. Hence, with (1.689),

$$\frac{dx^i}{d\eta} = \frac{p^i}{p^0} = \gamma^i(1 - D + A),$$

$= \gamma^i$ in zeroth order. Since $\partial f/\partial x^i$ is of first order, we obtain to first order

$$\frac{df}{d\eta} = \frac{\partial f}{\partial \eta} + \gamma^i \frac{\partial f}{\partial x^i} + \frac{\partial f}{\partial q}\frac{dq}{d\eta}. \tag{1.690}$$

Computation of $dq/d\eta$.

We start from the $\mu = 0$ component of (1.663). The left-hand side is with (1.689)

$$\frac{dp^0}{d\lambda} = \frac{dp^0}{d\eta}\frac{d\eta}{d\lambda} = \frac{d}{d\eta}\left[\frac{q}{a^2}(1-A)\right]\frac{q}{a^2}(1-A),$$

thus,

$$\frac{d}{d\eta}\left[\frac{q}{a^2}(1-A)\right] = -\frac{a^2}{q}(1+A)\Gamma^0{}_{\alpha\beta}p^\alpha p^\beta.$$

This gives

$$\frac{dq}{d\eta} = q(\partial_\eta A + \gamma^i \partial_i A - \frac{1+2A}{q}a^4\Gamma^0{}_{\alpha\beta}p^\alpha p^\beta + 2\mathscr{H}q. \tag{1.691}$$

For the metric (1.687), we obtain

$$\begin{aligned}(1+2A)\Gamma^0{}_{\alpha\beta}p^\alpha p^\beta &= -\frac{1}{2a^2}\left\{-\partial_0[a^2(1+2A)](p^0)^2 - 2\partial_i[a^2(1+2A)]p^0p^i\right\}\\ &\quad + \frac{1}{2a^2}\partial_0[a^2(1+2D)]\delta_{ij}p^ip^j.\end{aligned}$$

One readily verifies that $dq/d\eta$ vanishes in zeroth order (as expected). Therefore, we have to work out the first order of the last expression, i.e., of

$$\begin{aligned}\frac{1}{2a^2}\partial_0[a^2(1+2A)]\frac{q^2}{a^4}(1-2A) + \frac{1}{a^2}\partial_i[a^2(1+2A)]\frac{q^2}{a^4}\gamma^i(1-A)(1-D)\\ + \frac{1}{2a^2}\partial_0[a^2(1+2D)]\frac{q^2}{a^4}(1-2D).\end{aligned}$$

Here, the time derivatives of a^2 give no first-order contributions, so we obtain for the first-order part:

$$\frac{q^2}{a^4}[\partial_0 A + \partial_0 D + 2\gamma^i\partial_i A].$$

Using this in (1.691) gives to first order

$$\frac{dq}{d\eta} = -q(\gamma^i \partial_i A + D'). \tag{1.692}$$

Therefore,

$$\boxed{\frac{df}{d\eta} = f' + \gamma^i \partial_i f - q\frac{\partial f^{(0)}}{\partial q}(D' + \gamma^i \partial_i A).} \tag{1.693}$$

Note that $df/d\eta = 0$ in zeroth order for an equilibrium situation. According to the previous footnote, the right-hand side of (1.693), as a function of phase space, is equal to $L_{X_g} f$, up to the factor p^0, due to $d\eta/d\lambda = p^0$.

As a first application, we consider the collisionless Boltzmann equation for $m = 0$:

$$\boxed{(\partial_\eta + \gamma^i \partial_i)\delta f - \left[D' + \gamma^i \partial_i (A)\right] q\frac{\partial f^{(0)}}{\partial q} = 0.} \tag{1.694}$$

It is obvious how to write this in gauge invariant form

$$\boxed{(\partial_\eta + \gamma^i \partial_i)(\delta f)_\chi = \left[\Phi' + \gamma^i \partial_i (\Psi)\right] q\frac{\partial f^{(0)}}{\partial q}.} \tag{1.695}$$

(From this, the collisionless Boltzmann equation follows in any gauge; write this out.)

In terms of the Fourier amplitudes, we get, with $\mu := \hat{k} \cdot \gamma$,

$$\boxed{(\delta f)'_\chi + i\mu k(\delta f)_\chi = \left[\Phi' + ik\mu\Psi\right] q\frac{\partial f^{(0)}}{\partial q}.} \tag{1.696}$$

This equation can be used for neutrinos as long as their masses are negligible (the generalization to the massive case is easy).

1.9.5 Boltzmann Equation for Photons

The collision term for photons due to Thomson scattering on electrons will be derived later (Sect. 1.9.7). We shall find that in the longitudinal gauge, ignoring polarization effects[29],

$$C[f] = x_e n_e \sigma_T p \left[\langle \delta f \rangle - \delta f - q\frac{\partial f^{(0)}}{\partial q}\gamma^i \partial_i v_b + \frac{3}{4} Q_{ij}\gamma^i \gamma^j\right]. \tag{1.697}$$

---

[29] I refer to [8] for a detailed treatment that includes the polarization dependence of Thomson scattering; see also [27], especially Appendix E.

On the right, $x_e n_e$ is the unperturbed free electron density (x_e = ionization fraction), σ_T is the Thomson cross section, and v_b is the scalar velocity perturbation of the baryons. Furthermore, we have introduced the spherical averages

$$\langle \delta f \rangle = \frac{1}{4\pi} \int_{S^2} \delta f \, d\Omega_\gamma, \tag{1.698}$$

$$Q_{ij} = \frac{1}{4\pi} \int_{S^2} [\gamma_i \gamma_j - \frac{1}{3}\delta_{ij}] \delta f \, d\Omega_\gamma. \tag{1.699}$$

(Because of the tight coupling of electrons and ions, we can take $v_e = v_b$.)

The linearized Boltzmann equation thus becomes

$$\begin{aligned}(\partial_\eta + \gamma^i \partial_i)\delta f \; - \; \left[D' + \gamma^i \partial_i A\right] q \frac{\partial f^{(0)}}{\partial q} \\ = a x_e n_e \sigma_T \left[\langle \delta f \rangle - \delta f - q \frac{\partial f^{(0)}}{\partial q} \gamma^i \partial_i v_b + \frac{3}{4} Q_{ij} \gamma^i \gamma^j \right].\end{aligned} \tag{1.700}$$

This can immediately be written in a gauge invariant form, by replacing

$$\delta f \to (\delta f)_\chi, \;\; v_b \to V_b, \;\; A \to \Psi, \;\; D \to \Phi. \tag{1.701}$$

In our applications to the CMB, we work with the gauge invariant *(integrated) brightness temperature* perturbation

$$\Theta_\chi(\eta, x^i, \gamma^j) = \int (\delta f)_\chi q^3 dq \Big/ \, 4 \int f^{(0)} q^3 dq. \tag{1.702}$$

(The factor 4 is chosen because of the Stephan–Boltzmann law, according to which $\delta\rho/\rho = 4\delta T/T$.) It is simple to translate the Boltzmann equation for $\mathscr{F}_s$ to a kinetic equation for Θ_s. Using

$$\int q \frac{\partial f^{(0)}}{\partial q} q^3 dq = -4 \int f^{(0)} q^3 dq,$$

we obtain for the convective part (from the left-hand side of the Boltzmann equation for $(\delta f)_\chi$)

$$\Theta'_\chi + \gamma^i \partial_i \Theta_\chi + \Phi' + \gamma^i \partial_i \Psi.$$

The collision term gives

$$\dot{\tau}(\theta_0 - \Theta_\chi + \gamma^i \partial_i V_b + \frac{1}{16} \gamma^i \gamma^j \Pi_{ij}),$$

with $\dot{\tau} = x_e n_e \sigma_T a / a_0$, $\theta_0 = \langle \Theta_\chi \rangle$ (spherical average), and

$$\frac{1}{12} \Pi_{ij} := \frac{1}{4\pi} \int [\gamma_i \gamma_j - \frac{1}{3}\delta_{ij}] \Theta_s \, d\Omega_\gamma. \tag{1.703}$$

The basic equation for Θ_χ is thus

$$(\Theta_\chi + \Psi)' + \gamma^i \partial_i (\Theta_\chi + \Psi) =$$
$$(\Psi' - \Phi') + \dot{\tau}(\theta_0 - \Theta_\chi + \gamma^i \partial_i V_b + \frac{1}{16}\gamma^i \gamma^j \Pi_{ij}). \quad (1.704)$$

This equation clearly also holds for the (unintegrated) brightness temperature fluctuation, $(\delta T/T)(\eta, x^i, q, \gamma^i)$, defined by

$$\delta f = -q \frac{\partial f^{(0)}}{\partial q}(\delta T/T),$$

since the Thomson cross section is energy independent.

In a mode decomposition, we get (I drop from now on the index χ on Θ):

$$\boxed{\Theta' + ik\mu(\Theta + \Psi) = -\Phi' + \dot{\tau}[\theta_0 - \Theta - i\mu V_b - \frac{1}{10}\theta_2 P_2(\mu)]} \quad (1.705)$$

(recall $V_b \to -(1/k)V_b$). The last term on the right comes about as follows. We expand the Fourier modes $\Theta(\eta, k^i, \gamma^j)$ in terms of Legendre polynomials

$$\Theta(\eta, k^i, \gamma^j) = \sum_{l=0}^{\infty} (-i)^l \theta_l(\eta, k) P_l(\mu), \quad \mu = \hat{k} \cdot \gamma, \quad (1.706)$$

and note that

$$\frac{1}{16}\gamma^i \gamma^j \Pi_{ij} = -\frac{1}{10}\theta_2 P_2(\mu) \quad (1.707)$$

(Exercise). The expansion coefficients $\theta_l(\eta, k)$ in (1.706) are the *brightness moments*[30]. The lowest three have simple interpretations. We show that in the notation of Sect. 1.5:

$$\theta_0 = \frac{1}{4}\Delta_{s\gamma}, \quad \theta_1 = V_\gamma, \quad \theta_2 = \frac{5}{12}\Pi_\gamma. \quad (1.708)$$

Derivation of (1.708)

We start from the general formula (see Sect. 1.9.1)

$$T^\mu_{(\gamma)\nu} = \int p^\mu p_\nu f(p) \frac{d^3 p}{p^0} = \int p^\mu p_\nu f(p) p dp \, d\Omega_\gamma. \quad (1.709)$$

According to the general parametrization (1.520), we have

$$\delta T^0_{(\gamma)0} = -\delta\rho_\gamma = -\int p^2 \delta f(p) p dp \, d\Omega_\gamma. \quad (1.710)$$

Similarly, in zeroth order

[30] In the literature, the normalization of the θ_l is sometimes chosen differently: $\theta_l \to (2l+1)\theta_l$.

$$T^{(0)0}_{(\gamma)\ 0} = -\rho^{(0)}_\gamma = -\int p^2 f^{(0)}(p) p dp \, d\Omega_\gamma. \tag{1.711}$$

Hence,

$$\frac{\delta\rho_\gamma}{\rho^{(0)}_\gamma} = \frac{\int q^3 \delta f \, dq \, d\Omega_\gamma}{\int q^3 f^{(0)} dq \, d\Omega_\gamma}. \tag{1.712}$$

In the longitudinal gauge, we have $\Delta_{s\gamma} = \delta\rho_\gamma/\rho^{(0)}_\gamma$, $\mathscr{F}_s = \delta f$, and thus by (1.702) and (1.706)

$$\Delta_{s\gamma} = 4\frac{1}{4\pi}\int \Theta \, d\Omega_\gamma = 4\theta_0.$$

Similarly,

$$T^i_{(\gamma)0} = -h_\gamma v^{|i}_\gamma = \int p^i p_0 \delta f p dp \, d\Omega_\gamma$$

or

$$v^{|i}_\gamma = \frac{3}{4\rho^{(0)}_\gamma}\int \gamma^i \delta f p^3 dp \, d\Omega_\gamma. \tag{1.713}$$

With (1.711) and (1.702), we get

$$V^{|i}_\gamma = \frac{3}{4\pi}\int \gamma^i \Theta \, d\Omega_\gamma. \tag{1.714}$$

For the Fourier amplitudes, this gauge invariant equation gives ($V_\gamma \to -(1/k)V_\gamma$)

$$-iV_\gamma \hat{k}^i = \frac{3}{4\pi}\int \gamma^i \Theta \, d\Omega_\gamma$$

or

$$-iV_\gamma = \frac{3}{4\pi}\int \mu\Theta \, d\Omega_\gamma.$$

Inserting here the decomposition (1.706) leads to the second relation in (1.708).

For the third relation, we start from (1.520) and (1.709)

$$\delta T^i_{(\gamma)j} = \delta p_\gamma \delta^i{}_j + p^{(0)}_\gamma \left(\Pi^{|i}_{\gamma|j} - \frac{1}{3}\delta^i{}_j \nabla^2 \Pi_\gamma\right) = \int p^i p_j \delta f p \, dp \, d\Omega_\gamma.$$

From this and (1.710), we see that $\delta p_\gamma = \frac{1}{3}\delta\rho_\gamma$, thus $\Gamma_\gamma = 0$ (no entropy production with respect to the photon fluid). Furthermore, since $p^{(0)}_\gamma = \frac{1}{3}\rho^{(0)}_\gamma$ we obtain with (1.703)

$$\Pi^{|i}_{\gamma|j} - \frac{1}{3}\delta^i{}_j \nabla^2 \Pi_\gamma = 4\cdot 3\frac{1}{4\pi}\int [\gamma^i\gamma_j - \frac{1}{3}\delta^i{}_j]\Theta \, d\Omega_\gamma = \Pi^i{}_j.$$

In momentum space ($\Pi_\gamma \to (1/k^2)\Pi_\gamma$) this becomes

$$-(\hat{k}^i\hat{k}_j - \frac{1}{3})\Pi_\gamma = \Pi^i{}_j$$

or, contracting with $\gamma_i\gamma^j$ and using (1.707), the desired result.

Hierarchy for Moment Equations

Now, we insert the expansion (1.706) into the Boltzmann equation (1.705). Using the recursion relations for the Legendre polynomials,

$$\mu P_l(\mu) = \frac{l}{2l+1} P_{l-1}(\mu) + \frac{l+1}{2l+1} P_{l+1}(\mu), \tag{1.715}$$

we obtain

$$\sum_{l=0}^{\infty}(-i)^l \theta_l' P_l + ik\sum_{l=0}^{\infty}(-i)^l \theta_l \left[\frac{l}{2l+1}P_{l-1} + \frac{l+1}{2l+1}P_{l+1}\right] + ik\Psi P_1$$
$$= -\Phi' P_0 - \dot{\tau}\left[\sum_{l=1}^{\infty}(-i)^l \theta_l P_l - iV_b P_1 - \frac{1}{10}\theta_2 P_2\right].$$

Comparing the coefficients of P_l leads to the following hierarchy of ordinary differential equations for the brightness moments $\theta_l(\eta)$:

$$\theta_0' = -\frac{1}{3}k\theta_1 - \Phi', \tag{1.716}$$

$$\theta_1' = k\left(\theta_0 + \Psi - \frac{2}{5}\theta_2\right) - \dot{\tau}(\theta_1 - V_b), \tag{1.717}$$

$$\theta_2' = k\left(\frac{2}{3}\theta_1 - \frac{3}{7}\theta_3\right) - \dot{\tau}\frac{9}{10}\theta_2, \tag{1.718}$$

$$\theta_l' = k\left(\frac{l}{2l-1}\theta_{l-1} - \frac{l+1}{2l+3}\theta_{l+1}\right), \quad l > 2. \tag{1.719}$$

A lot could be said in connection with this hierarchy. Here, we consider only Eq. (1.717) and rewrite it with (1.708) as

$$V_\gamma' = k\Psi + \frac{k}{4}\Delta_{s\gamma} - \frac{1}{6}k\Pi_\gamma + \mathcal{H}F_\gamma, \tag{1.720}$$

where

$$\boxed{\mathcal{H}F_\gamma = -\dot{\tau}(V_\gamma - V_b).} \tag{1.721}$$

This agrees with (1.554), apart from the last term $\mathcal{H}F_\gamma$ that describes the momentum transfer due to the interaction with the baryon fluid (Thomson scattering). For the baryons, we have to add in (1.554) also a term $\mathcal{H}F_b$ to account for the back-reaction. This momentum transfer can immediately be obtained from the requirement that the sum of the two momentum equations for γ and b has to agree with the total momentum equation (1.230). This implies that

$$h_\gamma F_\gamma + h_b F_b = 0, \quad F_b = -\frac{4\rho_\gamma}{3\rho_b}F_\gamma. \tag{1.722}$$

The momentum equation of the baryon fluid then becomes

$$V_b' = -aHV_b + kc_b^2\delta_b + k\Psi + \dot{\tau}(\theta_1 - V_b)/R, \tag{1.723}$$

with $R = 3\rho_b/4\rho_\gamma$.

1.9.6 Tensor Contributions to the Boltzmann Equation

Considering again only the case $K = 0$, the metric (1.455) for tensor perturbations becomes

$$g_{\mu\nu} = a^2(\eta)[\eta_{\mu\nu} + 2H_{\mu\nu}], \tag{1.724}$$

where the $H_{\mu\nu}$ satisfy the TT gauge conditions (1.456):

$$H_{00} = H_{0i} = H^i{}_i = H_i{}^j{}_{|j} = 0. \tag{1.725}$$

We introduce again an adapted orthonormal tetrad

$$e_{\hat{0}} = \frac{1}{a}\,\partial_\eta, \quad e_{\hat{k}} = \frac{1}{a}\,(\partial_k - H^l{}_k\partial_l). \tag{1.726}$$

From $p^{\hat{\mu}} e_{\hat{\mu}} = p^\mu \partial_\mu$, we get

$$p^0 = \frac{p^{\hat{0}}}{a}, \quad p^k = \frac{1}{a}(p^{\hat{k}} - p^{\hat{l}} H^k{}_l)$$

or in terms of the comoving momentum q:

$$p^0 = \frac{q}{a^2}, \quad p^k = \frac{q}{a^2}(\gamma^k - \gamma^l H^k{}_l). \tag{1.727}$$

The first equation implies $d\eta/d\lambda = q/a^2$. To first order we have again (1.690) for $df/d\eta$, and we proceed as for scalar perturbations in computing $dq/d\lambda$. Instead of (1.691), we now obtain

$$\frac{dq}{d\eta} = -\frac{1}{q}a^4\Gamma^0{}_{\alpha\beta}p^\alpha p^\beta + 2\mathscr{H}q. \tag{1.728}$$

As before, one verifies that this vanishes in first order, and a straightforward calculation gives now

$$\frac{dq}{d\eta} = -qH_{ij}'\gamma^i\gamma^j. \tag{1.729}$$

For tensor perturbations, we thus obtain

$$\boxed{\frac{df}{d\eta} = f' + \gamma^i\partial_i f - q\frac{\partial f^{(0)}}{\partial q}H_{ij}'\gamma^i\gamma^j.} \tag{1.730}$$

For the temperature (brightness) perturbation this gives, if we neglect collisions,

$$\boxed{(\partial_\eta + \gamma^j\partial_i)\Theta = -H'_{ij}\gamma^i\gamma^j.} \tag{1.731}$$

(Collisions can be included without much effort; Exercise.)

This equation describes the influence of tensor modes on Θ. The evolution of these tensor modes is described according to (1.457) by

$$H''_{ij} + 2\mathscr{H}H'_{ij} - \nabla^2 H_{ij} = 0, \tag{1.732}$$

if we neglect tensor perturbations of the energy-momentum tensor. We decompose H_{ij} as in Sect. 1.7.2:

$$H_{ij}(\eta,k) = \sum_{\lambda=\pm 2} h_\lambda(\eta,k)\varepsilon_{ij}(k,\lambda), \tag{1.733}$$

where the polarization tensor satisfies (1.463). The mode functions $h_\lambda(\eta,k)$ satisfy the homogeneous linear differential equation

$$h'' + 2\frac{a'}{a}h' + k^2 h = 0. \tag{1.734}$$

At very early times, when the modes are still far outside the Hubble horizon, we can neglect the last term in (1.734), whence h is *frozen*. For this reason we solve (1.734) with the initial condition $h'(\eta_i,k) = 0$. Moreover, we are only interested in growing modes. For the matter or the radiation dominated eras, one can solve (1.734) analytically in terms of Bessel functions (Exercise). It is, however, more instructive to discuss this mode equation approximately.

On scales smaller than the Hubble horizon, we can use a WKB approximation. Without the damping term, due to the cosmological expansion, $h(\eta)$ is a linear combination of $\cos(k\eta)$ and $\sin(k\eta)$. In the WKB ansatz, we multiply this with a slowly varying amplitude $A(\eta)$. Neglecting A'', the differential equation shows that $A \propto 1/a$. We, therefore, expect that the tensor contributions to the CMB power spectra fall off rapidly on scales smaller than the Hubble horizon.

1.9.7 Collision Integral for Thomson Scattering

The main goal of this appendix is the derivation of Eq. (1.697) for the collision integral in the Thomson limit.

When we work relative to an orthonormal tetrad the collision integral has the same form as in special relativity. So lets first consider this case.

Collision Integral for Two-Body Scattering

In SR, the Boltzmann equation (1.683) reduces to

$$p^\mu \partial_\mu f = C[f] \tag{1.735}$$

or

$$\partial_t f + v^i \partial_i f = \frac{1}{p^0} C[f]. \tag{1.736}$$

In order to find the explicit expression for $C[f]$ things become easier if the following nonrelativistic normalization of the one-particle states $|p,\lambda\rangle$ is adopted:

$$\langle p',\lambda'|p,\lambda\rangle = (2\pi)^3 \delta_{\lambda,\lambda'} \delta^{(3)}(\mathbf{p}' - \mathbf{p}). \tag{1.737}$$

(Some readers may even prefer to discretize the momenta by using a finite volume with periodic boundary conditions.) Correspondingly, the one-particle distribution functions f are normalized according to

$$\int f(p) \frac{g d^3 p}{(2\pi)^3} = n, \tag{1.738}$$

where g is the statistical weight (= 2 for electrons and photons), and n is the particle number density.

The S-matrix element for a two-body collision $p,q \to p',q'$ has the form (suppressing polarization indices)

$$\langle p',q'|S-1|p,q\rangle = -i(2\pi)^4 \delta^{(4)}(p'+q'-p-q)\langle p',q'|T|p,q\rangle. \tag{1.739}$$

Because of our noninvariant normalization, we introduce the Lorentz invariant matrix element M by

$$\langle p',q'|T|p,q\rangle = \frac{M}{(2p^0 2q^0 2p'^0 2q'^0)^{1/2}}. \tag{1.740}$$

The transition probability per unit time and unit volume is then (see, e.g., Sect. 64 of [42])

$$dW = (2\pi)^4 \frac{1}{2p^0 2q^0} |M|^2 \delta^{(4)}(p'+q'-p-q) \frac{d^3 p'}{(2\pi)^3 2p'^0} \frac{d^3 q'}{(2\pi)^3 2q'^0}. \tag{1.741}$$

Since we ignore in the following polarization effects, we average $|M|^2$ over all polarizations (helicities) of the initial and final particles. This average is denoted by $\overline{|M|^2}$. Per polarization, we still have the formula (1.741), but with $|M|^2$ replaced by $\overline{|M|^2}$. From time reversal invariance, we conclude that $\overline{|M|^2}$ remains invariant under $p,q \leftrightarrow p',q'$.

With the standard arguments, we can now write down the collision integral. For definiteness, we consider Compton scattering $\gamma(p)+e^-(q)\to\gamma(p')+e^-(q')$ and denote the distribution functions of the photons and electrons by $f(p)$ and $f_{(e)}(q)$, respectively. In the following expression, we neglect the Pauli suppression factors $1-f_{(e)}$, since in our applications the electrons are highly nondegenerate. Explicitly, we have

$$\frac{1}{p^0}C[f]=\frac{1}{2p^0}\int\frac{2d^3q}{(2\pi)^3 2q'^0}\frac{2d^3q'}{(2\pi)^3 2q'^0}\frac{2d^3p'}{(2\pi)^3 2p'^0}(2\pi)^4\overline{|M|^2}\delta^{(4)}(p'+q'-p-q)$$
$$\times\left\{(1+f(p))f(p')f_{(e)}(q')-(1+f(p'))f(p)f_{(e)}(q)\right\}. \tag{1.742}$$

At this point, we return to the normalization of the one-particle distributions adopted in Sect. 1.9.1. This amounts to the substitution $f\to 4\pi^3 f$. Performing this in (1.735) and (1.742), we get for the collision integral

$$C[f]=\frac{1}{16\pi^2}\int\frac{d^3q}{q^0}\frac{d^3q'}{q'^0}\frac{d^3p'}{p'^0}\overline{|M|^2}\delta^{(4)}(p'+q'-p-q)$$
$$\times\left\{(1+4\pi^3 f(p))f(p')f_{(e)}(q')-(1+4\pi^3 f(p'))f(p)f_{(e)}(q)\right\}. \tag{1.743}$$

The invariant function $\overline{|M|^2}$ is explicitly known and can for instance be expressed in terms of the Mandelstam variables s,t,u (see Sect. 86 of [42]).

The integral with respect to d^3q' can trivially be done

$$C[f]=\frac{1}{16\pi^2}\int\frac{d^3q}{q^0}\frac{1}{q'^0}\frac{d^3p'}{p'^0}\delta(p'^0+q'^0-p^0-q^0)\overline{|M|^2}\times\{\cdots\}. \tag{1.744}$$

The integral with respect to $\mathbf{p}'$ can most easily be evaluated by going to the rest frame of q^μ. Then,

$$\int d^3p'\frac{1}{p'^0q'^0}\delta(p'^0+q'^0-p^0-q^0)\cdots=\int d\Omega_{\hat{\mathbf{p}}'}\int d|\mathbf{p}'|\frac{|\mathbf{p}'|}{q'^0}\delta(m+q'^0-p^0-q^0)\cdots.$$

We introduce the following notation: With respect to the rest system of q^μ, let $\omega:=p^0=|\mathbf{p}|$, $\omega':=p'^0=|\mathbf{p}'|$, $E'=\sqrt{\mathbf{q}'^2+m^2}$. Then, the last integral is equal to

$$\frac{\omega'}{E'}\frac{1}{|1+\partial E'/\partial\omega'|}=\frac{\omega'^2}{m\omega}.$$

In getting the last expression, we have used energy and momentum conservation.

So far, we are left with

$$C[f]=\frac{1}{16\pi^2 m}\int\frac{d^3q}{q^0}\int d\Omega_{\hat{\mathbf{p}}'}\frac{\omega'^2}{\omega}\overline{|M|^2}\times\{\cdots\}. \tag{1.745}$$

In the rest system of q^μ, the following expression for $\overline{|M|^2}$ can be found in many books (for a derivation, see [43])

$$\overline{|M|^2} = 3\pi m^2 \sigma_T \left[\frac{\omega'}{\omega} + \frac{\omega}{\omega'} - \sin^2 \vartheta \right], \tag{1.746}$$

where ϑ is the scattering angle in that frame. For an arbitrary frame, the combination $d\Omega_{\hat{\mathbf{p}}'} \frac{\omega'^2}{\omega} \overline{|M|^2}$ has to be treated as a Lorentz invariant object.

At this point, we take the nonrelativistic limit $\omega/m \to 0$, in which $\omega' \simeq \omega$ and $C[f]$ reduces to the simple expression

$$C[f] = \frac{3}{16\pi} \sigma_T \omega n_e \int d\Omega_{\hat{\mathbf{p}}'} (1 + \cos^2 \vartheta)[f(p') - f(p)]. \tag{1.747}$$

Derivation of (1.697)

In Sect. 1.9.4, the components p^μ of the four-momentum p refer to the tetrad e_μ defined in (1.688). Relative to this, we introduced the notation $p^\mu = (p, p\gamma^i)$. The electron four-velocity is according to (1.520) given to first order by

$$u_{(e)} = \frac{1}{a}(1 - A)\partial_\eta + \frac{1}{a} v^i_{(e)} \partial_i = e_0 + v^i_{(e)} e_i; \;\; v^i_{(e)} = v_{(e)i} = \partial_i v_{(e)}. \tag{1.748}$$

Now ω in (1.747) is the energy of the four-momentum p in the rest frame of the electrons, thus

$$\omega = -\langle p, u_{(e)} \rangle = p[1 - \partial_i v_{(e)} \gamma^i]. \tag{1.749}$$

Similarly,

$$\omega' = -\langle p', u_{(e)} \rangle = p'[1 - \partial_i v_{(e)} \gamma'^i]. \tag{1.750}$$

Since in the nonrelativistic limit $\omega' = \omega$, we obtain the relation

$$p'[1 - \hat{e}_i(v_{(e)}) \gamma'^i] = p[1 - \partial_i v_{(e)} \gamma^i]. \tag{1.751}$$

Therefore, to first order

$$\begin{aligned} f(p', \gamma'^i) &= f^{(0)}(p') + \delta f(p', \gamma'^i) \\ &= f^{(0)}(p) + \frac{\partial f^{(0)}}{\partial p}(p' - p) + \delta f(p, \gamma'^i) \\ &= f^{(0)}(p) + p \frac{\partial f^{(0)}}{\partial p} \partial_i v_{(e)} (\gamma'^i - \gamma^i) + \delta f(p, \gamma'^i). \end{aligned} \tag{1.752}$$

Remember that the surface element $d\Omega_{\hat{\mathbf{p}}'}$ in (1.747) also refers to the rest system. This is related to the surface element $d\Omega_{\gamma'}$ by[31]

$$d\Omega_{\hat{\mathbf{p}}'} = \left(\frac{p'}{\omega'} \right)^2 d\Omega_{\gamma'} = [1 + 2\partial_i v_{(e)} \gamma'^i] d\Omega_{\gamma'}. \tag{1.753}$$

[31] Under a Lorentz transformation, the surface element for photons transforms as

$$d\Omega = (\omega'/\omega)^2 d\Omega'$$

(Exercise).

Inserting (1.752) and (1.753) into (1.747) gives to first order, with the notation of Sect. 1.9.5,

$$C[f] = n_e \sigma_T p \left[\langle \delta f \rangle - \delta f - p \frac{\partial f^{(0)}}{\partial p} \partial_i v_{(e)} \gamma^i + \frac{3}{4} Q_{ij} \gamma^i \gamma^j \right], \tag{1.754}$$

that is the announced equation (1.697).

Acknowledgments I thank Vittorio Gorini, Sabino Matarrese, and Ugo Moschella for organizing such an interesting school at the beautiful lake of Como and for the warm hospitality during the school.

References

1. P.J.E. Peebles, *Principles of Physical Cosmology*. Princeton University Press 1993.
2. J.A. Peacock, *Cosmological Physics*. Cambridge University Press 1999.
3. A.R. Liddle and D.H. Lyth, *Cosmological Inflation and Large Scale Structure*. Cambridge University Press 2000.
4. S. Dodelson, *Modern Cosmology*. Academic Press 2003.
5. G. Börner, *The Early Universe*. Springer-Verlag 2003 (4th edition).
6. V.S. Mukhanov, *Physical Foundations of Cosmology*. Cambridge University Press 2005.
7. S. Weinberg, *Cosmology*, Oxford University Press (2008).
8. R. Durrer, *The Cosmic Microwave Background*, Cambridge Unive4rsity Press (2008).
9. N. Straumann, *General Relativity, With Applications to Astrophysics*, Texts and Monographs in Physics, Springer Verlag, 2004.
10. H. Nussbaumer and L. Bieri, *Discovering the Expanding Universe*, Cambridge University Press (2009).
11. N. Straumann, Helv. Phys. Acta **45**, 1089 (1972).
12. N. Straumann, *Allgemeine Relativitätstheorie und Relativistische Astrophysik*, 2. Auflage, Lecture Notes in Physics, Volume 150, Springer-Verlag (1988); Kap. IX.
13. N. Straumann, *General Relativity and Relativistic Astrophysics*, Springer-Verlag (1984).
14. N. Straumann, *Kosmologie I*, Vorlesungsskript, http://web.unispital.ch/neurologie/vest/homepages/straumann/norbert
15. G. Steigman, Phys. Scripta, **T121**, 142 (2005); arXiv: hep-ph/0501100.
16. E. Mörtsell and Ch. Clarkson, arXiv:0811.0981.
17. P. Astier, et al., Astron. Astrophys. **447**, 31 (2006)
18. R. Kantowski and R.C. Thomas, Astrophys. J. **561**, 491 (2001); astro-ph/0011176.
19. M. Sasaki, Mon. Not. R. Astron. Soc. **228**, 653 (1987).
20. N. Sigiura, N. Sugiyama, and M. Sasaki, Prog.Theo. Phys. **101**, 903 (1999).
21. C. Bonvin, R. Durrer and M.A. Gasparini, astro-ph/0511183.
22. V.A. Belinsky, L.P. Grishchuk, I.M. Khalatnikov, and Ya.B. Zeldovich, Phys. Lett. **155B**, 232 (1985).
23. A. Linde, *Lectures on Inflationary Cosmology*, hep-th/9410082.
24. H. Kodama and M. Sasaki, Prog. Theor. Phys. Suppl. **78**, 1-166 (1984); abreviated as KS,84.
25. V.F. Mukhanov, H.A. Feldman and R.H. Brandenberger, Physics Reports **215**, 203 (1992).
26. R. Durrer, Fund. of Cosmic Physics **15**,209 (1994).
27. N. Straumann, *From primordial quantum fluctuations to the anisotropies of the cosmic microwave background radiation*, Ann. Phys. (Leipzig) **15**, No. 10-11, 701-847 (2006); [hep-ph/0505249]. For an updated and expanded version, see: www.vertigocenter.ch/straumann/norbert

28. N. Straumann, Ann.Phys. (Berlin) **17**, 609 (2008).
29. J. Hwang, Astrophys. J. **375**, 443 (1991).
30. R. Durrer and N. Straumann, Helvetica Phys. Acta **61**, 1027 (1988).
31. J. Hwang and H. Noh, Gen. Rel. Grav. **31**, 1131 (1999); astro-ph/9907063.
32. A.A. Starobinski, S. Tsujikawa and J. Yokoyama, astro-ph/0107555.
33. C. Armendariz-Picon, J. Cosm. & Astropart. Phys. **0702**, 031 (2007) [astro-ph/0612288].
34. M. Abramowitz and I. Stegun, *Handbook of Mathematical Functions, with Formulas, Graphs, and Mathematical Tables*, Dover (1974).
35. W.H. Kinney, E. Kolb, A. Melchiorri and A. Riotto, Phys. Rev. **D74**, 023502 (2006); astro-ph/0605338.
36. M.S. Turner, M. White and J.E. Lidsey, Phys. Rev. **D48**, 4613 (1993).
37. Y. Watanable and E. Komatsu, Phys. Rev. **D73**, 123515 (2006).
38. M. Giovannini, arXiv:0901.3026 [astro-ph.Co].
39. H. Kodama and H. Sasaki, Internat. J. Mod. Phys. **A2**, 491 (1987).
40. J.M. Bardeen, J.R. Bond, N. Kaiser, and A.S. Szalay, Astrophys. J. **304**, 15 (1986).
41. J.E. Marsden and T.S. Ratiu, *Introduction to Mechanics and Symmetry*, Second Edition, Springer-Verlag (1999).
42. L.D. Landau and E.M. Lifshitz, *Quantum Electrodynamics*, Vol. 4, second edition, Pergamon Press (1982).
43. N. Straumann, *Relativistische Quantentheorie, Eine Einfuehrung in die Quantenfeldtheorie*, Springer-Verlag (2005).

Chapter 2
Cosmology with Cosmic Microwave Background and Large-Scale Structure Observations

Licia Verde

2.1 Introduction

Cosmology aims at obtaining a physical description of the Universe including its global dynamics and content. Having a standard cosmological model provides a good framework to do that. In this framework, we aim to measure cosmological parameters (the parameters of the model) and to develop a fundamental understanding of them. We also want to understand the origin and evolution of cosmic structure and to probe the physics of the early Universe. Typical energies of the early universe were very high: the physics of the early Universe had deep links with fundamental particle physics.

The basic framework of the cosmological model relies on the following:

- The Cosmological principle: the Universe on average is described by the Friedman–Robertson–Walker metric
- Hot big bang and big bang nucleosynthesis
- Structure formation is described by a perturbed Friedman–Robertson–Walker metric
- Initial perturbations were seeded by inflation
- Physical laws as we know them are valid throughout the Universe: general relativity + atomic, particle, thermal, radiation physics

These can be considered as "assumptions" underlying the standard cosmological model. They are supported by observations: the Hubble diagram, the light elements abundances, the Cosmic Microwave Background (CMB) etc., however, sometimes it is good to challenge the underlying assumptions. There is work in the literature about this, but we will not discuss it in this Chapter.

Licia Verde
ICREA &ICCUB, e-mail: liciaverde@icc.ub.edu

The parameters of the standard model can be grouped into different types.

Parameters that describe the smooth Universe. These are parameters that govern the global geometry of space-time and parameters that govern the expansion rate the current expansion rate; H_0 and the matter density parameter Ω_m, which is the total of the cold dark matter component Ω_c and the baryonic component Ω_b and radiation Ω_{rad} which is unimportant for the recent history but not in the early Universe. There is also a neutrino component, given that neutrino oscillations indicated that neutrinos have mass Ω_{nu}, although in the standard cosmological model it is assumed to be zero as current cosmological observations have –reliably – only given upper limits to Ω_{nu}. In the standard model, all these components are sub-dominant as the Universe is dominated by dark energy in the form of a cosmological constant Ω_Λ. Finally, Ω_k describes the global geometry, which in the standard cosmological model is 0. Often explored deviations from the standard cosmological models are the nonflat case $\Omega_k \neq 0$ and dark energy deviating from a cosmological constant, i.e., a cosmic fluid with negative equation of state parameter w, but not necessarily $w \equiv -1$. w is often assumed to be constant, but in many cases, its redshift evolution is also considered.

Parameters that describe the inhomogeneous Universe. These are parameters that characterize the properties of the inhomogeneities: σ_8, the power spectrum spectral slope n for scalar perturbations. The primordial power spectrum is assumed to be a power law, $n = 1$ will give a so-called Harrison–Zeldovich or scale-invariant power spectrum. Deviations from a power law, in the form of a running of the spectral index $dnd\ln k$ are also considered as deviations from the standard cosmological model and are expected to be small in simple inflation models. In the standard cosmological model, perturbation are expected to be scalar. However, inflation produces a background of gravity waves; for cosmology, these give tensor perturbations. The parameter r describes the tensor to scalar perturbation amplitude, and in simple inflationary model, the tensor perturbations also have a power-law power spectrum with slope given by the tensors consistency relation $n_t = -r/8$. Parameters parameterizing our ignorance. Typical examples τ optical depth to the last scattering surface. In principle if we knew exactly how reionization happens, we should be able to compute τ from all the other parameters, but we can't do that. Another example is galaxy bias.

A similar grouping can be made about observations: there are observations that probe the global space-time geometry and expansion rate such as Supernovae luminosity distance diagrams (see Tsujikawa's Chapter) and baryon acoustic oscillations (BAO), observations that probe the inhomogeneities such as large-scale structures -see below- and, in part, gravitational lensing –see chapter 3 by Heavens.

In this chapter, I will review what we have learned recently from CMB and large-scale structure results and what we hope to learn with forthcoming experiments. I will concentrate on some of the challenges these data sets offer and how the cosmological information is extracted from the survey data. I will then move on to the subject of large-scale structure where I will pay special attention to the relatively new subject of BAO.

2.2 Cosmic Microwave Background and Other Data Sets: What have we Learned About Cosmology?

The standard cosmological model is extremely successful: with only six parameters, it can fit observations of the Universe from $z = 1100$ to today. The first three parameters are energy density of cold dark matter, energy density of baryons (the Universe is spatially flat with dark energy in the form of a cosmological constant that makes it flat), and present-day expansion rate. In this model, the initial conditions are Gaussian, the perturbations are scalar and generated in the inflationary scenario: amplitude and slope of the primordial power spectrum fully characterize the statistics of the initial perturbations. The sixth parameter is the optical depth to the last scattering surface. This model was first proposed more than a decade ago but has survived, virtually unchanged, the avalanche of high-precision cosmological data of the past decade. Interestingly, we can determine the parameters of this model from Cosmic Microwave Background (CMB) data alone and then extrapolate the prediction for this model to observations of the late Universe (large-scale structure clustering, supernovae data etc.). We find that the extrapolated model provides an extremely good fit to the late Universe observations (See Fig. 2.1 and 2.2). When external data sets are added to WMAP5 data, the recovered value of the cosmological parameter do not shift significantly, also indicating consistency. This is summarized in the following table (Table 2.1).

2.2.0.1 Polarization

Although most of the cosmological information on cosmological parameters, such as Ω_b or Ω_m, comes from the CMB temperature anisotropy, the CMB light is polarized and the CMB polarization can open a new window into the early Universe. Shortly after the CMB was first detected, M. Rees [40] showed that the CMB should

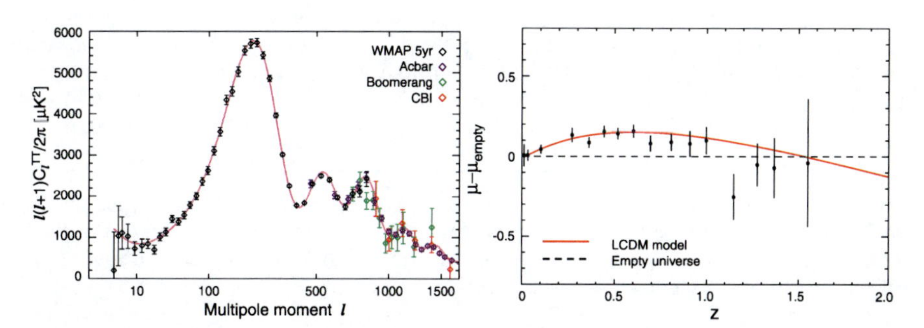

Fig. 2.1 Left:WMAP5 temperature angular power spectrum and best-fit LCDM model. This is consistent with data from recent small-scale CMB experiments. Right: luminosity-distance relationship predicted for the best-fit WMAP5-only model. The points show binned supernova observations from the Union compilation. Figures reproduced from [6]

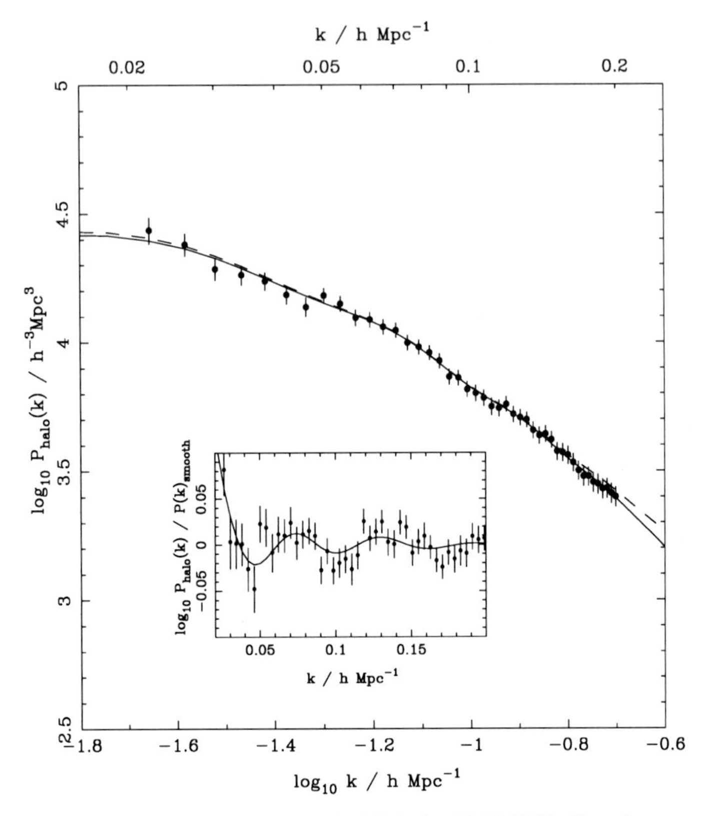

Fig. 2.2 Power spectrum from the Luminous Red Galaxies SDSS DR7. Also, shown are the best-fit LCDM model for WMAP5 data with large-scale structure data (solid line) and the underlying linear dark matter power spectrum. In the inset, the power spectrum is divided by a smooth fit to enhance the BAO feature.

Parameter	WMAP5	WMAP5 + CMB	WMAP5 + LRG	WMAP5 + SN	WMAP5 + BAO
H_0	71.9 ± 2.7	72.5 ± 2.6	69.4 ± 1.6	69.6 ± 1.7	70.1 ± 1.5
$\Omega_b \times 10^2$	4.40 ± 0.30	4.30 ± 0.30	4.71 ± 0.12	4.672 ± 0.20	4.6 ± 0.12
Ω_m	0.214 ± 0.027	0.207 ± 0.026	0.289 ± 0.019	0.238 ± 0.018	0.278 ± 0.018
τ	0.087 ± 0.017	0.086 ± 0.017	0.087 ± 0.017	0.083 ± 0.016	0.086 ± 0.016
n_s	0.963 ± 0.014	0.960 ± 0.014	0.961 ± 0.013	0.959 ± 0.014	0.961 ± 0.013
σ_8	0.796 ± 0.036	0.783 ± 0.035	0.824 ± 0.025	0.820 ± 0.028	0.813 ± 0.028

Table 2.1 Cosmological parameters constraints for a flat, power-law LCDM model, for the following data sets and data sets combinations WMAP5 [20, 6], WMAP5 + CBI, VSA, and ACBAR [39, 5, 23], WMAP5 + power spectrum of SDSS LRG galaxies [41], WMAP5 + Union Supernovae sample [21], WMAP5 + Baryon Acoustic Oscillations (BAO)[38].

be polarized; but it was only in 2002 that the DASI collaboration[22, 24] first statistically detected the CMB anisotropies in polarization, although WMAP in 2003 had detected the cross correlation between temperature and polarization [2, 19]. In 2006, WMAP provided the first full sky map of CMB polarization [32]. The temperature quadrupole at the surface of last scatter generates polarization through Thompson scattering off free electrons. During the evolution of the Universe, free electrons are available to polarize the CMB radiations twice. The first time is at the last scattering surface (because recombination is not instantaneous). The second time in the evolution of the Universe where free electrons can generate polarization is at the end of the dark ages (during reionization by the first stars).

Keeping this in mind, the observed properties of the polarization pattern are the result of different physical processes, depending on scale (see Fig. 2.3). On small sub-horizon scales, the local dipole seen by the free electrons at the last scattering surface is given by local (primordial) density perturbations, yielding a radial (tangential) pattern around hot(cold) spots.

On super-horizon scales, the dipole is created by velocities generated from adiabatic perturbations, thus polarization anisotropies on these scales should be anticorrelated to the temperature anisotropies (the extrema of density corresponds to minimum of velocity fluctuations and vice versa [47]). This was clearly seen in the WMAP data [19], and although this anticorrelation is not useful to constrain cosmological parameters in the standard cosmological model, it is an important confirmation of the assumed paradigm [33].

On very large scales $\ell < 10$, the polarization signal is dominated by the reionization signature. Recall that in the standard LCDM cosmology, all reionization physics effects on the CMB are summarized by a single parameter τ, the optical depth to

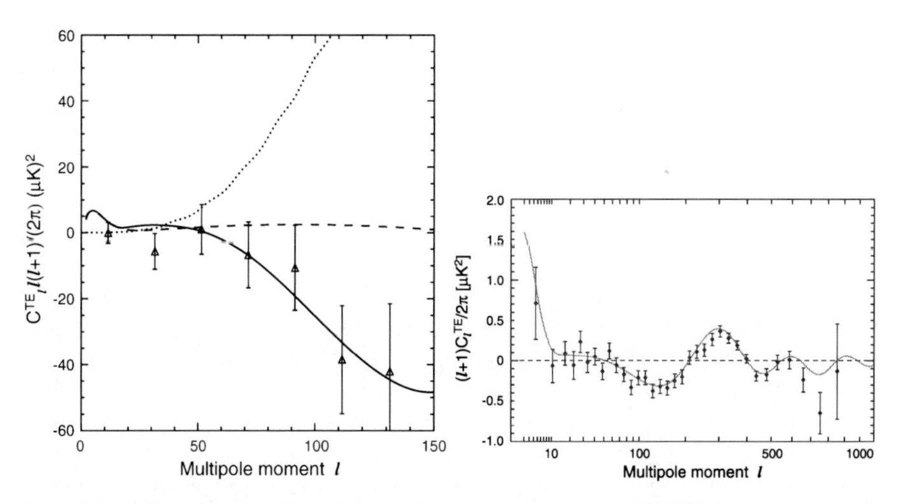

Fig. 2.3 Left: Temperature-polarization angular power spectrum from WMAP 1st year data (symbols, plotted in bandpowers of $\Delta\ell = 10$. The large-angle TE power spectrum predicted in primordial adiabatic models (solid line), primordial isocurvature models (dashed line), and causal scaling seed models (dotted line) is shown. Right:The WMAP 5-year-TE power spectrum. The green curve is the best-fit theory spectrum from the WMAP5 data. The clear anticorrelation between the primordial plasma density (corresponding approximately to T) and velocity (corresponding approximately to E) in causally disconnected regions of the sky indicates that the primordial perturbations must have been on a superhorizon scale. The enhanced correlation at large angular scales is the reionization feature.

the last scattering surface, and quantities are computed using that assumption of instantaneous (step) reionization. Even if we are not interested in reionization, this parameter cannot be neglected: in fact from temperature data alone, the determination of the primordial power spectrum spectral slope, n_s, is severely degenerate with value for τ. While for WMAP 1st year data, the polarization signal was not good enough to disentangle the two parameters, with WMAP 3 and 6-years-data, this degeneracy is lifted.

While in the standard LCDM model, primordial perturbations are scalar perturbations (vorticity modes are expected to decay rapidly), the inflationary paradigm is also expected to generate a stochastic background of gravity waves. The amplitude, such as the gravity waves in the background, is closely related to the energy scale of inflation. As CMB photons travel in a metric perturbed by gravity waves, the stretching of space-time caused by gravity waves cause the photons to produce a quadrupole intensity distribution. Thompson scattering thus leads to polarization.

It has been shown that the CMB polarization pattern on the sphere can be decomposed in two modes called E and B modes that are analogous to the E and B field in electromagnetism, with E being the curl-free and B being the curl components [16, 48]. Density perturbations, to linear order in perturbation theory[1], cannot produce B-mode polarization, but gravity waves can. This can be understood by considering a single Fourier mode of the density perturbation: being the density perturbation, a scalar field has some symmetries: a rotational symmetry around the **k** vector of the Fourier mode and reflection along any plane that include the **k** vector (no curl). So the polarization pattern must also satisfy this symmetry. This is not the case for the gravity waves (tensor-model) perturbations. The primordial gravity waves amount is parameterized by r, the so-called tensor to scalar ratio. Since most models predict power-law power spectra for tensor and scalar perturbations and also a relation between the scalar and the tensor spectral slope, r is defined as the ratio between the two power spectra at a given k (usually 0.002 Mpc/h).

Unfortunately, the galaxy around us (and our detectors) emits strong polarized light, and it emits roughly equally E and B modes. Any (primordial) polarization measurement must take into account the polarized emission from foregrounds, even when applying a galactic cut to the maps. The effect of this on the angular power spectrum of CMB polarization is shown in Fig. 2.4. Foreground emission has a frequency dependence distinct from that of the CMB, palso, for galactic foregrounds, it is expected to have a spatial pattern somewhat related to the structure and properties of our own galaxy. Ultimately, foreground subtraction relies on these two features, and it often relies on templates constructed from a combination of data at different frequencies and from modeling and our knowledge about the structure of our galaxy (see also Sec. 2.3.5).

[1] Perturbations at the CMB epoch are very small and linear theory is an excellent approximation.

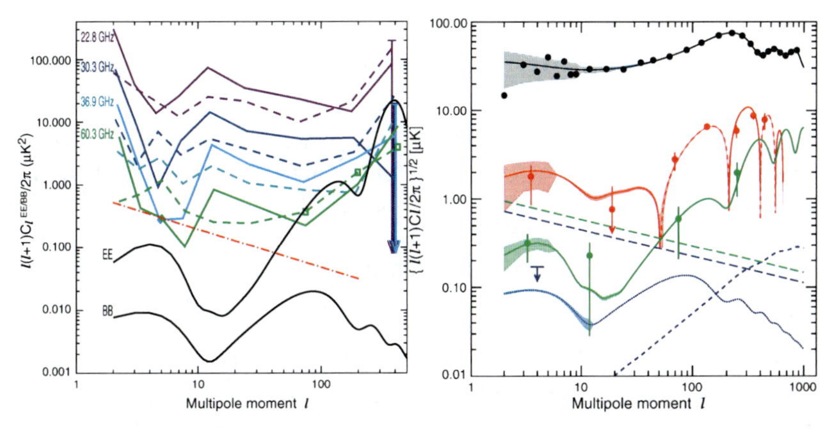

Fig. 2.4 Left: (From WMAP3 years data) The absolute value of the EE (solid, violet through green) and BB (dashed, violet through green) polarization spectra as s function of frequency for the region outside the Galactic mask. The best-fit LCDM model with $\tau = 0.09$ and an additional tensor contribution with $r = 0.3$ is shown in black. Right: Plots of signal for TT (black), TE (red), and EE (green) for the best-fit LCDM model. The dashed line for TE indicates areas of anticorrelation. The cosmic variance is shown as a light swath around each model. The blue line shows the BB contribution for a model with $r = 0.3$, and the blue short-dashed line show the contribution from gravitational lensing which moves power from E to B. The foreground contribution from synchrotron and dust at 65 GHz (close to the minimum emission) is shown as long-dashed lines: green for EE and blue for BB. From [32].

2.2.1 *Testing Inflation: Status and the Prospects*

The inflation paradigm was introduced and formulated in the 80s it postulates a period of accelerated expansion in the very early Universe, driven by a slowly rolling -potential dominated- scalar field; with this, inflation solves the problems of the classical big bang theory: the horizon problem and the flatness problem. In addition, it gives a natural set up for cosmological perturbations: they arise from quantum fluctuations stretched by the expansion which then evolve classically. Although the Universe becomes opaque beyond the last scattering surface ($z \sim 1100$), there are ways of seeing indirectly beyond redshift 1100. Information on the shape of the inflaton potential is enclosed in the shape and amplitude of the primordial power spectrum of perturbations. Information about the energy scale of inflation (the height of the inflaton potential), however, can only be obtained by the addition of B-modes polarization amplitude.

In general, the flatness, horizon, and homogeneity of the Universe gives an "observational" constraint of a number of inflationary e-foldings greater than 50. This requires the potential to be flat enough to generate enough e-foldings, implying that not every scalar field can be the inflaton. Detailed measurements of the shape of the primordial power spectrum can rule in or out different potentials.

Let us start by reviewing the generic predictions of inflation and the current observational status on these, we will then move on to more quantitative tests. Generic predictions of inflation are as follows:

- Flat Universe. The latest measurements using WMAP 5-years-data and H_0 constraints are [20] $-0.052 < \Omega_k < 0.013$ at the 95% confidence level and from WMAP 5 year + clustering of LRG galaxies for SDSS DR5[41] $-0.027 < \Omega_k < 0.003$ at the 95% confidence level.
- Gaussianity: [20] reports constraints on the local non-Gaussianity parameter f_{NL} consistent with Gaussianity: $-9 < f_{\rm NL} < 111$ (for equilateral type of non-Gaussianity, the constraint is $-151 < f_{NL}^{EQ} < 253$).
- Power spectrum nearly scale invariant: WMAP 5-years-data yields: $n_s = 0.963 \pm 0.015$ and the combination with LRG yields: 0.959 ± 0.014 [41].
- Adiabatic initial conditions.
- Super-horizon fluctuations. The large-scale anticorrelation of the temperature-E-mode polarization cross correlation indicates that initial conditions are consistent with being adiabatic and super-horizon (upper limits can be imposed on extra nonadiabatic contributions).

We, therefore, conclude that WMAP5 observations (and in general CMB observations alone and in combination with external data sets) are consistent with the simplest inflationary models. The near-scale invariance was evident since the time of COBE, indications for flatness were present in TOCO, Boomerang, Maxima, Archeops data well before WMAP. Interesting constraints on Gaussianity and on adiabatic and super-horizon fluctuations, however, became available only with WMAP data. The precision of current data sets enables us to go beyond this and to critically test specific inflationary models. It is interesting to note that the constraints one obtains depend on how the reconstruction is performed and on the chosen parameterization. The two main approaches used thus far are (1) Since simple inflationary models predict that the primordial power spectrum should be close to a power law, parameterize the primordial power spectrum shape as [2]

$$P(k) = A\left(\frac{k}{k_0}\right)^{n_s(k_0)+\frac{1}{2}\frac{dn_s}{d\ln k}\ln\left(\frac{k}{k_0}\right)}, \tag{2.1}$$

where A denotes the primordial amplitude, k_0 is called the *pivot point*, $n_s(k_0)$ is the spectral slope at the pivot, and $dn_s/d\ln k \equiv d^2 \ln P(k)/d\ln k^2$ is called running of the spectral index and it is assumed not to depend on scale. It is customary to use $k_0 = 0.002 Mpc^{-1}$, but sometimes $k_0 = 0.05 Mpc^{-1}$ is also used. However, depending on the data set used, there will be a k_* where the errors on amplitude and n_s

[2] Sometimes, the primordial amplitude is rescaled and related to the primordial curvature perturbations as follows: $\Delta^2_R(k_0) = 2.95 \times 10^{-9} A(k_0)$, where $\Delta^2_R = 1/(2\pi^2)k^3 P_R(k)$ and the curvature perturbation R is related to the potential perturbation Φ by $\Phi = -3/5R$. Also, recall that the density power spectrum and the potential power spectrum are related by $P_\Phi \propto 1/k^4 P_\delta$.

are decorrelated, thus some authors do not stick with the two "customary" options above. To convert between any two pivot conventions, say, k_1 and k_0:

$$A(k_1) = A(k_0)\left(\frac{k_1}{k_0}\right)^{n_s(k_0)+\frac{1}{2}\frac{dn_s}{d\ln k}\ln\left(\frac{k_1}{k_0}\right)} \tag{2.2}$$

and, since $n_s \equiv d\ln P(k)/d\ln k$

$$n(k_1) = n(k_0) + \frac{dn_s}{d\ln k}\ln(k_1/k_0). \tag{2.3}$$

The parameters n_s, $dn_s/d\ln k$, and r can be related to the so-called potential slow-roll parameters as outlined in e.g., [28, 27, 26]

$$\Delta_R^2 = \frac{V}{M_P^4}\frac{1}{24\pi^2\varepsilon_V}, \tag{2.4}$$

$$r = 16\varepsilon_V, \tag{2.5}$$

$$n_s - 1 = -16\varepsilon_v + 2\eta_V, \tag{2.6}$$

$$\frac{dn_s}{d\ln k} = -\frac{2}{3}[(n_s-1)^2 - 4\eta_V^2] - 2\xi_V, \tag{2.7}$$

where $m_P = 1.22\times 10^{19} GeV$ denotes the Planck mass, $M_P = m_P/\sqrt{8\pi}$ and

$$\varepsilon_V \equiv \frac{M_P^2}{2}\left(\frac{V'}{V}\right)^2, \tag{2.8}$$

$$\eta_V \equiv M_P^2\frac{V''}{V}, \tag{2.9}$$

$$\xi_V \equiv M_P^4\left(\frac{V'V'''}{V^2}\right). \tag{2.10}$$

In single-field inflationary models, there is a consistency relation relating the tensor tilt to the tensor amplitude itself $n_t = -r/8$.

(2) The other approach is to use the so-called Hubble slow-roll parameters. This approach goes under the name of slow-roll reconstruction. For inflation driven by a single, minimally coupled scalar field, the equation of motion can be written as a function of the inflaton field ϕ e.g., [17, 9, 33]:

$$\dot{\phi} = -\frac{m_P^2}{4\pi}H(\phi) \tag{2.11}$$

$$\left[H'(\phi)\right]^2 - \frac{12\pi}{m_P^2}H^2(\phi) = -\frac{32\pi^2}{m_P^4}V(\phi) \tag{2.12}$$

from where the Hubble slow-roll parameters can be defined by the infinite hierarchy of differential equations:

$$\varepsilon_H(\phi) \equiv \frac{m_P^2}{4\pi}\left[\frac{H'(\phi)}{H(\phi)}\right]^2, \tag{2.13}$$

$${}^{\ell}\lambda_H \equiv \left(\frac{m_P^2}{4\pi}\right)\frac{(H')^{\ell-1}}{H^{\ell}}\frac{d^{(\ell+1)}H}{d\phi^{(l+1)}} : \ell \geq 1. \tag{2.14}$$

This hierarchy can be truncated at order N so that ${}^{\ell}\lambda_H = 0$ for all $\ell > N$; then, the hierarchy can be solved explicitly and an exact expression for the inflaton potential can be obtained. The Hubble slow-roll parameters then are ε_H as defined above, $\eta_H =^1 \lambda_H$ and $\xi_H =^2 \lambda_H$. [25] stated that this is then equivalent to parameterize $H(\phi)$ as a polynomial of order $N+1$ $H(\phi) = H_o(1 + A_1\phi + A_2\phi^2 + ...)$, with the ${}^{\ell}\lambda_H$ related to the A_ℓ coefficients and that thus the corresponding potential $V(\phi)$ can also be written in term of these coefficients. Still it should be clear that while there are relations between the potential slow-roll parameters and the Hubble slow-roll parameters, the two parameterizations are different and using flat priors on one parameterization does not correspond to using flat priors on the other. If a bayesian analysis is performed (as it is customary in cosmology) and if simple priors on the minimum required number of e-folding is then applied, the constraints on observable quantities n_s and $dn_s/d\ln k$ need not coincide as illustrated in Fig. 2.5.

Different inflationary models "live" in different parts of this parameter space. However, note that by measuring the shape of the scalar perturbations spectrum, one cannot extract information about the absolute scale of inflation. In order to do that, constraints on the tensor to scalar ratio r are needed. This is the main goal of the

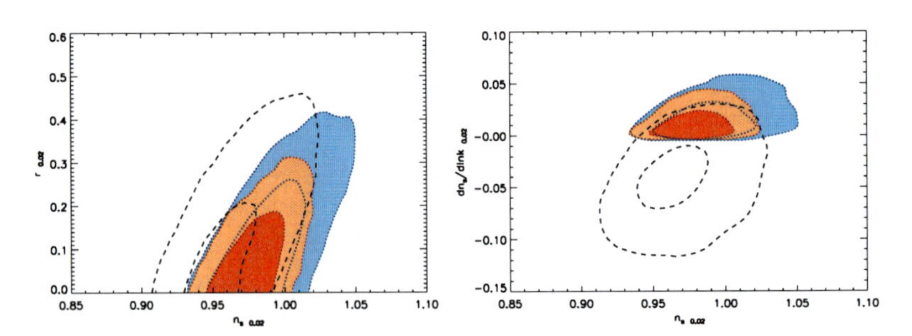

Fig. 2.5 The joint 68% (inner) and 95% (outer) bounds on the power law spectral parameters at the fiducial scale $k_0 = 0.02$ Mpc^{-1} obtained by transforming the constraints on the Hubble slow-roll parameters into this parameter space. The blue constraints are derived from WMAP 5-year-data alone, and the red constraints are derived from the WMAP5 + SN data combination. The dotted contours come from slow-roll reconstruction applying an e-fold prior assuming a reheating temperature $T_{reh} > 10$ TeV. For comparison, the dashed contours show an analysis using the empirical power-law prescription in terms of n_s , r, and $dn_s/d\ln k$ at a pivot scale of 0.02 Mpc^{-1}, using WMAP5 data. From [34].

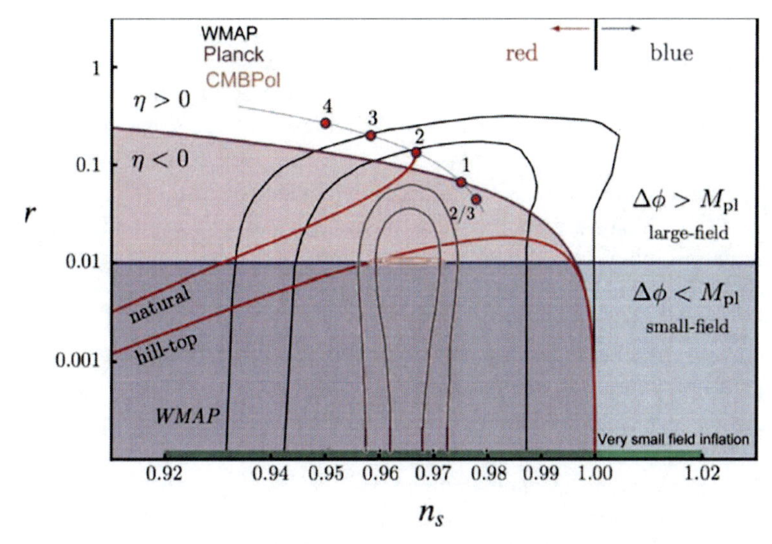

Fig. 2.6 Present and future constraints on the n_s r plane along with classification of inflationary models. The green contours show the WMAP current constraints, the dark red curves show the forecasted Planck constraints, and the light red oval curves show the forecasted CMBPol constraints for a fiducial model of $r = 0.01$ and $n_s = 0.97$. $r > 0.01$ requires a displacement of the field during inflation larger than M_P (large field models), small field models need a smaller displacement (smaller r). Also shown are the predictions of few representative models of single-field slow-roll inflation: chaotic inflation $\lambda\phi^p$ for general p (thin solid line) and for few representative values (bullets); natural inflation where the inflaton potential has to form $V_0[1 - cos(\phi/\mu)]$, hill top inflation with potential $V_0[1-(\phi/\mu)^2]+\ldots$, and models with very small tensor amplitudes (very small filed inflation).

next generation of CMB-polarization missions. A good introduction and theoretical motivation for such an experimental effort can be found e.g., in [1]. Here, in Fig. 2.6, we report an adapted figure from that paper with present and future constraints on the r n plane and some inflationary models.

2.2.2 Beyond the Standard Cosmological Model

While the standard LCDM, six-parameters model is extremely successful; cosmological data can be (and should be) used also to test for deviations from this minimal model. In the previous section, we have seen that an a couple of extra parameters were introduced: the running of the spectral index and the tensor to scalar ratio. Thus far, there are only upper limits on the tensor-to-scalar ratio: $r < 0.43$ for WMAP5 years alone at 95% CL, $r < 0.22$ from WMAP + BAO + SN also at the 95% CL. On the other hand, there have been claims for detection of nonzero running though not at high statistical significance. Clearly, a large running for a nearly scale-invariant spectrum implies that ξ is nonnegligible while ε and η are still relatively small (or fine tuned so that the cancel out in the expression for n_s). In any case, it implies that the potential is not flat (or, equivalently that the Hubble parameter during inflation is not nearly constant) and so that slow roll cannot be sustained and such model, if it is a standard, single field, inflation, will not have many e-foldings. At least 50 e-foldings are needed to explain the flatness problem and the horizon problem: a large running would imply troubles for the simplest standard slow-roll inflationary scenarios.

The flatness assumption can also be relaxed to obtain constraints on the curvature parameter Ω_k, which were briefly discussed in §2.2.1, as well as possible deviations from Gaussian initial conditions. Note that the CMB alone is not sensitive to Ω_k and to H_0 at the same time: a CMB constraint on H_0 can be obtained assuming a flat universe and a constraint on Ω_k can be obtained only by combining CMB with H_0 measurement (or other measurements of the low redshift expansion history such as the supernovae luminosity distance).

There are also other additional cosmological parameters that can be considered. Let us begin by relaxing the assumption that the dark energy is a cosmological constant. The simplest (and probably dummest) way to do that is to assume that the dark energy has an equation of state parameter $p_{DE}/\rho_{DE} = w \neq -1$, where p_{DE} denotes the dark energy pressure and ρ_{DE} its density; a cosmological constant has by definition $w = -1$. w is then assumed to be constant. In this case, the dark energy density evolves slightly in redshift so that $\rho_{DE} = \rho_0(1+z)^{3(1+w)}$. Again, the CMB alone is not sensitive to the w parameter, but constraints can be obtained in combination with low redshift probes such as supernovae, BAO, and H_0 determination.

In the standard model, one assumes that there are three massless neutrino species, so limits can be imposed from cosmological data on the sum of neutrino masses or on the effective number of neutrino species. Cosmology is sensitive to the *physical* energy density in relativistic particles in the early Universe ω_{rel} which, in the standard model (for cosmology), includes only photons and neutrinos $\omega_{rel} = \omega_\gamma + N_{eff}\omega_\nu$, where ω_γ is the energy density in photons and $\omega_\nu$is the energy density in one active neutrino. As ω_γ is extremely well constrained by CMB observations, ω_{rel} can be used to constrain neutrino properties; deviations from $N_{eff} = 3.046$ would signal nonstandard neutrino features or additional relativistic relics. Free-streaming relativistic particles affect the CMB mainly through their relativistic energy density, which alters the epoch of matter-radiation equality. Because the redshift of matter-radiation equality is well constrained by the ratio of the third to first CMB peak height, this effect defines a degeneracy between $\Omega_m h^2$ and N_{eff}: $1+z_{eq} = (\Omega_m h^2)/(\Omega_\gamma h^2)[1+0.227N_{eff}]^{-1}$. Any additional constraints on Ω_m and or h from external data set will the break such degeneracy. Neutrinos with mass ≤ 1 eV become nonrelativistic after the epoch of recombination probed by the CMB so that allowing massive neutrinos alters matter-radiation equality for fixed $\Omega_m h^2$. Their radiation-like behaviour at early times changes the expansion rate, shifting the peak positions, but this is somewhat degenerate with other cosmological parameters. Therefore, WMAP5 alone constrains $\sum m_\nu < 1.3$ eV at the 95% confidence interval in a flat LCDM universe. After the neutrinos become nonrelativistic, their free-streaming damps power on small scales and therefore modifies the matter power spectrum in the low-redshift universe. This effect can be tested by low-redshift data such as galaxy clustering.

Parameter	WMAP5	WMAP5 + H_0	WMAP5 + LRG	WMAP5 + SN
Ω_k (95% CL)	–	$0.014 < \Omega_k < 0.009$	$0.027 < \Omega_k < 0.003$	$-0.0316 < \Omega_k < 0.0078$
w (68% CL)	–	-1.12 ± 0.12	0.75 ± 0.15	$-1.04^{+0.066}_{-0.064}$
N_{eff} (68% CL)	–	4 ± 2	$4.8^{+1.8}_{-1.7}$	–
$\sum m_\nu$ (95% CL)	1.3 eV	< 0.6 eV	< 0.62 eV	–

Table 2.2 Cosmological parameters constraints for one-parameter deviations from the standard LCDM model for the following data sets and data sets combinations WMAP5 [20, 6], WMAP5 + H_0 determination from [42], WMAP5 + Power spectrum of SDSS LRG galaxies [41], WMAP5 + Union Supernovae sample [21].

Constraints on these additional parameters from a combination of several data sets is reported in Table 2.2.

It is clear that the combination of CMB and large-scale structure data is very powerful not only in constraining cosmological parameters but also in testing the standard cosmological model.

In the next section, I will try to explain or at least give an idea of how this information is extracted from the data. Textbooks and classical cosmology courses concentrate in illustrating how the cosmological information is encoded in the CMB maps and power spectra. There are also publicly available codes –CMBFAST, CAMB, CMBEASY– that given a set of cosmological parameters will provide the theory angular temperature power spectrum expected. I will, therefore, take the complementary approach and try to illustrate the challenges and outline solution for the issue of how to estimate the angular power spectrum of the primary CMB and how to constrain cosmological parameters from it.

2.3 CMB: How is the Information Extracted?

In the standard cosmological model and in standard models of inflation, the initial perturbations were nearly Gaussian. For Gaussian initial conditions, the power spectrum completely characterizes the statistical properties of the CMB temperature fluctuations. Therefore, the information enclosed in the megapixel CMB maps is *compressed* into a CMB angular power spectrum:

$$C_\ell = \frac{1}{(2\ell+1)} \sum_m |a_{\ell m}|^2, \tag{2.15}$$

where $a_{\ell m}$ denote the coefficients of the spherical harmonics expansion of the temperature fluctuations. Unfortunately, one can never measure this directly. Several real-world effects come into play, which we will be briefly outlined below. Once these effects can be accounted for, the cosmological information is thus encoded in CMB the angular power spectrum. Note that there are temperature spectra, C_ℓ^{TT}, polarization spectra C_ℓ^{EE} (in principle also C_ℓ^{BB}), and cross spectra C_ℓ^{TE}. C_ℓ^{TB} and C_ℓ^{EB} should vanish in a universe that conserves parity [16, 49, 48]; thus, these are used mostly to check for systematic effects (but see e.g., discussion in [20]). There are publicly available codes –CMBFAST, CAMB, CMBEASY– that given a set of cosmological parameters will provide the theory angular temperature power spectrum expected. So here, I will try to illustrate the challenges and outline solution of how to estimate the angular power spectrum of the primary CMB and I will illustrate how to constrain cosmological parameters from it.

Note that in principle, it is also possible to extract cosmological information directly from CMB maps without going through the C_ℓ measurement. In fact, the likelihood function for the temperature fluctuation observed by a noiseless experiment with full sky coverage has the form:

$$\mathscr{L}(\mathbf{T}|\theta) \propto \frac{\exp[-(\mathbf{T}\mathbf{S}^{-1}\mathbf{T}/2]}{\sqrt{det\mathbf{S}}} \tag{2.16}$$

where $\mathbf{T}$ denotes the temperature map, θ denotes the array of cosmological parameters and the elements of the signal matrix $\mathbf{S}$ are given by $S_{ij} = \sum_\ell (2\ell+1) C_\ell^{th}(\theta) P_\ell(n_i n_j)/(4\pi)$. However, in practice, this procedure is slow and prohibitively expensive for high-resolution maps, and it has thus far been used only in exceptional cases (see e.g., [6] and references therein).

2.3.1 Real-World Effects

A realistic CMB map is affected by instrumental noise, finite resolution, and, because of the galactic foregrounds, not the full sky can be used for cosmology. Of course, additional real-world effects may as well be present (correlated inhomogeneous noise, anisotropic beam etc.,), but the effects of noise (uniform) of beam and of sky cut need to be included even when forecasting the performance of an idealized experiments with some given (ideal) characteristics. Here, we start by follwing [18].

2.3.1.1 Noise

Every detector has a noise that is then superimposed to the signal. Today's detectors are so good that they are photon-noise limited, thus the only way to reduce the noise is to use multiple detectors and even to make detectors arrays. In the presence of noise, the measured temperature in a given direction in the sky T_{meas} is given by $T_{meas} = T + noise$, thus $a_{\ell m} = a_{\ell m}^{signal} + a_{lm}^{noise}$. Since the noise is not expected to correlate with the signal, this is a superposition of two independent processes, thus the resulting power spectrum is the sum of the two power spectra: $C_\ell^{meas} = C_\ell^{signal} + C_\ell^{noise}$. Thus, this is a biased estimator of the signal because, in general, $\langle C_\ell^{noise} \rangle \neq 0$: $\langle a_{lm}^{noise} \rangle = 0$ but $\langle a_{lm}^{noise\,2} \rangle \neq 0$. Here and hereafter, $\langle . \rangle$ denotes ensamble average. There is one way to make $C_\ell^{noise} = 0$: computing C_ℓ^{meas} from the cross correlation of different (uncorrelated) detectors. We call these cross-power spectra, in contrast to auto-power spectra. One needs to subtract the noise from the auto-power spectra. For an experiment with a detector sensitivity of s (usually expressed in $\mu k\sqrt{s}$), the rms per sky (or map) pixel is given by $\sigma_{pix} = s/\sqrt{t_{pix}}$, where t_{pix} is the observing time spent on each pixel. Note that for detecting a polarized signal, the instrument needs to "split the photons" and so the sensitivity s is at least a factor $sqrt2$ worst for a polarization-sensitive detector than for one sensitive to T only (all other characteristics being equal). Thus, for an experiment with negligible beam smearing (i.e., beam smearing much smaller than the pixel size), the noise spectrum per multipole becomes $w = (\sigma_{pix}^2 \Omega_{pix})^{-1}$, where Ω_{pix} s the pixel solid angle: $\Omega_{pix} = 4\pi f_{sky}/N_{pix}$, where N_{pix} is the number of pixels in the observed map. Thus, $C_\ell^{noise} = w^{-1}$.

2.3.2 Beam

Every experiment has a finite resolution, in optical astronomy this is called point spread function; in radio, it is called beam and is characterized by its solid angle Ω_{beam}. The resulting maps are then pixelized and the pixel size (or Ω_{pix}) must be matched to the beam size. The effect of this, if we neglect the effect of noise, is that the temperature in the sky at a given position (pixel) is given by $T_i = \int d\Omega' T(\hat{n})$ $b(|\hat{n} - \hat{n'}|)$, where b denotes the beam profile. This is often assumed to be (or is actually very close to a) Gaussian, in this case its full width at half maximum (FWHM) specifies $\sigma_b = 0.425FWHM$, thus $C_\ell^{meas} = C_\ell^{signal} \exp[-\ell 2\sigma_b^2]$.

Note that the noise is intrinsic to the detector and thus does not "see" the beam, therefore: $C_\ell^{meas} = C_\ell^{signal} \exp[-\ell 2\sigma_b^2] + C_\ell^{noise}$; but if one now deconvolves for the beam to estimate the signal, one obtains: $C_\ell^{meas} = C_\ell^{signal} + C_\ell^{noise} \exp[\ell 2\sigma_b^2]$. That is, the noise appears to "explode" exponentially at high ℓ, as shown in Fig. 2.7.

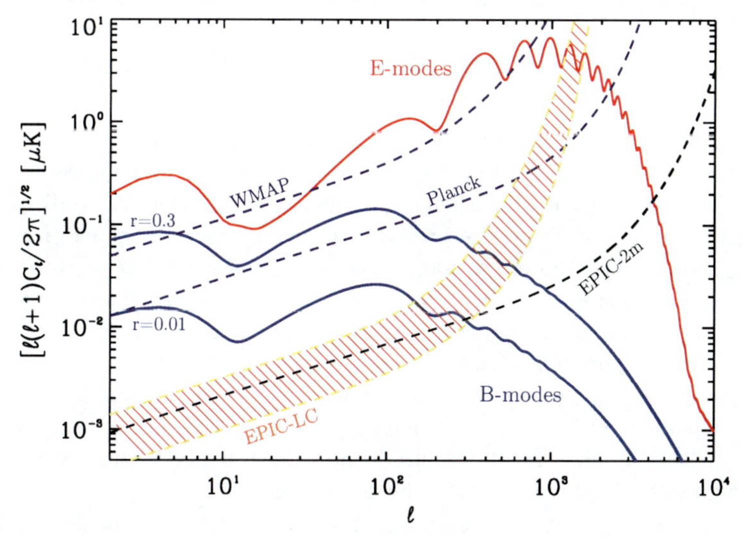

Fig. 2.7 E- and B-mode power spectra for a tensor-to-scalar ratio saturating current bounds, r = 0.3 and for r = 0.01. Also, shown are the experimental sensitivities for WMAP, Planck, and two different realizations of CMBPol (EPIC-LC and EPIC-2m). (Figure from[1]).

2.3.3 Sky Cut

The galactic emission dominates the CMB signal, thus even full sky experiments cannot use the entire sky for cosmological analysis. In the presence of a sky cut (i.e., regions of the sky with pixels set to zero "by hand" either because of contamination or for lack of data), the measured spherical harmonic coefficients are related to the underlying temperature fluctuation filed by a convolution with the mask: $\widetilde{a_{\ell m}} = \int d\Omega_n' T(\hat{n}) W(\hat{n}) Y^*_{\ell m}(\hat{n})$ and for a pixelized map this becomes $\widetilde{a_{\ell m}} = \Omega_p \sum_p T_p W_p Y^*_{\ell m}(\hat{n}_p)$. If one ignores the effect and computes the C_ℓ using $\widetilde{a_{\ell m}}$ instead of $a_{\ell m}$, one obtains the so-called pseudo-C_ℓ ([13]). Clearly, $\widetilde{C_\ell} \neq C_\ell$, however, it can be shown that $\langle \widetilde{C_\ell} \rangle = \sum_\ell G_{\ell\ell'} \langle C_{\ell'} \rangle$. The function G describes the mode coupling introduced by the sky cut and is related to the coefficients of the spherical harmonic transform of the mask $W_{\ell m}$ and to the mask power spectrum W_ℓ by $W_\ell = (2\ell+1)^{-1} \sum_m |W_{\ell m}|^2$;

$$G_{\ell\ell'} = \frac{2\ell'+1}{4\pi} \sum_{\ell''} (2\ell''+1) W_{\ell''} \begin{pmatrix} \ell & \ell' & \ell'' \\ 0 & 0 & 0 \end{pmatrix} . \tag{2.17}$$

If one could invert G and if one could say that $\langle C_\ell \rangle \equiv C_\ell$, then one could estimate $C_\ell = \sum_{\ell'} G^{-1}_{\ell\ell'} \widetilde{C_{\ell'}}$. This is the gist of the paper of [13]. Note that the same description is valid also if the original mask $W(\hat{n})$ does not only take values 0 or 1 but also if it takes other continuous values in between: usually a *weight* is used, which could be related for example to the number of time a pixel was observed or to the pixel noise in case of nonhomogenous noise etc.

2.3.4 How Do You Make a CMB Map in the First Place?

The first question one may ask before starting on a cosmological analysis with CMB data is: how do the raw data look like and do you get them in the form of a map? The "raw" data are usually collected as time-ordered data (TOD). A TOD is shown in Fig. 2.8, which is taken from reference [30] from the Archeops experiment. In the x-axis, there is time t and in the y-axis, there is the signal recorded by the detector d. As the beam scans the sky with time, following what is called a scanning strategy, spatial frequencies along the scan map into temporal variations in the detector (shown in the y-axis). One needs also to store some extra information (pointing) telling where in the sky one was pointing at each time t to be able to then reconstruct a sky map. So how does one makes a map from a TOD like the one shown in Fig. 2.8? I will give here a brief introduction, this is a broad subject where many smart algorithms are continuously being developed.

The goal here is to make maps from TOD. The problem can be recast in terms of operation of matrices on vectors.

$$d = g[M(T+T_{fg})] + g_{det} n + c, \tag{2.18}$$

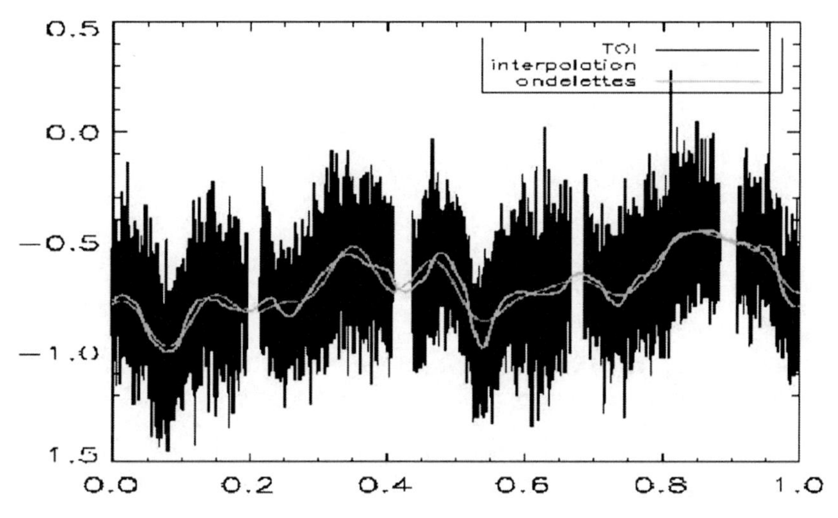

Fig. 2.8 Example of a TOD: In the x-axis, there is time t and in the y-axis, there is the signal recorded by the detector d. As the beam scans the sky with time, following what is called a scanning strategy, spatial frequencies along the scan map into temporal variations in the detector (shown in the y-axis). Note the low-frequency signal, which is mostly due to variations in gain, atmospheric loading, calibration drift, etc., and high-frequency signal (CMB and noise). The low-frequency signal should be taken out: in the figure, the low-frequency signal estimated by two different procedures are shown.[A&A 459, 987-1000 (2006)]Macias Perez, Bourachot; fig 11]

d is the raw TOD vector, its length is N_d, it can have additional information associated, such as temperature, humidity (for ground-based observations), etc. Note that element i, d_i and element $i+1$ are usually separated by a tiny fraction of a second. It is easy to figure out the sheer size of the problem by thinking that observations may last for years.

g denotes the gain. This is also an N_d-long vector. The gain is expected to vary much more slowly than the signal. In g, the total gain, there are several contributions: detector, receiver, etc. ($g = g_{det} \times g_{rec.} \times ...$), g_{det}, e.g., denotes the gain of the detector only.

c denotes the baseline vector also N_d-long; this also varies more slowly than the signal, and it depends on the details of the instrument.

T is the CMB temperature map vector. It is a vector N_t-long, where N_t is the number of pixels in the map. For completeness, I have included a foreground component to the temperature signal t_{fg}, which has a frequency dependence that differs from that of the CMB blackbody. Some map-making approaches use this information at this stage to separate out foregrounds (see Sec. 2.3.5)

M is the pointing matrix, its dimensions are $N_t \times N_d$ and it is a sparse matrix.

n is the noise vector also N_d-long; it should be that $\langle n \rangle = 0$, but one can characterize the noise via the noise matrix: $N = \langle nn^T \rangle$. The noise is often gaussian and piecewise stationary in the time domain (it varies much more slowly than the signal, so this should be a good approximation). It can thus be described by its power spectrum of correlation function N.

One solves for the noise correlation as follows:

$$N(\Delta t) \begin{array}{ll} = const & if\ \Delta t = 1 \\ = F(t) & if\ 1 < \Delta t < \Delta t_{max} \\ = 0 & if\ \Delta t > \Delta t_{max}, \end{array} \quad (2.19)$$

$$(2.20)$$

where Δt_{max} is a time interval such that the noise is uncorrelated on scales larger than Δt_{max}. $F(t)$ is some smooth-fitting function where parameters can be fit iteratively. The same procedure can be followed for g, c etc.

Note that if Eq (2.18) were simpler: $d = Mt + n$, then one could immediately define a *maximum likelihood* estimator:

$$\hat{T} = (M^T N^{-1} M)^{-1} (M^T N^{-1} d). \quad (2.21)$$

Although $(M^T N^{-1} d)$ can be computed directly once N is characterized, $(M^T N^{-1} M)^{-1}$ can be solved by conjugate gradient with preconditioner. In practice, define $t_o = M^t N^{-1} d$ and note that $T_o = (M^T N^{-1} M)\hat{t} \equiv \Sigma^{-1}\hat{t}$. Here, Σ is the pixel-to-pixel noise correlation matrix, and by conjugate gradient, one can find $\hat{t}$ iteratively. The *preconditioner* makes the iterative process to converge faster: imagine there exist a matrix S such that $S\Sigma^{-1}$ is diagonal. Then, if you solve for $S\Sigma^{-1}\hat{T} = ST_o$, the calculation can be performed much faster.

Unfortunately, one does not start with Eq. 2.21, one starts with 2.18, so the map making (finding the solution of Eq. 2.18) must be done as an iterative process.

It is at this point that cross linking (i.e., an observing pattern with interlocking scans) really helps the process of mapmaking.

There is also the issue of calibration. Let us consider that the CMB average temperature is about 3K, but the signal we are after are variations of one part in 1000000. The response (gain) of the detector is expected to be exquisitely linear in this tiny range, but it needs to be known much more precisely than it can be calibrated in the laboratory. Any small systematic change will give a calibration uncertainty. Full sky experiments like WMAP calibrate on the dipole: the dipole (due to the combined motion of the local group toward the great attractor, the motion of our galaxy in the local group, the motion of the sun around the galaxy and of the Earth and/or the satellite) is a large signal of a well-determined dependence on the position on the sky. Partial sky experiments could calibrate on well-known point sources or planets, although now they tend to calibrate on WMAP. Calibration can be done best directly from the time-ordered data.

2.3.5 Foregrounds

Outside the galactic plane, the CMB is the dominant signal in the radio part of the spectrum. Synchrotron emission resulting from the acceleration of cosmic ray electrons in the magnetic field of the Galaxy increases at low frequencies and is highly polarized. The other dominant foreground is thermal emission from dust. This component increases with frequency and is also polarized. Other sub-dominant (but nonnegligible) foregrounds, are free-free (bremsstrahlung emission from thermal electrons) and a possible component of spinning dust. Among extragalactic foregrounds, we find point sources (i.e., radio galaxies): while the known ones and the brightest ones can be masked out, there always remain an unresolved population contaminating the signal. As CMB experiments push to higher resolution, secondary anisotropies could contaminate the primary signal. Although the CMB follows a blackbody, foregrounds show a different frequency dependence. This is the key information used to clean foregrounds from CMB maps. The cleaning can be done either at the map-making stage (see T_{fg} term in Eq. 2.18) exploiting their frequency dependence or at the map stage using foregrounds templates. Foreground cleaning is particularly crucial for polarization: the temperature signal exceeds the galactic foregrounds outside the galactic cut over a wide range of frequencies (~ 50 to ~ 100 GHz) however, the polarization signal is dominated by foregrounds at all frequencies and on all scales. Foregrounds have their specific scale and frequency dependent and their specific sky pattern: foreground cleaning, especially for polarization should be done either at the TOD stage or at the map stage but NOT at the power spectrum stage. The only exception for this may be the unresolved point sources contribution. Bright, resolved point sources can be taken out by masking the relevant pixels. If a point source catalog is available, it can be used to build a better point source mask. Eventually, for numerous and dim sources, either external catalogs are not available or the sources large number along with the experiment pixel size may mean that to mask then one would need to masks most of the surveyed sky. In this case, one may want to leave the unresolved sources signal in the maps and subtract it at the power spectrum level. If the point sources statistics clustering can be modeled by a Poisson distribution (i.e., they are assumed to be unclustered), their contribution to the power spectrum is readily computed. In addition their frequency dependence helps in modeling and subtracting their contribution. Note, however, that not all point sources are well modeled by a population without intrinsic clustering. This may well be a major limitation for high-resolution CMB experiments that try to observe into the damping tail and/or to observe secondary anisotropies.

Only to give an idea of the expected level of galactic foregrounds contamination, below we report the parameters describing the foregrounds power spectra as characterized by early 2009, see [7, 1] for details and references therein. The relevant equations are

For synchrotron:

$$C_\ell^{S,XY}(\nu) = A_s \left(\frac{\nu}{\nu_o}\right)^{2\alpha_S} \left(\frac{\ell}{\ell_o}\right)^{\beta_S}, \tag{2.22}$$

where X,Y denotes T, E B; $\alpha_s = -3$, $\beta_S = -2.6$, $\nu_o = 30$GHz, $\ell_o = 350$, and $A_s = 4.7 \times 10^{-5}\,\mu\text{K}^2$.

For dust:

$$C_\ell^{D,XY}(\nu) = p^2 A_D \left(\frac{\nu}{\nu_o}\right)^{2\alpha_D} \left(\frac{\ell}{\ell_o}\right)^{\beta_D} \left[\frac{e^{\frac{h\nu_o}{kT}} - 1}{e^{\frac{h\nu}{kT}} - 1}\right]^2, \quad (2.23)$$

where $\alpha_D = 2.2$, $\nu_o = 94$GHz, $\ell_o = 10$, $A_D = 1\mu$ K^2, $\beta_D^{XY} = -2.5$ and the dust polarization fraction $p = 5\%$. Note that in principle the various parameters A, α, and β do not need to be the same for T E and B.

2.3.5.1 Atmosphere

Sub-orbital experiments have the additional complication of the atmosphere. The atmosphere is opaque to electromagnetic radiation except for few atmospheric windows (in the frequency range of interest around 30 and 90 to 140 GHz), where transmission can be higher or lower depending on the water vapor presence in the atmosphere (less water vapor corresponds to better transparency). This is why a good location for ground-based CMB experiments is the South Pole: the South Pole is high (3000 m) and, despite all the ice, because of the low temperatures, the air is very dry. The high plateau in the Atacama desert is also a good place for CMB observations. Even for observations done from high and dry places and in the atmospheric windows, spatial fluctuations in the atmospheric transparency are introduced by wind, turbulence etc., and they cause cause variations in the detector timestreams, referred to as "atmospheric noise." This acts as extra noise contribution, which is expected to dominate all other sources of noise at large angular scales (low ℓ but to become increasingly less important at high ℓ). Sophisticated techniques are under developement to "solve" for the atmospheric noise contribution to the signal and subtract it out as much as possible from the TOD.

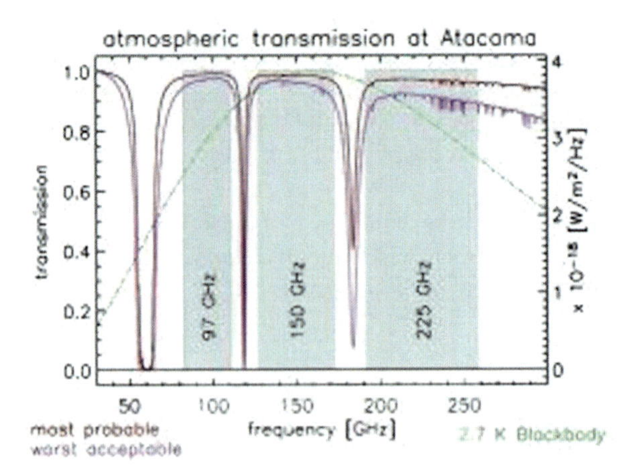

Fig. 2.9 Atmospheric transition from the Clover experiment: note the atmospheric "windows" (below 50 GHz and around 100, 150, and 220 GHz) well suited to observe the CMB.

2.3.6 Estimation of the C_ℓ

For an ideal experiment (full sky, no noise), the C_ℓ can be obtained as

$$C_\ell = \frac{\sum_m |a_{\ell m}|^2}{2\ell+1}. \tag{2.24}$$

This is the maximum likelihood estimator. In practical applications (if the noise is Gaussian), the covariance matrix of the data $\mathscr{C} = \langle \mathbf{dd}^T \rangle$, where d denotes the data (vector of the temperature fluctuation in the pixels map). $\mathscr{C} = S+N$, where N stands for the noise covariance and S is the signal covariance. Its harmonic transform is related to the C_ℓ as follows:

$$S_{\ell m}^{\ell' m'} = \langle a_{\ell m}^{meas} a_{\ell' m'}^{meas*} \rangle = \tilde{C}_\ell \delta^D(\ell m \ell' m'). \tag{2.25}$$

Note that $\tilde{C}_\ell$ here includes the effect of the beam, then

$$S_{ij} = \sum_{\ell m \ell' m'} T_\ell^m S_{\ell m}^{\ell' m'} Y_{\ell'}^{*m'}. \tag{2.26}$$

The angular power spectrum can then in principle be estimated from the map by maximizing the likelihood function: a multivariate Gaussian of the **d** data with the $\mathscr{C}$ matrix as the covariance. This operation, however, is computationally prohibitive for large maps. Various approaches have been proposed. They can be characterized in two classes:

- Exploit possible symmetries in the observational strategy and still maximize the likelihood as explained above. e.g., [31].
- Use an approximated algorithm that filters the data, compute the spectra on the masked sky, and de-bias the resulting estimate via Monte-carlo simulations, e.g., [13].

The second method is sub-optimal especially at low ℓ. A third approach has recently been proposed and has demonstrated to be very promising, see [10].

2.3.7 Likelihoods

Assuming that the CMB multipoles are Gaussainly distributed, the likelihood for an ideal, noiseless CMB experiment with full sky coverage is thus (see Eq. 2.16 and transform to spherical harmonics)

$$\mathscr{L} = \Pi_{\ell m} \frac{\exp[-1/2 \mathbf{s} \mathbf{C}^{-1} \mathbf{s}]}{\sqrt{det \mathbf{C}}}, \tag{2.27}$$

where $\mathbf{s}_\ell = (a_\ell^T, a_\ell^E, a_\ell^B)$ and a_{ell} denotes the spherical harmonic coefficients for temperature, E and B model polarization. The covariance matrix $\mathbf{C}_\ell$ is then given by

$$\mathbf{C}_\ell = \begin{pmatrix} C_\ell^{TT} & C_\ell^{TE} & 0 \\ C_\ell^{TE} & C_\ell^{EE} & 0 \\ 0 & 0 & C_\ell^{BB} \end{pmatrix}, \tag{2.28}$$

where C_ℓ denotes the angular CMB power spectrum.

Since the Universe is assumed to be isotropic, the likelihood function is independent of m and summing over it, one obtains, up to a constant:

$$-2\ln\mathscr{L} = \sum_\ell (2\ell+1)\left\{\ln\left(\frac{C_\ell^{BB}}{\hat{C_\ell}^{BB}}\right) + \ln\left(\frac{C_\ell^{TT}C_\ell^{EE}-(C_\ell^{TE})^2}{\hat{C_\ell}^{TT}\hat{C_\ell}^{EE}-(\hat{C_\ell}^{TE})^2}\right) + \right. \tag{2.29}$$

$$\left. \frac{C_\ell^{\hat{T}T}C_\ell^{EE}+C_\ell^{TT}C_\ell^{\hat{E}E}-2C_\ell^{TE}C_\ell^{\hat{T}E}}{C_\ell^{TT}C_\ell^{EE}-(C_\ell^{TE})^2} + \frac{C_\ell^{\hat{B}B}}{C_\ell^{BB}} - 3\right\}, \tag{2.30}$$

where the likelihood has been normalized with respect to the maximum likelihood value, where $C_\ell^{XY} = \hat{C_\ell}^{XY}$, $X, Y = T, E, B$.

In case of an experiment with partial sky coverage (of a sky fraction f_{sky}) and noise

$$\ln\mathscr{L} \longrightarrow f_{sky}\ln\mathscr{L}, \; C_\ell^{XY} \longrightarrow C_\ell^{XY} + \mathscr{N}_\ell^{XY}, \tag{2.31}$$

where $\mathscr{N}_\ell^{XY}$ denotes the noise power spectrum and it is added to both C_ℓ^{XY} and $\hat{C_\ell}^{XY}$.

2.4 The Dark Side of Large-Scale Structures

The primordial perturbations seeded from inflation are believed to grow by gravitational instability to form the large-scale structures, LSS, (as traced by galaxies and galaxy clusters) observed at low redshift. The structure of the Universe on large scales is largely determined by the force of gravity, which we believe we know well, and not too much by complex mechanisms (baryonic physics, galaxy formation), which we do not know well. At this point, we should bear in mind that there are theories of nonstandard gravity to explain e.g., the accelerated expansion of the Universe, and in these theory the growth of large-scale structure is modified. (See S. Carroll contribution and references therein).

From this, we can infer that "precision cosmology" has different meaning when talking about the CMB (where the underlying physics is simple, extremely well known and robust) and when talking about LSS at lower redshift. The systematic effect inherent to the imperfect modeling of the observations are much larger for LSS; while gravitational lensing circumvents a lot of this (see the chapter by

A. Heavens on page xxx.), the direct observation of clustering is much more sensitive to these uncertainties. Nevertheless, LSS and CMB are highly complementary probes since most of the statistical power from primary CMB temperature anisotropies has basically been harvested already, it is important to try to do the best in pursuing "precision cosmology" from LSS data.

2.4.1 Basic Tools for Large-Scale Structure

Let us start by defining the (dark matter) overdensity field $\delta(\mathbf{r}) = (\rho(\mathbf{r} - \bar{\rho})/\bar{\rho}$.

The assumption of homogeneity and isotropy implies that the expectation values of the field are the same in all volumes. Each volume is a realization of the global statistical process and, therefore, will have properties that differ from the global average. To recover the global properies, there are two options: (1) average over many realizations (not easily doable as we have only one observable universe, but doable with simulations) and (2) average over many (large enough) volumes. The "axiom" of ergodicity tell us that (1) and (2) are equivalent.

The basic tool of large scale structure studies is the power spectrum for two reasons: for a Gaussian random field, it specifies completely the statistical properties if the field and because it is what is predicted from theory. Even if initial conditions were Gaussian, the late time overdensity field will not be Gaussian, but theory (e.g., inflation) still gives clean predictions for the linear power spectrum. So still, one can try to map a linear power spectrum to a nonlinear one and thus compare with observation. Even in this case, the power spectrum is a very, very useful tool. The Fourier transform of the overdensity field is

$$\delta_{\mathbf{k}} = A \int d^3 r \delta(\mathbf{r}) \exp[-i\mathbf{k}\dot{\mathbf{r}}], \tag{2.32}$$

and its inverse is

$$\delta(\mathbf{r}) = B \int d^3 k \delta_{\mathbf{k}} \exp[i\mathbf{k}\dot{\mathbf{r}}], \tag{2.33}$$

with Dirac delta function

$$\delta_{\mathbf{k}}^D = AB \int d^3 r \exp[\pm i\mathbf{k}\dot{\mathbf{r}}]. \tag{2.34}$$

Different Fourier transform convention uses different values for A and B; we adopt $A = 1$, $B = 1/(2\pi)^3$. Recall that the two-point correlation function is $\xi(x) = \langle \delta(\mathbf{r})\delta(\mathbf{r} + \mathbf{x}) \rangle$ and dependent only on the modulus of x because of isotropy and $\langle\rangle$ denotes the ensamble average.

The power spectrum $P(k)$ is thus

$$\langle \delta_{\mathbf{k}} \delta_{\mathbf{k}'} \rangle = (2\pi)^3 P(k) \delta^D(\mathbf{k} + \mathbf{k}'). \tag{2.35}$$

Since $\delta(\mathbf{r})$ is real, we have $\delta^*_{\mathbf{k}} = -\delta_{-\mathbf{k}}$, where the superscript $*$,

$$\langle \delta_{\mathbf{k}} \delta^*_{\mathbf{k}'} \rangle = (2\pi)^3 P(k) \delta^D(\mathbf{k}-\mathbf{k}') = (2\pi)^3 \int d^3x \xi(x) \exp[-i\mathbf{k}\cdot\mathbf{x}] \delta^D(\mathbf{k}-\mathbf{k}')] \quad (2.36)$$

and power spectrum and correlation function are Fourier transform pairs and related by

$$\xi(x) = \frac{1}{(2\pi)^3} \int d^3kP(k) \exp[i\mathbf{k}\cdot\mathbf{x}] \quad (2.37)$$

$$P(k) = \int d^3x \xi(x) \exp[-i\mathbf{k}\cdot\mathbf{x}]. \quad (2.38)$$

$$(2.39)$$

Thus, the same amount of information is enclosed in $P(k)$ and $\xi(x)$, at least at this stage.

It is useful to define the variance of the overdensity field:

$$\sigma^2 = \langle \delta^2(x) \rangle = \xi(0) = \frac{1}{(2\pi)^3} \int d^3kP(k) \equiv \int \Delta^2(k) d\ln k, \quad (2.40)$$

where in the last equality, we have defined the variance per log-k $\Delta^2(k) = (2\pi)^{-1}k^3P(k)$. Note that the numerical value of $\Delta^2(k)$ does not depend on the Fourier transform convention. However, σ^2 depends on the filtering scale. When reporting σ^2, the density filed is first convolved with a smoothing filter; two typical choices are

- Gaussian filter:

$$f = \frac{1}{(2\pi)^{3/2}R_G^3} \exp[-x^2/(2R_G^2)]\,; f_k = \exp[-k^2R_G^2/2] \quad (2.41)$$

- Top hat

$$f = \frac{1}{4\pi R_T^3} \Theta(x/R_T)\,; f_k = \frac{3}{(kR_T)^3} [sin(kR_T) - kR_T cos(kR_T)]. \quad (2.42)$$

The relation between the two smoothing radii is $R+t \simeq \sqrt{5}R_G$. Also, note the oscillatory scale of f_k in the top hat case. The quantity σ_8 defined as the *rms* fluctuation of the field smoothed with a top hat window on scales of $R_T = 8Mpc/h$ has contribution from relatively large k, but most of the signal comes from scales $k \sim 0.14$ h/Mpc.

So, the initial conditions generated by inflation are characterized by a given power spectrum, but how do they evolve after that? In short, in linear theory, when $\delta \ll 1$, different Fourier modes evolve independently and thus a Gaussian filed remains Gaussian; $P(k)$ changes in amplitude but not in shape. There are two exceptions: in the radiation dominated era and for the imprint of the BAO on the dark matter distribution. These two effects are taken in account by the transfer function; otherwise, the linear growth can be derived as follows.

Start by assuming that dark matter is pressureless that only sub-horizon scales are considered well after recombination (and radiation drag), then one can write the following set of equations:

$$\frac{D\rho}{Dt} = -\rho\nabla\cdot\mathbf{u}\,\text{continuity equation,} \tag{2.43}$$

$$\frac{D\mathbf{u}}{Dt} = -\nabla\Phi \text{ Euler equation,} \tag{2.44}$$

$$\nabla^2\Phi = 4\pi G\rho \text{ Poisson equation,} \tag{2.45}$$

$$\tag{2.46}$$

where $\mathbf{u}$ denotes the velocity field and Φ the gravitational potential. Since

$$\frac{D}{Dt} = \frac{\partial}{\partial t} + \mathbf{u}\cdot\nabla, \tag{2.47}$$

this can be rewritten like

$$\frac{\partial\rho}{\partial t} + \nabla\cdot\rho\mathbf{u} = 0, \tag{2.48}$$

$$\frac{\partial\mathbf{u}}{\partial t} + (\mathbf{u}\cdot\nabla)\cdot\mathbf{u} - \nabla\Phi = 0, \tag{2.49}$$

$$\nabla^2\Phi - 4\pi G\rho = 0, \tag{2.50}$$

$$\tag{2.51}$$

If there are no perturbations and using Birkhoff's theorem, we find, as expected, that

$$\rho(t) = \rho_0 a(t)^{-3}, \tag{2.52}$$

$$\mathbf{r} = a(t)\mathbf{x}, \tag{2.53}$$

$$\mathbf{u} = \frac{\dot{a}}{a}\mathbf{r} = H\mathbf{r}, \tag{2.54}$$

where x denotes the comoving coordinate and a is the scale factor, usually normalized to be 1 at present time.

Let us now introduce small perturbations: $\rho(\mathbf{x},t) = \bar{\rho}(t)[1+\delta(\mathbf{x},t)]$; $\mathbf{u} = \dot{a}\mathbf{x} + a\dot{\mathbf{x}}$. In the last equation, the first term of the RHS is the Hubble flow and the second term is the *peculiar velocity*, $\mathbf{v}$. By substituting these equations in 2.51, using Eq. 2.54 and assuming that $\delta \ll 1$, we obtain the following *linearized* second-order differential equation:

$$\frac{\partial^2\delta}{\partial t^2} + 2\frac{\dot{a}}{a}\frac{\partial\delta}{\partial t} - 4\pi G\bar{\rho}\delta = 0. \tag{2.55}$$

The solution is (you can verify it)

$$\delta^{(1)} = D(t)\delta(x, t = 0) + \text{decaying mode}, \tag{2.56}$$

where $D(z) = a(z)g(z)$ and $g = 1$ in an Einstein de Sitter universe. You can also see that (perhaps unsurprisingly) perturbations do not grow in an empty universe.

In general, the linear growth function $g(a)$ can be well fitted by (e.g.,[29]):

$$g(a) = \exp[\int_0^a \frac{da'}{a'}(\Omega_m(a)^\gamma - 1)], \tag{2.57}$$

where $\gamma = 0.55$ for a standard LCDM model; $\gamma = 0.55 + 0.022(1 + w)$ for a dark energy model and it deviated from these values for nonstandard models.

2.4.1.1 Transfer Function

The primordial power spectrum is well approximated by a power law: then why is it that the large scale structure power spectrum is not and in particular, has a characteristic scale as clearly seen in Fig. 2.2? The main effect (enclosed in the transfer function and not described by the approach of the previous section) is due to a suppression of growth at small scales. A perturbation of wavelenght λ enters the horizon when $\lambda < d_H$ (with d_H being close to the Hubble radius). For scales that enter the horizon before matter radiation equality $\lambda < d_H(a_{eq})$, the expansion time-scale is shorter than the collapse time (i.e., they are larger than the Jeans length at that time). Take δ to be the matter (nonrelativistic) overdensity and define $y = \rho_{nr}/\rho_r$ the ratio between relativistic and nonrelativistic densities. Since $\rho_{nr} \propto a^{-3}$ and $\rho_r \propto a^{-4}$, we have $y = a/a_{eq}$. The linearized perturbation equation 2.55 can be rweritn as

$$\frac{d^2\delta}{dy} + \frac{2 + 3y}{2y(1 + y)}\frac{d\delta}{dy} - \frac{3\delta}{2y(1 + y)} = 0. \tag{2.58}$$

Try to find a solution of the type $a + by^n$ to obtain a growing mode $\delta = 1 + 3/2y$. Thus, the growing solution between $y = 0$ (i.e., initial conditions) and $y = 1$(i.e., matter radiation equality) is only a factor of 5/2: effective perturbations that enter the horizon in radiation domination are frozen; in other words and as illustrated in Fig 2.10, perturbations that are of small enough scales to enter the horizon before a_{eq} get their growth suppressed by a factor

$$f = \left(\frac{a_{enter}}{a_{eq}}\right)^2 = \left(\frac{k_0}{k}\right)^2, \tag{2.59}$$

thus, the power spectrum is suppressed by a factor k^4. Any effect that changes the matter radiation equality will, therefore, leave its imprint in the shape of the matter power spectrum on large scales.

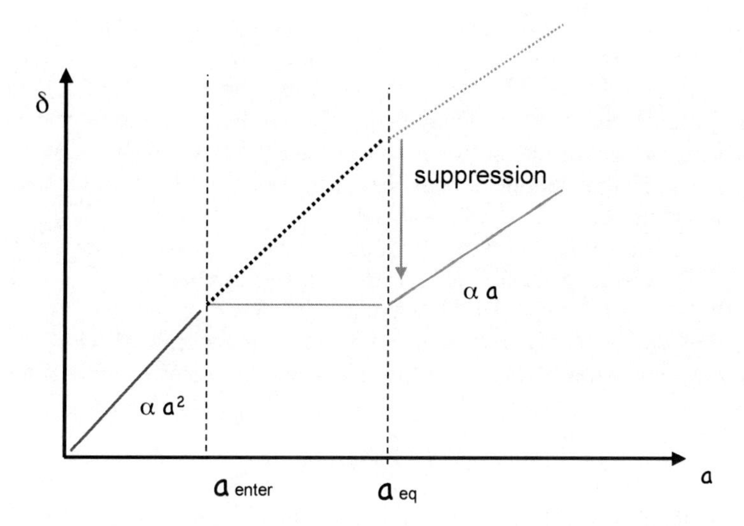

Fig. 2.10 Sketch of the effect of the matter-radiation equality on the growth of perturbation of a particular scale. This effect gives the transfer function and its characteristic shape.

Another effect comes in if some of the dark matter is hot that it is has some nonzero intrinsic velocity v. In this case, perturbations can be washed out on scales smaller than the free streaming length, defined as

$$\lambda_{FS} = a(t) \int_0^t \frac{v(t')}{a(t')} dt' . \tag{2.60}$$

This is the case for neutrinos: since neutrino oscillations indicate that neutrinos have mass, they are a (subdominant) contribution to the dark matter, but they are "hot." They, thus, suppress fluctuations on scales smaller than

$$k_{m_\nu} \simeq 0.026 \left(\frac{m_\nu}{1eV}\right)^{1/2} \Omega_m^{1/2} h\mathrm{Mpc}^{-1} \tag{2.61}$$

and give a small-scale suppression to the power spectrum of

$$\frac{\Delta P}{P} \simeq -8\frac{\Omega_\nu}{\Omega_m} = -\frac{0.085}{\Omega_m h^2} \frac{\sum m_{nu}}{1\mathrm{eV}} . \tag{2.62}$$

Although the two effects described thus far yield broadband effects on the power spectrum shape, on scales $0.01 < k < 0.3$ Mpc/h the photon-baryon coupling in the early Universe leaves its characteristic acoustic imprint in the dark matter power spectrum. This effect goes under the name "Baryon Acoustic Oscillations," and it is an extremely hot subject. We will get back to this in Sect. 2.4.6.

2.4.1.2 Real-World Effects

Thus far, we have considered the matter overdensity field δ and the matter power spectrum. In reality, large-scale galaxy surveys measure the galaxy overdensity field which may not be identical to the matter one. This is an extremely wide subject, so here we just introduce the basic idea and motivation of why the galaxy field may be a biased tracer of the dark matter. To do so, we sketch the classic demonstration of [14] that in a Gaussian random field, high peaks are more clustered than the underlying distribution: they are biased tracers. If one then makes the identification of high peaks of the initial (Gaussian) field with dark matter halos, which are the structure that host galaxies..., the conclusion is that galaxies are biased tracers. The proof is simple: compute the correlation function of regions of a Gaussian field above a threshold. Consider a Gaussian field smoothed on some scale R, with correlation function ξ and rms σ. The probability that a random point is above a threshold $\nu\sigma$ [3] is

$$P_1 = \int_{\nu\sigma}^{\infty} P_{Gauss}(y)dy, \tag{2.63}$$

and the probability that two random points are above the threshold is

$$P_{12} = \int_{\nu\sigma}^{\infty} P_{Gauss}(y_1,y_2)dy_1dy_2, \tag{2.64}$$

where $P(y_1,y_2)$ is the bivariate Gaussian distribution.

$$P(y_1,y_2 > \nu\sigma) = P_1P_2[1+\xi_{>\nu\sigma}], \tag{2.65}$$

where $\xi_{>\nu\sigma}$ denotes the correlation function of peaks above the threshold and

$$[1+\xi_{>\nu\sigma}] = \sqrt{\frac{2}{\pi}} erfc\left(\frac{\nu}{\sqrt{2}}\right)^{-2} \int_{\nu}^{\infty} \exp[-y^2/2] erfc\left(\frac{\nu - y\xi(r)/\xi(0)}{\sqrt{2[1-\xi^2(r)/\xi^2(0)]}}\right) dy. \tag{2.66}$$

Despite the apparent complication of the expression, it is easy to see that for $\xi << 1$ and $\nu >> 1$

$$\xi_{>\nu\sigma} = \frac{\nu^2}{\sigma^2}\xi\,. \tag{2.67}$$

At this point, a dubious step can be made to interpret Eq. 2.67 that is to relate the galaxy overdensity to the matter overdensity as $\delta_g = b\delta$, which is the equation defining *linear bias* with b being the linear bias factar. This relation implies that $\xi_g = b^2\xi$ with the interpretation $b = \nu/\sigma$. Note, however, that while linear bias implies Eq. 2.67 the converse is not true.

---

[3] Please do not confuse this ν with neutrinos: in keeping with the literature, we use the same symbol, the meaning should be clear from the context.

Of course, identifying the peaks of a Gaussian field with halos and galaxies may be a bit of a stretch. Although the modeling of galaxy bias is still a open issue, the modeling of halo bias is under control. The halo-model approach to clustering developed in the late 1990s–early 2000s along with the extended Press–Shecther theory offers a really good modeling of it. For more details, see [4].

2.4.1.3 Shot Noise

Thus far, when characterizing inhomogeneities via the power spectrum, we have considered a continuous dark matter field. In practice, a discrete distribution of galaxies is observed: how does this affect the clustering? We have seen above that if galaxies form on high peaks, the field in linearly bias and that in general galaxy bias will be more complicated. However, there is an additional contribution coming from discreteness. The number of galaxies at a given point is space is given by

$$n(\mathbf{x}) = \bar{n}[1+\delta(\mathbf{x})] = \sum_i \delta^D(\mathbf{x}-\mathbf{x}_i) \text{ where } \bar{n} = \langle \sum_i \delta^D(\mathbf{x}-\mathbf{x}_i)\rangle, \tag{2.68}$$

where the index i runs through the galaxies. To describe discreteness, one usually makes the assumption of Poisson sampling that is galaxies are a Poisson sampling of the underlying dark matter distribution. This is only an approximation, however. It has become clear recently that the Poisson assumption cannot hold in details. For example, dark matter halos exclusion (halos tend to avoid each other on scales comparable to the halo Lagrangian radius) gives a shot noise that cannot be fully Poisson, it will be sub-Poisson. On the other hand, if galaxies preferentially occupy halos and many galaxies occupy the same halo, this will give a super-Poisson contribution. The Poisson assumption, however, works quite well on large scales and is a useful work-horse tool to describe discreteness. Thus, divide the volume in infinitesimal cells so that each cell contains either one or no galaxies, but the probability for one cell to include two galaxies is fully negligible. Thus, the probability of having one galaxy in a volume δV is $\delta p P_1 = \bar{n}\delta V$, and the probability of finding no galaxy is $\delta P_0 = 1 - \bar{n}\delta V$. In addition, when averaging over nonempty volume elements, $\langle n_i\rangle = \langle n_i^2\rangle = \bar{n}\delta V$. To obtain the two-point correlation function, we start by computing the probability of finding galaxies in cells 1 and 2 as function of their distance:

$$\langle n_1 n_2\rangle = \langle \sum_{ij} \delta^D(\mathbf{x}_1-\mathbf{x}_1)\delta^D(\mathbf{x}_2-\mathbf{x}_j)\rangle = \bar{n}^2(1+\xi_{12}) + \bar{n}\delta^D(\mathbf{x}_1-\mathbf{x}_2), \tag{2.69}$$

where the second term comes from the zero-lag pairs. So, if we were to define a correlation function for our discrete field ξ^d as $\langle n_1 n_2\rangle \equiv \bar{n}^2(1+\xi^d)$, then $\xi^d = \xi + 1/\bar{n}\delta^D(\mathbf{x}_1-\mathbf{x}_2)$. In Fourier space, this becomes

$$\langle \delta_{\mathbf{k}_1}\delta_{\mathbf{k}_2}\rangle = (2\pi)^3\left(P(k_1) + \frac{1}{\bar{n}}\right)\delta^D(\mathbf{k}_1+\mathbf{k}_2). \tag{2.70}$$

Before we leave shot noise aside, let us note that this contribution is not Gaussian; it is only n the large numbers limit that a Poisson distribution is well approximated by a Gaussian. Therefore, as long as the galaxy density is high enough and the shot noise correction is small, the modeling of discreteness can be that of a superposition of two independent random processes, the clustering one and the white noise contribution coming from the Poisson sampling, which amplitude depends on $\bar{n}$. For any practical application, the goodness of these assumptions will have to be tested.

2.4.2 Window and Selection Function

Galaxy surveys are usually magnitude limited, which means that as you look further away you start missing some galaxies. The selection function tells you the probability for a galaxy at a given distance (or redshift z) to enter the survey. It is a multiplicative effect along the line of sight in real space. On the other hand, you can never observe a perfect (or even better infinite) squared box of the Universe, and in CMB studies, you can never have a perfect full sky map (we live in a galaxy...). The mask (sky cut in CMB jargon) is a function that usually takes values of 0 or 1 and is defined on the plane of the sky (i.e., it is constant along the same line of sight). The mask is also a real space multiplication effect. In addition, sometimes in CMB studies, different pixels may need to be weighted differently, and the mask is an extreme example of this where the weights are either 0 or 1. Also, this operation is a real space multiplication effect.

Let's recall that a multiplicaton in real space (where $W(\mathbf{x})$ denotes the effects of window and selection functions)

$$\delta^{true}(\mathbf{x}) \longrightarrow \delta^{obs}(\mathbf{x}) = \delta^{true}(\mathbf{x})W(\mathbf{x}) \tag{2.71}$$

is a convolution in Fourier space:

$$\delta^{true}(\mathbf{k}) \longrightarrow \delta^{obs}(\mathbf{k}) = \delta^{true}(\mathbf{k}) * W(\mathbf{k}) \tag{2.72}$$

the sharper $W(\mathbf{r})$ is the messier and delocalized $W(\mathbf{k})$ is. As a result, it will couple different k-modes even if the underlying ones were not correlated!

Thus (if there is no shot noise),

$$\langle|\delta^{obs}(k)|^2\rangle = (2\pi)^3 \int d^3\mathbf{k}' P(\mathbf{k}-\mathbf{k}')|W(\mathbf{k}')|^2, \tag{2.73}$$

where the kernel W has the effect of coupling modes, if the survey has a typical size L, then $W(k)$ has a width $\Delta k \sim 1/L$. Another consequence of this is that one cannot measure the $P(k)$ directly, one will end up with an estimator of it which includes effects from the mask and the selection function. Deconvolution is always a nast business. While in CMB, one often tries to deconvolve the effects of the mask from the C_ℓ; in large-scale structure, one usually decides to keep a $P(k)$ estimator that includes the mode coupling and apply the same operation to the theory before comparing it to observations. However, note that this mode coupling does not only affect the signal but also the covariance (error) estimation. In practice, these effects are estimated from mock realizations of the survey.

2.4.3 Weighting Schemes to Account for all that and More

In the presence of window and selection function, one can devise a weighting scheme that optimizes the performance of your estimator of the power spectrum. Here, I will review the approach of [11]: it is not the only one but it is useful and simple enough so that the gist of it can be explained in a page.

Instead of working with $\delta(\mathbf{x})$ or $n(\mathbf{x})$, one defines a new quantity

$$F(\mathbf{x}) = \frac{w(\mathbf{x})[n_g(\mathbf{x}) - \alpha n_{syn}(\mathbf{x})]}{[\int d^3\mathbf{x} n^2(\mathbf{x}) w^2(\mathbf{x})]^{1/2}}. \tag{2.74}$$

Here n_g denotes the number density of galaxies and n_{syn} denotes a synthetic catalog with the same angular and selection function as the real survey but without clustering. This synthetic catalog needs to have as many "points" as possible in order not to introduce additional shot noise. α is a dilution factor: the number density of the synthetic catalog is $1/\alpha$ that of the survey. As it will be clear below, α needs to be small (well below 0.01) so that the number density of the synthetic catalog is much higher than that of the survey. w is the weighting: it can be chosen to minimize variance, to give an unbiased estimator etc. Thus, we can write

$$\langle n_g(\mathbf{x}) n_g(\mathbf{x}')\rangle = \bar{n}(\mathbf{x})\bar{n}(\mathbf{x}')[1+\xi] + \bar{n}(\mathbf{x})\delta^D(\mathbf{x}-\mathbf{x}'), \tag{2.75}$$

$$\langle n_{syn}(\mathbf{x}) n_{syn}(\mathbf{x}')\rangle = \alpha^{-2}\bar{n}(\mathbf{x})\bar{n}(\mathbf{x}') + \alpha^{-1}\bar{n}(\mathbf{x})\delta^D(\mathbf{x}-\mathbf{x}'), \tag{2.76}$$

$$\langle n_g(\mathbf{x}) n_{syn}(\mathbf{x}')\rangle = \alpha^{-1}\bar{n}(\mathbf{x})\bar{n}(\mathbf{x}'). \tag{2.77}$$

By substituting this in the equation for F and taking its Fourier transform, we see that

$$\begin{aligned}\langle |F(k)^2| \rangle &= \\ &= \frac{\int d^3\mathbf{x} d^3\mathbf{x}' w(\mathbf{x}) w(\mathbf{x}') \langle [n_g(\mathbf{x}) - \alpha n_{syn}(\mathbf{x})][n_g(\mathbf{x}') - \alpha n_{syn}(\mathbf{x}')]\rangle \exp[i\mathbf{k}\cdot(\mathbf{x}-\mathbf{x}')}{\int d^3\mathbf{x} \bar{n}^2(\mathbf{x}) w^2(\mathbf{x})} \\ &= \int \frac{d^3\mathbf{k}'}{(2\pi)^3} P(k') |G(\mathbf{k}-\mathbf{k}')|^2 + (1+\alpha)\frac{\int d^3\mathbf{x}\bar{n}(\mathbf{x}) w^2(\mathbf{x})}{\int d^3\mathbf{x}\bar{n}^2(\mathbf{x}) w^2(\mathbf{x})},\end{aligned} \tag{2.78}$$

where we have assumed that there is no correlation between the mask and the clustering. This equation defines the function G:

$$G(\mathbf{k}) = \frac{\int d^3\mathbf{x}\bar{n} w \exp(i\mathbf{k}\cdot\mathbf{x})}{[\int d^3\mathbf{x}\bar{n}^2 w^2]^{1/2}}. \tag{2.79}$$

If the window function of the survey is well behaved then the width of G is $\sim 1/L$ for a survey of typical size L. Thus, on scales where $k \gg 1/L$, we see that

$$\langle |F(k)|^2 \rangle \simeq P(k) + P_{shot\ noise} \tag{2.80}$$

from where one can trivially estimate $P(k)$. The expression for the coupling between modes is

$$\begin{aligned} \langle F(k)F^*(k') \rangle &= \frac{1}{(2\pi)^3} \int P(k'')G(\mathbf{k}-\mathbf{k}'')G(\mathbf{k}'-\mathbf{k}'')d^3\mathbf{k}'' \\ &+(1+\alpha)\frac{\int d^3\mathbf{x}\bar{n}(\mathbf{x})w^2(\mathbf{x})}{\int d^3\mathbf{x}\bar{n}^2(\mathbf{x})w^2(\mathbf{x})}. \end{aligned} \tag{2.81}$$

If one takes the weight $w \propto 1/\bar{n_g}$, [11] points out, weight is given to distant parts of the survey where $\bar{n}$ is low. They show that the optimum weight is

$$w(\mathbf{x}) = [1 + \bar{n}(\mathbf{x})P(k)]^{-1}. \tag{2.82}$$

If the survey includes different galaxy populations with different values of bias, then one may want to select the weight, not to minimize the variance on $P(k)$ but to minimize instead any possible distortion of the recovered $P(k)$ shape, as shown by [36].

2.4.4 Redshift-Space Distortions

The issue of galaxy bias could be bypassed if one could observe directly the dark matter distribution. This would not bypass the nonlinearities, window, selection function, etc. But we are more confident that these effects can be reliably modeled if not analytically at least with n-body simulations. Beyond weak lensing, the next best thing is offered by the velocity field. If galaxies can be treated as test particles in the total gravitational field, their peculiar velocities should be directly related to the total matter distribution. Although peculiar velocities are difficult to observe directly, they leave a clear imprint in the observed galaxy distribution. In fact when a galaxy redshift is interpreted as its distance, the resulting galaxy maps will be distorted in the line of sight direction by peculiar velocities. From the linearized continuity equation

$$\frac{\partial \delta}{\partial t} = -\frac{1}{a}\nabla \cdot \mathbf{v} = \frac{1}{a} i\mathbf{k} \cdot \mathbf{v}, \tag{2.83}$$

where

$$\mathbf{v} = a\frac{\partial \delta}{\partial t}\frac{\mathbf{k}}{ik^2} = \frac{Ha\mathbf{k}}{ik^2}\left(\frac{a}{\delta}\frac{\partial \delta}{\partial a}\right)\delta = f(\Omega,z)\frac{Ha\mathbf{k}}{ik^2}\delta, \tag{2.84}$$

where

$$f = \frac{d\ln\delta}{d\ln a} \simeq \Omega^{\gamma}. \tag{2.85}$$

Recall that this function is crucial to constrain dark energy models and deviations from the standard cosmology. Clearly, velocities, if they can be measured, offer a direct handle to it. The issue is that (1) velocities are very difficult to measure directly and (2) that f comes in the relation between (dark matter) density and velocity, leaving us to deal with the bias issue. Nevertheless let's push on. Let us recall that $\mathbf{u} = H\mathbf{r} + v_{rad}$, where the line of sight peculiar velocity is $v_{rad} = [\mathbf{v}(\mathbf{r}) - \mathbf{v}(0)] \cdot \mathbf{r}$. In galaxy redshift surveys, the redshift coordinate is used as distance

$$\mathbf{s} = \frac{\mathbf{u}}{H} = \mathbf{r}[1 + v_{rad}/H]. \tag{2.86}$$

One also need to take into account that density in redshifts space (indicated by subscript s) is related to that in real space by $\rho_s d^3 s - \rho d^3 r$, thus

$$d^3 s = d^3 r \left(1 + \frac{r_{rad}}{Hr}\right)^2 \left(1 + \frac{1}{H}\frac{\partial v_{rad}}{\partial r}\right). \tag{2.87}$$

If at cosmological distances $v_{rel}/(Hr) \ll \partial v_{rad}/\partial r$, we can write that $\rho_s(\mathbf{r}) = \rho(\mathbf{r})(1 - 1/H\partial v_{rad}\partial r)$. For small overdensities, we obtain

$$\delta_s = \delta - \frac{1}{H}\frac{\partial v_{rad}}{\partial r}. \tag{2.88}$$

At this point, it is useful to go to Fourier space. Recall that $v_{\mathbf{k}} = iH\mathbf{k}/k^2 f \delta_{\mathbf{k}}$, where $delta_{\mathbf{k}} \propto \exp[-i\mathbf{k}\cdot\mathbf{r}]$. We obtain that $\partial v_{rad}\partial r \simeq -Hcos^2(\theta_{kr}) f\delta$ with θ_{kr} being the angle between $\mathbf{k}$ and the line of sight and therefore

$$\delta_s(\mathbf{k}) = \delta(\mathbf{k})(1 + f\mu^2), \tag{2.89}$$

where $\mu = \cos(\theta_{kr})$. This implies that on large scales, the power spectrum gets enhanced by a factor $(1 + f\mu^2)^2$. This is called "Kaiser effect" from the paper [15]. Note that this factor depends on the direction with respect to the line of sight: clustering does not appear isotropic any more. Now, using a Fourier expansion in this context makes not much sense because redshift-space distortions are radial: the natural decomposition should be in spherical harmonics. In this case, the maths are not for the faint of heart [12, 35]. However, one can (at least for illustration purposes) adopt the distant observer approximation: if the survey volume considered is far enough from the observer so that it subtends a small angle, then the line of sight direction can be considered constant across the survey and a Fourier expansion works well. In this case, the angle-averaged effect of the redshift-space boosting of power becomes

$$P^s(k) = P(k)\,\text{average}(1 + 2f\mu^2 + f^2\mu^4) = P(k)(1 + 2/3f + 1/5f^2). \tag{2.90}$$

Of course, δ for the dark matter cannot be directly observed. In the case of linear, deterministic bias where gelaxy and dark matter overdensities are connected the simple relation $\delta_g = b\delta$, it is easy to see that redshift-space distortions expressions get modified as

$$\delta_s(\mathbf{k}) = b\delta(\mathbf{k})(1 + \frac{f}{b}\mu^2). \tag{2.91}$$

The combination f/b is often indicated by the letter β. Clearly [46] the combination $\sigma_{8,g}\beta$ is independent of galaxy bias in the case of local linear deterministic bias.

2.4.5 Nonlinearities etc.

Thus far, we have assumed that $\delta \ll 1$; what happens when this assumption does not hold? First, let us notice that the linear, local, deterministic bias assumption even mathematically and for a Gaussian field can hold only in the regime where $\delta \ll 1$. In fact by definition $\delta \geq -1$, but if $\delta_g = b\delta$ for high b it may make $\delta_g < -1$, which is nonsense.

For the same reason, as δ grows by gravitational evolution when δ is no longer $<< 1$, the field cannot remain Gaussian: it must generate some skewness where positive overdensities can grow at will but negative underdensities are always ≥ -1. Thus, while in the linear regime, different Fourier modes evolved independently, in the nonlinear regime, modes gets coupled. As a consequence, the power spectrum will change its shape. One can go quite a long way in describing nonlinearites with analytical approaches, such as perturbation theory or variations of thereof, but ultimately the process can be modeled quite accurately resorting to n-body simulations (see Moscardini contribution). An extremely promising approach was pioneered by [45]; it consists in finding a possibly physically motivated model to interpolate results from N-body simulations.

A word of warning is, however, necessary regarding nonlinear redshift-space distortions. Although nonlinearities in real space give a scale-dependent boost of power, nonlinearites in redshift space give a scale dependent and anisotropic suppression of power. In fact, imagine to observe a fairly massive cluster with velocity dispersion of the order of 1000 km$/s$. If galaxies in the cluster have this kind of random velocities, their positions in redshift space will be smeared out to scales larger than 10 Mpc/h i.e., on scales that are naively assumed to be linear or quasi-linear. This effect is called "fingers-of-God" for obvious reasons as it can be seen in Fig. 2.11. To model the small-scale smearing effect on the power spectrum, it is common to multiply the redshift space $P(k)$ by a scale and μ-dependent suppression factor $D(k\sigma_p\mu)$, where σ_p is a parameter quantifying the strength of the fingers-of-God effect, two typical choices are exponential or Gaussian velocity distributions:

$$D = \frac{1}{1+(k\sigma_p\mu)^2/2} \text{ or } D = \exp[-k^2\sigma_p^2\mu^2/2]. \tag{2.92}$$

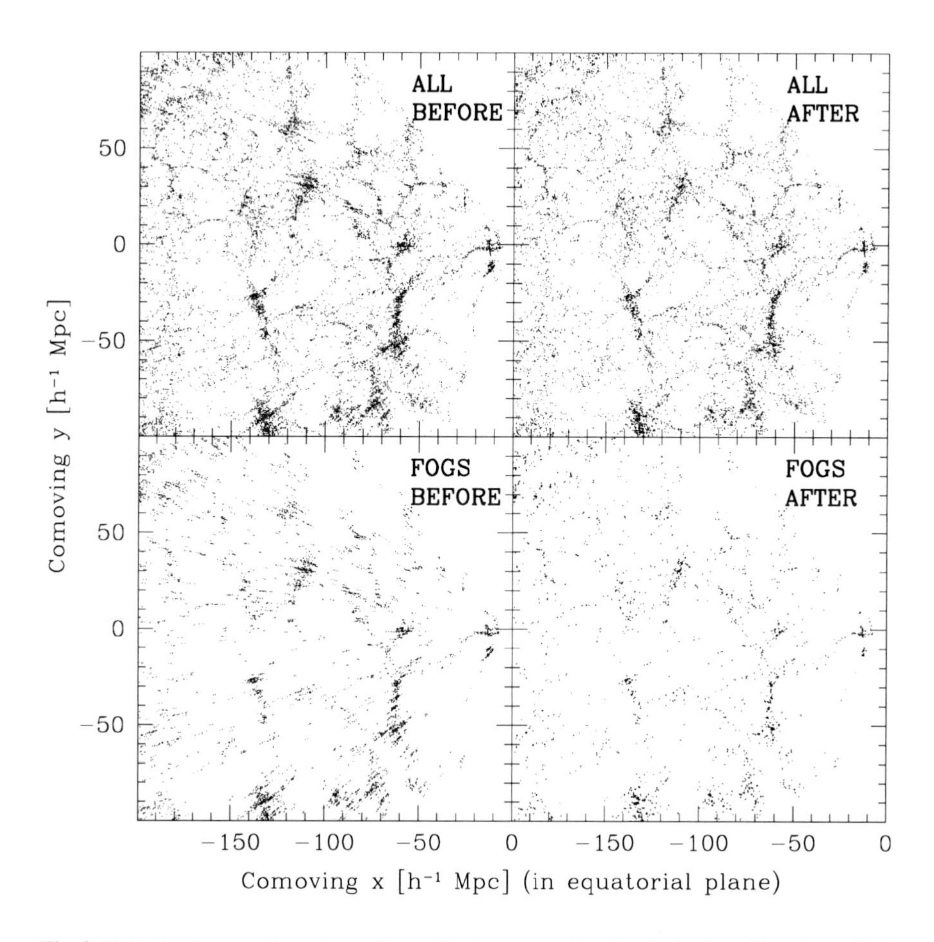

Fig. 2.11 Redshift-space distortions effects. On the top panels, all galaxies in a slice of the SDSS DR5 main survey catalog are shown, whereas while on the bottom panel, only galaxies identified as being part of clusters. On the left-hand side, galaxies are shown as the appear in redshift space: on the right-hand side, the fingers-of-God have been collapsed to a spherically symmetric structure. Note that after the fingers-of-God have been collapsed, clustering is still enhanced along the line of sight on large scales (note the appearance of "great walls"). This is the "Kaiser" effect. [Tegmark et al 2004, fig 7]

In this case, σ_p is interpreted as the line-of-sight pairwise velocity dispersion. Note that this effect is very difficult to model in Fourier space. In this particular example, only few clusters will display this effect: the line-of-sight smearing will be dependent on environment, type of galaxies selected, etc. As an aside, let us note here that any error in determining the galaxies redshifts can be modeled in a similar way, especially if the redshift errors are Gaussianly distributed. Both these effects (non-linear redshift-space distortions and redshift errors) are shown in Fig. 2.13 along with the large-scale boost of power. As before, as long as galaxies can be considered

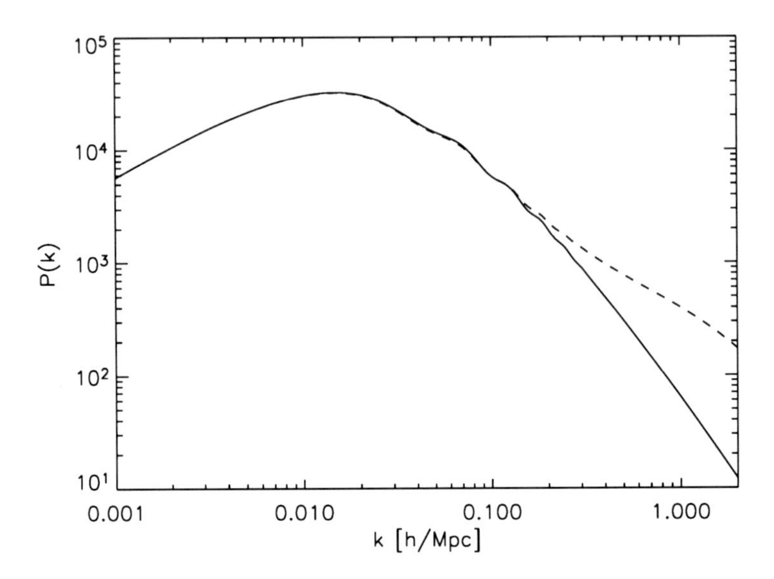

Fig. 2.12 Left: linear (solid) and halofit nonlinear (dashed) power spectra for the dark matter distribution at $z = 0$ in a LCDM model. At small, $k > 0.1$ Mpc$/h$, scales nonlinearities modify the shape of the power spectrum giving a scale-dependent boost.

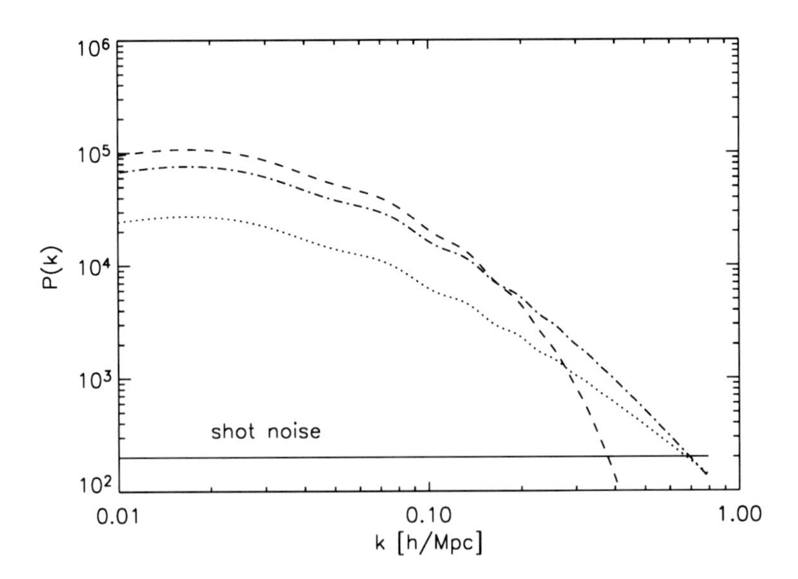

Fig. 2.13 Power spectra at $z = 0$ for a LCDM model: Linear dark matter power spectrum (dotted), angle-averaged galaxy redshift space power spectrum assuming a linear bias of $b = 1.5$ (dot-dashed), linear dark matter underlying power spectrum, and $\sigma_p = 600$ km/s; the same but only for modes along the line of sight (dashed); typical shot noise level for a densely sampled galaxy population (solid line).

as test masses tracing the dark matter velocity field, all these effects can, in principle, be accurately modeled: one can go a long way with perturbation theory-based approaches or simulate it with n-body simulation or use some hybrid approach. Possibly, then the fundamental limitation remaining in interpreting large-scale structure galaxy surveys will be nonlinear bias. Two date, the simultaneous modeling of nonlinearities and nonlinear bias has been attempted in two different ways.

- A phenomenological model. This was first used in [3] for the 2dFGRS galaxy survey. In this model, calibrated on n-body simulations, the relation between the galaxy redshift space power spectrum and the underlying dark matter linear power spectrum is given by

$$P_{gal}(k) = \frac{1+Qk^2}{1+ak} P_{linear\ matter}(k). \tag{2.93}$$

  It is crucial to calibrate the numbers a and Q in this expression to the particular survey under consideration keeping in mind both the survey redshift and the galaxy population considered. For the 2dF galaxies (effective redshift ~ 0.15) in redshift space, $A = 1.4$ and $Q = 4.0$. If one is to apply this prescription to different type of galaxies (e.g., LRG's), a new calibration will be needed.
- Reconstructing the halo density field. The idea is to single out what galaxies belong to the same dark matter halo and instead of collapsing their finger-of-God into a spherically symmetric overdensity, count all of them as one single object (a dark matter halo). This has been pioneered in [41].

The details of this modeling are important, especially if one is interested in extracting cosmological information enclosed in the broadband shape of the power spectrum. Although the BAO signal yields information on the expansion history of the Universe, the broadband shape can also offer information on, e.g., the mechanism that generated the perturbations in the early Universe.

In the remainder of this contribution, we will concentrate on the relatively new subject of BAO.

2.4.6 Baryon Acoustic Oscillations (BAO)

Cosmological perturbations in the early Universe excite sound waves in the photon-baryon fluid. After recombination, these BAO became frozen into the distribution of matter in the Universe imprinting a preferred scale, the sound horizon. This defines a standard ruler whose length is the distance sound can travel between the Big Bang and recombination. The BAO are directly observed in the CMB angular power spectrum and have been observed in the spatial distribution of galaxies by the 2dFGRS survey and the SDSS survey [8, 3, 37]. The BAO, observed at different cosmic epochs, act as a powerful measurement tool to probe the expansion of the Universe, which in turns is a crucial handle to constrain the nature of dark energy. The underlying physics which sets the sound horizon scale ($\sim$150 Mpc comoving) is well understood and involves only linear perturbations in the early Universe. The BAO scale is measured in surveys of galaxies from the statistics of the three-dimensional galaxy positions. Only recently have galaxy surveys such as SDSS grown large enough to allow for this detection. The existence of this natural standard measuring rod allows us to probe the expansion of the Universe. The angular size of the oscillations in the CMB revealed that the Universe is close to being flat. Measurement of the change of apparent acoustic scale in a statistical distribution of galaxies over a large range of redshift can provide stringent new constraints on the nature of dark energy. The acoustic scale depends on the sound speed and the propagation time. These depend on the matter to radiation ratio and the baryon-to-photon ratio. CMB anisotropy measures these and hence fixes the oscillation scale. A BAO survey measures the acoustic scale along and across the line of sight. At each redshift, the measured angular (transverse) size of oscillations, $\Delta\theta$, corresponds with the physical size of the sound horizon, where the angular diameter distance D_A is an integral over the inverse of the evolving Hubble parameter, $H(z)$. $r_\perp = (1+z)D_A(z)\delta\theta$. In the radial direction, the BAO directly measures the instantaneous expansion rate $H(z)$, through $r_\parallel = (c/H(z))\Delta z$, where the redshift interval (Δz) between the peaks is the oscillation scale in the radial direction. As the true scales, $r_\perp$ and $r_\parallel$ are known (given by r_s the sound horizon at radiation drag, well measured by the CMB) this is not an Alcock–Paczynsky test but a "standard ruler" test. Note that in this standard ruler test, the cosmological feature used as the ruler is not an actual object but a statistical property: a feature in the galaxy correlation function (or power spectrum). An unprecedented experimental effort is undergoing to obtain galaxy surveys that are deep, larger, and accurate enough to trace the BAO feature as a function of redshift. Before these surveys can even be designed, it is crucial to know how well a survey with given characteristic will do. This was illustrated very clearly in [43], which we follow closely here. To forecast the cosmological constraints achievable from a survey of given characteristics, we will adopt the Fisher matrix approach. To start, we need to compute the statistical error associated to a determination of the galaxy power spectrum $P(k)$. In what follows, we will ignore effects of nonlinearities and complicated biasing between galaxies and dark matter: we will assume that galaxies, at least on large scales, trace the linear matter power spectrum in such a way that their power spectrum is directly proportional to the dark matter one: $P(k) = b^2 P_{DM}(k)$, where b stands for galaxy bias. At a given wavevector k, the statistical error of the power spectrum is a sum of a cosmic variance term and a shot noise term:

$$\frac{\sigma_P(k)}{P(k)} = \frac{P(k) + 1/\bar{n}}{P(k)}. \tag{2.94}$$

Here $\bar{n}$ denotes the average density of galaxies and $1/\bar{n}$ is the white noise contribution from the fact that galaxies are assumed to be a Poisson sampling of the underlying distribution. When written in this way, this expression assumes that $\bar{n}$ is constant with position. While in reality, this is not true for forecasts one assumes that the survey can be divided in shells in redshifts and that the selection function is such that n is constant within a given shell. Since $P(k)$ is also expected to change in redshift, then one should really implicitly assume that there is a z dependence in Eq. 2.94. In general, $P(k,z) = b(z)^2 G^2(z) P_{DM}(k)$, where $G(z)$ denotes the linear growth factor: i.e., the bias is expected to evolve with redshift as well as clustering does, not only because galaxy bias changes with redshift but also because at different redshifts one may be seeing different type of galaxies that may have different bias parameter. We do not know a priori the form of $b(z)$, but given a fiducial cosmological model, we know $G(z)$. Preliminary observations seem to indicate that the z evolution of b tends to cancel that of $G(z)$, so it is customary to assume that $b(z)G(z) \sim constant$, but we should bear in mind that this is an assumption. Also let us recall that this is somewhat a naive modeling as halos (and this galaxies) may not necessarily be related to the dark matter by a linear, deterministic bias!

An extra complication arises because of redshift space distortions. Note that redshift space distortions only affect the line-of-sight clustering (it is a perturbation to the distances) not the angular clustering. Since these distortions are created by clustering they carry, in principle, important cosmological information. To write this dependence explicitly (assuming linear theory):

$$P(k,\mu,z) = b(z)^2 G(z)^2 P_{DM}(k)(1+\beta\mu)^2 \tag{2.95}$$

In the linear regime, the cosmological information carried by the redshift space distortions is enclosed in the $f(z) = \beta(z)b(z)$ combination.

For finite surveys, $P(k)$ at nearby wavenumbers are highly correlated, the correlation length is related to the size of the survey volume: for large volumes the cell size over which modes are correlated is $(2\pi)^3/V$, where V denotes the comoving survey volume. Only over distances in k-space larger than that modes can be considered independent. Therefore, if one wants to count over all the modes anyway (e.g., by transforming discrete sums into integrals in the limit of large volumes), then each k needs to be downweighted, to account the fact that all k are not independent. In addition, one should keep in mind that Fourier modes $\mathbf{k}$ and $-\mathbf{k}$ are not independent (the density field is real-valued!), giving an extra factor of 2 in the weighings. We can thus write the error on a band power centered around k,

$$\frac{\sigma_P}{P} = 2\pi\sqrt{\frac{2}{Vk^2\delta k\Delta\mu}}\left(\frac{1+nP}{nP}\right). \tag{2.96}$$

In the spirit of the Fisher approach, we now assume that the Likelihood function for the band-powers $P(k)$ is Gaussian, thus we can approximate the Fisher matrix by

$$F_{ij} = \int_{k_{min}}^{k_{max}} \frac{\partial \ln P(\mathbf{k})}{\partial\theta_i}\frac{\partial \ln P(\mathbf{k})}{\partial\theta_j} V_{eff}(\mathbf{k})\frac{d\mathbf{k}}{2(2\pi)^3}. \tag{2.97}$$

The derivatives should be evaluated at the fiducial model and V_{eff} denotes the effective survey volume given by

$$V_{eff}(\mathbf{k}) = V_{eff}(k,\mu) = \int\left[\frac{n(z)P(k,\mu)}{n(z)P(k,\mu)+1}\right]^2 dz = \left[\frac{nP(k,\mu)}{nP(k,\mu)+1}\right]^2 V, \tag{2.98}$$

where $n = \langle n(z) \rangle$. Equation 2.97 can be written explicitely as a function of k and μ as

$$F_{ij} = \int_{-1}^{1} \int_{k_{min}}^{k_{max}} = \frac{\partial \ln P(k,\mu)}{\partial \theta_i} \frac{\partial \ln P(k,\mu)}{\partial \theta_j} V_{eff}(k,\mu) \frac{k^2 dk d\mu}{2(2\pi)^2}. \tag{2.99}$$

In writing this equation, we have assumed that over the entire survey extension the line-of-sight direction does not change: in other words, we made the flat sky approximation. For forecasts, this encloses all the statistical information anyway; but for actual data analysis application, the flat sky approximation may not hold. In this equation, k_{min} is set by the survey, volume: for future surveys, where the survey volume is large enough to sample, the first BAO wiggle the exact value of k_{min} does not matter, however, recall that for surveys of typical size L (where $L \sim V^{1/3}$), the largest scale probed by the survey will be corresponding to $k = 2\pi/L$. Keeping in mind that the first BAO wiggle happens at ~ 150 Mpc, the survey size needs to be $L \gg 150$ Mpc for k_{min} to be unimportant and for the "large volume approximation" made here to hold. As anticipated above, one may want to sub-divide the survey in independent redshift shells, compute the Fisher matrix for each shell and then combine the constraints. In this case, L will be set by the smallest dimension of the volume (typically the width of the shell), so one needs to make sure that the width of the shell still guarantees a large volume and large L. k_{max} denotes the maximum wavevector to use. One could, for example, impose a sharp cut to delimit the range of validity of linear theory. In [44], this is improved as we will see below.

Before we do that, let us note that there are two ways to interpret the parameters θ_{ij} in Eq. (2.99). One could simply assume a cosmological model, say e.g., a flat quintessence model where the equation of state parameter $w(z)$ is parameterized by $w(z) = w(0) + w_a(1-a)$ and take derivatives of $P(k,\mu)$ with respect to these parameters. Alternatively, one could simply use as parameters the quantities $H(z_i)$ and $D_A(z_i)$, where z_i denote the survey redshift bins. These are the quantities that govern the BAO location and are more general: they allow one not to choose a particular dark energy model until the very end. Then, one must also consider the cosmological parameters that govern the $P(k)$ shape $\Omega_m h^2$, $\Omega_b h^2$, and n_s. Of course, one can also consider $G(z_i)$ as free parameters and constrain these either through the overall $P(k)$ amplitude (although one would have to assume that $b(z)$ is known, which is dicey) or through the determination of $G(z)$ and $\beta(z)$. The safest and most conservative approach, however, is to ignore any possible information coming from $G(z)$, $\beta(z)$, or n_s and to only try to constrain expansion history parameters.

The piece of information still needed is how the expansion history information is extracted from $P(k,\mu)$. When one converts **ra**, **dec**, and redshifts into distances and positions of galaxies of a redshift survey, one assumes a particular reference cosmology. If the reference cosmology differs from the true underlying cosmology, the inferred distances will be wrong and so the observed power spectrum will be distorted:

$$P(k_\perp, k_\parallel) = \frac{D_a(z)^2_{ref} H(z)_{true}}{D_A(z)^2_{true} H(z)_{ref}} P_{true}(k_\perp, k_\parallel). \tag{2.100}$$

Note that since distances are affected by the choice of cosmology and k vectors, $k_{ref,\|} = H(z)_{ref}/H(z)_{true}k_{true,\|}$ and $k_{ref,\perp} = D_A(z)_{true}/D_A(z)_{ref}k_{true,\perp}$, Eq. 2.100 can be written as

$$P_{true}(k_\perp, k_\|, z) = b(z)^2\left(1+\beta(z)\frac{k^2_{true,\|}}{k^2_{true,\perp}+k^2_{true,\|}}\right)^2\left[\frac{G)(z)}{G(z_o)}\right]^2 P_{DM}(k, z_o), \quad (2.101)$$

where z_o is some reference redshift where to normalize $P(k)$ typical choices can be the CMB redshift or redshift $z = 0$. Note that from these equations, it should be clear that what the BAO actually measure directly is $H(z)r_s$ and D_A/r_s, where r_s is the BAO scale; the advantage is that r_s is determined exquisitely from the CMB.

How would then one convert these constraints on those on a model parameter? Clearly, one then projects the resulting Fisher matrix on the dark energy parameters space. In general, if you have a set of parameters θ_i with respect to which the Fisher matrix has been computed, but you would like to have the Fisher matrix for a different set of parameters ϕ_i, where the θ_i are functions of the ϕ_i, the operation to implement is

$$F_{\phi_i,\phi_j} = \sum_{mn}\frac{\partial\theta_n}{\partial\phi_i}F_{\theta_n,\theta_m}\frac{\partial\theta_m}{\partial\phi_j}. \quad (2.102)$$

The full procedure for the BAO survey case is illustrated in Fig. 2.14. The slight complication is that one starts off with a Fisher matrix (for the original parameter set θ_i), where some parameters are nuisance and need to be marginalized over, so some matrix inversions are needed.

Thus far, non-linearities have been just ignored. It is, however, possible to include then at some level in this description. [44] proceeds by introducing a distribution of Gaussianly distributed random displacements parallel or perpendicular to the line-of-sight coming from nonlinear growth (in all directions) and from nonlinear redshift space distortions (only in the radial direction). The publicly available code that implements all this (and more) is available at the web site http://cmb.as.arizona.edu/ eisenste/acousticpeak/bao_forecast.html. In order to use the code, keep in mind that [44] models the the effect of nonlinearities is to convolve the galaxy distribution with a redshift-dependent and μ-dependent smoothing kernels. The effect on the power spectrum is to multiply $P(k)$ by $\exp[-k^2\Sigma(k,\mu)/2]$, where $\Sigma(k,\mu) = \Sigma^2_\perp - \mu^2(\Sigma^2_\| - \Sigma^2_\perp)$. As a consequence, the integrand of the Fisher matrix expression of Eq. (2.99) is multiplied by

$$\exp[-k^2\Sigma^2_\perp - k^2\mu^2(\Sigma^2_\| - \Sigma^2_\perp)], \quad (2.103)$$

where, to be conservative, the exponential factor has been taken outside the derivatives, which is equivalent to marginalize over the parameters $\Sigma_\|$ and $\Sigma_\perp$ with large uncertainties.

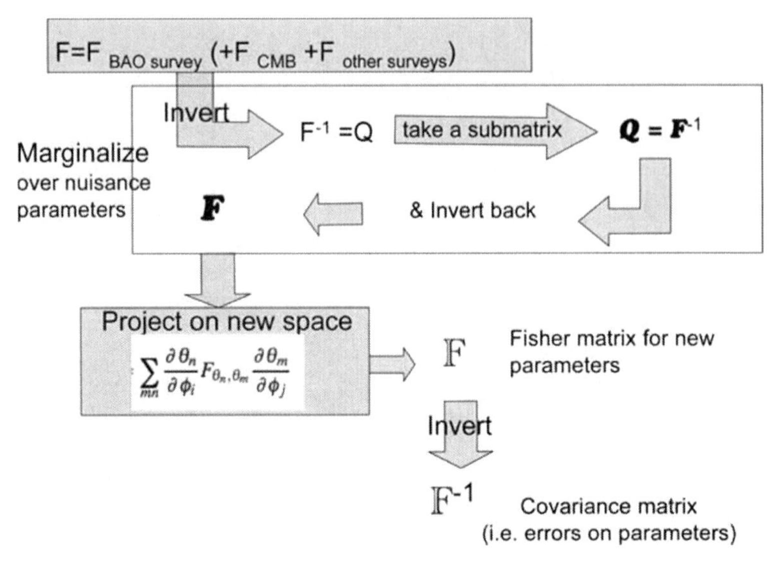

Fig. 2.14 Steps to implement once the Fisher matrix of Eq.2.99 has been computed to obtain error on dark energy parameters.

Note that $\Sigma_{\parallel}$ and $\Sigma_{\perp}$ depends on redshift and on the chosen normalization for $P_{DM}(k)$. In particular,

$$\Sigma_{\perp}(z) = \Sigma_0 G(z)/G(z_0), \tag{2.104}$$

$$\Sigma_{\parallel}(z) = \Sigma_0 G(z)/G(z_0)(1+f(z)), \tag{2.105}$$

$$\Sigma_0 \propto \sigma_8 \,. \tag{2.106}$$

If in your convention $z_0 = 0$, then $\Sigma_0(z=0) = 8.6h^{-1}\sigma_{8,DM}(z=0)/0.8$.

As an example of an application of this approach for survey design, it may be interesting to ask the question of what is the optimal galaxy number density for a given survey. Taking redshifts is expensive and for a given telescope time allocated, only a certain number of redshifts can be observed. Thus, it is better to survey more volume but have a low number density or survey a smaller volume with higher number density? You can try to address this issue using the available code. For a cross check, Fig. 2.15 shows what you should obtain. Here, we have assumed $\sigma_8 = 0.8$ at $z = 0$, $b(z=0) = 1.5$ and we have assumed that $G(z)b(z) = constant$. To interpret this figure note that with the chosen normalizations, $P(k)$ in real space at the BAO scale $k \sim 0.15$ h/Mpc is 6241(Mpc/h)^3, boosted up by large-scale redshift space distortions to roughly 10^4(Mpc/h)^3, so $n = 10^{-4}$ corresponds to $nP(k=0.15) = 1$. Note that the "knee" in this figure is, therefore, around $nP = 1$. This is where this "magic number" of reaching $nP \sim> 1$ in a survey comes from. Of course, there are other considerations (mainly related to the fast that galaxies may not be exactly a

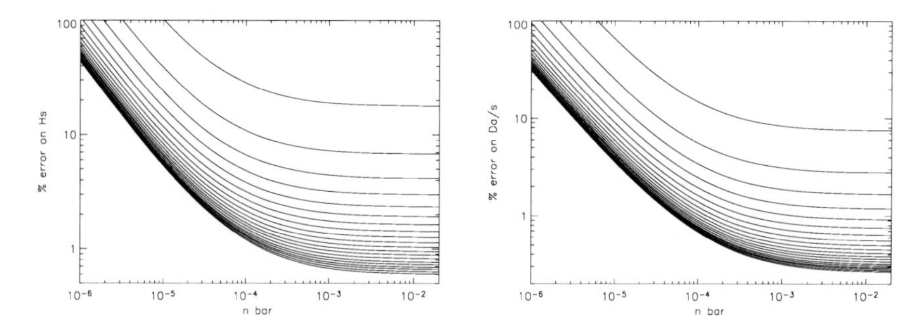

Fig. 2.15 Percent error on $H(z)r_s$ and D_a/r_s as a function of the galaxy number density of a BAO survey. This figure assumes full sky coverage $f_{sky} = 1$ (errors will scale like $1/\sqrt{f_{sky}}$) and redshift range from $z = 0$ to $z = 2$ in bins of $\Delta z = 0.1$.

Poisson sampling of the dark matter distribution) that would tend to yield an optimal nP bigger than unity and of order of few.

2.5 Conclusions

Although certainly not complete and exhaustive, this chapter aimed at giving an overview of what has been learned by combining CMB and large-scale structure observations and the potential of this combination for future surveys. I have also tried to explain, in part, how the cosmological information is encoded in these data sets and how it can be extracted. I have concentrated on some of the aspects that are not those more frequently discussed in summer schools. These lecture notes are no substitute for reading and studying the relevant papers, but hopefully it would help putting them in context of the big picture.

References

1. Baumann, D., et al. 2009, American Institute of Physics Conference Series, 1141, 10
2. Bennett, C. L., et al. 2003, ApJS, 148, 1
3. Cole, S., et al. 2005, MNRAS, 362, 505
4. Cooray, A., & Sheth, R. 2002, PysRep., 372, 1
5. Dickinson, C., et al. 2004, MNRAS, 353, 732
6. Dunkley, J., et al. 2009, ApJS, 180, 306
7. Dunkley, J. et al. 2009, American Institute of Physics Conference Series, 1141, 222
8. Eisenstein, D. J., et al. 2005, ApJ, 633, 560
9. Easther, R., & Kinney, W. H. 2003, PRD, 67, 043511
10. Eriksen, H. K., et al. 2007, ApJ, 656, 641
11. Feldman, H. A., Kaiser, N., & Peacock, J. A. 1994, ApJ, 426, 23
12. Heavens, A. F., & Taylor, A. N. 1995, MNRAS, 275, 483

13. Hivon, E., Górski, K. M., Netterfield, C. B., Crill, B. P., Prunet, S., & Hansen, F. 2002, ApJ, 567, 2
14. Kaiser, N. 1984, ApJLett, 284, L9
15. Kaiser, N. 1987, MNRAS, 227, 1
16. Kamionkowski, M., Kosowsky, A., & Stebbins, A. 1997, PRD, 55, 7368
17. Kinney, W. H. 1997, PRD, 56, 2002
18. Knox, L. 1995, PRD, 52, 4307
19. Kogut, A., et al. 2003, ApJS, 148, 161
20. Komatsu, E., et al. 2009, ApJS, 180, 330
21. Kowalski, M., et al. 2008, ApJ, 686, 749
22. Kovac, J. M., Leitch, E. M., Pryke, C., Carlstrom, J. E., Halverson, N. W., & Holzapfel, W. L. 2002, Nature, 420, 772
23. Kuo, C. L., et al. 2007, ApJ, 664, 687
24. Leitch, E. M., et al. 2002, Nature, 420, 763
25. Liddle, A. R. 2003, PRD, 68, 103504
26. Liddle, A. R., & Lyth, D. H. 2000, Cosmological Inflation and Large-Scale Structure, by Andrew R. Liddle and David H. Lyth, pp. 414. ISBN 052166022X. Cambridge, UK: Cambridge University Press, April 2000.,
27. Liddle, A. R., & Lyth, D. H. 1993, Phys.Rep., 231, 1
28. Liddle, A. R., & Lyth, D. H. 1992, Physics Letters B, 291, 391
29. Linder, E. V. 2005, PRD, 72, 043529
30. Macías-Pérez, J. F., & Bourrachot, A. 2006, A&A, 459, 987
31. Oh, S. P., Spergel, D. N., & Hinshaw, G. 1999, ApJ, 510, 551
32. Page, L., et al. 2007, ApJS, 170, 335
33. Peiris, H. V., et al. 2003, ApJS, 148, 213
34. Peiris, H. V., & Easther, R. 2008, Journal of Cosmology and Astro-Particle Physics, 7, 24
35. Percival, W. J., et al. 2004, MNRAS, 353, 1201
36. Percival, W. J., Verde, L., & Peacock, J. A. 2004, MNRAS, 347, 645
37. Percival, W. J., Cole, S., Eisenstein, D. J., Nichol, R. C., Peacock, J. A., Pope, A. C., & Szalay, A. S. 2007, MNRAS, 381, 1053
38. Percival, W. J., et al. 2009, arXiv:0907.1660
39. Readhead, A. C. S., et al. 2004, ApJ, 609, 498
40. Rees, M. J. 1968, ApJLett, 153, L1
41. Reid, B. A., et al. 2009, arXiv:0907.1659
42. Riess, A. G., et al. 2009, ApJ, 699, 539
43. Seo, H.-J., & Eisenstein, D. J. 2005, ApJ, 633, 575
44. Seo, H.-J., & Eisenstein, D. J. 2007, ApJ, 665, 14
45. Smith, R. E., et al. 2003, MNRAS, 341, 1311
46. Song, Y.-S., & Percival, W. J. 2009, Journal of Cosmology and Astro-Particle Physics, 10, 4
47. Spergel, D. N., & Zaldarriaga, M. 1997, Physical Review Letters, 79, 2180
48. Seljak, U., & Zaldarriaga, M. 1997, Physical Review Letters, 78, 2054
49. Zaldarriaga, M., & Seljak, U. 1997, PRD, 55, 1830

Chapter 3
Cosmology with Gravitational Lensing

Alan Heavens

Abstract In this chapter, I give an overview of gravitational lensing, concentrating on theoretical aspects, including derivations of some of the important results. Topics covered include the determination of surface mass densities of intervening lenses, as well as the statistical analysis of distortions of galaxy images by general inhomogeneities (cosmic shear), both in 2D projection on the sky and in 3D where source distance information is available. 3D mass reconstruction and the shear ratio test are also considered, and the sensitivity of observables to Dark Energy is used to show how its equation of state may be determined using weak lensing. Finally, the article considers the prospect of testing Einstein's General Relativity with weak lensing, exploiting the differences in growth rates of perturbations in different models.

3.1 Introduction

Gravitational lensing has emerged from being a curiosity of Einstein's General Relativity to a powerful cosmological tool. The reasons are partly theoretical, partly technological. The traditional tool of observational cosmology, the galaxy redshift survey, has become so large that statistical errors are very small, and questions about the fundamental limitations of this technique were raised. Fundamentally, studies of galaxy surveys will be limited by uncertain knowledge of galaxy formation and evolution—we will probably never know with high accuracy where galaxies should exist, even statistically, in a density field. Since it is the density field that is most directly predictable from fundamental theories (with some caveats), gravitational lensing is attractive as it is a direct probe. Furthermore, statistical analysis shows that large weak lensing surveys covering large fractions of the sky to a median redshift of around unity are in principle extremely powerful and can lead to error bars

Alan Heavens
Institute for Astronomy, University of Edinburgh, Blackford Hill, Edinburgh EH9 3HJ, U.K., e-mail: afh@roe.ac.uk

on cosmological parameters that are very small. In particular, the subtle effects of dark energy and even modifications to Einstein's General Relativity are potentially detectable with the sort of surveys that are being planned for the next decade. The limitations of lensing are likely to be in systematic errors arising from the ability to measure accurately shapes of galaxy images, physical effects aligning galaxies themselves, uncertain distances of sources, and uncertainties in the theoretical distribution of matter on small scales where baryon physics becomes important. Specially designed instrumentation and telescopes with excellent image quality help the situation a lot, and understanding of the physical systematics is now quite good. These notes will not be concerned very much with the practical issues, but more with the theoretical aspects of how lensing works, how it can be used to determine surface densities of clusters of galaxies (testing the dark matter content), how it can be used statistically on large scales (testing the dark energy properties) and how the combination of geometrical measurements and the growth rate of perturbations can probe the gravity law (testing so-called dark gravity). It is not a comprehensive review, and in particular does not cover the observational developments, which have seen the typical size of lensing surveys increase from 1 (past) $\to$ 100 (now) $\to 10^4$ (near future) square degrees. For more comprehensive recent reviews, see e.g., [62, 31] or the excellent SAAS-FEE lecture notes [65]. The structure of this review is as follows: section 2 covers basic lensing results, section 3 (Dark) Matter mapping, section 4 lensing on a cosmological scale, section 5 3D lensing and Dark Energy, and section 6 Dark Gravity.

3.2 Basics of Lensing

We begin with some basic results on light bending and derive results for the bending of light by an intervening lens with an arbitrary surface density.

3.2.1 The Bend Angle

The basic mechanism for gravitational lensing is that a point mass M will deflect a light beam through an angle

$$\tilde{\alpha}(r) = \frac{4GM}{rc^2} \qquad \text{point mass,} \tag{3.1}$$

where G is Newton's gravitation constant, c is the speed of light, and r is the distance of closest approach of the ray to the mass. This deflection angle is calculated using Einstein's General Theory of Relativity (GR), but is simply a factor two larger than Newtonian theory would predict if one treats the photon as a massive particle. The light path is curved, but if we are looking at a source which is far beyond the lens, the light path can be approximated by two straight lines, with a bend angle $\tilde{\alpha}$ between them. This is the *thin-lens* approximation.

The bend angle for a point mass equation (3.1) can be used to calculate the bend angle for an arbitrary mass distribution. For example, the bend angle for a thin lens whose mass distribution projected on the sky is circularly symmetric depends on the projected mass enclosed within a circle centred on the middle of the lens and extending out to the projected radius R of the ray:

$$\tilde{\alpha}(R) = \frac{4GM(<R)}{Rc^2} \qquad \text{circularly symmetric mass,} \qquad (3.2)$$

where $M(<R)$ is the mass enclosed within a projected radius R.

3.2.2 The Lens Equation

The lens equation relates the true position on the sky of a source to the position of its image(s). We can get the basic idea of gravitational lensing by considering a point mass or a circularly symmetric mass distribution. Figure 3.1 shows the geometry of the situation. The lens equation comes simply by noting that $PS_1 = PS + SS_1$ (where P is the unmarked point directly above L). If we denote the observer-source distance by D_S, the lens-source distance by D_{LS}, the position on the sky of the image by an angle θ, and the position on the sky of the unlensed source by β, then, for small angles, this translates to

$$D_S\theta = D_S\beta + D_{LS}\tilde{\alpha}, \qquad (3.3)$$

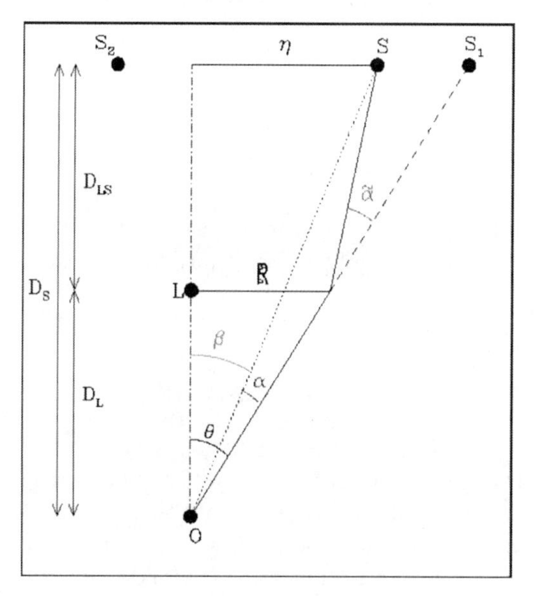

Fig. 3.1 The geometry for a thin lens (adapted from J. Wambsganss).

and we note that the bend angle is a function of the distance of the impact parameter of the ray R, $\tilde{\alpha} = \tilde{\alpha}(R = D_L\theta)$, where D_L is the distance to the lens. For cosmological applications, the D_i are angular diameter distances.

For convenience, we define the scaled bend angle by

$$\alpha = \frac{D_{LS}}{D_S}\tilde{\alpha}. \tag{3.4}$$

This gives us the *lens equation*

$$\beta = \theta - \alpha. \tag{3.5}$$

Note that for a given source position β, this is an implicit equation for the image position(s) θ, and for general circularly symmetric mass distributions, the lens equation cannot be solved analytically. There may indeed be more than one solution leading to multiple images of the same source. The lens equation is, however, an explicit equation for the source position given an image position. This is a useful feature that can be exploited for ray-tracing simulations of thin-lens systems, when one can ray-trace backwards from the observer to the sources.

3.2.2.1 Point Mass Lens: Multiple Images and Einstein Rings

The generic possibility of multiple images can be illustrated nicely by a point mass lens, where the implicit lens equation can be solved analytically for the image positions. For a point mass M, the bend angle is $\tilde{\alpha}(R) = 4GM/(Rc^2) = 4GM/(D_L\theta c^2)$, so the lens equation is

$$\beta = \theta - \frac{4GM}{c^2\theta}\frac{D_{LS}}{D_L D_S}, \tag{3.6}$$

which is a quadratic for θ:

$$\theta^2 - \beta\theta - \theta_E^2 = 0, \tag{3.7}$$

where we have defined the *Einstein angle*

$$\theta_E \equiv \sqrt{\frac{4GM}{c^2}\frac{D_{LS}}{D_L D_S}}. \tag{3.8}$$

There are evidently two images of the source for a point mass lens, one either side of the lens, at angles of

$$\theta_\pm = \frac{\beta}{2} \pm \sqrt{\frac{\beta^2}{4} + \theta_E^2}. \tag{3.9}$$

The solutions are the intersections of the line $\alpha(\theta)$ and the straight line $\theta - \beta$. For $\theta < 0$, the bend is in the opposite direction, so $\alpha < 0$. We see that, with the point mass, there are inevitably two solutions for $\beta \neq 0$.

There is a special case when the observer, lens, and source are lined up ($\beta = 0$), when one singular solution is the straight-line path, and the other, at $\theta = \theta_E$, is an

Einstein ring. Clearly, from the rotational symmetry of the arrangement, the image will appear at θ_E in all directions round the lens. A near-perfect Einstein ring is shown in the colour plate section.

Notice that if $|\beta| \gg \theta_E$, then the main image is perturbed only slightly, at $\beta + \theta_E^2/\beta + O(\beta^{-3})$, and the second image is very close to the lens at $\theta_- \simeq -\theta_E^2/\beta$. As we will see shortly, this second image is very faint. θ_E is a useful ballpark angle for determining whether the deflection of the source rays is significant or not. Typically, it is about an arcsecond for lensing by large galaxies at cosmological distances and microarcseconds for lensing of stars in nearby galaxies by stars in the Milky Way.

3.2.2.2 Magnification and Amplification

As we have seen, simple lenses alter the positions of the image of the source and may indeed produce multiple images. Another important, and detectable, effect is that the apparent size and the brightness of the source will change when its light undergoes a lensing event. The gravitational bending of light preserves surface brightness, so the change in apparent size is accompanied by a similar change in brightness.

For the circularly symmetric lenses, the ratio of the solid angle of the source and that of the image gives the amplification of an infinitesimally small source:

$$A = \frac{\theta}{\beta}\frac{d\theta}{d\beta}. \tag{3.10}$$

For a point lens, the amplifications of the two images are obtained straightforwardly by differentiating Eq. (3.9):

$$A_\pm = \frac{1}{2}\left(1 \pm \frac{\beta^2 + 2\theta_E^2}{\beta\sqrt{\beta^2 + 4\theta_E^2}}\right). \tag{3.11}$$

Note that the amplification can be negative. This corresponds to an image that is flipped with respect to the source.

In some cases, such as microlensing, where the two images are unresolved, one can only measure the total amplification,

$$A_{\text{total}} = |A_+| + |A_-| = \frac{\beta^2 + 2\theta_E^2}{\beta\sqrt{\beta^2 + 4\theta_E^2}} \tag{3.12}$$

and we note that the difference of the amplifications of images by a point lens is unity. In the limit $|\beta| \gg \theta_E$, $A_+ \to 1 + (\theta_E/\beta)^4$ and $A_- \to -(\theta_E/\beta)^4$, so the inner image is extremely faint in this limit.

3.2.3 General Thin Lens Mass Distributions

For a general distribution of mass, the bend angle is a 2D vector on the sky, $\boldsymbol{\alpha}$. The lens equation then generalises naturally to a vector equation

$$\boldsymbol{\beta} = \boldsymbol{\theta} - \boldsymbol{\alpha}. \tag{3.13}$$

For a thin lens with surface density $\Sigma(\boldsymbol{\theta}')$, the bend due at $\boldsymbol{\theta}$ to a small element of mass $dM = \Sigma(\boldsymbol{\theta}')d^2\boldsymbol{\theta}'$ is evidently in the direction $\boldsymbol{\theta} - \boldsymbol{\theta}'$ and has a magnitude $4GdM/(c^2|\boldsymbol{\theta} - \boldsymbol{\theta}'|)$, so the total bend angle is

$$\boldsymbol{\alpha}(\boldsymbol{\theta}) = \frac{4GD_LD_{LS}}{D_Sc^2} \int d^2\boldsymbol{\theta}' \frac{\Sigma(\boldsymbol{\theta}')(\boldsymbol{\theta} - \boldsymbol{\theta}')}{|\boldsymbol{\theta} - \boldsymbol{\theta}'|^2}. \tag{3.14}$$

The bend angle can be expressed as the (2D) gradient of the (thin lens) *lensing potential*, $\boldsymbol{\alpha} = \nabla\phi$, where

$$\phi(\boldsymbol{\theta}) = \frac{4GD_LD_{LS}}{c^2D_S} \int d^2\boldsymbol{\theta}'\Sigma(\boldsymbol{\theta}')\ln(|\boldsymbol{\theta} - \boldsymbol{\theta}'|) \qquad \text{(thin lens)}. \tag{3.15}$$

This potential satisfies the 2D Poisson equation

$$\nabla^2\phi(\boldsymbol{\theta}) = 2\kappa(\boldsymbol{\theta}), \tag{3.16}$$

where ∇ is a 2D gradient, with respect to angle. κ is the *convergence* $\kappa(\boldsymbol{\theta}) \equiv \Sigma(\boldsymbol{\theta})/\Sigma_{\text{crit}}$, with

$$\Sigma_{crit} \equiv \frac{c^2D_S}{4\pi GD_LD_{LS}} \tag{3.17}$$

being the *critical surface density*. The 2D Poisson equation is extremely useful, as it allows us to estimate the surface mass density of an intervening lens from lensing measurements, as we shall see later.

3.2.3.1 Convergence, Magnification, and Shear for General Thin Lenses

For a general distribution of mass within a thin lens, the magnification and distortion of an infinitesimal source are given by the transformation matrix from the source position $\boldsymbol{\beta}$ to the image position(s) $\boldsymbol{\theta}$. From the vector lens Eq. (3.13). The (inverse) amplification matrix is

$$A_{ij} \equiv \frac{\partial\beta_i}{\partial\theta_j} = \delta_{ij} - \phi_{ij}, \tag{3.18}$$

where $\phi_{ij} \equiv \partial^2\phi/\partial\theta_i\partial\theta_j$. We see that A is symmetric, and it can be decomposed into an isotropic expansion term and a shear. A general amplification matrix also includes a rotation term (the final degree of freedom being the rotation angle), but we see that weak lensing does not introduce rotation of the image and has only

3 degrees of freedom, rather than the four possible in a 2×2 matrix. We decompose the amplification matrix as follows:

$$A_{ij} = \begin{pmatrix} 1-\kappa & 0 \\ 0 & 1-\kappa \end{pmatrix} + \begin{pmatrix} -\gamma_1 & -\gamma_2 \\ -\gamma_2 & \gamma_1 \end{pmatrix}, \tag{3.19}$$

where κ is called the *convergence* and

$$\gamma = \gamma_1 + i\gamma_2 \tag{3.20}$$

is the *complex shear*. For weak lensing, both $|\kappa|$ and $|\gamma_i|$ are $\ll 1$. A nonzero κ represents an isotropic expansion or contraction of a source; $\gamma_1 > 0$ represents an elongation of the image along the x-axis and contraction along y. $\gamma_1 < 0$ stretches along y and contracts along x. $\gamma_2 \neq 0$ represents stretching along $x = \pm y$ directions. The effects are shown in Fig. 3.2.

Making the decomposition, we find that

$$\begin{aligned} \kappa &= \frac{1}{2}(\phi_{11} + \phi_{22}) \\ \gamma_1 &= \frac{1}{2}(\phi_{11} - \phi_{22}) \equiv D_1\phi \\ \gamma_2 &= \phi_{12} \equiv D_2\phi. \end{aligned} \tag{3.21}$$

which defines the two operators $D_{1,2}$, and it is straightforward to prove that

$$D_i D_i = (\nabla^2)^2, \tag{3.22}$$

where the summation is over $i = 1, 2$.

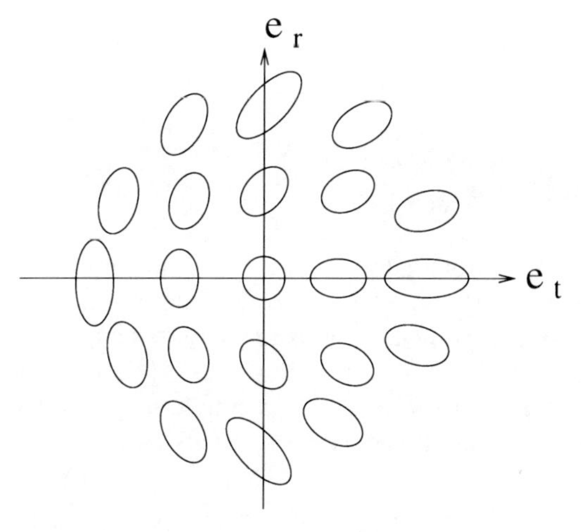

Fig. 3.2 The effect of shear distortions on a circular source. In the notation of the current chapter, e_t and e_r are the real and imaginary parts of the ellipticity (or shear). From [79].

Note that $\kappa > 0$ corresponds to *magnification* of the image. Lensing preserves surface brightness, so this also amounts to *amplification* of the source flux. The magnification is

$$A = \frac{1}{\det A_{ij}} = \frac{1}{(1-\kappa)^2 - |\gamma|^2}. \tag{3.23}$$

We see that we may have infinite amplifications if $\kappa \geq 1$. Such effects apply only for infinitesimal sources, and places in the source plane, which lead to infinite magnifications, are referred to as *caustics*, and it leads to highly distorted images along lines called *critical lines* in the lens plane. The giant arcs visible in images of some rich clusters lie on or close to critical lines. For cosmic shear, due to the general inhomogeneities along the line-of-sight, κ and $|\gamma|$ are typically 0.01, and the lensing is weak.

It is worth noting that the amplification matrix may be written

$$A_{ij} = (1-\kappa)\begin{pmatrix} 1-g_1 & -g_2 \\ -g_2 & 1+g_1 \end{pmatrix}, \tag{3.24}$$

where $g \equiv \gamma/(1-\kappa)$ is called the *reduced shear*. Since the $1-\kappa$ multiplier affects only the overall size (and hence brightness) of the source, but not its shape, we see that shear measurements can determine only the reduced shear and not the shear itself. For weak lensing, $\kappa \ll 1$, so the two coincide to linear order.

Note that a rotation of $\pi/4$ in the coordinate system changes a real shear into an imaginary shear - i.e., the complex shear rotates by $\pi/2$, twice the angle of rotation of the coordinate system. This behaviour is characteristic of a *spin-weight* 2 field, and is encountered also in microwave background polarisation and gravitational wave fields.

The shear field is in principle observable, so let us see how it can be used to estimate the surface mass density of a lens.

3.2.3.2 Estimating Shear

The estimation of shear from galaxy image data is a complex business, and I will give no more than a highly simplified sketch. For more details, see reviews such as [62, 79, 31]. One way to estimate shear is to measure the *complex ellipticity* e of a galaxy, which can be defined in terms of moments [46] even if the galaxy image is not elliptical in shape. Figure 3.2 shows how simple shapes map onto ellipticity. In the limit of weak distortions, the observed ellipticity is

$$e = \frac{e^s + \gamma}{1 + \gamma^* e^s}, \tag{3.25}$$

where e^s is the undistorted source ellipticity. The source ellipticity has a dispersion $\langle |e^s|^2 \rangle \simeq 0.3$, whereas γ itself is usually much smaller, 0.01–0.1. If we average over some sources,

$$\langle e \rangle = \gamma \tag{3.26}$$

so we can use the average ellipticity as an estimator of shear.

Two points to note:

1. The estimator is noisy, dominated by the intrinsic dispersion of e, so γ has a variance $\sim \langle |e^s|^2 \rangle / N$ if N sources are averaged.

2. In these notes, the formulae will often refer to γ; it should be noted that in reality it will be e, a noisy estimate of γ, which is used in practice.

The main practical difficulty of lensing experiments is that the atmosphere and telescope affect the shape of the images. These modifications to the shape may arise due to such things as the point spread function, or poor tracking of the telescope. The former needs to be treated with great care. Stars (whose images should be round) can be used to remove image distortions to very high accuracy, although a possibly fundamental limitation may arise because of the finite number of stars in an image. Interpolation of the anisotropy of the PSF needs to be done carefully, and examples of how this can be done in an optimal way are given in [80]. Bayesian methods are beginning to be used, with *lensfit* [61, 51] showing very promising results.

3.3 Dark Matter

Theoretical work with numerical simulations indicates that in the absence of the effects of baryons, virialised Dark Matter haloes should follow a uniform "NFW" profile, $\rho(r) = \rho_s(r_s/r)(1+r/r_s)^{-2}$ [63], if the Dark Matter is cold (CDM). Simulations also predict how the physical size of the clusters should depend on mass, characterised by the concentration index $c_s \equiv r_{vir}/r_s$, where r_{vir} is defined as the radius within which the mean density is 200 times the background density. Roughly, $c_s \propto M^{-0.1}$. This can be tested by measuring the shear signal and stacking the results from many haloes to increase signal-to-noise.

3.3.1 2D Mass Surface Density Reconstruction

We wish to take a map of estimated shear (ellipticities) and estimate the surface mass density of the intervening lens system. For simplicity here, we assume the sources are at the same distance. The classic way to do this was given by [45].

In some respects, it is easier to work in Fourier space. Expanding

$$\kappa_{\mathbf{k}} \equiv \int d^2\theta \, \kappa(\boldsymbol{\theta}) \exp(i\mathbf{k}.\boldsymbol{\theta}) \tag{3.27}$$

etc, then evidently

$$\begin{aligned}\kappa_{\mathbf{k}} &= -\frac{1}{2}k^2\phi_{\mathbf{k}} \\ \gamma_{1\mathbf{k}} &= -\frac{1}{2}(k_1^2 - k_2^2)\phi_{\mathbf{k}} \\ \gamma_{2\mathbf{k}} &= -k_1k_2\phi_{\mathbf{k}},\end{aligned} \tag{3.28}$$

where $k^2 = k_1^2 + k_2^2 = |\mathbf{k}|^2$.

We see that the following are estimators of $\kappa_{\mathbf{k}}$:

$$\gamma_{1\mathbf{k}}\frac{k^2}{k_1^2 - k_2^2}; \qquad \gamma_{2\mathbf{k}}\frac{k^2}{2k_1k_2}. \tag{3.29}$$

In practice, we use the shapes of galaxies to estimate shear, so our estimates of $\gamma_{i\mathbf{k}}$ are dominated by noise in the form of scatter in the intrinsic shapes of galaxies. Thus, the variance of each of these estimators is determined by the variances in $|e_{i\mathbf{k}}|^2$. The variance of the first estimator is, therefore, proportional to $k^4/(k_1^2 - k_2^2)^2$, and the second estimator to $k^4/(2k_1k_2)^2$. The optimal estimator for $\kappa_{\mathbf{k}}$ is given by the standard inverse variance weighting, giving

$$\kappa_{\mathbf{k}} = \left(\frac{k_1^2 - k_2^2}{k^2}\right)\gamma_{1\mathbf{k}} + \left(\frac{2k_1k_2}{k^2}\right)\gamma_{2\mathbf{k}}. \tag{3.30}$$

Since this is the sum of two terms, each of which is a multiplication in **k** space, it represents a convolution in real space. (An exercise for the enthusiastic reader is to Fourier transform to find the real-space convolution). In fact, it is easier to find the convolution by noting that

$$\begin{aligned}\gamma_i &= D_i\phi \\ &= 2D_i\nabla^{-2}\kappa \\ \Rightarrow \kappa &= 2D_i\nabla^{-2}\gamma_i,\end{aligned} \tag{3.31}$$

where the last line follows from Eq. 3.22. Now, we know the solution to the 2D Poisson equation—it is given in Eq. 3.15:

$$\nabla^{-2}\gamma_i(\boldsymbol{\theta}) = \frac{1}{2\pi}\int d^2\theta'\,\gamma_i(\boldsymbol{\theta}')\ln|\boldsymbol{\theta}' - \boldsymbol{\theta}|. \tag{3.32}$$

Upon differentiation with D_i and summation, we find

$$\kappa(\boldsymbol{\theta}) = \frac{2}{\pi}\int d^2\theta'\,\frac{[\gamma_1(\boldsymbol{\theta}')\cos(2\psi) + \gamma_2(\boldsymbol{\theta}')\sin(2\psi)]}{|\boldsymbol{\theta}' - \boldsymbol{\theta}|^2}, \tag{3.33}$$

where ψ is the angle between $\boldsymbol{\theta}$ and $\boldsymbol{\theta}'$.

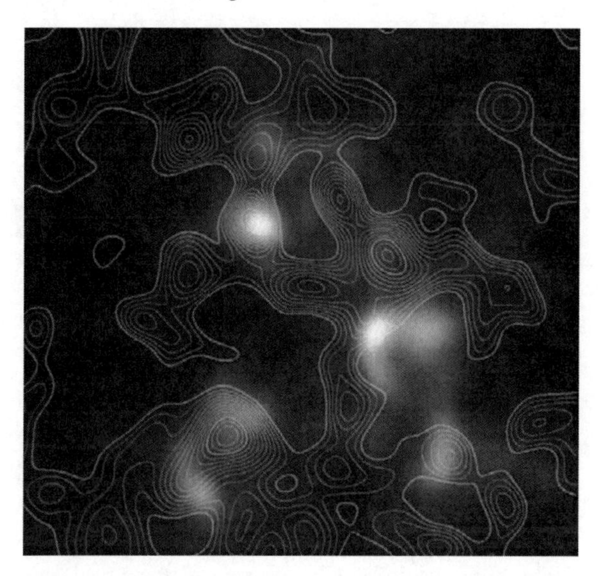

Fig. 3.3 2D reconstruction of surface matter density from the COMBO17 A901 study (contours; [21]) superimposed on galaxy surface density.

It is quite tempting to work with Eq. 3.33 in its discrete form, replacing the integral by a sum over shear estimates at the galaxy locations g, i.e., considering the following estimator for κ:

$$\hat{\kappa}(\boldsymbol{\theta}) = \frac{1}{\bar{n}} \sum_g \frac{[\gamma_1(\boldsymbol{\theta}_g)\cos(2\psi) + \gamma_2(\boldsymbol{\theta}_g)\sin(2\psi)]}{|\boldsymbol{\theta}_g - \boldsymbol{\theta}|^2}, \tag{3.34}$$

where $\bar{n}$ is the mean surface density of sources. This estimator would, however, be a mistake. It is unbiased, but it has the awkward and undesirable property of having infinite noise. This can be seen as follows:

The fourier transform of Eq. 3.34 is

$$\hat{\kappa}_{\mathbf{k}} = \frac{1}{\bar{n}} \left[\left(\frac{k_1^2 - k_2^2}{k^2} \right) \sum_g \gamma_1(\boldsymbol{\theta}_g) \exp(i\mathbf{k}.\boldsymbol{\theta}_g) + \left(\frac{2k_1 k_2}{k^2} \right) \sum_g \gamma_2(\boldsymbol{\theta}) \exp(i\mathbf{k}.\boldsymbol{\theta}_g) \right]. \tag{3.35}$$

If we assume (not quite correctly—see a later discussion on intrinsic alignments; but this will do here) that the galaxy shapes are uncorrelated, then

$$\langle \hat{\kappa}_{\mathbf{k}} \hat{\kappa}^*_{\mathbf{k}} \rangle = \frac{\langle |e^2| \rangle}{2\bar{n}}, \tag{3.36}$$

where $\langle |e^2| \rangle$ is the dispersion in galaxy ellipticities. Note that as usual, the estimate of the shear is dominated by the intrinsic ellipticity. If we use Parseval's theorem, we see that the variance in the recovered convergence, being the integral of Eq. 3.36 over $\mathbf{k}$, diverges.

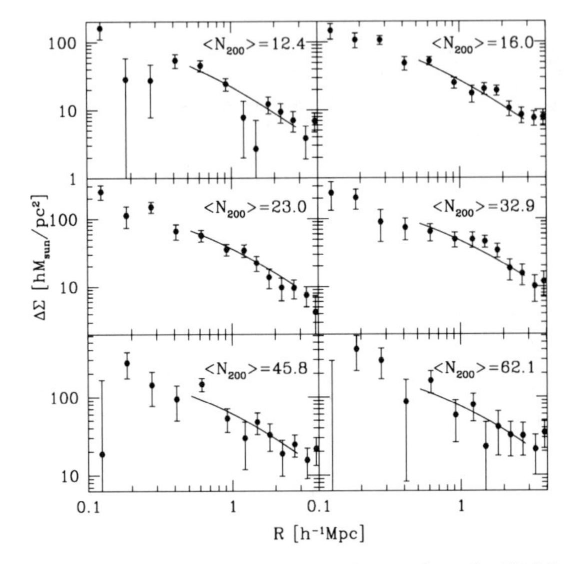

Fig. 3.4 Excess surface density from stacked galaxy clusters from the SDSS survey, with best-fitting NFW profiles. N_{200} is a measure of the richness of the clusters. From [58].

The solution is to smooth the field, which can be done at any stage—either by smoothing the estimated shears (or averaging them in cells) or by working in Fourier space and introducing a filter, or even by smoothing the estimated noisy convergence field. The amount of smoothing required can prevent high-resolution maps of all but the richest clusters, but the advent of high-surface densities with space telescopes has improved this situation considerably.

Finally, we note that the estimate of the zero-wavenumber coefficient κ_0 is undetermined by Eq. 3.35. We cannot determine the mean surface density this way. This feature is an example of the so-called "mass-sheet degeneracy". Other measurements (amplification/magnification or by observing sources at different distances behind the lens) can alleviate this problem or one can assume that $\kappa \to 0$ as one goes far from the lens centre.

3.3.2 Testing the Navarro–Frenk–White Profile of CDM

Figure 3.4 shows the average radial surface density profiles for clusters identified in the sloan digital sky survey (SDSS), grouped by number of cluster galaxies, and Navarro–Frenk–White (NFW) fits superimposed [58]. Figure 3.5 shows that the observed concentration indices are close to the theoretical predictions, but some tension exists. Broadly, weak lensing data on clusters, therefore, support the CDM model. We will consider 3D mapping later, but unfortunately, the limited accuracy of photo-zs ($\sim 0.03 \simeq 100h^{-1}$ Mpc typically) means the 3D mass map is smoothed heavily in the radial direction, and this limits the usefulness of 3D mapping for testing the NFW profile.

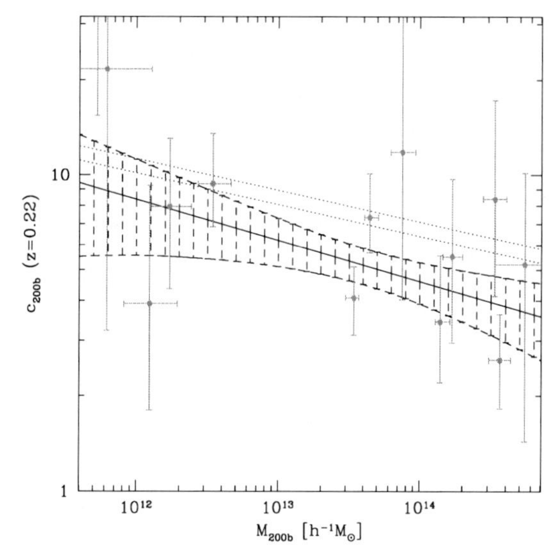

Fig. 3.5 Concentration indices from SDSS clusters as a function of mass, compared with simulation (dotted, for different cosmologies). Dashed regions show range assuming a power law $c_s - M$ relation. From [58].

A more radical test of theory has been performed with the Bullet Cluster [15], actually a pair of clusters which have recently passed through each other. There are two clear peaks in the surface density of galaxies, and X-ray emission from hot shocked gas in between. In the standard cosmological model, this makes perfect sense, as the galaxies in the clusters are essentially collisionless. If the (dominant) Dark Matter is also collisionless, then we would expect to see surface mass concentrations at the locations of the optical galaxy clusters, and this is exactly what is observed. In MOND or TeVeS models without Dark Matter, one would expect the surface mass density to peak where the dominant baryon component is—the X-ray gas. This is not seen. A caveat is that it is not quite the surface density which is observed, rather the *convergence*, which is related to the distortion pattern of the galaxy images and which is proportional to surface density in GR, but not in MOND/TeVeS. However, no satisfactory explanation of the bullet cluster has been demonstrated without Dark Matter.

3.4 Cosmological Lensing

Gravitational lensing is strong if the distortions to the images are substantial and weak if the distortions are small. Weak lensing of background images can be used to learn about the mass distribution in individual objects, such as galaxy clusters, but we concentrate now on weak lensing on a cosmological scale, which is usually analysed in a statistical way. Reviews of this area include [62, 31, 6, 69, 79].

The basic effect of weak lensing is for the clumpy matter distribution to perturb slightly the trajectories of photon paths. By considering how nearby light paths are perturbed, one finds that the shapes of distant objects are changed slightly. Associated with the size change is a change in the brightness of the source. Size and magnitude changes can, in principle, be used to constrain the properties of the matter distribution along the line-of-sight (and cosmological parameters as well), but it is the change in shape of background sources that has almost exclusively been used in cosmological weak lensing studies. The reason for this is simply that the signal-to-noise is better. These notes will concentrate on shear (=shape changes), but the magnification and amplification of sources can also be used and will probably be used in future when the surveys are larger. The great promise of lensing is that it acts as a direct probe of the matter distribution (whether dark or not) and avoids the use of objects that are assumed to trace the mass distribution in some way, such as galaxies in large-scale structure studies. Theoretically, lensing is very appealing, as the physics is very simple and very robust, direct connections can be made between weak lensing observables and the statistical properties of the matter distribution. These statistical properties are dependent on cosmological parameters in a known way, so weak lensing can be employed as a cosmological tool. The main uncertainties in lensing are observational – it is very challenging to make images of the necessary quality. In this section, we concentrate here on the weak effects on a cosmological scale of the nonuniform distribution of matter all the way between the source and observer, an effect often referred to as *cosmic shear*.

3.4.1 Distortion of Light Bundles

The distortion of a light bundle has to be treated with GR, but if one is prepared to accept one modification to Newtonian physics, one can do without GR.

In an expanding universe, it is usual to define a *comoving coordinate* $\mathbf{x}$, such that "fundamental observers" retain the same coordinate. Fundamental observers are characterised by the property of seeing the Universe as isotropic; the Earth is not (quite) a fundamental observer, as from here the Cosmic Microwave Background looks slightly anisotropic. The equation of motion for the transverse coordinates (about some fiducial direction) of a photon in a flat universe is

$$\frac{d^2 x_i}{d\eta^2} = -\frac{2}{c^2}\frac{\partial \Phi}{\partial x_i}. \qquad i = 1,2 \tag{3.37}$$

We assume a flat universe throughout, but the generalisation to nonflat universes is straightforward (there is an extra term in the equation above, and some r symbols need to be changed to an angular diameter distance).

The equation of motion can be derived using General Relativity(GR) see Appendix for details). We will use $x_i, i = 1,2$ for coordinates transverse to the line-of-sight, and r to indicate the radial coordinate. η is the conformal time, related to the coordinate t by $d\eta = cdt/R(t)$ and $R(t)$ is the cosmic scale factor, equal to R_0 at the present time. $\Phi(x_i, r)$ is the peculiar gravitational potential related to the matter overdensity field $\delta \equiv \delta\rho/\rho$ by Poisson's equation

$$\nabla^2_{3D}\Phi = \frac{3H_0^2\Omega_m}{2a(t)}\delta, \tag{3.38}$$

where H_0 is the Hubble constant, Ω_m is the present matter density parameter, and $a(t) = R(t)/R_0 = (1+z)^{-1}$, where z is redshift.

The equation of motion is derived in the Appendix from General Relativity, in a (nearly flat) metric given in the Newtonian gauge by

$$ds^2 = (1+2\Phi/c^2)c^2dt^2 - (1-2\Phi/c^2)R^2(t)(dr^2 + r^2d\theta^2 + r^2\sin^2\theta d\varphi^2). \tag{3.39}$$

From a Newtonian point-of-view, Eq. 3.37 is understandable if we note that time is replaced by η (which arises because we are using comoving coordinates), and there is a factor 2 that does not appear in Newtonian physics. This same factor of two gives rise to the famous result that in GR the angle of light bending round the Sun is double that of Newtonian theory.

The coordinates $\mathbf{x}_i$ are related to the (small) angles of the photon to the fiducial direction $\boldsymbol{\theta} = (\theta_x, \theta_y)$ by $\mathbf{x}_i = r\boldsymbol{\theta}_i$.

3.4.2 Lensing Potential

The solution to (3.37) is obtained by first noting that the zero-order ray has $ds^2 = 0 \Rightarrow dr = -d\eta$, where we take the negative root because the light ray is incoming. Integrating twice and reversing the order of integration gives

$$x_i = r\theta_i - \frac{2}{c^2}\int_0^r dr' \frac{\partial\Phi}{\partial x_i'}(r-r'). \tag{3.40}$$

We now perform a Taylor expansion of $\partial\Phi/\partial x_i'$ and find the deviation of two nearby light rays is

$$\Delta x_i = r\Delta\theta_i - \frac{2}{c^2}\Delta\theta_j \int_0^r dr' r'(r-r')\frac{\partial^2\Phi}{\partial x_i'\partial x_j'}, \tag{3.41}$$

which we may write as

$$\Delta x_i = r\Delta\theta_j(\delta_{ij} - \phi_{ij}), \tag{3.42}$$

where δ_{ij} is the Kronecker delta ($i = 1,2$) and we define

$$\phi_{ij}(\mathbf{r}) \equiv \frac{2}{c^2}\int_0^r dr' \frac{(r-r')}{rr'}\frac{\partial^2\Phi(\mathbf{r}')}{\partial\theta_i\partial\theta_j}. \tag{3.43}$$

The integral is understood to be along a radial line (i.e. $\mathbf{r} \parallel \mathbf{r}'$); this is the *Born approximation*, which is a very good approximation for weak lensing [8, 66, 78]. In reality, the light path is not quite radial.

It is convenient to introduce the *(cosmological) lensing potential*, which controls the distortion of the ray bundle:

$$\phi(\mathbf{r}) \equiv \frac{2}{c^2} \int_0^r dr' \frac{(r-r')}{rr'} \Phi(\mathbf{r}') \qquad \text{(cosmological; flat universe)}. \tag{3.44}$$

Note that $\phi_{ij}(\mathbf{r}) = \partial^2 \phi(\mathbf{r})/\partial\theta_i \partial\theta_j$. So, remarkably, we can describe the distortion of an image as it passes through a clumpy universe in a rather simple way. The shear and convergence of a source image is obtained from the potential in the same way as the thin lens case (Eq. 3.22), although the relationship between the convergence and the foreground density is more complicated, as we see next.

3.4.2.1 Relationship to Matter Density Field

The gravitational potential Φ is related to the matter overdensity field $\delta \equiv \delta\rho/\rho$ by Poisson's equation (3.38). The convergence is then

$$\kappa(\mathbf{r}) = \frac{3H_0^2\Omega_m}{2c^2} \int_0^r dr' \frac{r'(r-r')}{r} \frac{\delta(\mathbf{r}')}{a(r')}. \tag{3.45}$$

Note that there is an extra term $\partial^2\Phi/\partial r'^2$ in ∇_{3D}^2, which integrates to zero to the order to which we are working.

3.4.2.2 Averaging Over a Distribution of Sources

If we consider the distortion averaged over a distribution of sources with a radial distribution $p(r)$ (normalised such that $\int dr\, p(r) = 1$), the average distortion is again obtained by reversing the order of integration:

$$\Delta x_i = r\Delta\theta_j \left(\delta_{ij} - \frac{2}{c^2} \int_0^r \frac{dr'}{r'} g(r') \frac{\partial^2\Phi(\mathbf{r}')}{\partial\theta_i \partial\theta_j} \right), \tag{3.46}$$

where

$$g(r) \equiv \int_r^\infty dr'\, p(r') \frac{r'-r}{r'}. \tag{3.47}$$

In order to estimate $p(r)$, surveys began to estimate distances to source galaxies using photometric redshifts. This has opened up the prospect of a full 3D analysis of the shear field, which we will discuss briefly later in this article.

3.4.2.3 Convergence Power Spectrum and Shear Correlation Function

The average shear is zero, so the most common statistics to use for cosmology are two-point statistics, quadratic in the shear. These may be in "configuration" ("real") space or in transform space (using Fourier coefficients or similar). I will focus on two quadratic measures, the convergence power spectrum and the shear–shear correlation function.

To find the expectation value of a quadratic quantity, it is convenient to make use of the matter density power spectrum, $P(k)$, defined by the following relation between the overdensity Fourier coefficients:

$$\langle \delta_{\mathbf{k}} \delta^*_{\mathbf{k}'} \rangle = (2\pi)^3 \delta^D(\mathbf{k}-\mathbf{k}')P(k), \tag{3.48}$$

where δ^D is the Dirac delta function. $P(k)$ is evolving, so we write it as $P(k;r)$ in future, where r and t are related through the lookback time. (This r-dependence may look strange; there is a subtlety: (3.48) holds if the field is homogeneous and isotropic, which the field on the past light cone is not, since it evolves. In the radial integrals, one has to consider the homogeneous field at the same cosmic time as the time of emission of the source). The trick is to get the desired quadratic quantity into a form, which includes $P(k;r)$.

For the convergence power spectrum, we first transform the convergence in a 2D Fourier transform on the sky, where ℓ is a 2D dimensionless wavenumber:

$$\kappa_\ell = \int d^2\boldsymbol{\theta}\, \kappa(\boldsymbol{\theta}) e^{-i\ell.\boldsymbol{\theta}}, \tag{3.49}$$

$$= A\int_0^\infty dr\, r\frac{g(r)}{a(r)}\int d^2\boldsymbol{\theta}\, \delta(r\boldsymbol{\theta},r) e^{-i\ell.\boldsymbol{\theta}}, \tag{3.50}$$

where $A \equiv 3H_0^2\Omega_m/2c^2$. We expand the overdensity field in a Fourier transform,

$$\delta(r\boldsymbol{\theta},r) = \int \frac{d^3\mathbf{k}}{(2\pi)^3}\, \delta_{\mathbf{k}} e^{ik_\parallel r} e^{i\mathbf{k}_\perp .r\boldsymbol{\theta}} \tag{3.51}$$

and substitute into (3.50). We form the quantity $\langle \kappa_\ell \kappa^*_{\ell'} \rangle$, which, by analogy with (3.48), is related to the (2D) *convergence power spectrum* by

$$\langle \kappa_\ell \kappa^*_{\ell'} \rangle = (2\pi)^2 \delta^D(\ell-\ell')P_\kappa(|\ell|). \tag{3.52}$$

Straightforwardly,

$$\langle \kappa_\ell \kappa^*_{\ell'} \rangle = A^2\int_0^\infty dr\, G(r)\int_0^\infty dr'\, G(r')\int d^2\boldsymbol{\theta} d^2\boldsymbol{\theta}'\frac{d^3\mathbf{k}}{(2\pi)^3}\frac{d^3\mathbf{k}'}{(2\pi)^3} \tag{3.53}$$
$$\langle \delta_{\mathbf{k}} \delta^*_{\mathbf{k}'} \rangle \exp(ik_\parallel r - ik_\parallel{}' r')\exp(i\mathbf{k}_\perp .\boldsymbol{\theta} - i\mathbf{k}'_\perp .\boldsymbol{\theta}')\exp(-i\ell.\boldsymbol{\theta} + i\ell'.\boldsymbol{\theta}'),$$

where $G(r) \equiv rg(r)/a(r)$. Using (3.48), we remove the $\mathbf{k}'$ integration, introducing the power spectrum $P(k) = P(\sqrt{k_\parallel^2 + |\mathbf{k}_\perp|^2})$. For small-angle surveys, most of the signal comes from short wavelengths, and the $k_\parallel$ is negligible, so $P(k) \simeq P(|\mathbf{k}_\perp|)$. The only $k_\parallel$ term remaining is the exponential, which integrates to $(2\pi)\delta^D(r-r')$. The integrals over $\boldsymbol{\theta}$ and $\boldsymbol{\theta}'$ give $(2\pi)^2\delta^D(\boldsymbol{\ell}-r\mathbf{k}_\perp)$ and $(2\pi)^2\delta^D(\boldsymbol{\ell}'-r\mathbf{k}_\perp{}')$, respectively, so the whole lot simplifies to give the convergence power spectrum as

$$P_\kappa(\ell) = \left(\frac{3H_0^2\Omega_m}{2c^2}\right)^2 \int_0^\infty dr \left[\frac{g(r)}{a(r)}\right]^2 P(\ell/r;r). \tag{3.54}$$

An exercise for the reader is to show that the power spectrum for the complex shear field γ is the same: $P_\gamma = P_\kappa$. The shear correlation function, for points separated by an angle θ is

$$\begin{aligned} \langle\gamma\gamma^*\rangle_\theta &= \int \frac{d^2\ell}{(2\pi)^2} P_\gamma(\ell)\, e^{i\boldsymbol{\ell}.\boldsymbol{\theta}} \\ &= \int \frac{\ell d\ell}{(2\pi)^2} P_\kappa(\ell) e^{i\ell\theta\cos\varphi} d\varphi \\ &= \int \frac{d\ell}{2\pi} \ell P_\kappa(\ell) J_0(\ell\theta), \end{aligned} \tag{3.55}$$

where we have used polar coordinates, with φ the angle between $\boldsymbol{\ell}$ and $\boldsymbol{\theta}$, and we have exploited the isotropy (P_κ depends only on the modulus of $\boldsymbol{\ell}$). J_0 is a Bessel function.

Other quadratic quantities (examples are shear variances on different scales, Aperture Mass (squared)) can be written similarly as integrals over the power spectrum, with different kernel functions.

3.4.3 Matter Power Spectrum

As we have seen, the two-point statistics of the shear and convergence fields depend on the power spectrum of the matter, $P(k;t)$. The power spectrum grows in a simple way when the perturbations in the overdensity are small, $|\delta| \ll 1$, and when the power spectrum grows in amplitude while keeping the same shape as a function of k. However, gravitational lensing can still be weak, even if the overdensity field is nonlinear. Poisson's equation still holds provided we are in the weak-field limit as far as GR is concerned, and this essentially always holds for cases of practical interest. In order to get as much statistical power out of lensing, one must probe the nonlinear regime, so it is necessary for parameter estimation to know how the power spectrum grows. Through the extensive use of numerical simulations, the growth of dark matter clustering is well understood down to quite small scales, where uncertainties in modelling, or uncertain physics, such as the influence of baryons on the dark matter [81], make the predictions unreliable. Accurate fits for the nonlinear power spectrum have been found [71] up to $k > 10hMpc^{-1}$, which is far beyond the

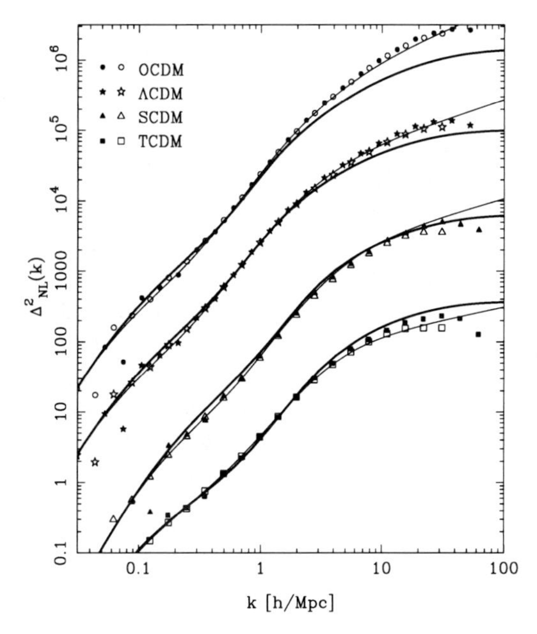

Fig. 3.6 The nonlinear power spectrum from numerical simulations, along with fitting functions (from [71].

linear/nonlinear transition $k \sim 0.2hMpc^{-1}$. Figure 3.6 shows fits for a number of CDM models. For precision use, one must make sure that the statistics do not have substantial contributions from the high-k end where the nonlinear power spectrum is uncertain. This can be explored by looking at the kernel functions implicit in the quantities, such as the shear correlation function (3.56).

3.4.4 Intrinsic Alignments

The main signature of weak lensing is a small alignment of the images, at the level of a correlation of ellipticities of $\sim 10^{-4}$. One might be concerned that physical processes might also induce an alignment of the galaxies themselves. In the traditional lensing observations, the distances of individual galaxies are ignored, and one simply uses the alignment on the sky of galaxies, and one might hope that the galaxies will typically be at such large separations along the line-of-sight that any physical interactions would be rare and can be ignored. However, the lensing signal is very small, so the assumption that intrinsic alignment effects are sufficiently small needs to be tested. This was first done in a series of papers by a number of groups in 2000–2001 e.g., [23, 18, 16, 13], and the answer is that the effects may not be negligible. The contamination by intrinsic alignments is highly depth-dependent. This is easy to see, since at fixed angular separation, galaxies in a shallow survey will be physically closer together in space, and hence more likely to experience tidal interactions, which might align the galaxies. In addition to this, the shallower the survey, the smaller the lensing signal. In a pioneering study, the alignments of nearby galaxies in the SuperCOSMOS survey were investigated [11]. This survey is so shallow (median redshift ~ 0.1) that the expected lensing signal is tiny. A nonzero alignment was found, which agrees with at least some of the theoretical estimates of the effect. The main exception is the numerical study of Jing [44], which predicts a contamination so high that it could dominate even deep surveys. For deep surveys, the effect, which is sometimes called the "II correlation," is expected to be rather small, but if one wants to use weak lensing as a probe of subtle effects, such as the effects of altering the equation of state of dark energy, then one has to do something. There are essentially two options – either one tries to calculate the intrinsic alignment signal and subtract it or one tries to remove it altogether. The former approach is not practical, as, although there is some agreement as to the general level of the contamination, the details are not accurately enough known. The latter approach is becoming possible, as lensing surveys are now obtaining estimates of the distance to each galaxy, via photometric redshifts (spectroscopic redshifts are difficult to obtain because one needs a rather deep sample, with median redshift at least 0.6 or so, and large numbers, to reduce shot noise due to the random orientations of ellipticities). With photometric redshifts, one can remove physically close galaxies from the pair statistics (such as the shear correlation function)[26, 48]. Thus, one removes a systematic error in favour of a slightly increased statistical error. The analysis in [27] is the only study that has explicitly removed close pairs.

Another effect that is more subtle is the correlation between the orientation or a foreground galaxy and the orientation of the lensed image of a background galaxy. The latter is affected gravitationally by the tidal field in the vicinity of the former (as well as the tidal fields all the way along the line-of-sight), and if the orientation of the foreground galaxy is affected by the local tidal field, as it surely must at some level, then there can be a contamination of the cosmological lensing signal by what is sometimes referred to as the "GI correlation". This was first pointed out in [30], and it seems to be likely to be a significant effect [29]. It is less easy to deal with than the second correlation, but modelling and nulling methods exist, at the price of diminished signal-to-noise [10, 40].

3.4.5 E/B Decomposition

Weak gravitational lensing does not produce the full range of locally linear distortions possible. These are characterised by translation, rotation, dilation, and shear, with six free parameters. Translation is not readily observable, but weak lensing is specified by three parameters rather than the four remaining degrees of freedom permitted by local affine transformations. This restriction is manifested in a number of ways: e.g., the transformation of angles involves a 2×2 matrix that is symmetric, so it is not completely general, see Eq. (3.18). Alternatively, a general spin-weight 2

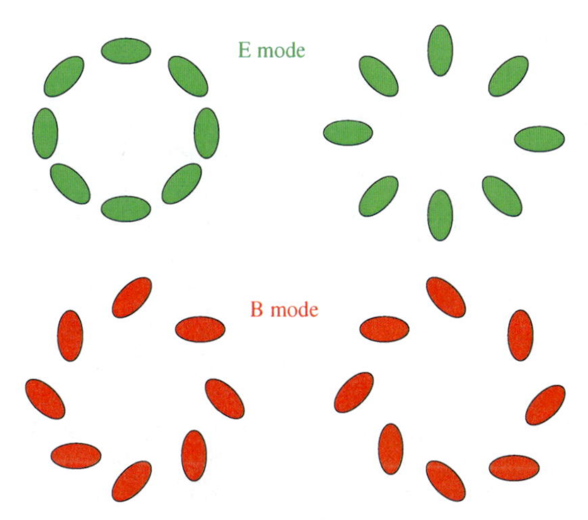

Fig. 3.7 Example patterns from E-mode and B-mode fields (from [79]). Weak lensing only produces E-modes at any significant level, so the presence of B-modes can indicate systematic errors.

field can be written in terms of second derivatives of a *complex* potential, where as the lensing potential is real. There are many other consistency relations that have to hold if lensing is responsible for the observed shear field. In practice, the observed ellipticity field may not satisfy the expected relations, if it is contaminated by distortions not associated with weak lensing. The most obvious of these is optical distortions of the telescope system, but it could also involve physical effects such as intrinsic alignment of galaxy ellipticities, which we will consider later.

A convenient way to characterise the distortions is via E/B decomposition, where the shear field is described in terms of an "E-mode," which is allowed by weak lensing, and a "B-mode," which is not. These terms are borrowed from similar decompositions in polarisation fields. In fact, weak lensing can generate B-modes, but they are expected to be very small [67], so the existence of a significant B-mode in the observed shear pattern is indicative of some nonlensing contamination. The easiest way to introduce a B-mode mathematically is to make the lensing potential complex:

$$\phi = \phi_E + i\phi_B. \tag{3.56}$$

There are various ways to determine whether a B-mode is present. A neat way is to generalise a common statistic called the aperture mass to a complex $M = M_{ap} + iM_\perp$, where the real part picks up the E-modes, and the imaginary part the B-modes. Alternatively, the $\xi_\pm$ can be used [17, 68]:

$$P_{\kappa\pm}(\ell) = \pi \int_0^\infty d\theta\, \theta\, [J_0(\ell\theta)\xi_+(\theta) \pm J_4(\ell\theta)\xi_-(\theta)], \tag{3.57}$$

where the $\pm$ power spectra refer to E- and B-mode powers. In principle, this requires the correlation functions to be known over all scales from 0 to ∞. Variants of this (see e.g., [17]) allow the E/B-mode correlation functions to be written in terms of integrals of $\xi_\pm$ over a finite range:

$$\xi_E(\theta) = \frac{1}{2}\left[\xi_-(\theta) + \xi'_+(\theta)\right] \tag{3.58}$$
$$\xi_B(\theta) = -\frac{1}{2}\left[\xi_-(\theta) - \xi'_+(\theta)\right],$$

where

$$\xi'_+(\theta) = \xi_+(\theta) + 4\int_0^\theta \frac{d\theta'}{\theta'}\xi_+(\theta') - 12\theta^2\int_0^\theta \frac{d\theta'}{\theta'^3}\xi_+(\theta'). \tag{3.59}$$

This avoids the need to know the correlation functions on large scales, but it needs the observed correlation functions to be extrapolated to small scales; this was one of the approaches taken in the analysis of the CFHTLS data [32]. Difficulties with estimating the correlation functions on small scales have led others to prefer to extrapolate to large scales, such as in the analysis of the GEMS [28] and William Herschel data [59]. Note that without full sky coverage, the decomposition into E- and B-modes is ambiguous, although for scales much smaller than the survey it is not an issue.

3.4.6 Results

The first results from cosmic shear were published in 2000 [3, 77, 47, 82], so as an observational science, cosmological weak lensing is very young. Till date, the surveys have been able to show clear detections of the effect and reasonably accurate determination of some cosmological parameters, usually the amplitude of the dark

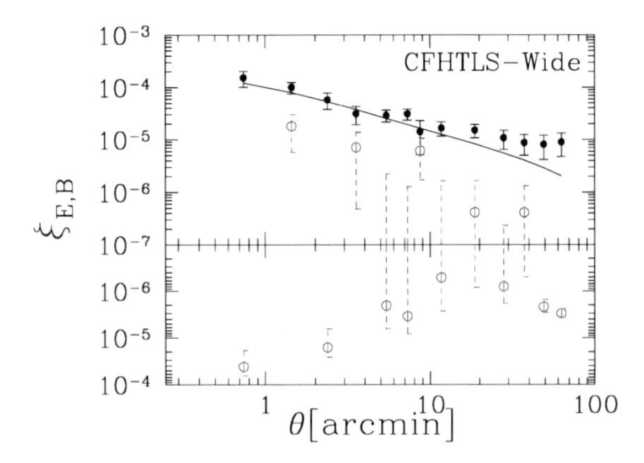

Fig. 3.8 E- and B-modes from an early analysis of CFHTLS data [7]. Top points are the E-modes and bottom points are the B-modes.

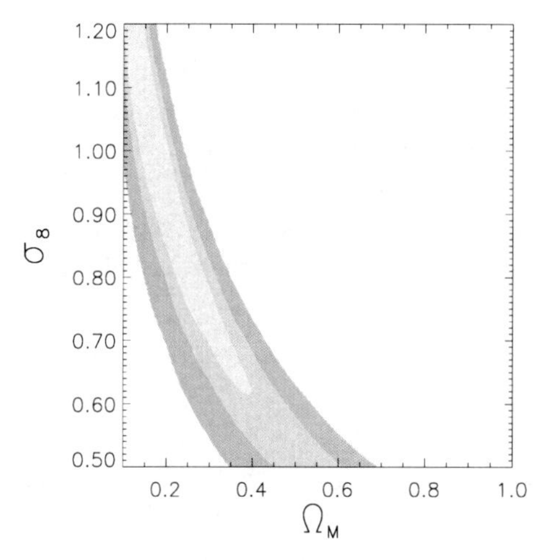

Fig. 3.9 Cosmological parameters from an early analysis of CFHTLS data [7]

matter perturbations (measured by the rms fractional fluctuations in an $8h^{-1}Mpc$ sphere and denoted σ_8), and the matter density parameter Ω_m. Current surveys cannot lift a near-degeneracy between these two, and usually a combination (typically $\sigma_8\Omega_m^{0.5}$) is quoted. This makes sense—it is difficult, but certainly not impossible, to distinguish between a highly clumped low-density universe and a modestly clumped high-density universe. There is no question that the surveys do not yet have the size or the careful control of systematics required to compete with the microwave background and other techniques used for cosmological parameter estimation. However, this situation is changing fast, particularly with the CFHT Legacy Survey, which is now complete, and Pan-STARRS 1, which is underway. Future surveys such as proposed by Euclid, JDEM, and LSST should be much more powerful. Early results from CFHTLS are shown in Figs. 3.8 and 3.9.

3.5 Lensing in 3D

Knowing the redshift distribution of sources is vital to interpret 2D lensing statistics, and to help with this, recent lensing surveys have obtained multicolour data to estimate distances using photo-zs. With photo-zs for sources, it makes sense to use the information individually, and this opens up the possibility of 3D lensing.

3.5.1 3D Potential and Mass Density Reconstruction

As we have already seen, it is possible to reconstruct the surface density of a lens system by analysing the shear pattern of galaxies in the background. An interesting question is then whether the reconstruction can be done in three dimensions, when distance information is available for the sources. It is probably self-evident that mass distributions can be *constrained* by the shear pattern, but the more interesting possibility is that one may be able to *determine* the 3D mass density in an essentially nonparametric way from the shear data.

The idea [73] is that the shear pattern is derivable from the lensing potential $\phi(\mathbf{r})$, which is dependent on the gravitational potential $\Phi(\mathbf{r})$ through the integral equation

$$\phi(\mathbf{r}) = \frac{2}{c^2}\int_0^r dr' \left(\frac{1}{r'} - \frac{1}{r}\right)\Phi(\mathbf{r}'), \tag{3.60}$$

where the integral is understood to be along a radial path (the Born approximation), and a flat universe is assumed in Eq. (3.60). The gravitational potential is related to the density field via Poisson's equation (3.38). There are two problems to solve here: one is to construct ϕ from the lensing data, the other is to invert Eq. (3.60). The second problem is straightforward: the solution is

$$\Phi(\mathbf{r}) = \frac{c^2}{2}\frac{\partial}{\partial r}\left[r^2 \frac{\partial}{\partial r}\phi(\mathbf{r})\right]. \tag{3.61}$$

From this and Poisson's equation $\nabla^2\Phi = (3/2)H_0^2\Omega_m\delta/a(t)$, we can reconstruct the mass overdensity field

$$\delta(\mathbf{r}) = \frac{a(t)c^2}{3H_0^2\Omega_m}\nabla^2\left\{\frac{\partial}{\partial r}\left[r^2\frac{\partial}{\partial r}\phi(\mathbf{r})\right]\right\}. \tag{3.62}$$

The construction of ϕ is more tricky, as it is not directly observable, but must be estimated from the shear field. This reconstruction of the lensing potential suffers from a similar ambiguity to the mass-sheet degeneracy for simple lenses. To see how, we first note that the complex shear field γ is the second derivative of the lensing potential:

$$\gamma(\mathbf{r}) = \left[\frac{1}{2}\left(\frac{\partial^2}{\partial\theta_x^2} - \frac{\partial^2}{\partial\theta_y^2}\right) + i\frac{\partial^2}{\partial\theta_x\partial\theta_y}\right]\phi(\mathbf{r}). \tag{3.63}$$

As a consequence, since the lensing potential is real, its estimate is ambiguous up to the addition of any field $f(\mathbf{r})$ for which

$$\frac{\partial^2 f(\mathbf{r})}{\partial\theta_x^2} - \frac{\partial^2 f(\mathbf{r})}{\partial\theta_y^2} = \frac{\partial^2 f(\mathbf{r})}{\partial\theta_x\partial\theta_y} = 0. \tag{3.64}$$

Since ϕ must be real, the general solution to this is

$$f(\mathbf{r}) = F(r) + G(r)\theta_x + H(r)\theta_y + P(r)(\theta_x^2 + \theta_y^2), \tag{3.65}$$

where F, G, H, and P are arbitrary functions of $r \equiv |\mathbf{r}|$. Assuming these functions vary smoothly with r, only the last of these survives at a significant level to the mass density and corresponds to a sheet of overdensity

$$\delta = \frac{4a(t)c^2}{3H_0^2\Omega_m r^2}\frac{\partial}{\partial r}\left[r^2\frac{\partial}{\partial r}P(r)\right]. \tag{3.66}$$

There are a couple of ways to deal with this problem. For a reasonably large survey, one can assume that the potential and its derivatives are zero on average, at each r, or that the overdensity has average value zero. For further details, see [4]. Note that the relationship between the overdensity field and the lensing potential is a linear one, so if one chooses a discrete binning of the quantities, one can use standard linear algebra methods to attempt an inversion, subject to some constraints such as minimising the expected reconstruction errors. With prior knowledge of the signal properties, this is the Wiener filter. See [35], for further details of this approach.

This method was first applied to COMBO-17 data [74] and recently to COSMOS HST data [60]—see Fig. 3.10.

3.5.2 Tomography

In the case where one has distance information for individual sources, it makes sense to use the information for statistical studies. A natural course of action is to divide the survey into slices at different distances and perform a study of the shear pattern on each slice. In order to use the information effectively, it is necessary to look at cross-correlations of the shear fields in the slices, as well as correlations within each

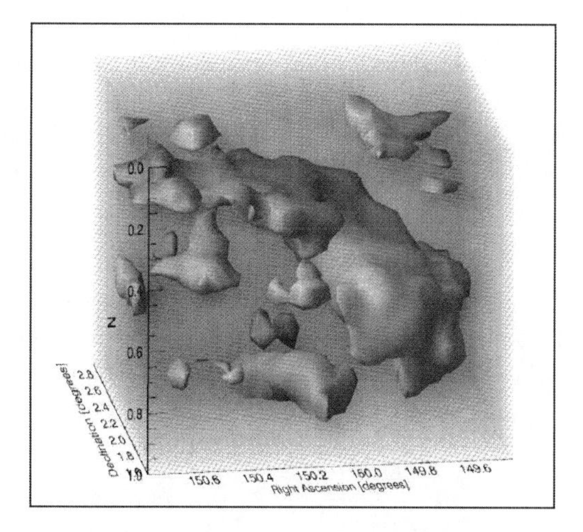

Fig. 3.10 3D reconstruction of matter density from the COSMOS ACS data [60].

slice [33]. This procedure is usually referred to as tomography, although the term does not seem entirely appropriate.

We start by considering the average shear in a shell, which is characterised by a probability distribution for the source redshifts $z = z(r)$, $p(z)$. The shear field is the second derivative of the lensing potential [12]

$$\gamma(\mathbf{r}) = \frac{1}{2}\tilde{\partial}\,\tilde{\partial}\,\phi(\mathbf{r}) \simeq \frac{1}{2}(\partial_x + i\partial_y)^2\phi(\mathbf{r}), \tag{3.67}$$

where the derivatives are in the angular direction, and the last equality holds in the flat-sky limit. If we average the shear in a shell, giving equal weight to each galaxy, then the average shear can be written in terms of an effective lensing potential

$$\phi_{\rm eff}(\theta) = \int_0^\infty dz\, p(z)\phi(\mathbf{r}), \tag{3.68}$$

where the integral is at fixed θ, and $p(z)$ is zero outside the slice (we ignore errors in distance estimates such as photometric redshifts; these could be incorporated with a suitable modification to $p(z)$). In terms of the gravitational potential, the effective lensing potential is

$$\phi_{\rm eff}(\theta) = \frac{2}{c^2}\int_0^\infty dr\, \Phi(\mathbf{r})g(r), \tag{3.69}$$

where reversal of the order of integration gives the lensing efficiency to be

$$g(r) = \int_{z(r)}^\infty dz'\, p(z')\left(\frac{1}{r} - \frac{1}{r'}\right), \tag{3.70}$$

$z' = z'(r')$ and we assume flat space. If we perform a spherical harmonic transform of the effective potentials for slices i and j, then the cross power spectrum can be related to the power spectrum of the gravitational potential $P_\Phi(k)$ via a version of Limber's equation:

$$\langle \phi^{(i)}_{\ell m}\phi^{*(j)}_{\ell' m'}\rangle = C^{\phi\phi}_{\ell,ij}\,\delta_{\ell'\ell}\delta_{m'm}, \tag{3.71}$$

where

$$C^{\phi\phi}_{\ell,ij} = \left(\frac{2}{c^2}\right)^2\int_0^\infty dr\, \frac{g^{(i)}(r)g^{(j)}(r)}{r^2}\, P_\Phi(\ell/r;r) \tag{3.72}$$

is the cross power spectrum of the lensing potentials. The last argument in P_Φ allows for evolution of the power spectrum with time or equivalently distance. The power spectra of the convergence and shear are related to $C^{\phi\phi}_{\ell,ij}$ by [34]

$$\begin{aligned} C^{\kappa\kappa}_{\ell,ij} &= \frac{\ell^2(\ell+1)^2}{4}\,C^{\phi\phi}_{\ell,ij} \\ C^{\gamma\gamma}_{\ell,ij} &= \frac{1}{4}\frac{(\ell+2)!}{(\ell-2)!}\,C^{\phi\phi}_{\ell,ij}. \end{aligned} \tag{3.73}$$

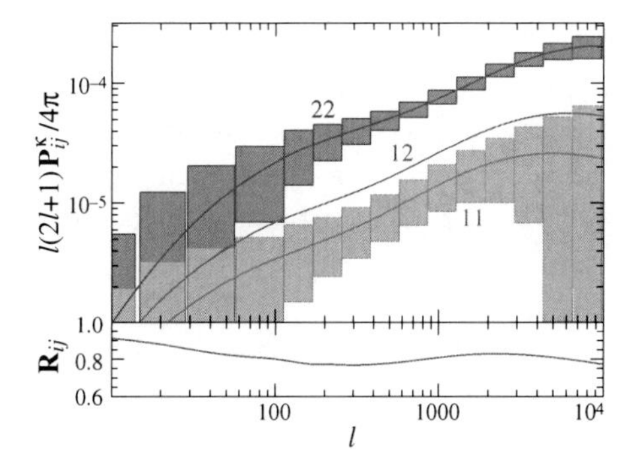

Fig. 3.11 The power spectra of two slices, their cross power spectrum, and their correlation coefficient. From [33]

The sensitivity of the cross power spectra to cosmological parameters is through various effects, as in 2D lensing: the shape of the linear gravitational potential power spectrum is dependent on some parameters, as is its nonlinear evolution; in addition, the $z(r)$ relation probes cosmology, via

$$r(z) = c\int_0^{\infty} \frac{dz'}{H(z}. \tag{3.74}$$

The reader is referred to standard cosmological texts for more details of the dependence of the distance-redshift relation on cosmological parameters.

Hu [33] illustrates the power and limitation of tomography, with two shells (Fig. 3.11). As expected, the deeper shell (2) has a larger lensing power spectrum than the nearby shell (1), but it is no surprise to find that the power spectra from shells are correlated, since the light from both passes through some common material. Thus, one does gain from tomography, but, depending on what one wants to measure, the gains may or may not be very much. For example, tomography adds rather little to the accuracy of the amplitude of the power spectrum, but far more to studies of dark energy properties. One also needs to worry about systematic effects, as leakage of galaxies from one shell to another, through noisy or biased photometric redshifts, can degrade the accuracy of parameter estimation [36, 57, 1, 50, 52].

3.5.3 The Shear Ratio Test

The shear contributed by the general large-scale structure is typically about 1%, but the shear behind a cluster of galaxies can far exceed this. As always, the shear of a background source is dependent on its redshift, and on cosmology, but also on the mass distribution in the cluster. This can be difficult to model, so it is attractive to consider methods that are decoupled from the details of the mass distribution of the cluster. Various methods have been proposed (e.g. [41, 9]). The method currently receiving the most attention is simply to take ratios of average tangential shear in different redshift slices for sources behind the cluster.

The amplitude of the induced tangential shear is dependent on the source redshift z, and on cosmology. via the angular diameter distance-redshift relation $S_k[r(z)]$ by [75]

$$\gamma_t(z) = \gamma_t(z=\infty)\frac{S_k[r(z) - r(z_l)]}{S_k[r(z)]}, \tag{3.75}$$

where $\gamma_{t,\infty}$ is the shear which a galaxy at infinite distance would experience and which characterises the strength of the distortions induced by the cluster, at redshift z_l. Evidently, we can neatly eliminate the cluster details by taking ratios of tangential shears, for pairs of shells in source redshift:

$$R_{ij} \equiv \frac{\gamma_{t,i}}{\gamma_{t,j}} = \frac{S_k[r(z_j)]\,S_k[r(z_i) - r(z_l)]}{S_k[r(z_i)]\,S_k[r(z_j) - r(z_l)]}. \tag{3.76}$$

In reality, the light from the more distant shell passes through an extra pathlength of clumpy matter, so suffers an additional source of shear. This can be treated as a noise term [75]. This approach is attractive in that it probes cosmology through the distance-redshift relation alone, being (at least to good approximation) independent of the growth rate of the fluctuations. Its dependence on cosmological parameters is, therefore, rather simpler, as many parameters (such as the amplitude of matter fluctuations) do not affect the ratio except through minor side effects. More significantly, it can be used in conjunction with lensing methods that probe both the distance-redshift relation and the growth rate of structure. Such a dual approach can in principle distinguish between quintessence-type dark energy models and modifications of Einstein gravity. This possibility arises because the effect on global properties (e.g., $z(r)$) is different from the effect on perturbed quantities (e.g., the growth rate of the power spectrum) in the two cases. The method has a signal-to-noise that is limited by the finite number of clusters, which are massive enough to have measurable tangential shear. In an all-sky survey, the bulk of the signal would come from the $10^5 - 10^6$ clusters above a mass limit of $10^{14}M_\odot$.

3.5.4 Full 3D Analysis of the Shear Field

An alternative approach to take is to recognise that, with photometric redshift estimates for individual sources, the data one is working with is a very noisy 3D shear field, which is sampled at a number of discrete locations, and for whom the locations are somewhat imprecisely known. It makes some sense, therefore, to deal with the data one has and to compare the statistics of the discrete 3D field with theoretical predictions. This was the approach of [22, 12, 24, 49]. It should yield smaller statistical errors than tomography, as it avoids the binning process that loses information.

In common with many other methods, one has to make a decision whether to analyse the data in configuration space or in the spectral domain. The former, usually studied via correlation functions, is advantageous for complex survey geometries, where the convolution with a complex window function implicit in spectral methods is avoided. However, the more readily computed correlation properties of a spectral analysis are a definite advantage for Bayesian parameter estimation, and we follow that approach here.

The natural expansion of a 3D scalar field (r,θ,ϕ) that is derived from a potential is in terms of products of spherical harmonics and spherical Bessel functions, $j_\ell(kr)Y_\ell^m(\boldsymbol{\theta})$. Such products, characterised by 3 spectral parameters (k,ℓ,m), are eigenfunctions of the Laplace operator, thus making it very easy to relate the expansion coefficients of the density field to that of the potential (essentially via $-k^2$ from the ∇^2 operator). Similarly, the 3D expansion of the lensing potential,

$$\phi_{\ell m}(k) \equiv \sqrt{\frac{2}{\pi}} \int d^3\mathbf{r}\, \phi(\mathbf{r}) k j_\ell(kr) Y_\ell^m(\boldsymbol{\theta}), \tag{3.77}$$

where the prefactor and the factor of k are introduced for convenience. The expansion of the complex shear field is most naturally made in terms of spin-weight 2 spherical harmonics ${}_2Y_\ell^m$ and spherical Bessel functions since $\gamma = \frac{1}{2}\tilde{\partial}\,\tilde{\partial}\,\phi$ and $\tilde{\partial}\,\tilde{\partial}\,Y_\ell^m \propto {}_2Y_\ell^m$:

$$\gamma(\mathbf{r}) = \sqrt{2\pi} \sum_{\ell m} \int dk\, \gamma_{\ell m} k\, j_\ell(kr)\; {}_2Y_\ell^m(\boldsymbol{\theta}). \tag{3.78}$$

The choice of the expansion becomes clear when we see that the coefficients of the shear field are related very simply to those of the lensing potential:

$$\gamma_{\ell m}(k) = \frac{1}{2}\sqrt{\frac{(\ell+2)!}{(\ell-2)!}}\, \phi_{\ell m}(k). \tag{3.79}$$

The relation of the $\phi_{\ell m}(k)$ coefficients to the expansion of the density field is readily computed, but more complicated as the lensing potential is a weighted integral of the gravitational potential. The details will not be given here, but relevant effects such as photometric redshift errors, nonlinear evolution of the power spectrum, and the discreteness of the sampling are easily included. The reader is referred to the original papers for details.

In this way, the correlation properties of the $\gamma_{\ell m}(k)$ coefficients can be related to an integral over the power spectrum, involving the $z(r)$ relation, so cosmological parameters can be estimated via standard Bayesian methods from the coefficients. Clearly, this method probes the dark energy effect on both the growth rate and the $z(r)$ relation.

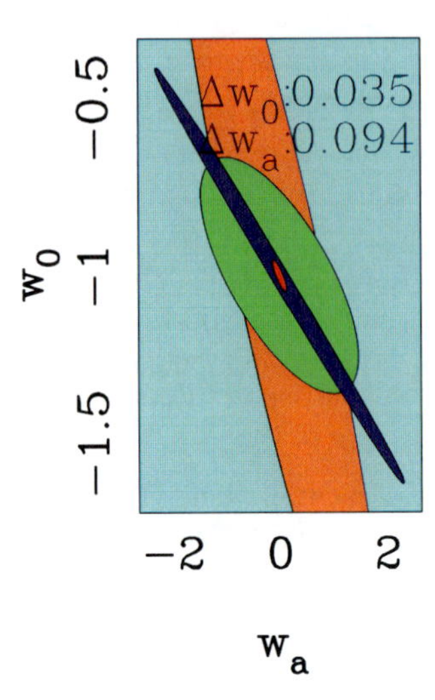

Fig. 3.12 The accuracy expected from the combination of experiments dedicated to studying Dark Energy properties. Marginal 1σ, 2-parameter regions are shown for the experiments individually and in combination. The supernova study fills the plot, the thin diagonal band is Planck, the near-vertical band is BAO, and the ellipse is the 3D lensing power spectrum method. The small ellipse is from combined. From [24].

3.5.5 Dark Energy with 3D Lensing Methods

In this section, we summarise some of the forecasts for cosmological parameter estimation from 3D weak lensing. We concentrate on the statistical errors that should be achievable with the shear ratio test and with the 3D power spectrum techniques. Tomography should be similar to the latter. We show results from 3D weak lensing alone, as well as in combination with other experiments. These include CMB, supernova, and baryon oscillation (BAO) studies. The methods generally differ not only in the parameters that they constrain well but also in terms of the degeneracies inherent in the techniques. Using more than one technique can be very effective at lifting the degeneracies, and very accurate determinations of cosmological parameters, in particular dark energy properties, may be achievable with 3D cosmic shear surveys covering thousands of square degrees of sky to median source redshifts of order unity.

Figures 3.5 and 3.13 show the accuracy that might be achieved with a number of surveys designed to measure cosmological parameters. We concentrate here on the capabilities of each method, and the methods in combination, to constrain the dark energy equation of state, and its evolution, parametrised by [14]

$$w(a) = \frac{p}{\rho c^2} = w_0 + w_a(1-a), \tag{3.80}$$

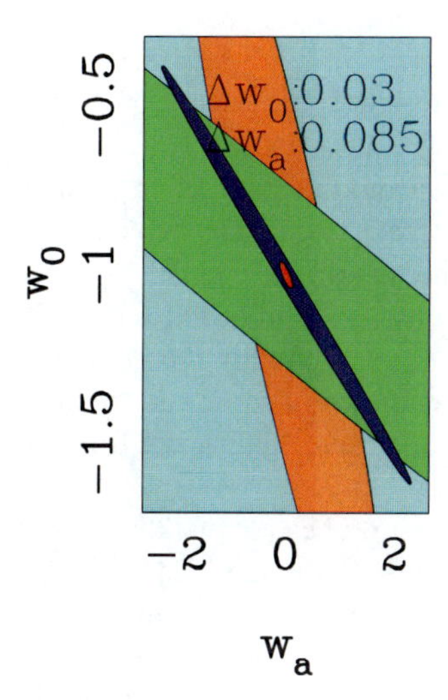

Fig. 3.13 As in Fig. 3.12, but with the shear ratio test as the lensing experiment. Supernovae fill the plot, Planck is the thin diagonal band, BAO the near-vertical band, and the shear ratio is the remaining 45 degree band. The combination of all experiments is in the centre. From [75].

where the behaviour as a function of scale factor a is, in the absence of a compelling theory, assumed to have this simple form. $w = -1$ would arise if the dark energy behaviour was actually a cosmological constant.

The assumed experiments are as follows: a 5-band 3D weak lensing survey, analysed either with the shear ratio test or with the spectral method, covering 10,000 square degrees to a median redshift of 0.7, similar to the capabilities of a ground-based 4 m-class survey with a several square degree field; the Planck CMB experiment (14-month mission); a spectroscopic survey to measure BAO in the galaxy matter power spectrum, assuming constant bias, and covering 2000 square degrees to a median depth of unity, and a smaller $z = 3$ survey of 300 square degrees, similar to WFMOS capabilities on Subaru; a survey of 2000 Type Ia supernovae to $z = 1.5$, similar to SNAPs design capabilities.

We see that the experiments in combination are much more powerful than individually, as some of the degeneracies are lifted. Note that the combined experiments appear to have rather smaller error bars than is suggested by the single-experiment constraints. This is because the combined ellipse is the projection of the product of several multidimensional likelihood surfaces, which intersect in a small volume. (The projection of the intersection of two surfaces is not the same as the intersection of the projection of two surfaces). The figures show that errors of a few percent on w_0 are potentially achievable, or, with this parametrisation, an error of w at a "pivot" redshift of $z \simeq 0.4$ of under 0.02. This error is essentially the minor axis of the error ellipses. These graphs include statistical errors only; systematic errors, from, e.g., bias in the photo-z distribution, are expected to degrade errors by $\sim \sqrt{2}$ [50].

3.6 Dark Gravity

In addition to the possibility that Dark Energy or the cosmological constant drives acceleration, there is an even more radical solution. As a cosmological constant, Einstein's term represents a modification of the gravity law, so it is interesting to consider whether the acceleration may be telling us about a failure of GR. Although no compelling theory currently exists, suggestions include modifications arising from extra dimensions, as might be expected from string-theory braneworld models. Interestingly, there are potentially measurable effects of such exotic gravity models which weak lensing can probe, and finding evidence for extra dimensions would of course signal a radical departure from our conventional view of the Universe.

Weak lensing is useful as it probes not just the distance-redshift relation but also the growth rate of perturbations (see, e.g., Eqs. 3.70 and 3.72). This is important because in principle measures which use only the $r(z)$ relation suffer from a degeneracy between a different gravity law and the equation of state of the contents of the Universe. To see this, first note that a modified gravity law will lead to some sort of Hubble relation $H(a)$. If we combine the Friedmann equation

$$H^2(a) + \frac{k}{a^2} = \frac{8\pi G\rho}{3} \tag{3.81}$$

and

$$\frac{d}{da}\left(\rho a^3\right) = -pa^2 = -w(a)\rho a^2, \tag{3.82}$$

we find that

$$w(a) = -\frac{1}{3}\frac{d}{d\ln a}\left[\frac{1}{\Omega_m(a)} - 1\right]. \tag{3.83}$$

So we see that any modified gravity law can be mimicked, as far as the distance-redshift relation is concerned, by GR with an appropriate equation of state.

To analyse other gravity laws, we consider scalar perturbations in the conformal Newtonian gauge (flat for simplicity), $ds^2 = a^2(\eta)\left[(1+2\psi)d\eta^2 - (1-2\phi)d\mathbf{x}^2\right]$, where ψ is the potential fluctuation, and ϕ the curvature perturbation, and η being the conformal time. Information on the gravity law is manifested in these two potentials. For example, in GR and in the absence of anisotropic stresses (a good approximation for epochs when photon and neutrino streaming are unimportant) $\phi = \psi$. More generally, the Poisson law may be modified, and the laws for ψ and ϕ may differ. This difference can be characterised [19] by the *slip*, ϖ. This may be scale- and time-dependent: $\psi(k,a) = [1+\varpi(k,a)]\,\phi(k,a)$, and the modified Poisson equation may be characterised by Q, an effective change in G [2]:

$$-k^2\phi = 4\pi G a^2 \rho_m \delta_m Q(k,a). \tag{3.84}$$

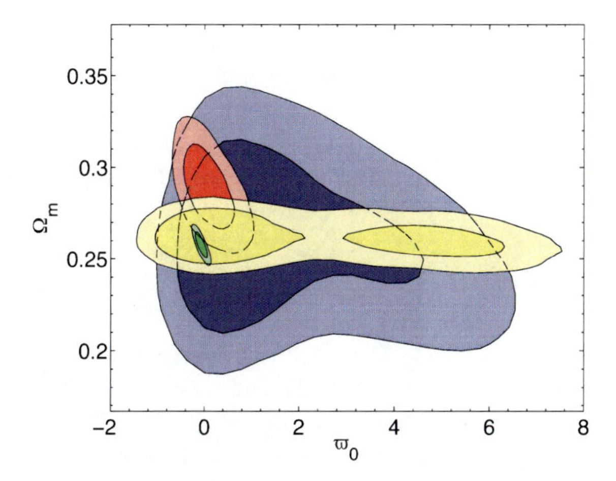

Fig. 3.14 The projected marginal 68% and 95% likelihood contours for the slip, ϖ, assuming $\varpi = \varpi_0(1+z)^{-3}$, for WMAP 5-year data (blue), adding current weak lensing and ISW data (red). Yellow is mock Planck CMB data, and green adds weak lensing from a 20,000 square degree survey [19].

Different observables are sensitive to ψ and ϕ in different ways [42]. For example, the integrated Sachs–Wolfe effect depends on $\dot{\psi}+\dot{\phi}$, but the effect is confined to large scales and cosmic variance precludes accurate use for testing modified gravity. Peculiar velocities are sourced by ψ. Lensing is sensitive to $\psi+\phi$, and this is the most promising route for next-generation surveys to probe beyond-Einstein gravity. The Poisson-like equation for $\psi+\phi$ is

$$-k^2(\psi+\phi) = 2\Sigma\frac{3H_0^2\Omega_m}{2a}\delta_m, \tag{3.85}$$

where $\Sigma \equiv Q(1+\varpi/2)$. For GR, $\Sigma = 1$, $\varpi = 0$. The DGP braneworld model [20] has $\Sigma = 1$, so mass perturbations deflect light in the same way as GR, but the growth rate of the fluctuations differs. Thus, we have a number of possible observational tests, including probing the expansion history, the growth rate of fluctuations, and the mass density-light bending relation. Future WL surveys can put precise constraints on Σ [2] and on ϖ (see Fig. 3.14) [19].

By probing the growth rate and the expansion history, weak lensing can lift a degeneracy that exists in methods, that consider the distance-redshift relation alone since the expansion history in a modified gravity model can always be mimicked by GR and Dark Energy with a suitable $w(a)$. In general, however, the growth history of cosmological structures will be different in the two cases (e.g., [53, 37], but see [54]).

3.6.1 Growth Rate

Although not the most general, the growth index γ [55] (not to be confused with the shear) is a convenient minimal extension of GR. The growth rate of perturbations in the matter density ρ_m, $\delta_m \equiv \delta\rho_m/\rho_m$, is parametrised as a function of scale factor $a(t)$ by

$$\frac{\delta_m}{a} \equiv g(a) = \exp\left\{ \int_0^a \frac{da'}{a'} \left[\Omega_m(a')^\gamma - 1\right] \right\}, \tag{3.86}$$

In the standard GR cosmological model, $\gamma \simeq 0.55$, whereas in modified gravity theories it deviates from this value. E.g. the flat DGP braneworld model [20] has $\gamma \simeq 0.68$ on scales much smaller than those where cosmological acceleration is apparent [56].

Measurements of the growth factor can in principle be used to determine the growth index γ, and it is interesting to know if it is of any practical use. In contrast to parameter estimation, this is an issue of model selection—is the gravity model GR or is there evidence for beyond-Einstein gravity? This question may be answered with the Bayesian evidence, B [70], which is the ratio of probabilities of two or more models, given some data. Following [25], Fig. 3.15 shows how the Bayesian evidence for GR changes with increasing true deviation of γ from its GR value for a combination of a future WL survey and *Planck*. From the WL data alone, one should be able to distinguish GR decisively from the flat DGP model at $\ln B \simeq 11.8$, or, in the frequentist view, 5.4σ [25]. The combination of WL + *Planck* + BAO + SN should be able to distinguish $\delta\gamma = 0.041$ at 3.41 sigma. This data combination should be able to decisively distinguish a dark energy GR model from a DGP modified-gravity model with expected evidence ratio $\ln B \simeq 50$. An alternative is to

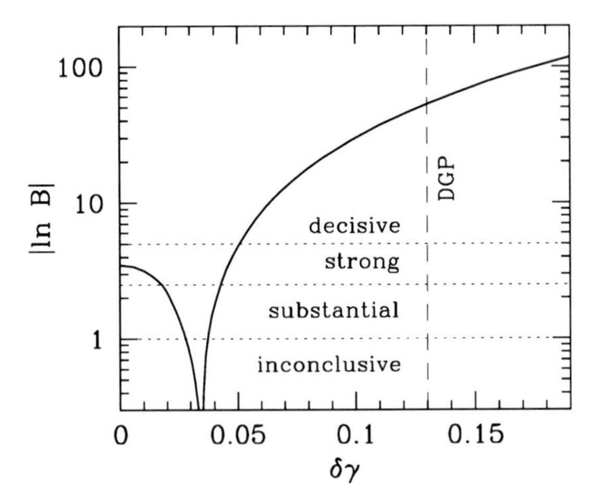

Fig. 3.15 Expected Bayesian evidence B vs. deviation of the growth index from GR, for a future WL survey + *Planck* [25]. If modified gravity is the true model, GR will still be favoured by the data to the left of the cusp. The Jeffreys scale of evidence [43] is labeled.

Survey	ν	$\|\ln B\|$	
DES+*Planck*+BAO+SN	3.5	1.28	substantial
DES+*Planck*	2.2	0.56	inconclusive
DES	0.7	0.54	inconclusive
PS1+*Planck*+BAO+SN	2.9	3.78	strong
PS1+*Planck*	2.6	2.04	substantial
PS1	1.0	0.62	inconclusive
WL_{NG}+*Planck*+BAO+SN	10.6	63.0	decisive
WL_{NG}+*Planck*	10.2	52.2	decisive
WL_{NG}	5.4	11.8	decisive

Table 3.1 The evidence ratio for the three weak lensing experiments considered with and without *Planck*, supernova, and BAO priors. WL_{NG} is a next-generation space-based imaging survey such as proposed for *DUNE* or *SNAP*. z_m is the median redshift, n_0 is the number of sources per square arcminute, and σ_z is the assumed photometric redshift error. For completeness, we also list the frequentist significance $\nu\sigma$ with which GR would be expected to be ruled out, if the DGP braneworld were the correct model.

parametrise geometry and growth with two separate effective values of w and look for (in)consistency [39, 72, 76, 83].

One caveat on all of these conclusions is that WL requires knowledge of the non-linear regime of galaxy clustering, and this is reasonably well understood for GR, but for other models, further theoretical work is required. This has already started [64]. The case for a large, space-based 3D weak lensing survey is strengthened, as it offers the possibility of conclusively distinguishing Dark Energy from at least some modified gravity models.

3.7 The Future

The main promise of weak lensing in the future will come from larger surveys with optics designed for excellent image quality. Currently, the CFHTLS is the state-of-the-art, covering ~ 170 square degrees to a median redshift in excess of one. In the near future, Pan-STARRS, VST, and DES promise very small PSF distortions and large areal coverage, and in the far future, LSST on the ground and satellites, such as Euclid or JDEM, may deliver extremely potent lensing surveys. In parallel with these developments, the acquisition of photometric redshifts for the sources has opened up the exciting possibility of analysing weak lensing surveys in 3D. Each source represents a noisy estimate of the shear field at a location in 3D space, and this extra information turns out to be extremely valuable, increasing substantially the statistical power of lensing surveys. In particular, it can lift the degeneracy between σ_8 and Ω_m, measure directly the growth of Dark Matter clustering [5] and, more excitingly still, it represents a powerful method to measure the equation of state of Dark Energy [22, 41, 24]—surely one of the most important remaining questions in cosmology. In addition, photometric redshifts allow the possibility of direct 3D Dark Matter mapping [73, 4, 74], thus addressing another of the unsolved problems. Finally, there is the interesting prospect that we may be able to find evidence for extra dimensions in the Universe, from braneworld models, or other cosmological models, from the effect they have on the gravity law and the growth rate of perturbations. Most excitingly, it seems that this may be within reach for ambitious lensing surveys planned for the next decade.

3.8 Appendix: The Propagation of Light through a Weakly Perturbed Universe

3.8.1 The Geodesic Equation

The geodesic equation governs the worldline x^λ ($\lambda = 0,1,2,3$) of a particle and is readily found in textbooks on General Relativity (e.g., [38]). It is

$$\frac{d^2x^\lambda}{dp^2} + \Gamma^\lambda_{\ \mu\nu}\frac{dx^\mu}{dp}\frac{dx^\nu}{dp} = 0, \tag{3.87}$$

where p is an affine parameter and $\Gamma^\lambda_{\ \mu\nu}$ is the affine connection, which can be written in terms of the metric tensor $g_{\mu\nu}$ as

$$\Gamma^\lambda_{\ \mu\nu} = \frac{1}{2}g^{\sigma\lambda}\left\{\frac{\partial g_{\mu\nu}}{\partial x^\sigma} + \frac{\partial g_{\sigma\nu}}{\partial x^\mu} - \frac{\partial g_{\mu\sigma}}{\partial x^\nu}\right\}. \tag{3.88}$$

For weak fields, the interval is given by

$$ds^2 = \left(1+\frac{2\Phi}{c^2}\right)c^2dt^2 - \left(1-\frac{2\Phi}{c^2}\right)R^2(t)\left[dr^2 + S_k^2(r)d\beta^2\right], \tag{3.89}$$

where Φ is the peculiar gravitational potential, $R(t)$ is the scale factor of the Universe, and r,θ,φ are the usual comoving spherical coordinates. The angle $d\beta = \sqrt{d\theta^2 + \sin^2\theta d\varphi^2}$. $S_k(r)$ depends on the geometry of the Universe, being given by

$$S_k(r) = \begin{cases} \sinh r, & \text{if } k < 0; \\ r, & \text{if } k = 0; \\ \sin r, & \text{if } k > 0. \end{cases}$$

The curvature $k = -1,0,1$ corresponds to open, flat, and closed universes, respectively. We are interested in the distortion of a small light bundle, so we can concentrate on a small patch of sky. If we choose the polar axis of the coordinate system to be along the centre of the light bundle, we can define angles $\theta_x \equiv \theta \cos\varphi$ and $\theta_y \equiv \theta \sin\varphi$. For convenience, we also use the *conformal time*, defined by $d\eta = cdt/R(t)$, in place of the usual time coordinate. With these definitions, the interval is more simply written as

$$ds^2 = R^2(t)\left\{\left(1+\frac{2\Phi}{c^2}\right)d\eta^2 - \left(1-\frac{2\Phi}{c^2}\right)\left[dr^2 + S_k^2(r)(d\theta_x^2 + d\theta_y^2)\right]\right\}.$$

The metric tensor for weakly perturbed flat Friedmann–Robertson–Walker metric is then

$$R^2(t)\begin{pmatrix} 1+2\Phi/c^2 & 0 & 0 & 0 \\ 0 & -\left(1-2\Phi/c^2\right) & 0 & 0 \\ 0 & 0 & -r^2\left(1-2\Phi/c^2\right) & 0 \\ 0 & 0 & 0 & -r^2\left(1-2\Phi/c^2\right) \end{pmatrix}.$$

We are interested in how the angles of the ray, (θ_x, θ_y) change as the photon moves along its path, responding to the varying gravitational potential. The unperturbed, radial, path is set by $0 = ds^2 \simeq d\eta^2 - dr^2$, i.e., For a radial incoming ray,

$$\frac{dr}{d\eta} = -1.$$

With $g^{\mu\nu}$ defined as the inverse of $g_{\mu\nu}$ (so $g_{\mu\nu}g^{\nu\alpha} = \delta_\mu{}^\alpha$), the affine connections are readily computed.

The parametrised equation for η is required only to zero-order in Φ and reduces to

$$\frac{d^2\eta}{dp^2} = -2\frac{\dot{R}}{R}\dot{\eta},$$

where a dot here denotes d/dp. By choosing the unit of p appropriately, we find

$$\frac{d\eta}{dp} = \frac{1}{R^2}.$$

We can also relate the radial coordinate to the conformal time, again to zero-order in Φ: The first-order equations governing θ_x and θ_y are obtained from the geodesic equation, or by the variational methods (see, e.g., d'Inverno (1992), Sect. 7.6 [38])

$$\frac{\partial L^2}{\partial x^\mu} - \frac{d}{dp}\left(\frac{\partial L^2}{\partial \dot{x}^\mu}\right) = 0,$$

where $L^2 \equiv (ds/dp)^2$. With $x^\mu = \theta_x$,

$$R^2\frac{2}{c^2}\frac{\partial\Phi}{\partial\theta_x}\dot{\eta}^2 + \frac{2}{c^2}R^2\frac{\partial\Phi}{\partial\theta_x}\left(\dot{r}^2 + r^2\theta_x^2 + r^2\theta_y^2\right) - \frac{d}{dp}\left[-2R^2r^2\left(1-\frac{2\Phi}{c^2}\right)\dot{\theta}_x\right] = 0.$$

With the zero-order solutions for $d\eta/dp$ and $dr/d\eta$, to first order this reduces simply to

$$\frac{d^2\theta_x}{d\eta^2} - \frac{2}{r}\frac{d\theta_x}{d\eta} = -\frac{2}{c^2r^2}\frac{\partial\Phi}{\partial\theta_x}.$$

It is convenient to write this as an equation for the comoving displacement of the ray from a fiducial direction,

$$x_i \equiv r\theta_i. \qquad i = 1,2$$

and the equation for θ_x and a similar one for θ_y simplify to

$$\frac{d^2\mathbf{x}}{d\eta^2} = -\frac{2}{c^2}\nabla\Phi,$$

where ∇ here is a comoving transverse gradient operator (∂_x, ∂_y).

We see that the propagation equation for the displacement looks similar to what one would guess from a Newtonian point-of-view; the presence of η (instead of t) in the acceleration term on the left is a result of the expansion of the Universe and the choice of comoving coordinates. The right-hand side looks like the gradient of the potential, but it is larger than the naıve gradient by a factor of two. This is the same factor of two which leads to the classic result of GR, famously tested by Eddington's 1919 solar eclipse observations, that the angle of light bending by the Sun is double what Newtonian theory predicted.

References

1. Amara A., Refregier A., MNRAS, 391, 228 (2008)
2. Amendola L., Kunz M., Sapone D., JCAP, 04, 13A (2008)
3. Bacon D.J., Refregier A., Ellis R.S., MNRAS, 318, 625 (2000)
4. Bacon D.J., Taylor A.N., MNRAS, 344, 1307 (2003)
5. Bacon D.,J., et al., MNRAS, 363, 723 (2005)
6. Bartelmann, M., Schneider, P., *Weak Gravitational Lensing*, Physics Reports 340, 291-472; astroph 9912508 (2001)
7. Benjamin J., et al., MNRAS, 281, 792 (2007)
8. Bernardeau F., van Waerbeke L., Mellier Y., A&A, 322, 1 (1997)
9. Bernstein G., Jain B., ApJ, 600, 17 (2004)
10. Bridle S., King L., NJPh, 9, 444 (2007)
11. Brown M., et al., MNRAS, 333,501 (2002)
12. Castro P.G., Heavens A.F., Kitching T., Phys. Rev. D (2006)
13. Catelan P., Kamionkowski M., Blandford R.D., MNRAS, 320, L7 (2001)
14. Chevallier M., Polarski D., IJMPD, 10, 213 (2001)
15. Clowe D., Gonzalez A., Markevitch M., ApJ, 604, 596 (2004)
16. Crittenden R.,Natarajan P., Pen U.-L., Theuns T., ApJ, 559, 552 (2001)
17. Crittenden R.,Natarajan P., Pen U.-L., Theuns T., ApJ, 568, 20 (2002)
18. Croft R.A.C., Metzler C.A., ApJ, 545, 561 (2000)
19. Daniel S., Caldwell R., Cooray A., Serra P., Melchiorri A., astroph/0901.0919 (2009)

20. Dvali G., Gabadaze G., Porrati M., Phys. Lett. B, 485, 208 (2000)
21. Gray M. et al., MNRAS, 347, 73 (2004)
22. Heavens A.F., MNRAS, 343, 1327 (2003)
23. Heavens A.F., Refregier A., Heymans C.E.C., MNRAS, 319, 649 (2000)
24. Heavens A.F., Kitching T., Taylor A.N., MNRAS, 373, 105 (2006)
25. Heavens A.F., Kitching T., Verde L., MNRAS, 380, 1029 (2007)
26. Heymans C., Heavens A.F., MNRAS, 337, 711 (2003)
27. Heymans C. et al., MNRAS, 347, 895, (2004)
28. Heymans C. et al., MNRAS, 361, 160 (2005)
29. Heymans C., White M., Heavens A.F., Vale C., van Waerbeke L., MNRAS, 371, 750 (2006)
30. Hirata C., Seljak U., Phys. Rev. D70, 063526 (2004)
31. Hoekstra H., Jain B., ARNPS, 58, 99 (2008)
32. Hoekstra H., Mellier Y., van Waerbeke L., Semboloni E., Fu L., Hudson M. J., Parker L. C., Tereno I., Benabed K., ApJ, 647, 116 (2006)
33. Hu W., ApJ, 522, 21 (1999)
34. Hu W., PRD, 62, 3007 (2000)
35. Hu W., Keeton C., PRD, 66, 3506 (2002)
36. Huterer D., Takada M., Bernstein G., Jain B., MNRAS, 366, 101 (2006)
37. Huterer D., Linder E.V., PRD, 75, 2, 3519 (2007)
38. d'Inverno R., *Introducing Einstein's Relativity*, OUP, Oxford, (1992)
39. Ishak M., Upadhye A., Spergel D., Phys. Rev. D., 74, 3513 (2006)
40. Joachimi B., Schneider P., A& A, 488, 829 (2008)
41. Jain B., Taylor A.N., PRL, 91, 141302 (2003)
42. Jain B., Zhang P., Phys. Rev. D, 78, 3503 (2008)
43. Jeffreys H., *Theory of Probability*, Oxford University Press, UK (1961)
44. Jing Y.P., MNRAS, 335, L89 (2002)
45. Kaiser N., Squires G., ApJ, 404, 441 (1993)
46. Kaiser N., Squires G., Broadhurst T., ApJ, 449, 460 (1995)
47. Kaiser N., Wilson G., Luppino G., astroph/0003338 (2000)
48. King L., Schneider P., A&A, 396, 411 (2002)
49. Kitching T. D., Heavens A. F., Taylor A. N., Brown M. L., Meisenheimer K., Wolf C., Gray M. E., Bacon D. J., MNRAS, 376, 771 (2007)
50. Kitching T.D., Taylor A.N., Heavens A.F., MNRAS, 389, 173 (2008)
51. Kitching T.D., Miller L., Heymans C., van Waerbeke L., Heavens A.F., MNRAS, 390, 149 (2008)
52. Kitching T.D., Amara A., Rassat A., Refregier A., astroph/0901.3143 (2009)
53. Knox L., Song Y.-S., Tyson J.A., Phys. Rev. D, 74, 3512 (2006)
54. Kunz M., Sapone D., PRL, 98, 12, 121301 (2007)
55. Linder E.V., Phys. Rev. D, 72, 043529 (2005)
56. Linder E.V., Cahn R.N. 2007, Astroparticle Physics, 28, 481 (2007)
57. Ma Z., Hu W., Huterer D., ApJ, 636, 21 (2006)
58. Mandelbaum R., Seljak U., Hirata C.M., JCAP 8, 6 (2008).
59. Massey R., Refregier A., Bacon D.J., Ellis R., Brown M.L., MNRAS, 359, 1277 (2005)
60. Massey R., et al., Nature, 445, 286 (2007)
61. Miller L., Kitching T.D., Heymans C., Heavens A.F., van Waerbeke L., MNRAS, 382, 315 (2007)
62. Munshi D., Valageas P., van Waerbeke L., Heavens A.F., Phys. Rev., 462, 67 (2008)
63. Navarro J., Frenk C.S, White S.D.M., ApJ, 490, 493 (1997)
64. Schmidt F., Lima M., Oyaizu H., Hu W., astroph/0812.0545 (2008)
65. Schneider P., Kochanek C., Wambsganss J., SAAS-FEE Advanced Course 33, Springer, Berlin (2006) (http://www.astro.uni-bonn.de/ peter/SaasFee.html)
66. Schneider P. et al., MNRAS, 296, 893 (2002)
67. Schneider P., van Waerbeke L., Mellier Y., A& A, 389, 729 (2002)
68. Schneider P.,van Waerbeke L., Kilbinger M., Mellier Y., A& A, 396, 1 (2002)

69. Schneider, P., *Gravitational lensing as a probe of structure*, to appear in the proceedings of the XIV Canary Islands Winter School of Astrophysics *Dark Matter and Dark Energy in the Universe* Tenerife. astroph 0306465 (2003)
70. Skilling J., 2004, avaliable at http://www.inference.phy.cam.ac.uk/bayesys
71. Smith R.E. et al., MNRAS, 341, 1311 (2003)
72. Song Y.-S., Doré O., JCAP, 1208, 039 (2009)
73. Taylor A.N., astroph/0111605 (2001)
74. Taylor A.N. et al., MNRAS, 353, 1176 (2004)
75. Taylor A.N., Kitching T.D., Bacon D.J., Heavens A.F., 2007, MNRAS, 374, 1377 (2007)
76. Wang S., Khoury J., Haiman Z., May M., PRD, 70, 13008 (2004)
77. van Waerbeke L., et al., A& A, 358, 30 (2000)
78. van Waerbeke L. et al., A&A, 393, 369 (2002)
79. van Waerbeke L., Mellier Y., *Gravitational Lensing by Large Scale Structures: A Review*. Proceedings of Aussois Winter School, astroph/0305089 (2003)
80. van Waerbeke L., Mellier Y., Hoekstra H., A& A, 429, 75 (2005)
81. White M., Astroparticle Physics, 22, 211-217 (2004)
82. Wittman D. et al., Nature, 405, 143 (2000)
83. Zhang J., Hui L., Stebbins A., ApJ, 635, 806 (2005)

Chapter 4
Cosmology with Numerical Simulations

Lauro Moscardini and Klaus Dolag

Abstract The birth and growth of cosmic structures is a highly nonlinear phenomenon that needs to be investigated with suitable numerical simulations. The main goal of these simulations is to provide robust predictions, which, once compared to the present and future observations, allows us to constrain the main cosmological parameters. Different techniques have been proposed to follow both the gravitational interaction inside cosmological volumes and the variety of physical processes acting on the baryonic component only. In this chapter, we review the main characteristics of the numerical schemes most commonly used in the literature, discuss their pros and cons, and summarize the results of their comparison.

4.1 Introduction

Numerical simulations have become in the last years one of the most effective tools to study and to solve astrophysical problems. The computation of the mutual gravitational interaction between a large set of particles is a problem that cannot be investigated with analytical techniques only. Therefore, it represents a good example of a problem where computational resources are absolutely fundamental.

Thanks to the enormous technological progress in the recent years, the available facilities allow us now to afford the problem of gravitational instability, which is the basis of the accepted model of cosmic structure formation, with a very high mass and space resolution. Moreover, the development of suitable numerical techniques permits to include a realistic treatment of the majority of the complex physical

Lauro Moscardini
Dipartimento di Astronomia, Università di Bologna, via Ranzani 1, I-40127 Bologna, Italy
e-mail: lauro.moscardini@unibo.it

Klaus Dolag
Max-Planck-Institut für Astrophysik, P.O. Box 1317, D-85741 Garching, Germany
e-mail: kdolag@mpa-garching.mpg.de

processes acting on the baryonic component, which is directly related to observations. For this reason, the numerical simulations are now used not only to better understand the general picture of structure formation in the Universe but also as a tool to validate cosmological models and to investigate the possible presence of biases in real data. In some sense, they substitute the laboratory experiments which are in practice impossible for cosmology, given the uniqueness of the Universe.

The chapter is organized as follows. In Section 4.2, we will discuss the schemes proposed to follow the formation and evolution of cosmic structures when the gravity only is in action: in particular, after presenting the model equations in Section 4.2.1, we will introduce the Particle–Particle method (Section 4.2.2), the Particle–Mesh method (Section 4.2.3), the Tree code (Section 4.2.4), and the so-called Hybrid methods (Section 4.2.5). Section 4.3 is devoted to the presentation of the numerical codes used to solve the hydrodynamical equations related to the baryonic component, introduced in Section 4.3.1. More in detail, Section 4.3.2 discusses the characteristics of the most used Lagrangian code, the Smoothed Particle Hydrodynamics, while Section 4.3.3 introduces the bases of the methods based on grids, the Eulerian codes.

4.2 N-Body Codes

4.2.1 The Model Equations

In order to write the equations of motion determining the gravitational instability leading to the formation and evolution of cosmic structures, it is necessary to choose the underlying cosmological model, describing the expanding background universe, where $a = 1/(1+z)$. In the framework of General Relativity, this means to assume a Friedmann–Lemaître model, with its cosmological parameters, namely the Hubble parameter H_0 and the various contributions coming from baryons, dark matter, and dark energy/cosmological constant to the total density parameter Ω_0.

Many different observations are now giving a strong support to the idea that the majority of the matter in the Universe is made by cold dark matter (CDM), i.e., nonrelativistic collisionless particles, which can be described by their mass m, comoving position $\mathbf{x}$, and momentum $\mathbf{p}$. The time evolution of the phase-space distribution function $f(\mathbf{x}, \mathbf{p}, t)$ is given by the coupled solution of the Vlasov equation

$$\frac{\partial f}{\partial t} + \frac{\mathbf{p}}{ma^2}\nabla f - m\nabla\Phi\frac{\partial f}{\partial \mathbf{p}} = 0 \tag{4.1}$$

and of the Poisson equation

$$\nabla^2\Phi(\mathbf{x},t) = 4\pi \mathrm{G} a^2 \left[\rho(\mathbf{x},t) - \bar{\rho}(t)\right]. \tag{4.2}$$

Here, Φ represents the gravitational potential and $\bar{\rho}(t)$ represents the mean background density. The proper mass density

$$\rho(\mathbf{x},t) = \int f(\mathbf{x},\mathbf{p},t) d^3 p \tag{4.3}$$

is the integral over the momenta $\mathbf{p} = ma^2\dot{\mathbf{x}}$.

The solution of this high-dimension problem is standardly obtained by using a finite set of N_p particles to trace the global matter distribution. For these tracers, it is possible to write the usual equations of motion, which in comoving coordinates read:

$$\frac{d\mathbf{p}}{dt} = -m\nabla\Phi \tag{4.4}$$

and

$$\frac{d\mathbf{x}}{dt} = \frac{\mathbf{p}}{ma^2}. \tag{4.5}$$

Introducing the proper peculiar velocity $\mathbf{v} = a\dot{\mathbf{x}}$, these equations can be written as

$$\frac{d\mathbf{v}}{dt} + \mathbf{v}\frac{\dot{a}}{a} = -\frac{\nabla\Phi}{a}. \tag{4.6}$$

The time derivative of the expansion parameter, $\dot{a}$, is given by the Friedmann equation, once the cosmological parameters are assumed.

In the following subsections, we will present some of the standard approaches used to solve the N-body problem.

4.2.2 The Particle–Particle (PP) Method

This is certainly the simplest possible method because it makes direct use of the equations of motion plus the Newton's gravitational law. In this approach, the forces on each particle are directly computed by accumulating the contributions of all remaining particles.

At each time-step Δt, the following operations are repeated:

- clearing of the force accumulators: $F_i = 0$ for $i = 1,...,N_p$, where N_p is the number of particles;
- accumulation of the forces considering all $N_p(N_p - 1)$ pairs of particles: $F_{ij} \propto m_i m_j / r_{ij}^2$, where r_{ij} is the distance between two particles having masses m_i and m_j, respectively:

$$F_i = F_i + \sum F_{ij}; \quad F_j = F_j + \sum F_{ij}; \tag{4.7}$$

- integration of the equations of motion to obtain the updated velocities v_i and positions x_i of the particles:

$$v_i^{\text{new}} = v_i^{\text{old}} + F_i \Delta t / m_i \ , \tag{4.8}$$

$$x_i^{\text{new}} = x_i^{\text{old}} + v_i \Delta t \ ; \tag{4.9}$$

- updating of the time counter: $t = t + \Delta t$.

Even if implementing this code is extremely easy from a numerical point of view (see an example in the Appendix of [3]), its computational cost is quite large, being proportional to N_p^2: for this reason, its application is in practice forbidden for problems with a very large number of particles.

Notice that in principle this method would compute the exact Newtonian force, used then to estimate the particles' acceleration: for this reason, the PP method can be considered the most accurate N-body technique. However, in order to avoid the divergence at very small scales, the impact parameter must be reduced by introducing a softening parameter ε in the equation for the gravitational potential Φ:

$$\Phi = -Gm_p/(r^2 + \varepsilon^2)^{1/2}. \tag{4.10}$$

In some sense, this corresponds to assign a finite size to each particle, which can be considered as a statistical representation of the total mass distribution. Typical choices for ε range between 0.02 and 0.05 times the mean interparticle distance. As a consequence, the frequency of strong deflections is also reduced, decreasing the importance of the spurious two-body relaxation, which is generated by the necessarily small number of particles used in the simulations, many orders of magnitude smaller than the number of collisionless dark matter particles really exist in the Universe.

As said, the largest limitation of this method is its scaling as N_p^2. An attempt to overcome this problem has been done by building a special-purpose hardware, called GRAPE (GRAvity PipE) [13]. This hardware is based on custom chips that compute the gravitational force with a hardwired force law. Consequently, this device can solve the gravitational N-body problem adopting the direct sum with a computational cost that is extremely smaller than for traditional processors.

Few words, which are valid also for most of the following methods, must be spent about time-stepping and integration. In general, the accuracy obtained when evolving the system depends on the size of the time step Δt and on the integrator scheme used. Finding the optimum size of time step is not trivial. A possible choice is given by

$$\Delta t = \alpha \sqrt{\varepsilon / |\mathbf{F}/m_p|} \ , \tag{4.11}$$

where the force is the one obtained at the previous time step, ε is a length scale associated to the gravitational softening, and α is a suitable tolerance parameter. Alternative and more accurate criteria are discussed in [19]. To update velocities and then positions, it is necessary to integrate first-order ordinary differential equations, once the initial conditions are specified: many methods, classified as explicit or implicit, are available, ranging from the simplest Euler's algorithm to the more accurate Runge–Kutta method (see, e.g., [20] for an introduction to these methods).

4.2.3 The Particle–Mesh (PM) Method

In this method, a mesh is used to describe the field quantities and to compute their derivatives. Thanks to the structure of the Poisson equation and to the assumption that the considered volume is a fair sample of the whole universe, it is convenient to re-write all relevant equations in the Fourier space, where we can take advantage of the Fast Fourier Techniques: this will allow a strong reduction of the CPU time necessary for each time step. This improvement in the computational cost is, however, paid with a loss of accuracy and resolution: the PM method cannot follow close interactions between particles on small scales. In fact, using a grid to describe the field quantities (like density, potential and force) does not allow a fair representation on scales smaller than the intergrid distance.

Since the introduction of a computational mesh is equivalent to a local smoothing of the field, for the PM method, it is not necesssary to adopt the softening parameter in the expression of the force and/or gravitational potential.

Going in more detail, each time step for the PM method is composed of the following operations:

- computation of the density at each grid point starting from the particles' spatial distribution;
- solution of the Poisson equation for the potential;
- computation of the force on the grid points;
- estimation of the force at the positions of each particle using a suitable interpolation scheme;
- integration of the equations of motion.

If the computational volume is a cube of side L and N_p is the number of particles having equal mass m_p, a regular three-dimensional grid with M nodes per direction is built: therefore, the grid spacing is $\Delta \equiv L/M$. Each grid point can be identified by a term of integer numbers (i,j,k), such that its spacial coordinates are $\mathbf{x}_{i,j,k} = (i\Delta, j\Delta, k\Delta)$, with $i,j,k = 1,...,M$.

4.2.3.1 Density Computation

The particle mass decomposition at the grid nodes represents one of the critical steps for the PM method, both in terms of resolution and computational cost. The mass density ρ at the grid point $\mathbf{x}_{i,j,k}$ can be written as

$$\rho(\mathbf{x}_{i,j,k}) = m_p M^3 \sum_{l=1}^{N_p} W(\delta \mathbf{x}_l) \,, \qquad (4.12)$$

where W is a suitable interpolation function and $\delta \mathbf{x}_l = \mathbf{x}_l - \mathbf{x}_{i,j,k}$ is the distance between the position of the l-th particle and the considered grid point.

The choice of W is related to the accuracy of the required approximation: of course, the higher the number of grid points involved in the interpolation, the better the approximation. For problems in higher dimensions, it is possible to write the function W as product of more functions, each of them depending only on the displacement in one dimension: $W_{i,j,k} = w_i w_j w_k$. In order of increasing accuracy, the most commonly adopted interpolation functions are as follows:

- "nearest-grid-point" (NGP): in this case, the mass of each particle is totally assigned to the nearest grid point only. As a consequence, the density shows a discontinuity every time a particle crosses the grid borders. The NGP interpolation function reads

$$w_i = 1, \quad M|\delta x_i| \leq 1/2\,; \tag{4.13}$$

- "cloud-in-cell" (CIC): the mass of each particle is assigned to two (i.e., $2^3 = 8$ in the 3D case) nearest points in an inversely proportional way with respect to its distance from the grid point: in this way, the density varies with continuity when a particle crosses a cell border, but its gradient is still discontinuous. The CIC interpolation function can be written as

$$w_i = 1 - M|\delta x_i|\,, \quad M|\delta x_i| \leq 1\,; \tag{4.14}$$

- "triangular-shaped-cell" (TSC): the mass decomposition involves three (i.e., $3^3 = 27$ in the 3D case) nearest points. In this way, the density gradient also varies smoothly during the cell border crossing; on the contrary, the second derivative remains discontinuous. The TSC interpolatation function can be expressed as

$$w_i = \begin{cases} 3/4 - M^2|\delta x_i|^2 & M|\delta x_i| \leq 1/2 \\ (1/2)(3/2 - M|\delta x_i|)^2 & 1/2 \leq M|\delta x_i| \leq 3/2 \end{cases}. \tag{4.15}$$

Of course, the interpolating functions can be easily extended to higher orders, providing gradually continuous derivatives of higher orders and better accuracy. However, because of the increasing number of involved grid points, the resulting interpolation would be more computationally demanding.

Notice that, in order to conserve the momentum, the same interpolation schemes W adopted here for the density computation have to be used to obtain the components of the force at the position of each particle, once the force on the mesh will be obtained (see above). In this case,

$$F_x(\mathbf{x}_l) = m_p \sum_{i,j,k=1}^{M} W(\delta \mathbf{x}_l) F_x(\mathbf{x}_{i,j,k})\,; \tag{4.16}$$

$$F_y(\mathbf{x}_l) = m_p \sum_{i,j,k=1}^{M} W(\delta \mathbf{x}_l) F_y(\mathbf{x}_{i,j,k}) \; ; \tag{4.17}$$

$$F_z(\mathbf{x}_l) = m_p \sum_{i,j,k=1}^{M} W(\delta \mathbf{x}_l) F_z(\mathbf{x}_{i,j,k}) \; . \tag{4.18}$$

4.2.3.2 Poisson Equation

The application of the Fast Fourier Transforms (FFTs) allows to solve the Poisson equation in a much easier way in the Fourier space, where it becomes

$$\Phi_k = G_k \delta_k \; ; \tag{4.19}$$

here, Φ_k and δ_k are the Fourier transforms of the gravitational potential and of the density contrast, respectively. In the previous equation, G_k represents a suitable Green function for the Laplacian, for which, in the case of a continuous system, a good expression is given by $G_k \propto k^{-2}$; alternative expressions for the Laplacian giving a better approximation are also available. Using a discrete system on a grid introduces errors and anisotropies in the computation of the force. In order to reduce this problem, it is necessary to find an expression for G_k, which corresponds to the desired shape for the particles in the configuration space and which minimizes the errors with respect to a reference force: for this reason, the optimal Green function depends on the chosen shape and interpolation scheme.

4.2.3.3 Force Calculation

In a typical PM scheme, the computation of the forces at each grid point requires differentiating the potential Φ to derive the i-th force component:

$$F_i(\mathbf{x}) = -m_p \frac{d\Phi}{dx_i} \; . \tag{4.20}$$

This can be done using the finite difference schemes, largely adopted to numerically solve differential equations. The approximation depends on the number of grid points involved in the computation. As examples, at the lowest order ("two-point centered"), the x-component of the force can be obtained as

$$\frac{F_x(\mathbf{x}_{i,j,k})}{m_p} = \frac{\Phi(\mathbf{x}_{i-1,j,k}) - \Phi(\mathbf{x}_{i+1,j,k})}{2\Delta} \, , \tag{4.21}$$

while the four-point approximation is given by

$$\frac{F_x(\mathbf{x}_{i,j,k})}{m_p} = \frac{2\left(\Phi(\mathbf{x}_{i+1,j,k}) - \Phi(\mathbf{x}_{i-1,j,k})\right)}{3\Delta} - \frac{\Phi(\mathbf{x}_{i+2,j,k}) - \Phi(\mathbf{x}_{i-2,j,k})}{12\Delta} \; . \tag{4.22}$$

It is possible to increase the resolution of a PM scheme by computing the force on a grid shifted with respect to the one used for the computation of potential. For example, in the case of the "two-point centered" approximation (see Eq. 4.21), this "staggered mesh" technique reads

$$\frac{F_x(\mathbf{x}_{i,j,k})}{m_p} = \frac{\Phi(\mathbf{x}_{i-1/2,j,k}) - \Phi(\mathbf{x}_{i+1/2,j,k})}{\Delta} \,. \tag{4.23}$$

It is well known that the finite difference schemes introduce truncation errors in the solutions. A way to avoid this problem and to gain a better accuracy is to obtain the forces directly from the gravitational potential in Fourier space: $\mathbf{F}_k = -i\mathbf{k}\Phi_k$. In this case, it is necessary to inverse-transform separately each single component of the force, using more frequently the FFT routines.

4.2.3.4 Pros and Cons

The big advantage of the PM method is its high computational speed: in fact, the number of operations scales as $N_p + N_g \log(N_g)$, where N_p is the number of particles and $N_g = M^3$ is the number of grid points. The small dynamical range, strongly limitated by the number of grid points and by the corresponding memory occupation, represents the largest problem of the method: only adopting hybrid methods (see above), it is possible to reach the sufficient resolution necessary in cosmological simulations.

4.2.4 Tree Codes

Thanks to its computational performance and accuracy, this method represents today the favorite tool for cosmological N-body simulations. The idea to solve the N-body problem is based on the exploitation of a hierarchical multipole expansion, the so-called tree algorithm. The speed up is obtained by using, for sufficiently distant particles, a single multipole force, in spite of computing every single distance, as required for methods based on direct sum. In this way, the sum ideally reduces to $N_p \log(N_p)$ operations, even if, for gravitationally evolved structures, the scaling can be less efficient.

In practice, the multiple expansion is based on a hierarchical grouping that is obtained, in the most common algorithms, by subdividing in a recursive way the simulation volume. In the approach suggested by [2], a cubical root node is used to encompass the full mass distribution. This cube is then repeatedly subdivided into eight daughter nodes of half the side-length each, until one ends up with "leaf" nodes containing single particles. The computation of the force is then done by "descending" the tree. Starting from the root node, the code evaluates, by applying a suitable criterion, whether or not the multipole expansion of the node provides an accurate enough partial force: in case of a positive answer, the tree descent is stopped and the multipole force is used; in the negative case, the daughter nodes are considered in turn. Usually, the opening criterion is based on a given fixed angle, whose choice controls the final error: typically one assumes the angle to be ≈ 0.5 rad. Obviously, the smaller and the more distant the particles' groups, the more accurate the assumption of multipole expansion.

4.2.5 Hybrid Methods

Having the previous techniques quite different numerical properties with opposite pros and cons, it is possible to combine them to build new algorithms (called hybrid methods), possibly mantaining the positive aspects only. A first attempt has been done with the P^3M code, which combines the high accuracy of the direct sum implemented by the PP method at small scale with the speed of the PM algorithm to compute the large-scale interactions. An improved version of this code has also been proposed, where spatially adaptive mesh refinements are possible in regions at very high density. This algorithm, called AP^3M [6], has been used to run several cosmological simulations, including the Hubble Volume simulations [9].

In the last-generation codes like GADGET[23], hybrid codes are built by replacing the direct sum with tree algorithms: the so-called TreePM. In this case, the potential is explicitly split in Fourier space into a long-range and a short-range part according to $\Phi_k = \Phi_k^{\rm long} + \Phi_k^{\rm short}$, where

$$\Phi_k^{\rm long} = \Phi_k \exp(-k^2 r_{\rm s}^2)\ ; \tag{4.24}$$

here, $r_{\rm s}$ corresponds to the spatial scale of the force-split. The long-range potential can be computed very efficiently with mesh-based Fourier methods, like in the PM method. The short-range part of the potential can be solved in real space by noting that for $r_{\rm s} \ll L$, the short-range part of the real-space solution of the Poisson equation is given by

$$\Phi^{\rm short}(\mathbf{x}) = -G \sum_i \frac{m_i}{\mathbf{r}_i} \mathrm{erfc}\left(\frac{\mathbf{r}_i}{2r_{\rm s}}\right). \tag{4.25}$$

In the previous equation, $\mathbf{r}_i$ is the distance of any particle i to the point $\mathbf{x}$. Thus, the short-range force can be computed adopting the tree algorithm, except that the force law is modified by a long-range cut-off factor.

This approach allows to largely improve the computational performance compared with ordinary tree methods, maintaining all their advantages: the very large dynamical range, the insensitivity to clustering, and the precise control of the softening scale of the gravitational force.

4.2.6 Initial Conditions and Simulation Setup

As said, the N-body simulations are a tool often used to produce, once a given cosmological model is assumed, the predictions, which can be directly compared with real observational data to falsify the model itself. For this reason, it is necessary to have a robust method to assign to the N_p particles the correct initial conditions, corresponding to a fair realization of the desired cosmological model. The standard inflationary scenarios predict that the initial density fluctuations are a random field with an almost Gaussian distribution (for non-Gaussian models and the corresponding initial conditions, see e.g., [11]). In this case, the field is completely defined by its power spectrum $P(k)$, whose shape is related to the choices made about the underlying cosmological model and the nature of dark matter.

The generation of initial conditions with a Gaussian distribution and a given $P(k)$ can be easily obtained exploiting the characteristics of a Gaussian distribution. In fact, as discussed by [1], in order that a generic field $F(x)$ to be strictly Gaussian, all its different Fourier spatial modes F_k have to be reciprocally independent, to have random phases θ_k and to have amplitude distributed according to a Rayleigh distribution:

$$P(|F_k|,\theta_k)d|F_k|d\theta_k = \exp\left(-\frac{|F_k|^2}{2P(k)}\right)\frac{|F_k|d|F_k|}{P(k)}\frac{d\theta_k}{2\pi}\ . \tag{4.26}$$

The real and imaginary parts of F_k are then reciprocally independent and Gaussian distributed. Consequently, it is sufficient to generate complex numbers with a phase randomly distributed in the range $[0,2\pi]$ and with an amplitude normally distributed with a variance given by the desired spectrum.

To obtain the perturbation field corresponding to this density distribution, it is then necessary to multiply it by a suitable Green function to obtain the potential which can be then differentiated. The following application of the Zel'dovich approximation [26] enables us to find the initial positions and velocities for a given particle distribution. A more detailed description of the algorithms to create cosmological initial conditions can be found in [8]. Notice that the main limitations of this method are essentially due to the use of a finite computational volume: the wavelengths close to the box size are badly sampled while those larger are not present at all!

Two further complications should be mentioned. Using a perfectly regular grid to distribute the unperturbed particles can introduce on the resulting density power spectrum discreteness effects which can be reduced by starting from an amorphous fully relaxed particle distribution. As suggested by [25], this can be constructed by applying negative gravity to a system and evolving it for a long time until it reaches a relaxed state. Second, when interested to studies of individual objects like galaxy clusters, we have to recall that also large-scale tidal forces have an important role in determining their final properties. To include these effects, one can apply the so-called "zoom" technique [24]: a high-resolution region is self-consistently embedded in a larger scale cosmological volume at low resolution. This allows an increase of the dynamical range up to two orders of magnitude while keeping the full cosmological context. For galaxy simulations, it is even possible to apply this technique on several levels of refinements to further improve the dynamical range of the simulation.

As important as the initial conditions is the problem of the optimal setup for cosmological simulations. For example, the number of particles necessary to reach the numerical convergence in the description of a given region of interest depends on the astrophysical quantity of interest, ranging from $\approx 30-50$ for the halo mass function to some thousands for the density profile in the central regions. Recently, [19] presented a comprehensive series of convergence tests designed to study the effect of numerical parameters on the structure of simulated dark matter haloes. In particular, this paper discusses the influence of the gravitational softening, the time stepping algorithm, the starting redshift, the accuracy of force computations, and the number of particles in the spherically-averaged mass profile of a galaxy-sized halo. The results were summarized in empirical rules for the choice of these parameters and in simple prescriptions to assess the effective convergence of the mass profile of a simulated halo, when computational limitations are present.

In general, it is important to notice that both the size and the dynamical range or resolution of the N-body simulations have been increasing very rapidly over the last decades, in a way that, thanks to improvements in the algorithms, is faster than the underlying growth of the available CPU power.

4.2.7 Code Comparison

Thanks to the analysis of very recent data regarding the cosmic microwave background, the galaxy surveys, and the high-redshift supernovae, we entered the era of the so-called high-precision cosmology, with very stringent constraints (i.e., at the per cent accuracy level) on the main parameters. In the perspective of even better data as expected in upcoming projects, the theoretical predictions, in order to be a useful and complementary tool, must reach a similar level of precision, that, particularly in the highly nonlinear regime, represents a real challenge.

A first step in this direction is the launch of an extensive program of comparison between the different numerical codes available in the community. An example is the work presented in [12], where ten different codes adopting a variety of schemes (tree, APM, tree-PM, etc.) have been tested against the same initial conditions. In general, the comparison has been very satisfactory, even if the variance between the results is still too large to make predictions suitable for the next-generation of galaxy surveys. In particular, the accuracy in the determination of the halo mass function is always better than 5 per cent, while the agreement for the power spectrum in the nonlinear regime is at the $5-10$ per cent level, even on moderate spatial scales around $k = 10h\,\mathrm{Mpc}^{-1}$. Considering the internal structure of halos in the outer regions of $\sim R_{200}$, it also appears to be very similar between different simulation codes. Larger differences between the codes in the inner region of the halos occur only if the halo is not in a relaxed state.

4.3 Hydrodynamical Codes

4.3.1 The Model Equations

Even if the dark matter component represents the dominant contribution to the total matter distribution in the Universe, it is quite important to have also a fair representation of baryons: in fact, most of the astrophysical signals are originated by physical process related to them.

In general, the description of the baryonic component is based on the assumption of an ideal fluid, for which the time evolution is obtained by solving the following set of hydrodynamical equations: the Euler equation,

$$\frac{d\mathbf{v}}{dt} = -\frac{\nabla P}{\rho} - \nabla \Phi; \tag{4.27}$$

the continuity equation,

$$\frac{d\rho}{dt} + \rho \nabla \mathbf{v} = 0; \tag{4.28}$$

the first law of thermodynamics,

$$\frac{du}{dt} = -\frac{P}{\rho} \nabla \cdot \mathbf{v}. \tag{4.29}$$

In the previous equations, P and u represent the pressure and the internal energy (per unit mass), which under the assumption of an ideal monatomic gas, are related by the equation of state: $P = 2/3\rho u$. Notice that for the moment, we neglect the effect produced by radiative losses, usually encrypted by the cooling function $\Lambda(u,\rho)$ (see the discussion in Sect. 4.3.5).

In the recent years, different techniques have been proposed to solve the coupled equations of dark and baryonic matter. All of them treat the gravitational term related to $\nabla\Phi$ with the same kinds of technique described in the previous section. For the hydrodynamical part, they adopt different strategies that can be classified in two big categories: Lagrangian and Eulerian techiques. The former are methods based on a finite number of particles used to describe the mass distribution, while the latter are methods that adopt a mesh to discretize the simulation volume. Before reviewing in the following subsections the main characteristics of their most important prototypes, we notice here that considering self-gravity, as necessary in cosmological applications, introduces a much higher level of complexity, with respect to standard hydrodynamical simulations. In fact the formation of very evolved density structures induces motions, which are often extremely supersonic, with the presence of shocks and discontinuities. Moreover, in order to have reliable simulations, the process of structure formation must be accurately followed on a very large range of scales, covering many order of magnitudes in space, in time, and in the interesting physical quantities (temperature, pressure, energy, etc.). Finally, we need to include the expansion of the Universe, which modifies the previous set of equations as

$$\frac{\partial \mathbf{v}}{\partial t} + \frac{1}{a}(\mathbf{v} \cdot \nabla)\mathbf{v} + \frac{\dot{a}}{a}\mathbf{v} = -\frac{1}{a\rho}\nabla P - \frac{1}{a}\nabla \Phi, \tag{4.30}$$

$$\frac{\partial \rho}{\partial t} + \frac{3\dot{a}}{a}\rho + \frac{1}{a}\nabla \cdot (\rho \mathbf{v}) = 0 \tag{4.31}$$

and

$$\frac{\partial}{\partial t}(\rho u) + \frac{1}{a}\mathbf{v} \cdot \nabla(\rho u) = -(\rho u + P)\left(\frac{1}{a}\nabla \cdot \mathbf{v} + 3\frac{\dot{a}}{a}\right). \tag{4.32}$$

4.3.2 Smoothed Particle Hydrodynamics (SPH)

The most popular Lagrangian method is certainly SPH that has a very good spatial resolution in high-density regions, while its performance in underdense region is unsatisfactory. Another charcateristic of SPH is the fact that it adopts an artificial viscosity, which does not allow to reach a sufficiently high resolution in shocked region. Even if in general, it is not able to treat dynamical instabilities with the same accuracy of the Eulerian methods presented in the next subsection, thanks to its adaptive nature, SPH is still the preferred method for cosmological hydrodynamical simulations.

Here, we introduce the basic idea of the technique. We refer to [16] for a more extended presentation of the method. Unlike in the Eulerian algorithms, in SPH, the fluid is represented using mass elements, i.e., a finite set of particles. This choice originates the different performance in high- and low-density regions: in fact, where the mean interparticle distance is smaller, the fluid is less sparsely sampled, with a consequent better resolution; the opposite holds when the interparticle distance is large.

4.3.2.1 Basics of SPH

As said, in SPH, the fluid is discretized using mass elements (i.e., a finite set of particles), unlike in Eulerian codes where the discretization is made using volume elements. Thanks to an adaptive smoothing, the mass resolution is kept fixed. In more detail, a generic fluid quantity A is defined as

$$\langle A(\mathbf{x}) \rangle = \int W(\mathbf{x} - \mathbf{x}', h) A(\mathbf{x}') d\mathbf{x}', \tag{4.33}$$

where the smoothing kernel W is suitably normalized: $\int W(\mathbf{x},h)d\mathbf{x} = 1$. Moreover, W, assumed to depend only on the distance modulus, collapses to a delta function if the smoothing length h tends to zero.

Using a set of particles with mass m_j and position $\mathbf{x}_j$, the previous integral can be done in a discrete way, replacing the volume element of the integration with the ratio of the mass and density m_j/ρ_j of the particles:

$$\langle A_i \rangle = \sum_j \frac{m_j}{\rho_j} A_j W(\mathbf{x}_i - \mathbf{x}_j, h) . \tag{4.34}$$

If one adopts kernels with a compact support (i.e., $W(\mathbf{x},h) = 0$ for $|\mathbf{x}| > h$), it is not necessary to do the summation over the whole set of particles, but only over the neighbours around the i-th particle under consideration, namely the ones inside a sphere of radius h. The following kernel represents an optimal choice in many cases and it has been used often in the literature:

$$W(x,h) = \frac{8}{\pi h^3} \begin{cases} 1 - 6\left(\frac{x}{h}\right)^2 + 6\left(\frac{x}{h}\right)^3 & 0 \le x/h < 0.5 \\ 2\left(1 - \frac{x}{h}\right)^3 & 0.5 \le x/h < 1 \\ 0 & 1 \le x/h \end{cases} . \tag{4.35}$$

If the quantity A is the density ρ_i, Eq. 4.34 becomes simpler:

$$\langle \rho_i \rangle = \sum_j m_j W(\mathbf{x}_i - \mathbf{x}_j, h). \tag{4.36}$$

The big advantage of SPH is that the derivatives can be easily calculated as

$$\nabla \langle A_i \rangle = \sum_j \frac{m_j}{\rho_j} A_j \nabla_i W(\mathbf{x}_i - \mathbf{x}_j, h), \tag{4.37}$$

where ∇_i denotes the derivative with respect to $\mathbf{x}_i$. A pairwise symmetric formulation of derivatives can be obtained by exploiting the identity

$$(\rho \nabla) \cdot A = \nabla(\rho \cdot A) - \rho \cdot (\nabla A), \tag{4.38}$$

which allows one to re-write a derivative as

$$\nabla \langle A_i \rangle = \frac{1}{\rho_i} \sum_j m_j (A_j - A_i) \nabla_i W(\mathbf{x}_i - \mathbf{x}_j, h). \tag{4.39}$$

Another symmetric representation of the derivative can be alternatively obtained from the identity

$$\frac{\nabla A}{\rho} = \nabla \left(\frac{A}{\rho}\right) + \frac{A}{\rho^2} \nabla \rho, \tag{4.40}$$

which then leads to

$$\nabla \langle A_i \rangle = \rho_i \sum_j m_j \left(\frac{A_j}{\rho_j^2} + \frac{A_i}{\rho_i^2} \right) \nabla_i W(\mathbf{x}_i - \mathbf{x}_j, h). \tag{4.41}$$

By applying the previous identities, one can re-write the Euler equation as

$$\frac{d\mathbf{v}_i}{dt} = -\sum_j m_j \left(\frac{P_j}{\rho_j^2} + \frac{P_i}{\rho_i^2} + \Pi_{ij} \right) \nabla_i W(\mathbf{x}_i - \mathbf{x}_j, h), \tag{4.42}$$

while the term $-(P/\rho)\nabla \cdot \mathbf{v}$ in the first law of thermodynamics can be written as

$$\frac{du_i}{dt} = \frac{1}{2} \sum_j m_j \left(\frac{P_j}{\rho_j^2} + \frac{P_i}{\rho_i^2} + \Pi_{ij} \right) (\mathbf{v}_j - \mathbf{v}_i) \nabla_i W(\mathbf{x}_i - \mathbf{x}_j, h). \tag{4.43}$$

In order to have a good reproduction of shocks, in the previous equations it has been necessary to introduce the so-called artificial viscosity Π_{ij}, for which different forms have been proposed in the literature (see, e.g., [17]): most of them try to reduce its effects in the regions where there are no shocks, adopting a specific articial viscosity for each particle.

To complete the set of fluid equations, we notice that the continuity equation does not represent a problem for Lagrangian methods like SPH, being automatically satisfied thanks to the conservation of the number of particles.

4.3.2.2 The Smoothing Length

Usually, each individual particle i has its own smoothing length h, which is determined by finding the radius h_i of a sphere which contains N_{nei} neighbours. A large value for N_{nei} would allow better estimates for the density field but with larger systematics; vice versa small N_{nei} would lead to larger sample variances. In the literature standard choices for N_{nei} range between 32 and 100. The presence of a variable smoothing length in the hydrodynamical equations can break their conservative form: to avoid it, it is necessary to introduce a symmetric kernel $W(\mathbf{x}_i - \mathbf{x}_j, h_i, h_j) = \bar{W}_{ij}$, for which the two main variants used in the literature are the kernel average,

$$\bar{W}_{ij} = (W(\mathbf{x}_i - \mathbf{x}_j, h_i) + W(\mathbf{x}_i - \mathbf{x}_j, h_j))/2 , \tag{4.44}$$

and the average of the smoothing lengths

$$\bar{W}_{ij} = W(\mathbf{x}_i - \mathbf{x}_j, (h_i + h_j)/2) . \tag{4.45}$$

Note that when writing the derivatives, we assumed that h is independent of the position $\mathbf{x}_j$. Thus, if the smoothing length h_i is variable for each particle, one would formally introduce the correction term $\partial W/\partial h$ in all the derivatives. An elegant way to do that is using the formulation which conserves numerically both entropy and internal energy, described in the next subsection.

4.3.2.3 Conserving the Entropy

In adiabatic flows, the entropic function $A = P/\rho^\gamma$ is conserved. The quantity A is related to the internal energy per unit mass:

$$u_i = \frac{A_i}{\gamma - 1}\rho_i^{\gamma-1}\,. \tag{4.46}$$

Shocks, which can be captured in SPH thanks to the artificial viscosity Π_{ij}, can originate an evolution of A:

$$\frac{dA_i}{dt} = \frac{1}{2}\frac{\gamma-1}{\rho_i^{\gamma-1}}\sum_j m_j \Pi_{ij}\left(\mathbf{v}_j - \mathbf{v}_i\right)\nabla_i \bar{W}_{ij}. \tag{4.47}$$

The Euler equation can be derived starting by defining the Lagrangian of the fluid as

$$L(\mathbf{q},\dot{\mathbf{q}}) = \frac{1}{2}\sum_i m_i \dot{\mathbf{x}}_i^2 - \frac{1}{\gamma-1}\sum_i m_i A_i \rho_i^{\gamma-1}, \tag{4.48}$$

which represents the entire fluid and has the coordinates $\mathbf{q} = (\mathbf{x}_1,...,\mathbf{x}_N,h_1,...,h_N)$.

The next important step is to define constraints, which allow an unambiguous association of h_i for a chosen number of neighbours. This can be done by requiring that the kernel volume contains a constant mass for the estimated density,

$$\phi_i(\mathbf{q}) = \frac{4\pi}{3}h_i^3\rho_i - nm_i = 0. \tag{4.49}$$

The equation of motion can be obtained as the solution of

$$\frac{d}{dt}\frac{\partial L}{\partial \dot{\mathbf{q}}_i} - \frac{\partial L}{\partial \mathbf{q}_i} = \sum_j \lambda_j \frac{\partial \phi_j}{\partial \mathbf{q}_i}, \tag{4.50}$$

which, as demonstrated by [21], can be written as

$$\frac{d\mathbf{v}_i}{dt} = -\sum_j m_j \left(f_j \frac{P_j}{\rho_j^2}\nabla_i W(\mathbf{x}_i - \mathbf{x}_j, h_j) + f_i \frac{P_i}{\rho_i^2}\nabla_i W(\mathbf{x}_i - \mathbf{x}_j, h_i) + \Pi_{ij}\nabla_i \bar{W}_{ij}\right). \tag{4.51}$$

In the previous equation, we notice the additional term due to the artificial viscosity Π_{ij}, which is needed to capture shocks.

The coefficients f_i, defined as

$$f_i = \left(1 + \frac{h_i}{3\rho_i}\frac{\partial \rho_i}{\partial h_i}\right)^{-1} , \quad (4.52)$$

incorporate fully the variable smoothing length correction term. For a detailed derivation of this formalism and its conserving capabilities, see [21].

4.3.3 Eulerian Methods

These methods solve the hydrodynamical equations adopting a grid, which can be fixed or adaptive in time. The first attempts were based on the so-called central difference schemes, i.e., schemes where the relevant hydrodynamical quantities were only represented by their values at the center of the grid, and the various derivatives were obtained by the finite-difference representation. This approach was not satisfactory: in fact, the schemes were only first-order accurate and they made use of some artificial viscosity to treat discontinuities, similarly to SPH. In more modern codes (like, e.g., the piecewise parabolic method, PPM), the shape of the hydrodynamical quantities $f_{n,u}(x)$ is recovered with much higher accuracy thanks to the use of several neighbouring cells. This allows to calculate the total integral of the given quantity over the grid cell, divided by the volume of each cell (e.g., cell average, $\hat{u}_n$), rather than pointwise approximations at the grid centres (e.g., central variables, u_n):

$$\hat{u}_n = \int_{x_{n-0.5}}^{x_{n+0.5}} f_{n,u}(x)dx . \quad (4.53)$$

This global shape is also used to estimate the values at the cell boundaries (e.g., $u^l_{n\pm0.5}$,$u^r_{n\pm0.5}$, the so-called interfaces), which can be used later as initial conditions of a Riemann problem, e.g., the evolution of two constant states separated by a discontinuity. Once the Riemann is analytically solved (see, e.g., [7]), it is possible to compute the fluxes across these boundaries for the time interval and then to update the cell averages $\hat{u}_n$. Notice that for high-dimensional problems, this procedure has to be repeated for each coordinate direction separately.

In general, the Eulerian methods suffer from limited spatial resolution, but they work extremely well in both low- and high-density regions, as well as in treating shocks. In cosmological simulations, accretion flows with large Mach numbers are very common. Here, following the total energy in the hydrodynamical equations can produce inaccurate thermal energy, leading to negative pressure, due to discretisation errors when the kinetic energy dominates the total energy. In such cases, the numerical schemes usually switch from formulations solving the total energy to formulations based on solving the internal energy in these hypersonic flow regions.

4.3.4 Code Comparison

The hydrodynamical schemes presented in the previous sections are based on quite different approaches, having for construction very different characteristics (in practice, particles against grid). For this reason, in order to be validated, they must be tested against problems with known solutions and then one has to compare their results, once the same initial conditions are given. The first standard test is the reproduction of the shock tube problem: in this case, codes able to capture the presence of shocks, like the most sophisticated Eulerian methods, have certianly better performances. Another possible test is the evolution of an initially spherical perturbation.

In general, a code validation in a more cosmological context is difficult. A first attempt of comparison between grid-based and SPH-based codes can be found in [14]. More recently, [18] compared the thermodynamical properties of the intergalactic medium predicted by the Lagrangian code *GADGET* and by the Eulerian code *ENZO*.

In the case of simulations for single cosmic structures, the most complete code comparison is the one provided by the Santa Barbara Cluster Comparison Project [10], where 12 different groups ran the own codes (including both Eulerian and Lagrangian schemes), starting from the same identical initial conditions corresponding to a massive galaxy cluster. In general, the agreement for the dark matter properties was satisfactory, with a 20 per cent scatter around the mean density and velocity dispersion profiles. A similar agreement was also obtained for many of the gas properties, like the temperature profile or the ratio of the specific dark matter kinetic energy to the gas thermal energy; somewhat larger differences are found in the temperature or entropy profiles in the innermost regions. One of the most worrying discrepancies is certainly when the total X-ray luminosity is considered: in this case, the spread can be also a factor 10. Most of the problem originates from a too low resolution in the central core, where the gas density (which enters as squared in the estimate of the X-ray luminosity) reaches its maximum. Another large difference between the results of different hydro-codes is related to the predicted baryon fraction and its profile within the cluster.

In general, we can conclude that Lagrangian and Eulerian schemes are providing compatible results, with some spread due to their specific weaknesses. However, we have to remind that these comparisons have been done in the nonradiative regime, i.e., excluding a long list of complex physical processes acting on baryons (see the following section), processes whose implementation is mandatory for a complete description of the formation of cosmic structures.

4.3.5 Extra Gas Physics

We know very well that the formation and evolution of cosmic structures is not determined only by the gravity due to the total matter distribution and by the adiabatic behavior of gas: many other important processes are in action, in particular on the baryonic component, influencing the final (physical and observational) properties of the structures. Here, we briefly list these processes, discussing their importance and pointing out the related numerical issues.

- *Cooling.* Its standard implementation is made by adding the cooling function $\Lambda(u,\rho)$ in the first law of thermodynamics, under the assumptions that the gas is optically thin and in ionization equilibrium, and that only two-body cooling processes are important. Considering a plasma with primordial composition of H and He, these processes are collisional excitation of H I and He II, collisional ionization of H I, He I, and He II, standard recombination of H II, He II, and He III, dielectric recombination of He II, and free-free emission (Bremsstrahlung). Being the collisional ionization and recombination rates depending only on the temperature, when the presence of a ionizing background radiation is excluded, it is possible to solve the resulting rate equation analytically; alternatively, the solution is obtained iteratively (see a discussion in [15]). When also the metallicity is implemented in the code, the number of possible processes becomes so large that it is necessary to use pre-computed tabulated cooling function. Finally, we notice that for pratical reasons the gas cooling is followed as "sub time step" problem, decoupled from the hydrodynamical treatment.
- *Star formation.* The inclusion of the cooling process originates two numerical problems. First, since the cooling is a runaway process, in the central regions of clusters the typical cooling time becomes significantly shorter than the Hubble time, causing the so-called overcooling problem: the majority of baryons can cool down and condense out of the hot phase. Second, since cooling is proportional to the square of the gas density, its efficiency is strongly related to the number of the first collapsed structures, whose good representation in simulations depends on the assumed numerical resolution. To avoid these problems, it is necessary to include the process of star formation starting from the cold and dense phase of the gas. These stellar objects, once the phase of supernova explosion is reached, can inject a large amount of energy (the so-called feedback) in the gas, increasing its temperature and possibly counteracting the cooling catastrophe. From a numerical point of view, this process is implemented via simple recipes (see, e.g., [15]), based on the assumptions that the gas has a convergent flow, it is Jeans unstable and, most important, it has an overdensity larger than the threshold corresponding to collapsed regions. This procedure allows to compute the fraction of gas converted into stars. For computational and numerical reasons, the star formation is done only when a significant fraction of the gas particle mass is interested by the transformation: at this point, a new collisionless "star" particle is created from the parent star-forming gas particle, whose mass is reduced accordingly. This process takes place until the gas particle is entirely transformed into stars. In order to avoid spurious numerical effects, which arise from the gravitational interaction of particles with widely differing masses, one usually restricts the number of star particles spawned by a gas particle to be relatively small, typically $2-3$.

- *Supernova feedback.* Assuming a specific initial mass function, it is possible to estimate the number of stars ending their lifes as type II supernovae and then to compute the total amount of energy (typically 10^{51} erg per supernova) that each star particle can release to the surrounding gas. Assuming that the typical lifetime of massive stars is smaller than the typical simulation time step, the feedback energy is injected in the surrounding gas in the same step (instantaneous recycling approximation). Improvements with respect to this simple model include an explicit sub-resolution description of the multiphase nature of the interstellar medium, which provides the reservoir of star formation. Such a subgrid model tries to model the global dynamical behavior of the interstellar medium in which cold, star-forming clouds are embedded in a hot medium (see, e.g., the self-regulated model proposed by [22] and a general critical discussion in [4]).
- *Chemical enrichment.* A possible way to improve the previous models is the inclusion of a more detailed description of stellar evolution and of the corresponding chemical enrichment. In particular, it is possible to follow the release of metals also from type Ia supernovae and low-and intermediate-mass stars, avoiding the instantaneous recycling approximation (see the discussion in [5]).
- *Other processes.* The baryons present in the cosmic structures undergo other physical processes that can have an important role in their modeling a nonexaustive list includes the effects related to thermal conduction, radiative transfer, magnetic fields, relativistic particles, black holes, and extra sources of feedback (like, for instance, AGN). The first attempts at implementing the corresponding physics in cosmological simulations have been made, even if in some cases the results are not yet convergent. Having a robust implementation of all these phenomena represents the more difficult and changelling frontier for future numerical experiments.

References

1. Bardeen, J. M., Bond, J. R., Kaiser, N., & Szalay, A. S. 1986, ApJ, 304, 15
2. Barnes, J., & Hut, P. 1986, Nature, 324, 446
3. Binney, J., & Tremaine, S. 1987, Princeton, NJ, Princeton University Press, 1987
4. Borgani, S., et al. 2006, MNRAS, 367, 1641
5. Borgani, S., Fabjan, D., Tornatore, L., Schindler, S., Dolag, K., & Diaferio, A. 2008, Space Science Reviews, 134, 379
6. Couchman, H. M. P. 1991, ApJ, 368, L23
7. Courant, R., & Friedrichs, K. O. 1948, Pure and Applied Mathematics, New York: Interscience
8. Efstathiou, G., Davis, M., White, S. D. M., & Frenk, C. S. 1985, ApJS, 57, 241
9. Evrard, A. E., et al. 2002, ApJ, 573, 7
10. Frenk, C. S., et al. 1999, ApJ, 525, 554
11. Grossi, M., Dolag, K., Branchini, E., Matarrese, S., & Moscardini, L. 2007, MNRAS, 382, 1261
12. Heitmann, K., et al. 2008, Computational Science and Discovery, 1, 015003
13. Ito, T., Makino, J., Fukushige, T., Ebisuzaki, T., Okumura, S. K., & Sugimoto, D. 1993, PASJ, 45, 339

14. Kang, H., Ostriker, J. P., Cen, R., Ryu, D., Hernquist, L., Evrard, A. E., Bryan, G. L., & Norman, M. L. 1994, ApJ, 430, 83
15. Katz, N., Weinberg, D. H., & Hernquist, L. 1996, ApJS, 105, 19
16. Monaghan, J. J. 1992, ARA&A, 30, 543
17. Monaghan, J. J. 1997, Journal of Computational Physics, 136, 298
18. O'Shea, B. W., Nagamine, K., Springel, V., Hernquist, L., & Norman, M. L. 2005, ApJS, 160, 1
19. Power, C., Navarro, J. F., Jenkins, A., Frenk, C. S., White, S. D. M., Springel, V., Stadel, J., & Quinn, T. 2003, MNRAS, 338, 14
20. Press, W. H., Teukolsky, S. A., Vetterling, W. T., & Flannery, B. P. 1992, Cambridge: University Press
21. Springel, V., & Hernquist, L. 2002, MNRAS, 333, 649
22. Springel, V., & Hernquist, L. 2003, MNRAS, 339, 289
23. Springel, V. 2005, MNRAS, 364, 1105
24. Tormen, G., Bouchet, F. R., & White, S. D. M. 1997, MNRAS, 286, 865
25. White, S. D. M. 1996, Cosmology and Large Scale Structure, 349
26. Zel'Dovich, Y. B. 1970, A&A, 5, 84

Part II
Dark Matter

Chapter 5
Dark Matter Astrophysics

Guido D'Amico, Marc Kamionkowski, and Kris Sigurdson

Abstract This chapter is intended to provide a brief pedagogical review of dark matter for the newcomer to the subject. We begin with a discussion of the astrophysical evidence for dark matter. The standard weakly interacting massive particle (WIMP) scenario—the motivation, particle models, and detection techniques—is then reviewed. We provide a brief sampling of some recent variations to the standard WIMP scenario, as well as some alternatives (axions and sterile neutrinos). Exercises are provided for the reader.

5.1 Introduction

Dark matter is an essential ingredient in a good fraction of the literature on extragalactic astronomy and cosmology. Since dark matter cannot be made of any of the usual standard model particles (as we will discuss below), dark matter is also a central focus of elementary-particle physics. The purpose of this review is to provide a pedagogical introduction to the principle astrophysical evidence for dark matter and to some of the particle candidates.

Rather than present a comprehensive survey of the vast and increasingly precise measurements of the amount and distribution of dark matter, we will present very simple ("squiggly-line") arguments for the existence of dark matter in clusters and galaxies, as well as the arguments for why it is nonbaryonic. The aim will be to

Guido D'Amico
SISSA, via Beirut 2-4, 34014 Trieste, Italy; e-mail: damico@sissa.it

Marc Kamionkowski
California Institute of Technology, Mail Code 350-17, Pasadena, CA 91125, USA;
e-mail: kamion@caltech.edu

Kris Sigurdson
Department of Physics and Astronomy, University of British Columbia, Vancouver, BC V6T 1Z1, Canada; e-mail: krs@phas.ubc.ca

provide insight into the evidence and arguments, rather than to summarize results from the latest state-of-the-art applications of the techniques.

Likewise, construction of particle-physics models for dark matter has become a huge industry, accelerated quite recently, in particular, with anomalous cosmic-ray and diffuse-background results [1, 2]. Again, we will not attempt to survey these recent developments and focus instead primarily on the basic arguments for particle dark matter. In particular, there has developed in the theoretical literature over the past 20 years a "standard" weakly-interacting massive particle (WIMP) scenario, in which the dark matter particle is a particle that arises in extensions (e.g., supersymmetry [3] or universal extra dimensions [4]) of the standard model that are thought by many particle theorists to provide the best prospects for new-physics discoveries at the Large Hadron Collider (LHC). We, therefore, describe this basic scenario. More detailed reviews of WIMP, the main subject of this article, can be found in Refs. [3, 5, 6].

After describing the standard WIMP scenario, we provide a brief sampling of some ideas for "nonminimal" WIMPs, scenarios in which the WIMP is imbued with some additional properties, beyond simply those required to account for dark matter. We also briefly discuss some other attractive ideas (axions and sterile neutrinos) for WIMPs. Exercises are provided throughout.

5.2 Astrophysical Evidence

It has been well established since the 1930s that there is much matter in the Universe that is not seen. It has also been long realized, and particularly since the early 1970s, that much of this matter must be nonbaryonic. The evidence for a significant quantity of dark matter accrued from galactic dynamics, the dynamics of galaxy clusters, and applications of the cosmic virial theorem. The evidence that much of this matter is nonbaryonic came from the discrepancy between the total matter density $\Omega_m \simeq 0.2 - 0.3$ (in units of the critical density $\rho_c = 3H_0^2/8\pi G$, where H_0 is the Hubble parameter), obtained from such measurements, and the baryon density $\Omega_b \simeq 0.05$ required for the concordance between the observed light-element (H, D, ^{3}He, ^{4}He, ^{7}Li) abundances with those predicted by big-bang nucleosynthesis (BBN) [7], the theory for the assembly of light elements in the first minutes after the big bang.

Rather than review the historical record, we discuss the most compelling arguments for nonbaryonic dark matter today, as well as some observations, most relevant to astrophysical phenomenology of dark matter today.

5.2.1 Galactic Rotation Curves

The flatness of galactic rotation curves has provided evidence for dark matter since the 1970s. These measurements are particularly important now not only for establishing the existence of dark matter, but particularly for fixing the local dark-matter density, relevant for direct detection of dark matter. We live in a typical spiral galaxy, the Milky Way, at a distance ~ 8.5 kpc from its center. The visible stars and gas in the Milky Way extend out to a distance of about 10 kpc. From the rotation curve, the rotational velocity $v_c(r)$ of stars and gas as a function of Galactocentric radius r, we can infer the mass $M_<(r)$ of the Galaxy enclosed within a radius r. If the visible stars and gas provided all the mass in the Galaxy, one would expect that the rotation curve should decline at radii larger than the 10 kpc extent of the stellar disk according to the Keplerian relation $v_c^2 = GM_{obs}/r$. Instead, one observes that $v_c(r)$ remains constant (a flat rotation curve) out to much larger radii, indicating that $M_<(r) \propto r$ for $r \gg 10$ kpc and thus that the Galaxy must contain far more matter than contributed by the stars and the gas.

Assuming a spherically symmetric distribution of matter, the mass inside a radius r is given by

$$M_<(r) = 4\pi \int_0^r \rho(r')r'^2 dr'. \tag{5.1}$$

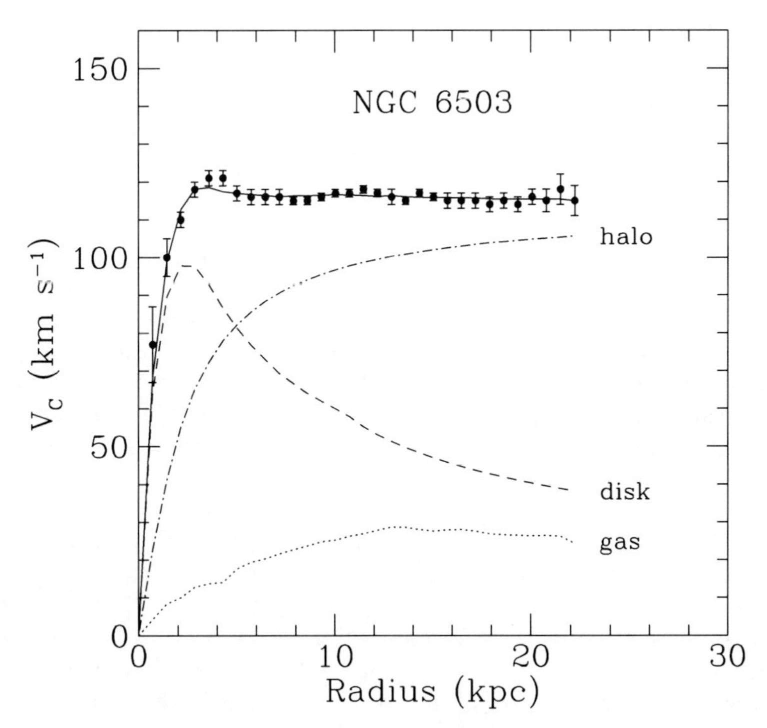

Fig. 5.1 Measured rotation curve of NGC6503 with best fit and contributions from halo, disk, and gas. From Ref. [8]

An estimate for the distribution of dark matter in the Galaxy can be obtained from the behavior of the rotation curve in the inner and outer galaxy. For example, the density distribution for the cored isothermal sphere, given by,

$$\rho(r) = \rho_0 \frac{R^2 + a^2}{r^2 + a^2}, \tag{5.2}$$

where $R \sim 8.5$ kpc is our distance from the Galactic center and ρ_0 is the local dark matter density, provides a qualitatively consistent description of the data. For large r, $\rho \sim r^{-2} \Rightarrow M(r) \propto r \Rightarrow v \sim$ const, while for small r, $\rho \sim$ const $\Rightarrow M(r) \propto r^3 \Rightarrow v \propto r$. Equation (5.2) describes a 2-parameter family of density profiles and by fitting the observed data one finds a scale radius $a \sim 3 - 5$ kpc and local matter density $\rho_0 \sim 0.4\,\mathrm{GeV\,cm^{-3}}$; the uncertainties arise from standard error in the rotation-curve measurements and from uncertainties in the contribution of the stellar disk to the local rotation curve. Because the dark matter is moving in the same potential well, the velocity dispersion of the dark matter can be estimated to be $\langle v_{\mathrm{dm}}^2 \rangle^{1/2} \sim 300$ km/sec. The simplest assumption is that the dark matter has a Maxwell–Boltzmann distribution with $f(\mathbf{v}) \sim e^{-v^2/2\bar{v}^2}$, where $\bar{v} \sim 220\,\mathrm{km/sec}$.

Exercise 1. Explain/estimate how ρ_0 would be affected if

- *(a) the halo were flattened, keeping the rotation curve unaltered;*
- *(b) the profile were of the Navarro–Frenk–White (NFW) type: $\rho(r) \propto \rho_c/[r(r + r_c)^2]$, keeping the local rotation speed the same;*
- *(c) the stellar contribution to the rotation curve was either increased or decreased.*

5.2.2 Galaxy Clusters

Galaxy clusters are the largest gravitationally bound objects in the Universe. They were first observed as concentrations of thousands of individual galaxies, and early application of the virial theorem $v^2 \sim GM/R$ (relating the observed velocity dispersion v^2 to the observed radius R of the cluster) suggested that there is more matter in clusters than the stellar component can provide [9]. It was later observed that these galaxies are embedded in hot x-ray–emitting gas, and we now know that clusters are the brightest objects in the x-ray sky. The x-rays are produced by hot gas excited to virial temperatures $T \sim$ keV of the gravitational potential well of the dark matter, galaxies, and gas. A virial temperature $T \sim$ keV corresponds to a typical velocity for the galaxies of $v \sim 10^3$ km/sec.

Observations of clusters come from optical and x-ray telescopes and more recently from the Sunyaev–Zeldovich effect [10]. Several independent lines of evidence from clusters indicate that the total mass required to explain observations is much larger than can be inferred by the observed baryonic content of galaxies and gas.

5.2.2.1 Lensing

Galaxy clusters exhibit the phenomenon of gravitational lensing [11, 12]. Because the gravitational field of the cluster curves the space around it, light rays emitted from objects behind the cluster travel along curved rather than straight paths on their way to our telescopes [13]. If the lensing is strong enough, there are multiple paths from the same object, past the cluster, that arrive at our location in the Universe; this results in multiple images of the same object (e.g., a background galaxy or active galactic nucleus). Furthermore, because the light from different sides of the same galaxy travels along slightly different paths, the images of strongly lensed sources are distorted into arcs. For instance, HST observations of Abell 2218 show arcs and multiple images as shown in Fig. 5.2. If the lensing is weak, the images may become slightly elongated, even if they are not multiply imaged.

For a lensing cluster with total mass M and impact parameter d the deflection angle is of order

$$\alpha \sim \left(\frac{GM}{dc^2}\right)^{1/2}. \tag{5.3}$$

Thus, from measurements of the deflection angle and impact parameter (which can be inferred by knowing the redshift to the lensing cluster and source), one can infer that the total mass M of a cluster is much larger than the observed baryonic mass M_b.

Exercise 2. Suppose a massive particle with velocity v is incident, with impact parameter b, on a fixed deflector of mass M. Calculate the deflection angle (using classical physics) due to scattering of this particle via gravitational interaction with the deflector. Show that you recover $\alpha = \left(GM/dc^2\right)^{1/2}$ in the limit $v \to c$,

Fig. 5.2 Image of the galaxy cluster Abel 2218. Credits: NASA, Andrew Fruchter, and the ERO team.

the velocity at which light rays propagate. Actually, the correct general relativistic calculation recovers this expression, but with an extra factor of 2.

Exercise 3. Estimate the deflection angle α for lensing by a cluster of $M \sim 10^{15} M_\odot$ and for an impact parameter of 1 Mpc.

5.2.2.2 Hydrostatic Equilibrium

In a relaxed cluster, the temperature profile $T(r)$ of gas, as a function of radius r, can be inferred using the strength of the emission lines, and the electron number density $n_e(r)$ can be inferred using the the x-ray luminosity $L(r)$. Combined, these observations give an estimate of the radial pressure profile $p(r) \propto n_e(r) k_B T(r)$. In steady state, a gravitating gas will satisfy the equation of hydrostatic equilibrium,

$$\frac{dp}{dr} = -G \frac{M_<(r)\, \rho_{\rm gas}(r)}{r^2} . \tag{5.4}$$

Here, $M_<(r)$ is the total (dark matter and baryonic gas) mass enclosed by a radius r and $\rho_{\rm gas}(r)$ is the density at radius r. Equation (5.4) can be used to determine the total mass M of the cluster. Comparison with the observed baryonic mass M_b again shows that $M \gg M_b$. In particular, observations using the x-ray satellites XMM-Newton and Chandra indicate that the ratio of baryonic matter to dark matter in clusters is $\Omega_b/\Omega_m \sim 1/6$. Additional constraints to the cluster-gas distribution can be obtained from the Sunyaev–Zeldovich (SZ) effect. This is the upscattering of cosmic microwave background (CMB) photons by hot electron gas in the cluster; the magnitude of the observed CMB-temperature change is then proportional to the integral of the electron pressure through the cluster (see, e.g., [10]).

Exercise 4. Estimate, in order of magnitude, the x-ray luminosity L_X for a cluster with total mass $M \sim 10^{15} M_\odot$ and a baryon fraction 1/6 in hydrostatic equilibrium with maximum radius $R \sim$ Mpc.

Exercise 5. Assume the cluster in Exer. 4 is isothermal ($T(r) = T =$ const.) with a dark matter distribution consistent with an NFW profile with $r_c \simeq R/10$. Neglecting the self-gravity of the gas:

- *(a) Show the properly normalized dark matter density profile is approximately $\rho(r) \simeq (233/45) M_c/[r(r+r_c)^2]$, where $M_c = M_<(r_c)$ is the mass enclosed within the scale radius r_c. Determine $M_<(r)$ and M_c and in terms of M for this cluster.*
- *(b) Using your results from (a) solve Eq. 5.4 and show that the gas density profile in such an NFW cluster takes the form $\rho_{gas}(r) \propto (1 + r/r_c)^{\Gamma r_c/r}$, where $\Gamma \propto (G M_c \mu m_p/r_c)/(k_B T)$.*

5.2.2.3 Dynamics

According to the virial theorem, the velocity dispersion of galaxies is approximately $v^2(r) \sim GM_<(r)/r$, where $M_<(r)$ is the mass enclosed within a radius r. Therefore, from measurements of the velocity dispersion and size of a cluster (which can be determined if the redshift and angular size of the cluster are known), one can infer the total mass M. Once again, the total mass is much larger than the baryonic mass $M \gg M_b$.

Cluster measurements are by now well established, with many well-studied and very well-modeled clusters, and there is a good agreement of estimates of M from dynamics, lensing, x-ray measurements, and the SZ effect. The current state of the art actually goes much further: one can now not only establish the existence of dark matter but also map its detailed distribution within the cluster.

Exercise 6. Following Zwicky [14], use the virial theorem to find an approximate formula relating the average mass of a galaxy in a galaxy cluster to the observed size and velocity dispersion of the cluster assuming that the system is self-gravitating (and assuming only that the observed galaxies contribute to the mass of the system). What answer would Zwicky have found for the Coma cluster with modern data?

5.2.3 Cosmic Microwave Background and Large-Scale Structure

Measurements of the CMB radiation and large-scale structure (LSS) of the Universe provide perhaps the most compelling evidence that the dark matter is nonbaryonic and the most precise measurements of its abundance.

One obtains from CMB maps the angular power spectrum C_ℓ of CMB temperature anisotropies as a function of multipole ℓ. If the temperature $T(\hat{\mathbf{n}})$ is measured as a function of position $\hat{\mathbf{n}}$ on the sky, then one can obtain the spherical-harmonic coefficients $a_{\ell m} = \int d\hat{\mathbf{n}} T(\hat{\mathbf{n}}) Y^*_{\ell m}(\hat{\mathbf{n}})$. The C_ℓ's are then simply the variance of the spherical-harmonic coefficients: $C_\ell = \langle |a_{\ell m}|^2 \rangle$. Theoretical predictions for the power spectrum depend on the values of cosmological parameters like the matter density $\Omega_m h^2$, the baryon density $\Omega_b h^2$, the cosmological constant Λ, the scalar spectral index n_s, the optical depth τ due to reionization, and the Hubble parameter H_0. One can thus determine these cosmological parameters by fitting precise measurements of the C_ℓs to the theoretical predictions [16]. Current measurements provide detailed information on C_ℓ over the range $2 < l < \mathcal{O}(1000)$, thus providing precise constraints to the cosmological parameters.

In 2000, data from the Boomerang and MAXIMA experiments (with supernova measurements) gave $\Omega_m h^2 = 0.13 \pm 0.05$ with error bars that shrink to ± 0.01 taking into account other measurements or assumptions (e.g., LSS, Hubble-constant, and supernova measurements, and/or the assumption of a flat Universe) [17]. Now, with WMAP, $\Omega_m h^2 = 0.133 \pm 0.006$ and $\Omega_b h^2 = 0.0227 \pm 0.0006$ [18].

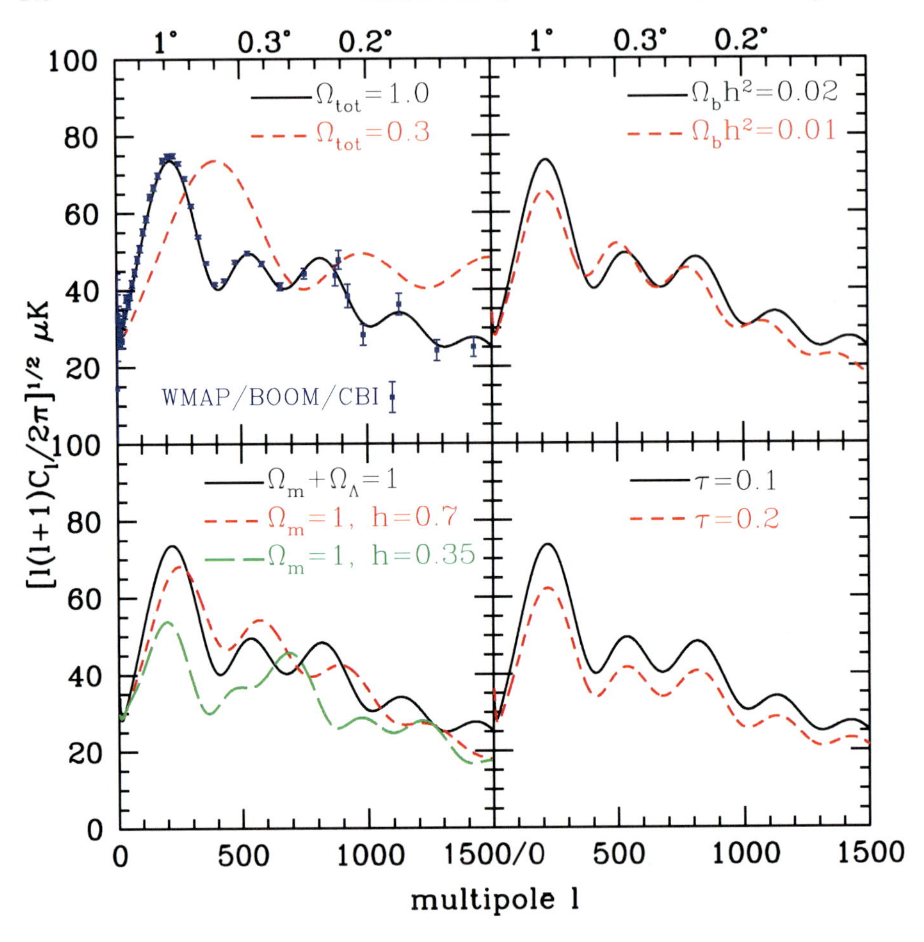

Fig. 5.3 Dependence of the CMB power spectrum on the cosmological parameters. From Ref. [15].

Exercise 7. Suppose that the temperature is measured with a Gaussian noise $\sigma_T \simeq 25\,\mu\mathrm{K}$ in $N_{\rm pix} \sim 10^6$ pixels on the sky. Estimate the rms temperature $\left\langle (\delta T/T)^2 \right\rangle^{1/2}$ that results.

5.3 Basic Properties of Dark Matter

Having established the existence of dark matter and presented the case that it is non-baryonic, we now consider the requirements for a dark-matter candidate and discuss some possibilities. Every dark-matter candidate should satisfy several requirements:

- Dark matter must be *dark*, in the sense that it must generically have no (or extremely weak) interactions with photons; otherwise, it might contribute to the dimming of quasars, create absorption lines in the spectra of distant quasars [19], or emit photons. One way to quantify this is by assuming that dark-matter particles have a tiny charge fe (where e is the electron charge and $f \ll 1$), which can be quantitatively constrained [20].

- Self-interactions of the dark matter should be small. We can estimate the cross section for DM–DM scattering in the following way: if DM particles scatter less than once in the history of the Universe, then the mean free path is less than $\lambda = v_{DM} H_0^{-1} \sim \left(3 \times 10^7 \mathrm{cm/sec}\right)\left(10^{17}\,\mathrm{sec}\right) \sim 3 \times 10^{24}\,\mathrm{cm}$. Then, if the galactic-halo density is $\rho_{DM} \sim 10^{-24}\mathrm{g/cm}^3$, the opacity for self-scattering in the galactic halo is $\kappa = (\rho_{DM}\lambda)^{-1} = \sigma/m \sim \mathrm{cm}^2/\mathrm{g}$. Thus, if the elastic-scattering cross section is $\sigma \gtrsim 10^{-24}\,(m/\mathrm{GeV})\,\mathrm{cm}^2$, then $\kappa \gtrsim 1$ and the typical halo–dark-matter particle scatters more than once during the history of the Universe. If dark matter self-scattered, it would suffer *gravothermal catastrophe*: that is, in binary interactions of two dark-matter particles, one particle can get ejected from the halo, while the other moves to a lower energy state at smaller radius. As this occurs repeatedly, much of the halo evaporates and the remaining halo shrinks. Although a variety of arguments can constrain dark-matter self-interactions, stringent and very transparent constraints come from observations of the Bullet Cluster, the merger of two galaxy clusters, in which it is seen (from gravitational-lensing maps of the projected matter density) that the two dark-matter halos have passed through each other while the baryonic gas has shocked and is located between the two halos [21].
- Interactions with baryons must also be weak. Suppose baryons and dark matter interact. As an overdense region collapses to form a galaxy, baryons and dark matter would fall together, with photons radiated from this baryon-DM fluid. This would result in a baryon-DM disk, in contradiction with the more diffuse and extended dark-matter halos that are observed. If DM interacted with baryons other than gravitationally in the early Universe, the baryon-photon fluid would be effectively heavier (have a higher mass loading relative to radiation pressure) even before recombination so that the baryon acoustic oscillations in the matter power spectrum and the CMB angular power spectrum would be modified [22].
- Dark matter cannot be made up of Standard Model (SM) particles since most leptons and baryons are charged. The only potentially suitable SM candidate is the neutrino, but it cannot be dark matter because of the celebrated Gunn-Tremaine bound [23], which imposes a lower bound on the masses of dark matter particles that decoupled when relativistic. The argument is the following: The momentum distribution in the galactic halo is roughly Maxwell–Boltzmann with a momentum uncertainty $\Delta p \sim m_\nu \langle v \rangle$ ($\langle v \rangle \sim 300\,\mathrm{km/sec}$), while the mean spacing between neutrinos is $\Delta x \sim n_\nu^{-1/3} \sim (\rho_\nu/m_\nu)^{-1/3}$. The Heisenberg uncertainty principle gives $\Delta x \Delta p \gtrsim \hbar$, which translates into a lower bound $m_\nu \gtrsim 50\,\mathrm{eV}$. (This Heisenberg bound can actually be improved by a factor of 2 by using arguments involving conservation of phase space.) Stronger bounds ($m_\nu \gtrsim 300\,\mathrm{eV}$) can be obtained from dwarf galaxies which have higher phase-space densities. As discussed below, there will be a cosmological density of neutrinos left over from the big bang, with a density $\Omega_\nu h^2 \sim 0.1\,(m_\nu/10\,\mathrm{eV})$. The neutrinos of mass

$m_\nu \gtrsim 300\,\mathrm{eV}$ consistent with the Gunn-Tremaine bound would overclose the Universe. Thus, neutrinos are unable to account for the dark matter.

5.4 Weakly Interacting Massive Particles (WIMPs)

Perhaps the most attractive dark matter candidates to have been considered are WIMPs. Many theories for new physics at the electroweak scale (e.g., supersymmetry, universal extra dimensions) introduce a new stable, WIMP, with a mass of order $M_\chi \sim 100\,\mathrm{GeV}$.

For example, in supersymmetric (SUSY) theories, the WIMP is the neutralino

$$\tilde{\chi} = \xi_\gamma \tilde{\gamma} + \xi_Z \tilde{Z}^0 + \xi_{h1} \tilde{h}_1^0 + \xi_{h2} \tilde{h}_2^0, \tag{5.5}$$

a linear combination of the supersymmetric partners of the photon, Z^0 boson, and neutral Higgs bosons. Neutralinos are neutral spin-1/2 Majorana fermions. In theories with universal extra dimensions there are Kaluza–Klein (KK) states γ_{KK}, Z^0_{KK}, H^0_{KK}, which are neutral KK bosons. The candidates are stable (or quasi-stable; i.e., lifetimes greater than the age of the Universe $\tau \gg t_U$) and particle-theory models suggest masses $M_\chi \sim 10 - 10^3\,\mathrm{GeV}$.

In typical theories, two WIMPs can annihilate to SM particles. For example, for a neutralino, we have the tree-level diagram in Fig. 5.4, where $m_{\tilde{q},\tilde{l}} \sim 100\,\mathrm{GeV}$ so that $\sigma \sim \alpha^2 m_{\tilde{q},\tilde{l}}^{-4} M_\chi^2 \sim 10^{-8}\,\mathrm{GeV}^{-2}$.

5.4.1 WIMP Freezeout in Early Universe

We now estimate the relic abundance of WIMPs in the standard scenario of thermal production (see, e.g., Ref. [24]). In the early Universe, at temperatures $T \gg M_\chi$, WIMPs are in thermal equilibrium and are nearly abundant as lighter particles, like

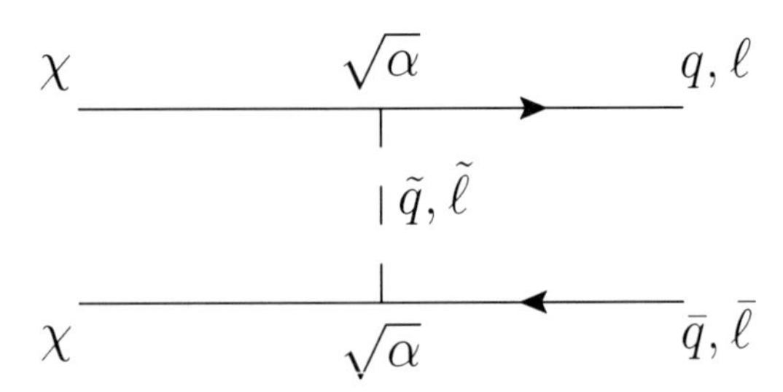

Fig. 5.4 An example of a Feynman diagram for annihilation of two WIMPs χ (neutralinos in this case) to fermion-antifermion pairs (where the fermions are either quarks q or leptons l) via exchange of an intermediate-state squark $\tilde{q}$ or slepton $\tilde{l}$.

photons, quarks, leptons, etc. Their equilibrium abundance is maintained via rapid interconversion of $\chi\chi$ pairs and particle-antiparticle pairs of SM particles. When the temperature falls below the WIMP mass, however, the WIMP abundances become Boltzmann suppressed, and WIMPs can no longer find each other to annihilate. The remaining WIMPs constitute a primordial relic population that still exists today.

We now step through a rough calculation. To do so, we assume that the WIMP is a Majorana particle, its own antiparticle (as is the case for the neutralino, for example), although the calculation is easily generalized for WIMPs with antiparticles (e.g., KK WIMPs).

The annihilation rate for WIMPs is $\Gamma(\chi\chi \leftrightarrow q\bar{q}, \ell\bar{\ell}, \ldots) = n_\chi \langle \sigma v \rangle$, where σ is the cross section for annihilation of two WIMPs to all lighter SM particles, v is the relative velocity, and the angle brackets denote a thermal average. The expansion rate of the Universe is $H = (8\pi G\rho/3)^{1/2} \sim T^2/M_{\rm Pl}$ during the radiation era, where $\rho \propto T^4$. In the spirit of "squiggly lines," we have neglected factors like the effective number of relativistic degrees of freedom g_* in the expansion rate, which the careful reader can restore for a more refined estimate.

By comparing these two rates, one can identify two different regimes:

- At early times, when $T \gg M_\chi$, $n_\chi \propto T^3$ and $\Gamma \gg H$: particles scatter and annihilate many times during an Hubble time and this maintains chemical equilibrium.
- At late times, when $T \ll M_\chi$, $n_\chi \propto T^{3/2} e^{-M_\chi/T}$ (note that the chemical potential $\mu_X = 0$ in the case of Majorana particles such as the neutralino) and $\Gamma \ll H$: there can be no annihilations, and the WIMP abundance freezes out (the comoving number density becomes constant).

This sequence of events is illustrated in Fig. 5.5, which shows the comoving number density of WIMPs as a function of the inverse temperature in equilibrium (solid curve) and including freezeout (dashed curves).

Freezeout occurs roughly when $\Gamma(T_f) \sim H(T_f)$. For nonrelativistic particles, $n_\chi = g_\chi \left(M_\chi T/2\pi\right)^{3/2} e^{-M_\chi/T}$, so the freezeout condition becomes

$$\left(M_\chi T_f\right)^{3/2} e^{-M_\chi/T_f} \sim \frac{T_f^2}{M_{Pl}} \quad \Rightarrow \quad \frac{T_f}{M_\chi} \sim \ln\left[\frac{M_{Pl} M_\chi^{3/2} \langle \sigma v \rangle}{T_f^{1/2}}\right], \tag{5.6}$$

where the mass parameters are in GeV. Taking $\langle \sigma v \rangle \sim \alpha^2/M_\chi^2$, and taking as a first guess $T_f \sim M_\chi$, we finally find

$$\frac{T_f}{M_\chi} \sim \left\{\ln\left[\frac{M_{Pl}\alpha^2}{(M_\chi T_f)^{1/2}}\right]\right\}^{-1} \sim \left\{\ln\left[\frac{10^{19}10^{-4}}{100}\right]\right\}^{-1} \sim \frac{1}{25} + \text{log corrections}, \tag{5.7}$$

where the numerical values are characteristic electroweak-scale parameters (i.e., $\sigma \sim 10^{-8}\,\text{GeV}^{-2}$, $M_\chi \sim 100\,\text{GeV}$).

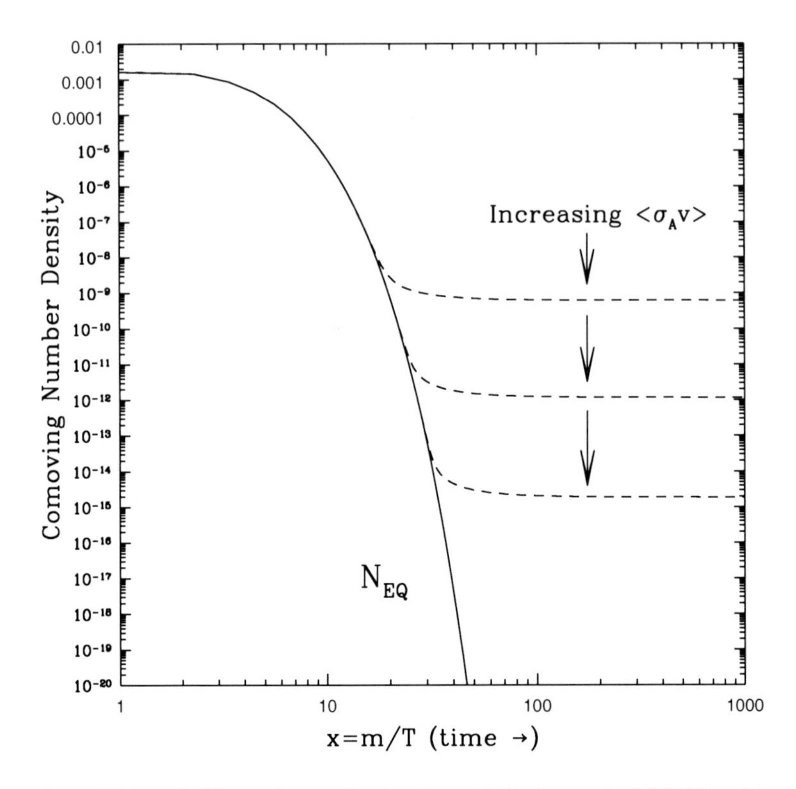

Fig. 5.5 Equilibrium (solid curve) and relic abundance (dashed curves) of WIMP particles. From Ref. [3].

At freezeout, the abundance relative to photons is

$$\frac{n_\chi}{n_\gamma} = \frac{\Gamma(T_f)/\langle\sigma v\rangle}{T_f^3} = \frac{H(T_f)/\langle\sigma v\rangle}{T_f^3} \sim \frac{T_f^2}{M_{Pl}\langle\sigma v\rangle T_f^3}$$
$$\sim \frac{1}{M_{Pl}\langle\sigma v\rangle T_f} \sim \frac{25}{M_{Pl}\langle\sigma v\rangle M_\chi}. \tag{5.8}$$

Today, we know that

$$\Omega_\chi = \frac{\rho_\chi}{\rho_c} \sim \frac{n_\chi^0}{n_\gamma^0}\frac{M_\chi n_\gamma^0}{\rho_c} \sim \frac{25}{M_{Pl}\langle\sigma v\rangle}\frac{400\,\mathrm{cm}^{-3}}{10^{-6}\,\mathrm{GeV\,cm}^{-3}}, \tag{5.9}$$

with no explicit dependence on the particle mass.

We thus obtain the observed abundance $\Omega_\chi h^2 \sim 0.1$ for $\sigma \sim 10^4\,(0.1 \times 10^{19} \times 10^{-6})^{-1}\,\mathrm{GeV}^{-2} \sim 10^{-8}\,\mathrm{GeV}^{-2}$, which turns out to be nearly exact, even though we have been a bit sloppy. A more precise calculation (including all the factors we have dropped) gives

$$\Omega_\chi h^2 \sim 0.1 \left(\frac{3 \times 10^{-26} \mathrm{cm}^3/\mathrm{sec}}{\langle \sigma v \rangle} \right) + \text{log corrections}, \qquad (5.10)$$

a remarkable result, as it implies that if there is a new stable particle at the electroweak scale, it is the dark matter.

As an aside, note that partial-wave unitarity of annihilation cross sections requires $\sigma \lesssim M_\chi^{-2}$, which means $\Omega_\chi h^2 \gtrsim (M_\chi/300\,\mathrm{TeV})^2$. This thus requires $\Omega_\chi h^2 \lesssim 0.1$, $M_\chi \lesssim 100\,\mathrm{TeV}$, without knowing anything about particle physics [25]. More precisely, this bound applies for point particles and does not apply if dark matter particles are bound states or solitons. If the interactions are strong, $\alpha \sim 1$, the bound is already saturated.

Although our arguments have been rough, one finds in SUSY and KK models that there are many combinations of reasonable values for the the SUSY or KK parameters that provide a WIMP with $\Omega_\chi h^2 \sim 0.1$ for $10\,\mathrm{GeV} \lesssim M_\chi \lesssim 1\,\mathrm{TeV}$.

Exercise 8. Equation (5.10) was derived assuming that the annihilation cross section $\langle \sigma v \rangle$ is temperature-independent. Redo the estimate for $\Omega_\chi h^2$ assuming that $\langle \sigma v \rangle \propto T^n$, where $n = 1,2,3,\cdots$.

5.4.2 Direct Detection

If WIMPs make up the halo of the Milky Way, then they have a local spatial density $n_\chi \sim 0.004\,(M_\chi/100\,\mathrm{GeV})^{-1}\mathrm{cm}^{-3}$ (roughly one per liter) and are moving with velocities $v \sim 200$ km sec^{-1}. Moreover, there is a crossing symmetry between the annihilation $\chi\chi \to q\bar{q}$ and the elastic scattering $\chi q \to \chi q$ processes—apart from some kinematic factors, the diagrams are more or less the same (as shown in Fig. 5.6)—so one expects roughly that the cross section $\sigma(\chi q \to \chi q) \sim \sigma(\chi\chi \to q\bar{q}) \sim 10^{-36}\,\mathrm{cm}^2$. One can, therefore, hope to detect a WIMP directly by observing its interaction with some target nucleus in a low-background detector composed, e.g., of germanium, xenon, silicon, sodium, iodine, or some other element.

At low energies, quarks are bound into nucleons and nucleons in turn are bound into nuclei, so the cross section one actually needs is $\sigma(\chi N \to \chi N)$ (where N here

Fig. 5.6 Crossing symmetry between annihilation and scattering diagrams.

stands for a nucleus). The calculation relating the χq interaction to the χN interaction requires both QCD and nuclear physics. It is complicated but straightforward. Here, we will simply assume, for illustration, that $\sigma(\chi N \to \chi N) \sim \sigma(\chi q \to \chi q)$.

The rate at which a nucleus in the detector is hit by halo WIMPs is then

$$R \sim n_\chi \sigma v \sim (0.004\,\mathrm{cm}^{-3})(10^{-36}\,\mathrm{cm}^2)\left(3\times 10^7 \frac{\mathrm{cm}}{\mathrm{sec}}\right) \sim 10^{-24}\mathrm{yr}^{-1}; \tag{5.11}$$

if there are $6\times 10^{23}\,M/(A\,\mathrm{g})$ nuclei in a detector, for an atomic number $A \sim 100$, we expect to see $R \sim 10/\mathrm{kg}/\ \mathrm{yr}$ events.

Let us estimate the recoil energy of a nucleus struck by a WIMP. If a WIMP of $M_\chi \sim 100\,\mathrm{GeV}$ runs into a nucleus with $A \sim 100$, the momentum change is $\Delta p \sim M_\chi v$, and the nucleus recoils with an energy of order $E \sim (\Delta p)^2/2m \sim (100\,\mathrm{GeV}\,10^{-3})^2(100\,\mathrm{GeV})^{-1} \sim 100\,\mathrm{keV}$.

To do things more carefully, one has to account for the fact that the cross section one actually needs the interaction cross sections with nuclei, and via the following steps,

$$\sigma(\chi q) \underset{\mathrm{QCD}}{\longrightarrow} \sigma(\chi n), \sigma(\chi p) \underset{\text{nuclear physics}}{\longrightarrow} \sigma(\chi N),$$

some theoretical uncertainties are introduced. One also finds that $\sigma(\chi N)$ is reduced relative to $\sigma(\chi q)$ by several orders of magnitude.

Qualitatively, there are two different types of interactions, axial and scalar (or spin-dependent and spin-independent). The first is described by the Lagrangian,

$$\mathscr{L}_{\mathrm{axial}} \propto \bar{\chi}\gamma^\mu\gamma_5\chi\,\bar{q}\gamma_\mu\gamma_5 q, \tag{5.12}$$

which couples χ to the spin of unpaired nucleons; this works only for nuclei with spin, and the coupling is different for unpaired protons or neutrons. Through this interaction one expects $\sigma \propto \bar{s}^2$, where $\bar{s}$ is the average spin $\sim 1/2$ of the unpaired proton or neutron in nuclei with odd atomic number.

The second interaction is described by the Lagrangian,

$$\mathscr{L}_{\mathrm{scalar}} \propto \bar{\chi}\chi\bar{q}q, \tag{5.13}$$

which couples χ to the mass of the nucleus, thus giving a cross section $\sigma \propto M^2 \propto A^2$ (where M and A are the nuclear mass and atomic number), which implies higher cross sections for larger A. However, this scaling is only valid up to a limit. In fact, the momentum exchanged is $\Delta p \sim (100\,\mathrm{GeV})(10^{-3}) \sim 0.1\,\mathrm{GeV}$, and the nuclear radius is roughly $r \sim A^{1/3}10^{-13}\,\mathrm{cm}$, so from the uncertainty principle one has $r\Delta p \gtrsim 1$, when

$$\frac{(0.1\,\mathrm{GeV})(10^{-13}\,\mathrm{cm})}{2\times 10^{-14}\,\mathrm{GeV\,cm}} A^{1/3} \gtrsim 1 \qquad \Longrightarrow \qquad A \gtrsim 10. \tag{5.14}$$

Detailed calculations show that the cross section for WIMP-nucleus elastic scattering does not increase much past $A \gtrsim 100$.

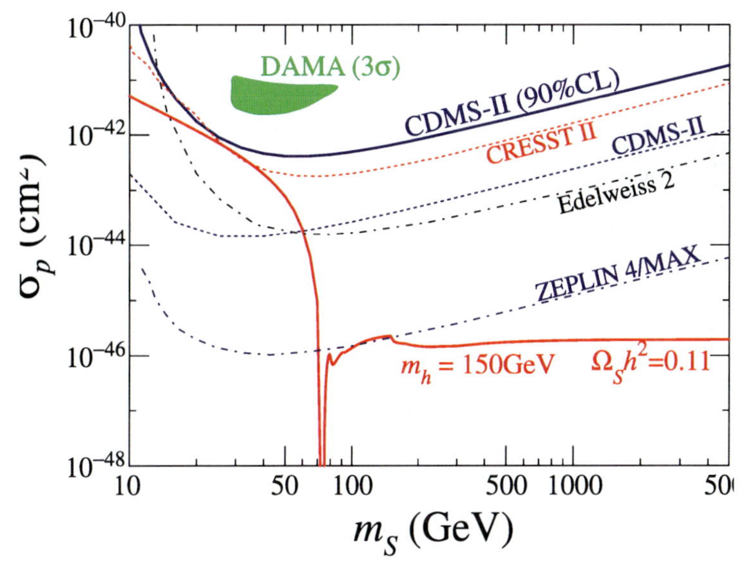

Fig. 5.7 Exclusion plot for the spin-independent dark-matter parameter space. The region favored by the DAMA annual modulation is inconsistent with the current bound (solid curve) from CDMS. The broken curves are forecasts for future experiments. We also show, for illustrative purposes only, predictions for a WIMP model with a lightest-Higgs-boson mass of $m_h = 150$ GeV.

In experiments, people usually draw exclusion curves for the WIMP-nucleon cross section versus the WIMP mass M_χ. The exclusion curves are less constraining both for low M_χ because of the low recoil energy, and for large M_χ because (for fixed local energy density ρ_χ) of the number density $n_\chi \propto M_\chi^{-1}$. Till date, only the DAMA experiment has reported a positive signal [26]. They used NaI, in which both nuclei have spin, one with an unpaired proton and the other with an unpaired neutron. The interpretation of their signal in terms of a WIMP with scalar interactions was ruled out by null results (at the time) from CDMS. An interpretation of their signal in terms of a spin-dependent WIMP-neutron interaction was ruled out by the null search in their Xe detector [27]. Although the interpretation in terms of spin-dependent WIMP-proton scattering was consistent with null results from other direct searches [27], it was ruled out by null searches for energetic neutrinos from the Sun (see Fig. 5.8), as we explain below. The interpretation in terms of spin-dependent scattering is now also ruled out directly by null results from the COUPP experiment [28].

5.4.3 Energetic ν's from the Sun

The escape velocity at the surface of the Sun is $v_s \sim 600\,\mathrm{km/sec}$, while at the center it is $v_c \sim 1300\,\mathrm{km/sec}$. If in passing through the Sun, a WIMP from the galactic halo scatters from a nucleus (most likely a proton) therein to a velocity less than the escape velocity, then it is gravitationally trapped within the Sun. As

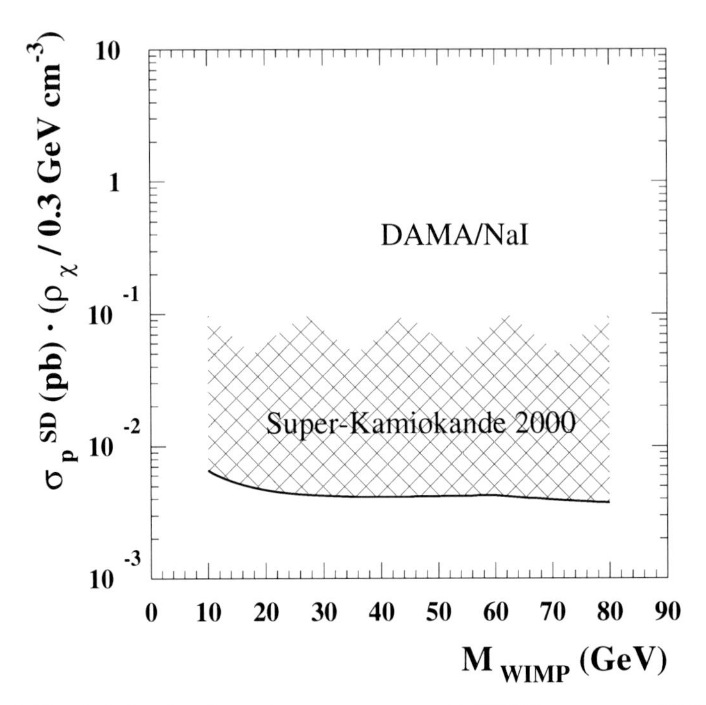

Fig. 5.8 The shaded region shows the parameter space (in WIMP mass versus SD WIMP-proton cross section) implied by the DAMA annual modulation for a WIMP with exclusively SD interactions with protons and no interaction with neutrons. The solid curve indicates the upper bound to the SD WIMP-proton cross section from null searches for neutrino-induced upward muons from the Sun; thus the cross-hatched region is excluded [27].

the gravitationally-trapped WIMP passes through the Sun subsequently, it loses energy in additional nuclear scatters and thus settles to the center of the Sun. In this way, the number of WIMPs in the center of the Sun is enhanced. These WIMPs can then annihilate to SM particles, through the same early-Universe processes that set their relic abundance [29]. Decays of the annihilation products (e.g., $W^+W^-, Z^0Z^0, \tau^+\tau^-, t\bar{t}, b\bar{b}, c\bar{c}, \ldots$) to neutrinos will produce energetic neutrinos that can escape from the center of the Sun. The neutrino energies are $E_\nu \sim [(1/3)-(1/2)]M_\chi \sim 100\,\text{GeV}$ and so cannot be confused with ordinary solar neutrinos, which have energies ~MeV. At night, these neutrinos will move up through the Earth. If the neutrino produces a muon through a charged-current interaction in the rock below a neutrino telescope (e.g., super-Kamiokande, AMANDA, or IceCube), the muon may be seen. In this way, one can search for these WIMP-annihilation–induced neutrinos from the Sun.

5.4.4 Cosmic Rays from DM Annihilation

In the galactic halo, one expects the annihilation processes $\chi\chi \to \cdots \to e^+e^-, p\bar{p}, \gamma\gamma$; detection of these products can be a signal of the presence of dark matter.

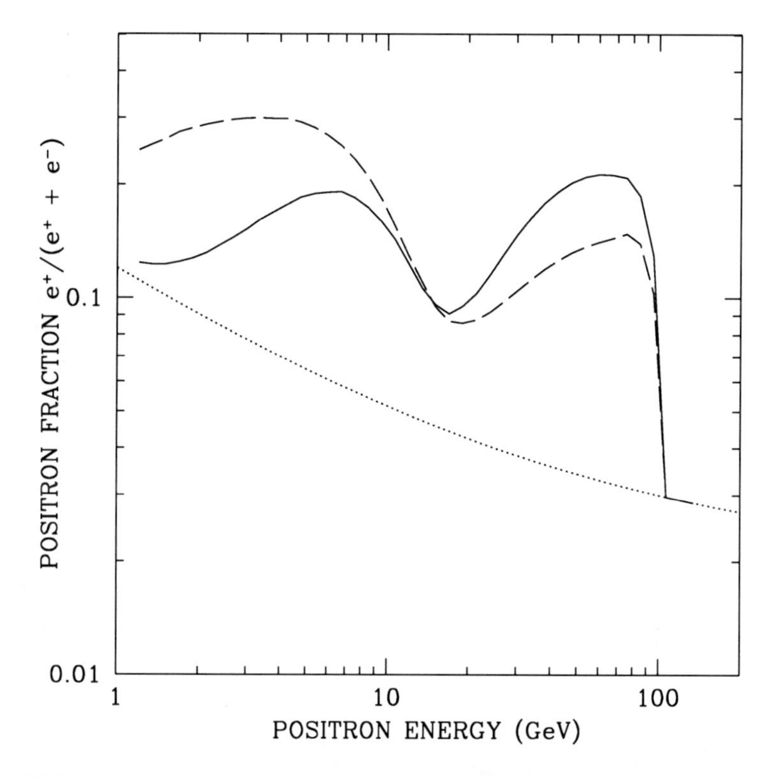

Fig. 5.9 The positron fraction, as a function of electron-positron energy, from annihilation of a 120 GeV neutralino WIMP to gauge bosons. The different curves are for different cosmic-ray-propagation models, and in both cases, the annihilation rate has been boosted by a factor of ten relative to the canonical (smooth-halo) value. From Ref. [30].

Exercise 9. Show that the annihilation process $\chi\chi \to e^+e^-$ is suppressed for Majorana WIMPs as the relative velocity $v \to 0$.

5.4.4.1 Positrons

Because of Galactic magnetic fields, cosmic-ray positrons and antiprotons do not propagate in straight lines and will thus appear to us as a diffuse background. Continuum e^+'s from WIMP annihilation are difficult to separate from ordinary cosmic-ray positrons. It has been argued that indirect processes, such as the annihilation $\chi\chi \to W^+W^- \to e^+\nu e^-\bar{\nu}$ [30], will produce a distinctive bump in the positron spectrum at energies $E_e \lesssim M_\chi$ (direct annihilation of Majorana WIMPs to electron-positron pairs is suppressed at galactic relative velocities), as illustrated in Fig. 5.9, and there has been tremendous excitement recently with the reported detection by the PAMELA experiment of such a bump [31]. However, it may be that nearby pulsars can also produce a bump in the positron spectrum [32], and more recent results from the Fermi telescope [33] call the PAMELA result into questions. It will thus be important to understand the possible pulsar signal, as well as the data, more carefully before the PAMELA excess can be attributed to WIMP annihilation.

5.4.4.2 Antiprotons

Likewise, it has also been argued that low-energy antiprotons from WIMP annihilation can be distinguished, through their energy spectrum, from the more prosaic cosmic-ray antiprotons produced by cosmic-ray spallation. Antiprotons can be produced by the decay of the standard WIMP-annihilation products, and the energy spectrum of such antiprotons is relatively flat at low energies. On the other hand, the energy spectrum of low-energy cosmic-ray antiprotons due to cosmic-ray spallation decreases at energies $E \lesssim$GeV. This is because the process $\bar{p} + p_{ISM} \rightarrow p + p + \bar{p} + \bar{p}$ has an energy threshold, in the center of mass, of $E_{\rm CM} > 4m_p$. This requires the primary cosmic-ray momentum to be very high. Production of an antiproton with $E_{\bar{p}} \lesssim$GeV, therefore, requires that the antiproton be ejected with momentum opposite to that of the initial cosmic-ray proton, and the phase-space for this ejection is small.

5.4.4.3 Gamma Rays

A final channel to observe WIMP annihilation is via gamma rays from WIMP annihilation. Direct annihilation of WIMPs to two photons, $\chi\chi \rightarrow \gamma\gamma$, through loop diagrams such as those shown in Fig. 5.10, produce monoenergetic photons, with energies equal to the WIMP mass. For $v \sim 10^{-3}c$, the photon energies would be $E_\gamma = E_\chi \left(1 \pm 10^{-3}\right)$, and one would see a narrow γ-ray line with $\Delta v/v \sim 10^{-3}$, superposed on a continuum spectrum produced by astrophysical processes; such a line would be difficult to mimic with traditional astrophysical sources. Decays of WIMP-annihilation products also produce a continuum spectrum of gamma rays at lower energies.

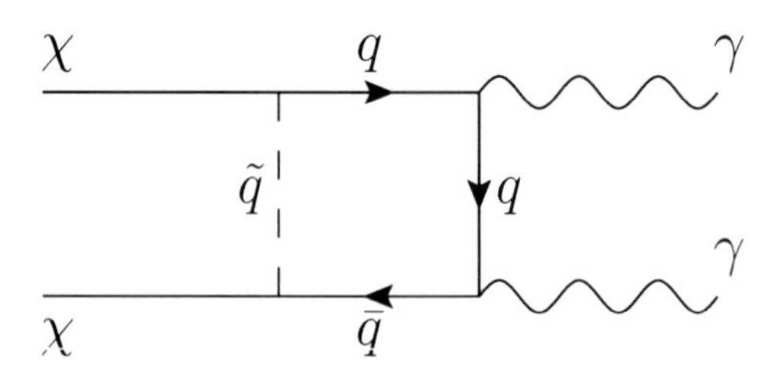

Fig. 5.10 Example of a Feynman diagram for annihilation of two neutralinos to two photons through a quark-squark loop.

The other advantage of gamma rays is that they propagate in straight lines. This opens up the possibility to distinguish gamma rays from WIMP annihilation from those from traditional sources through directionality—there should be a higher flux of WIMP-annihilation photons from places where WIMPs are abundant; e.g., the Galactic center. Another possibility is dwarf galaxies, which represent regions of high dark-matter density in the Milky Way halo. In general, the γ-ray flux (the number of photons per unit time-area–solid-angle) is given by

$$\frac{dF}{d\Omega} = \frac{\langle \sigma_{\chi\chi\to\gamma\gamma} v \rangle}{4\pi M_\chi^2} \int_0^\infty \rho^2(l)dl, \tag{5.15}$$

where the integral is taken along a given line of sight, l is the distance along that line of sight, and $\rho(l)$ is the dark-matter density at that distance. (Note that if $\rho(r) \propto r^{-1}$ with galactocentric radius r, as in an NFW profile, the intensity formally diverges, but the flux form any finite angular window around $r = 0$ is finite.)

Exercise 10. Estimate the γ-ray flux from WIMP annihilation, for a given annihilation cross section (times relative velocity) $\langle \sigma v \rangle_{\mathrm{ann}}$, in an angular window of radius ~ 5 degrees around the galactic center. Estimate a characteristic $\langle \sigma v \rangle$ for WIMPs and evaluate your result for the γ-ray flux for that value. How does it compare, in order of magnitude, with the sensitivity of the Fermi Gamma Ray Telescope?

5.4.4.4 Galactic Substructure and Boost Factors

The rate for annihilation, per unit volume, at any point in the Galactic halo is proportional to ρ^2, the square of the density at that point. The total annihilation rate in the halo, or in some finite volume of the halo, is then proportional to $\int dV \rho^2$, the integral, over that volume, of the density squared. In the canonical model, the halo density is presumed to vary smoothly with position in the galaxy with some density profile; e.g., the isothermal profile in Eq. (5.2).

However, a Galactic halo forms as part of a recent stage in a sequence of hierarchical structure formation. In this scenario, small objects undergo gravitational collapse first; they then merge to form more massive objects, which then merge to form even more massive objects, etc. If some of these substructures remain partially intact as they merge into more massive halos, then any given halo (in particular, the Milky Way halo) may have a clumpy distribution of dark matter. This is in fact seen in simulations. What this implies is that the annihilation rate in the halo may be enhanced by a "boost factor" $B \propto \langle \rho^2 \rangle / \langle \rho \rangle^2$, where the averages are over volume in the halo [34]. It may be possible to see angular variations in the γ-ray signal from WIMP annihilation, due to this substructure [35, 36]. It has even been suggested that proper motions of nearby substructures may be visible [37], although Ref. [38] disputed this claim.

As we will see below, the first gravitationally-collapsed objects in WIMP models have masses in the range $10^{-6} - 100$ Earth masses [39]. These objects may have densities several hundred times those of the mean halo density today. If so, and if these Earth-mass substructures survive intact through all subsequent generations of structure formation, then the boost factor B may be as large as several hundred, implying much larger cosmic-ray fluxes than the canonical model predicts.

Such large boost factors are, however, unlikely. Simulations of recent generations in the structure-formation hierarchy show that while the tightly bound inner parts of halos may survive during merging, the outer parts are stripped. Ref. [40] developed an analytic model, parametrized in terms of a halo-survival fraction, to describe the (nearly) scale-invariant process of hierarchical clustering. This model then provided the boost factor B in terms of that survival fraction. By comparing the results (cf., Fig. 5.11) of the analytic model for the local halo-density probability distribution function with subsequent measurements of the same distribution in simulations (Fig. 1 in Ref. [41]), one infers a small halo-survival fraction. The analytic model of Ref. [40] then suggests for this survival fraction no more than a small boost factor, $B \lesssim$few.

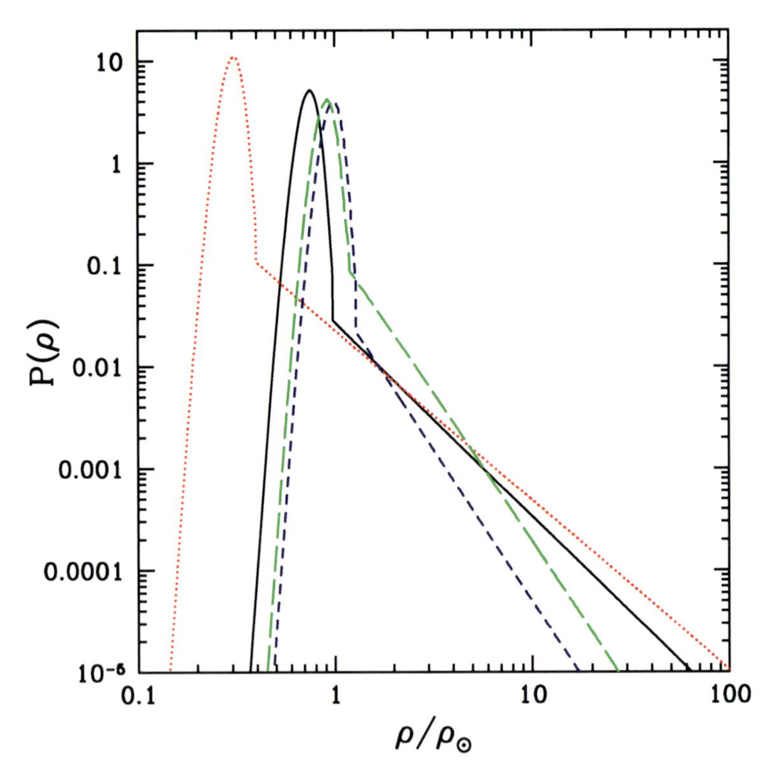

Fig. 5.11 The probability distribution function $P(\rho)$, due to substructure, for the local dark-matter density ρ, due to substructure, in units of the local halo density for a smooth halo. The different curves are for different substructure-survival fractions. The power-law tail is due to substructures. From Ref. [40].

5.5 Variations and Additions

What we have described this far may be referred to as the minimal-WIMP scenario. In this scenario, the dark matter is a thermal relic with electroweak-scale cross sections. It is neutral and scatters from baryons with cross sections $\sim 10^{-40}\,\text{cm}^2$ (to within a few orders of magnitude). It has no astrophysical consequences in the post-freezeout Universe beyond its gravitational effects. However, the recent literature is filled with a large number of astrophysical anomalies for which explanations have been proposed in terms of nonminimal WIMPs, WIMPs endowed with extra interactions or properties. This is a vast literature, far too large to review here. We, therefore, provide here only a brief sampling, focusing primarily on those that we have worked on.

5.5.1 Enhanced Relic Abundance

The calculation above of the freezeout abundance is the standard one in which it is assumed that the Universe is radiation-dominated at $T_f \sim 10-100\,\text{GeV}$. However, we have no empirical constraints to the expansion rate before big bang nucleosynthesis, which happens later, at $T_{BBN} \sim 1\,\text{MeV}$.

One can imagine other scenarios in which the WIMP abundance changes. For instance, suppose the pre-BBN Universe is filled with some exotic matter which has a stiff equation of state, $p_s = \rho_s$. This results in a scaling of the energy density of this stuff $\rho_s \propto a^{-6}$ with scale factor a [42]. Such an equation of state may arise if the energy density is dominated by the kinetic energy of some scalar field. The equation of motion of a scalar field with a flat potential is

$$\ddot{\varphi} + 3H\dot{\varphi} = 0 \qquad \Longrightarrow \qquad \ln\dot{\varphi} \propto -3\ln a, \tag{5.16}$$

which means that

$$\rho = \frac{1}{2}\dot{\varphi}^2 \propto a^{-6}. \tag{5.17}$$

A stiff equation of state, or something that behaves effectively like it, may also arise, e.g., in scalar-tensor theories of gravity or if there is anisotropic expansion in the early Universe.

Big-bang nucleosynthesis constrains the energy density of some new component of matter at a temperature $T \sim$MeV to be $(\rho_6/\rho_\gamma) \lesssim 0.1\,(T/\text{MeV})^2$. Since $\rho_s/\rho_{\text{rad}} \propto T^2$, the expansion rate with this new stiff matter will at earlier times be $H(T) \lesssim H_{\text{st}}(T)(T/\text{MeV})$, where $H_{\text{st}}(T)$ is the standard expansion rate. Neglecting the logarithmic dependence of the freezeout temperature $T_f \propto \ln[H\rho_6 n_\gamma]$ on the expansion, the WIMP abundance with this new exotic matter will be

$$\frac{n_\chi}{n_\gamma} = \frac{1}{n_\gamma}\frac{\Gamma}{\sigma v} = \frac{1}{n_\gamma}\frac{H}{\sigma v} \lesssim \left(\frac{n_\chi}{n_\gamma}\right)_{st}\left(\frac{T}{\text{MeV}}\right) \sim \left(\frac{n_\chi}{n_\gamma}\right)_{st}\left(\frac{M_\chi/25}{\text{MeV}}\right). \tag{5.18}$$

Thus, e.g., the relic abundance of an $M_\chi \sim 150\,\mathrm{GeV}$ WIMP can be increased by as much as $\sim 10^4$ in this way [42, 43].

Exercise 11. Show that anisotropic expansion gives rise to a Friedmann equation that looks like that for a universe with a new component of matter with $\rho \propto a^{-6}$. To do so, consider a universe with metric $ds^2 = dt^2 - [a_x(t)]^2 dx^2 - [a_y(t)]^2 dy^2 - [a_z(t)]^2 dz^2$, with $a_x(t)$, $a_y(t)$, and $a_z(t)$ different, and then derive the Friedmann equation for a universe filled with homogeneous matter of density ρ.

5.5.2 Kinetic Decoupling

There are two different kinds of equilibrium for WIMPs in the primordial bath. One is chemical equilibrium, which is maintained by the reactions

$$\chi\chi \leftrightarrow f\bar{f};$$

the other is kinetic equilibrium, maintained by the scattering

$$\chi f \leftrightarrow \chi f.$$

The first reaction freezes out before the second since $n_f \gg n_\chi$, where f is any kind of light degree of freedom. However, $\sigma(\nu\chi \leftrightarrow \nu\chi) \propto E_\nu^2$ since the ν's are Yukawa coupled, and $\sigma(\gamma\chi \leftrightarrow \gamma\chi) \propto E_\gamma^2$ since the photons are coupled by $\varepsilon_{\mu\nu\rho\sigma}k^\mu k^\nu \varepsilon^\rho \varepsilon^\sigma$ [44]. This means that $\Gamma(\chi f \leftrightarrow \chi f)$ drops rapidly and so kinetic freezeout happens not too much later than chemical freezeout.

Detailed calculations of the kinetic-decoupling temperature T_{kd} show that T_{kd} varies over 6 orders of magnitude in scans of the SUSY and UED parameter spaces [39]. During the time particles are chemically but not kinetically decoupled, they have the same temperature of the thermal bath, which scales as $T_\gamma \propto a^{-1}$, and after that, $T_\chi = p_\chi^2/2M_\chi \propto a^{-2}$. So, density perturbations $\delta\rho_\chi/\rho_\chi$ are suppressed on $\lambda_{phys} \sim H^{-1}$ while the WIMPs are kinetically coupled. The cutoff in the power spectrum $P(k)$ is at physical wavenumber $k_c = H(T_{kd})$, so if T_{kd} decreases, also k_c decreases. We expect power suppressed at mass scales $M < M_c$, where $M_c \sim 10^{-4} - 10^2 M_\oplus$ is the mass enclosed in the horizon at T_{kd}, as shown in Fig. 5.12 [39].

Exercise 12. Derive the mass M_{kd} enclosed within the horizon at a temperature T_{kd}.

5.5.3 Particle Decay and Suppression of Small-Scale Power

It might be the case that dark matter is produced by the decay of a metastable particle that was once in kinetic equilibrium with the thermal bath. For instance, although the dark matter cannot be a charged particle, it might be produced by the decay of

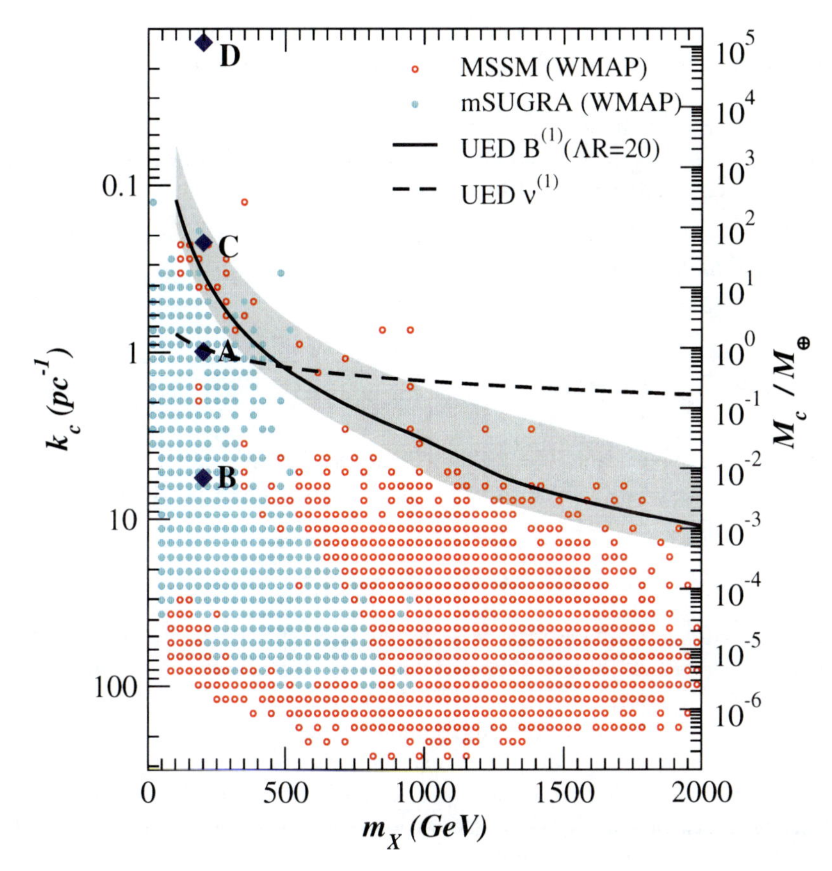

Fig. 5.12 The wavenumber and mass scale at which the primordial power spectrum is cut off due to kinetic decoupling of WIMPs in supersymmetric and UED models for WIMPs. From Ref. [39].

a charged particle. The growth of perturbation modes that enter the horizon prior to the decay of the charged particle will be suppressed relative to the standard case due to the coupling to the thermal bath: growth of charged-particle density perturbations is suppressed since charged particles cannot move through the baryon-photon fluid. If one has $\chi^+ \to \chi^0 + e^+$, with $\tau \sim 3.5$ yr ($z \sim 10^7$), then the matter power spectrum $P(k)$ is suppressed on $k \gtrsim \mathrm{Mpc}^{-1}$ [45], while for shorter lifetimes structure will be suppressed for larger k (smaller length scales). Models exhibiting charged-particle decay can be found in the parameter space of standard or minimal extensions of canonical WIMP (e.g., supersymmetric) scenarios [46]. While limits on energy injection and the formation of exotic bound states in big bang nucleosynthesis (BBN) constrain the fraction of the Universe bound up in charged particles [47], the suppression of power due to particle decay in the Universe remains a potentially observable effect of metastable particles. It is possible the metastable particle might remain in kinetic equilibrium via another interaction, or even if the particle is out of kinetic equilibrium the energy released in the decay process may impart the dark-matter particle with a velocity high enough to erase small-scale structure via free streaming [48].

Future measurements of high-redshift cosmic 21-cm fluctuations may provide a direct probe of modifications to the small-scale dark-matter power spectrum and other aspects of fundamental physics (see, e.g., [46, 49]).

Exercise 13. Derive the comoving wavenumber k that enters the horizon at the time a particle of lifetime τ decays.

5.5.4 Dipole Dark Matter

Although dark matter cannot be a charged particle, it may (via higher order interactions) be endowed with an electric or magnetic dipole moment interactions of the form [22, 19],

$$\mathscr{L}_{\rm dipole} \propto \bar{\chi}_i \sigma_{\mu\nu} \left(\mu_{ij} + \gamma_5 \mathscr{D}_{ij}\right) \chi_j F^{\mu\nu}, \tag{5.19}$$

Here, diagonal interaction terms ($i = j$) are the magnetic (μ) or electric ($\mathscr{D}$) dipole moments of a particle χ, while off-diagonal terms ($i \neq j$) are referred to as transition moments between the lightest WIMP state i and another, slightly heavier, WIMP state j. Such a dipole coupling to photons alters the evolution of dark-matter density perturbations and CMB anisotropies [22], although the strongest constraints to dipole moments comes from precision tests of the Standard Model for WIMP masses $M_\chi \lesssim 10$ GeV and direct-detection experiments for $M_\chi \gtrsim 10$ GeV [22, 50]; see Fig. 5.13 for the full constraints.

It may be possible to explain the results of the DAMA experiment using low-mass dipolar dark matter with a transition moment [50]. It may also be possible to look for the effects of a transition dipole moment in the absorption of high-energy photons from distant sources [19].

Exercise 14. Calculate the cross section for elastic scattering of a particle with an electric dipole moment of magnitude d from a nucleus with charge Ze.

5.5.5 Gravitational Constraints

It is generally assumed that although dark matter may involve new physics, the gravitational interactions of the dark matter are standard. In other words, it is generally assumed that the gravitational force between two DM particles and between a dark matter particle and a baryon is the same as that between two baryons. More precisely, the Newtonian gravitational force law between baryons that has been tested in the laboratory and in the Solar System reads $F_{b_1 b_2} = G m_1 m_2 / d^2$. We then usually assume that the force between baryons and DM is $F_{bd} = G m_b m_d / d^2$ and also that

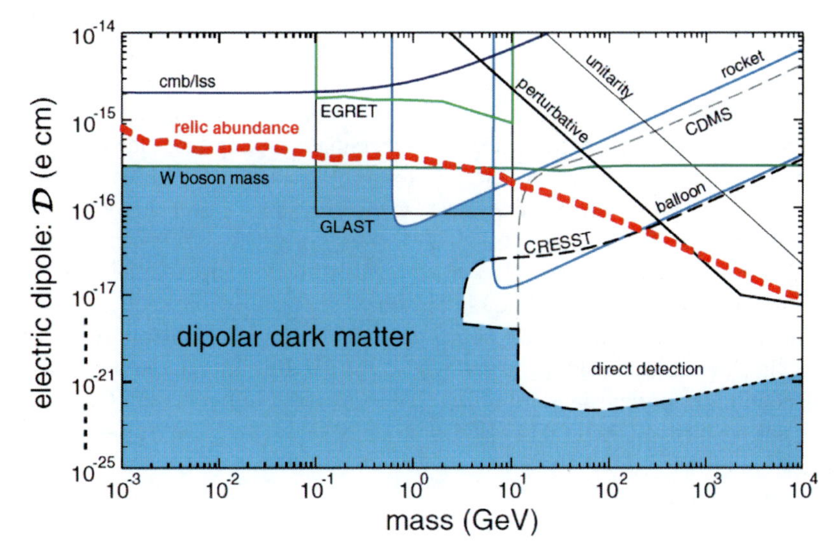

Fig. 5.13 Constraints to the dipole-mass parameter space for dark matter with an electric or magnetic dipole. From Ref. [22].

the gravitational DM–DM force law is $F_{d_1d_1} = G_d m_{d_1} m_{d_2}/d^2$ with $G_d = G$. However, there is no empirical evidence that this is true at more than the order-unity level [51], and it has even been postulated that $G_d = 2G$ in order to account for the void abundance [52]. A similar behavior (an increase in the DM–DM force law) could also arise if there were a new long-range interaction mediated by a nearly massless scalar field φ with Yukawa interactions $\varphi\bar{\psi}\psi$ with the DM field ψ. The difficulty in providing empirical constraints to this model is that measurements (e.g., gravitational lensing or stellar/galactic dynamics) of the dark-matter distribution determine only the gravitational potential Φ due to the dark-matter distribution, represented by some density $\rho_d(\mathbf{r})$, obtained through the Poisson equation $\nabla^2\Phi = 4\pi G\rho_d$. However, the same Φ can be obtained by replacing $\rho_d \to (1/2)\rho_d$ if we simultaneously replace $G \to 2G$.

It turns out, though, that this exotic interaction can be constrained by looking at substructures in the Milky Way halo [53, 54]. The Sagittarius dwarf galaxy, is dark-matter dominated, and it follows an elongated orbit around the Milky Way. When the dwarf reaches its point of closest approach to the Milky Way, the tidal forces it experiences in the Milky Way potential are largest. Stars are then stripped from the innermost and furthermost edge of the dwarf. Those from the innermost parts move at slightly larger velocities in the Galactic halo and at slightly smaller galactocentric radii; they thus subsequently run ahead of the Sagittarius dwarf and form the leading tidal tail of the Sagittarius dwarf that is observed. Conversely, those stripped from the outer edge subsequently lag behind forming the trailing tidal tail that is observed. Observationally, the leading and trailing tails have roughly the same brightness, as expected. Suppose now that the DM–DM force law were modified to

$G_d = fG$ with $f > 1$. The dark-matter halo of the Sagittarius dwarf would then be accelerated toward the Milky Way center more strongly than the stellar component of the Sagittarius dwarf. The stellar component would then slosh to the furthermost edge. Then, when the dwarf reaches its point of closest approach to the Milky Way, stars are still stripped from the outer edge, forming a trailing tail. However, there are now no stars in the innermost edge to form the leading tail. The evacuation of stars from the leading tail is inconsistent with observations, and this leads, with detailed calculations, to a bound $G_d = G(1 \pm 0.1)$ to Newton's constant for DM–DM interactions. In other words, dark matter and ordinary matter fall the same way, to within 10%, in a gravitational potential well.

Although Ref. [55] has more recently claimed to run a simulation of the tidal tails of the Sagittarius dwarf consistent with $G_d = 2G$, Ref. [56] has argued that the initial conditions for that simulation are self-inconsistent. References [57, 58] argue that a new long-range DM–DM force law implies, under fairly general conditions, a weaker long-range DM-baryon force law, and they discuss and compare possible tests of such a scenario.

5.5.6 Electromagnetic-Like Interactions for Dark Matter?

Another possibility is that dark matter experiences long-range electromagnetic-like forces mediated by a dark massless photon that couples only to gravity. Of course, if the fine-structure constant α_d associated with this dark $U(1)$ symmetry is too large, then long-range dark forces will induce the dark matter to be effectively collisional. This constrains $\alpha_d \lesssim 0.005\,(M_\chi/\mathrm{TeV})^{3/2}$ [59]. Far more restrictive constraints may arise from the development of plasma instabilities that may arise if there are (dark) positively- and negatively-charged dark-matter particles, but precise calculations of these effects remain to be done. See Refs. [59, 60] for more discussion of these models.

Exercise 15. Estimate the relic abundance of a dark-matter particle with dark charge α_d assuming that it annihilates to dark-photon pairs and assuming that the dark sector has the same temperature as the rest of the primordial plasma.

5.6 Some Other Particle Dark-Matter Candidates

WIMP models are interesting for a number of reasons: (1) The correct relic density arises naturally if there is new physics at the electroweak scale; (2) there are good prospects for detection of these particles, if they are indeed the dark matter; and (3) there is synergy with the goals of accelerator searches (especially at the LHC) for new electroweak-scale physics.

Still, there are a large number of other particle candidates for dark matter. Here we discuss two, the sterile neutrino and the axion, which may also arise in extensions of the standard model and for which there are clear paths toward detection if they make up the dark matter.

5.6.1 Sterile Neutrinos

A convenient mechanism to introduce neutrino masses and explain their smallness by a minimal extension of the Standard Model is to add 3 right-handed neutrinos, which are singlets under the SM gauge group. The mass matrix is taken to be of the form (for simplicity we consider only one family),

$$\begin{array}{c c} & \begin{array}{cc} \nu_L & \nu_R \end{array} \\ \begin{array}{c} \nu_L \\ \nu_R \end{array} & \left(\begin{array}{cc} 0 & M_D \\ M_D & M \end{array} \right), \end{array} \tag{5.20}$$

where the ν_L and ν_R are left-handed and right-handed (weakly interacting and sterile, respectively) fields. In the see-saw mechanism, the Dirac mass is assumed to be tiny compared with the Majorana mass: i.e., $M_D \ll M$. The mass eigenstates then have masses $M_1 \simeq M_D^2/M \ll M$ and $M_2 \simeq M$. For our purposes, it is advantageous to map the two-dimensional $M_D - M$ parameter space onto the $M_s - \theta$ parameter space, where M_s is the mass of the sterile (heavier) neutrino and θ is the mixing angle between the two states. The active and sterile mass eigenstates can then be written

$$|\nu_a\rangle = \cos\theta\,|\nu_L\rangle + \sin\theta\,|\nu_R\rangle, \tag{5.21}$$

$$|\nu_s\rangle = -\sin\theta\,|\nu_L\rangle + \cos\theta\,|\nu_R\rangle, \tag{5.22}$$

where $\theta = M_D/M$.

Sterile neutrinos can be produced in the early Universe and have both (1) a lifetime longer than the age of the Universe and (2) a cosmological density $\Omega_s \sim 0.2$ if the sterile-neutrino mass is in the ~keV regime [61].

The main decay mode of the sterile neutrino is then $\nu_S \to \nu\nu\bar{\nu}$, through the exchange of a Z^0 boson, as shown in Fig. 5.14. The decay rate and lifetime are

$$\Gamma = \frac{G_F^2 M_S^5}{96\pi^3}\theta^2 \qquad \Rightarrow \qquad \tau_S = \frac{\hbar}{\Gamma} \sim 10^{20}\,\mathrm{sec}\left(\frac{M_S}{\mathrm{keV}}\right)^5 \theta^{-2}. \tag{5.23}$$

If the sterile neutrinos constitute the dark matter, then it must be that $\tau_S \gg 10^{17}$ sec, which is possible if $M_S \sim O(1)$ keV. This mass cannot, however, be too small because of the Gunn–Tremaine limit from dwarf-spheroidal galaxies, which is $M_S \gtrsim 0.3$ keV. A stronger constraint to the model comes from the X-ray emission in the radiative decay $\nu_S \to \nu\gamma$, through the diagram in Fig. 5.15. This produces an x-ray line that can be sought in the spectrum of, e.g., a galaxy cluster. Although null searches for such lines (and from the diffuse cosmic x-ray

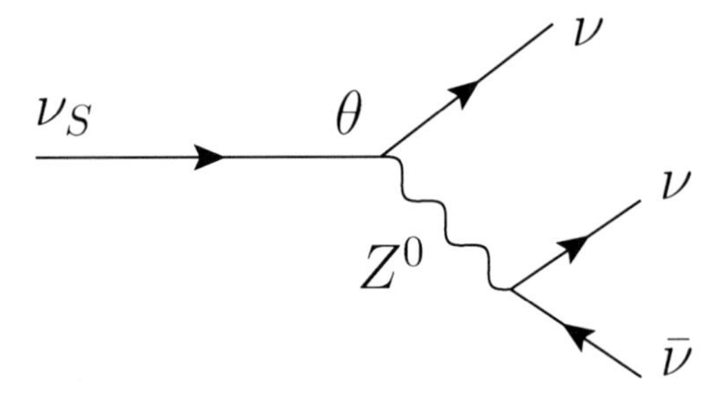

Fig. 5.14 Main decay channel for sterile neutrinos.

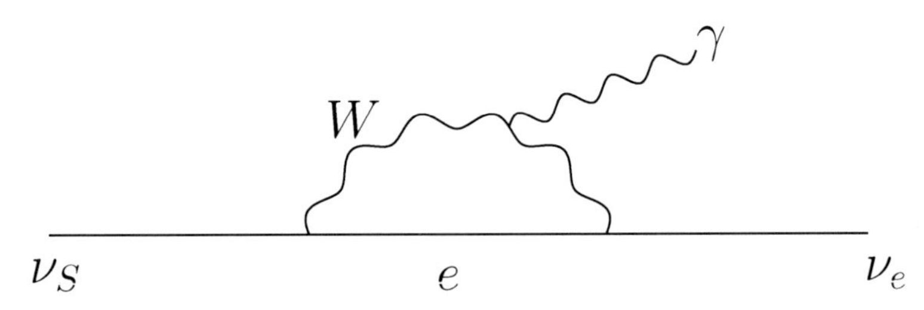

Fig. 5.15 Loop diagram for the decay $\nu_s \to \nu\gamma$.

background) provide [62, 63] stringent constraints to the model, there are still some regions in the $M_s - \theta$ parameter space that remain consistent with current constraints. This region may be probed, however, with future more sensitive x-ray searches. One interesting extended application of sterile neutrino dark matter was its use as a potential mechanism for generating momentum-anisotropy during supernova to drive pulsar kicks [64]. See, for instance, Refs. [65, 66], for the current status of sterile neutrino dark matter.

5.6.2 Axions

Axions arise in the Peccei-Quinn (PQ) solution to the strong-*CP* problem [67]. A global $U(1)_{PQ}$ symmetry is spontaneously broken at a scale f_a, and the CP-violating phase θ in the QCD Lagrangian becomes a dynamical field with a flat potential. At temperatures below the QCD phase transition, nonperturbative quantum effects break explicitly the symmetry and produce a nonflat potential that is minimized at $\theta \to 0$. The axion is the pseudo-Nambu–Goldstone boson of this near-global symmetry, the particle associated with excitations about the minimum at $\theta = 0$. The axion mass is $m_a \simeq \mathrm{eV}\,(10^7\,\mathrm{GeV}/f_a)$, and its coupling to ordinary matter is $\propto f_a^{-1}$.

The Peccei-Quinn solution works equally well for any value of f_a. However, a variety of astrophysical observations and laboratory experiments constrain the axion mass to be $m_a \sim 10^{-4}$ eV. Smaller masses would lead to an unacceptably large cosmological abundance. Larger masses are ruled out by a combination of constraints from supernova 1987A, globular clusters, laboratory experiments, and a search for two-photon decays of relic axions.

Curiously enough, if the axion mass is in the relatively small viable range, the relic density is $\Omega_a \sim 1$, and so the axion may account for the halo dark matter. Such axions would be produced with zero momentum by a misalignment mechanism in the early Universe and, therefore, act as cold dark matter. During the process of galaxy formation, these axions would fall into the Galactic potential well and would, therefore, be present in our halo with a velocity dispersion near 270 km sec^{-1}.

It has been noted that quantum gravity is generically expected to violate global symmetries, and unless these Planck-scale effects can be suppressed by a huge factor, the Peccei-Quinn mechanism may be invalidated [68]. Of course, we have at this point no predictive theory of quantum gravity, and several mechanisms for forbidding these global-symmetry violating terms have been proposed [69]. Therefore, discovery of an axion might provide much needed clues to the nature of Planck-scale physics.

There is a very weak coupling of an axion to photons through the triangle anomaly, a coupling mediated by the exchange of virtual quarks and leptons. The axion can, therefore, decay to two photons, but the lifetime is $\tau_{a\to\gamma\gamma} \sim 10^{50}\,\text{s}\,(m_a/10^{-5}\,\text{eV})^{-5}$ which is huge compared with the lifetime of the Universe and therefore unobservable. However, the $a\gamma\gamma$ term in the Lagrangian is $\mathcal{L}_{a\gamma\gamma} \propto a\mathbf{E}\cdot\mathbf{B}$, where $\mathbf{E}$ and $\mathbf{B}$ are the electric and magnetic field strengths. Therefore, if one immerses a resonant cavity in a strong magnetic field, Galactic axions that pass through the detector may be converted to fundamental excitations of the cavity, and these may be observable [70]. Such an experiment is currently underway [71] and has already begun to probe part of the cosmologically interesting parameter space (see the Figure in Reference. [72]), and it should cover most of the interesting region parameter space in the next few years.

Axions, or other light pseudoscalar particles, may show up astrophysically or experimentally in other ways. For example, the PVLAS Collaboration [73] reported the observation of an anomalously large rotation of the linear polarization of a laser when passed through a strong magnetic field. Such a rotation is expected in quantum electrodynamics, but the magnitude they reported was in excess of this expectation. One possible explanation is a coupling of the pseudoscalar $F\tilde{F}$ of electromagnetism to a low-mass axion-like pseudoscalar field. The region of the mass-coupling parameter space implied by this experiment violates limits for axions from astrophysical constraints, but there may be nonminimal models that can accommodate those constraints. Reference [74] reviews the theoretical interpretation and shows how the interactions of axions and other axion-like particles may be tested with x-ray re-appearance experiments. Although the original PVLAS results have now been called into question Ref. [75], variations of the model may still be worth investigating.

5.7 Conclusions

Here, we have reviewed briefly the basic astrophysical evidence for dark matter, some simple astrophysical constraints to its physical properties, and the canonical WIMP model for dark matter. We then discussed a number of variations of the canonical model, as well as some alternative particle dark-matter candidates. Still, we have only scratched the surface here, surveying only a small fraction of the possibilities for nonminimal dark matter. Readers who are interested in learning more are encouraged to browse the recent literature, where they will find an almost endless flow of interesting possibilities for dark matter, beyond those we have reviewed here.

Acknowledgments We thank Sabino Matarrese for initiating this collaboration during the Como summer school at which these lectures were given. We also thank the Aspen Center for Physics, where part of this review was completed. This work was supported at Caltech by DoE DE-FG03-92-ER40701 and the Gordon and Betty Moore Foundation, and at the University of British Columbia by a NSERC of Canada Discovery Grant.

References

1. D. P. Finkbeiner, Astrophys. J. **614**, 186 (2004) [arXiv:astro-ph/0311547]; G. Dobler and D. P. Finkbeiner, Astrophys. J. **680**, 1222 (2008) [arXiv:0712.1038 [astro-ph]]; D. Hooper, D. P. Finkbeiner, and G. Dobler, Phys. Rev. D **76**, 083012 (2007) [arXiv:0705.3655 [astro-ph]]; P. Jean *et al.*, Astron. Astrophys. **407**, L55 (2003) [arXiv:astro-ph/0309484]; J. Knodlseder *et al.*, Astron. Astrophys. **411**, L457 (2003) [arXiv:astro-ph/0309442]; G. Weidenspointner *et al.*, arXiv:astro-ph/0406178. A. W. Strong *et al.*, Astron. Astrophys. **444**, 495 (2005) [arXiv:astro-ph/0509290]; D. J. Thompson, D. L. Bertsch, and R. H. . O'Neal, arXiv:astro-ph/0412376.
2. O. Adriani *et al.* [PAMELA Collaboration], Nature **458**, 607 (2009) [arXiv:0810.4995 [astro-ph]].
3. G. Jungman, M. Kamionkowski and K. Griest, Phys. Rept. **267**, 195 (1996) [arXiv:hep-ph/9506380].
4. H. C. Cheng, J. L. Feng, and K. T. Matchev, Phys. Rev. Lett. **89**, 211301 (2002) [arXiv:hep-ph/0207125]; G. Servant and T. M. P. Tait, Nucl. Phys. B **650**, 391 (2003) [arXiv:hep-ph/0206071]; for a review, see D. Hooper and S. Profumo, Phys. Rep. **453**, 29 (2007) [arXiv:hep-ph/0701197].
5. L. Bergstrom, Rept. Prog. Phys. **63**, 793 (2000) [arXiv:hep-ph/0002126].
6. G. Bertone, D. Hooper and J. Silk, Phys. Rept. **405**, 279 (2005) [arXiv:hep-ph/0404175].
7. F. Iocco, G. Mangano, G. Miele, O. Pisanti and P. D. Serpico, Phys. Rept. **472**, 1 (2009) [arXiv:0809.0631 [astro-ph]].
8. K. G. Begeman, A. H. Broeils and R. H. Sanders, Mon. Not. Roy. Astron. Soc. **249** (1991) 523.
9. F. Zwicky, Helv. Phys. Acta **6**, 110 (1933).
10. R. A. Sunyaev and Y. B. Zeldovich, Ann. Rev. Astron. Astrophys. **18**, 537 (1980).
11. A. Einstein, Science **84**, 506 (1936).
12. F. Zwicky, Phys. Rev. **51**, 290 (1937).
13. R. D. Blandford and R. Narayan, Ann. Rev. Astron. Astrophys. **30**, 311 (1992).
14. F. Zwicky, Astrophys. J. **86**, 217 (1937).
15. M. Kamionkowski, arXiv:0706.2986 [astro-ph].

16. G. Jungman, M. Kamionkowski, A. Kosowsky and D. N. Spergel, Phys. Rev. Lett. **76**, 1007 (1996) [arXiv:astro-ph/9507080]; G. Jungman, M. Kamionkowski, A. Kosowsky and D. N. Spergel, Phys. Rev. D **54**, 1332 (1996) [arXiv:astro-ph/9512139].
17. A. H. Jaffe *et al.* [Boomerang Collaboration], Phys. Rev. Lett. **86**, 3475 (2001) [arXiv:astro-ph/0007333].
18. J. Dunkley *et al.* [WMAP Collaboration], Astrophys. J. Suppl. **180**, 306 (2009) [arXiv:0803.0586 [astro-ph]].
19. S. Profumo and K. Sigurdson, Phys. Rev. D **75**, 023521 (2007) [arXiv:astro-ph/0611129].
20. S. Davidson, S. Hannestad and G. Raffelt, JHEP **0005**, 003 (2000) [arXiv:hep-ph/0001179].
21. S. W. Randall, M. Markevitch, D. Clowe, A. H. Gonzalez and M. Bradac, arXiv:0704.0261 [astro-ph].
22. K. Sigurdson, M. Doran, A. Kurylov, R. R. Caldwell and M. Kamionkowski, Phys. Rev. D **70**, 083501 (2004) [Erratum-ibid. D **73**, 089903 (2006)] [arXiv:astro-ph/0406355].
23. S. Tremaine and J. E. Gunn, Phys. Rev. Lett. **42** (1979) 407.
24. E. W. Kolb and M. S. Turner, "The Early universe," (Addison-Wesley, Redwood City, 1990).
25. K. Griest and M. Kamionkowski, Phys. Rev. Lett. **64**, 615 (1990).
26. R. Bernabei *et al.* [DAMA Collaboration], Phys. Lett. B **480**, 23 (2000).
27. P. Ullio, M. Kamionkowski and P. Vogel, JHEP **0107** (2001) 044 [arXiv:hep-ph/0010036].
28. E. Behnke *et al.* [COUPP Collaboration], Science **319**, 933 (2008) [arXiv:0804.2886 [astro-ph]].
29. J. Silk, K. A. Olive and M. Srednicki, Phys. Rev. Lett. **55**, 257 (1985).
30. M. Kamionkowski and M. S. Turner, Phys. Rev. D **43** (1991) 1774.
31. O. Adriani *et al.* [PAMELA Collaboration], arXiv:0810.4995 [astro-ph]; J. Chang et al. (ATIC) (2005), prepared for 29th International Cosmic Ray Conferences (ICRC 2005), Pune, India, 31 Aug 03 - 10 2005.
32. S. Profumo, arXiv:0812.4457 [astro-ph]; D. Hooper, P. Blasi and P. D. Serpico, JCAP **0901**, 025 (2009) [arXiv:0810.1527 [astro-ph]].
33. A. A. Abdo *et al.* [The Fermi LAT Collaboration], Phys. Rev. Lett. **102**, 181101 (2009) [arXiv:0905.0025 [astro-ph.HE]].
34. L. Bergstrom, J. Edsjo, P. Gondolo and P. Ullio, Phys. Rev. D **59**, 043506 (1999) [arXiv:astro-ph/9806072]; L. Bergstrom, J. Edsjo and P. Ullio, Phys. Rev. D **58**, 083507 (1998) [arXiv:astro-ph/9804050].
35. S. Ando and E. Komatsu, Phys. Rev. D **73**, 023521 (2006) [arXiv:astro-ph/0512217]; S. Ando, E. Komatsu, T. Narumoto and T. Totani, Phys. Rev. D **75**, 063519 (2007) [arXiv:astro-ph/0612467]; A. Cuoco, S. Hannestad, T. Haugbolle, G. Miele, P. D. Serpico and H. Tu, JCAP **0704**, 013 (2007) [arXiv:astro-ph/0612559]; A. Cuoco, J. Brandbyge, S. Hannestad, T. Haugboelle and G. Miele, Phys. Rev. D **77**, 123518 (2008) [arXiv:0710.4136 [astro-ph]]; J. M. Siegal-Gaskins, arXiv:0807.1328 [astro-ph].
36. S. K. Lee, S. Ando and M. Kamionkowski, arXiv:0810.1284 [astro-ph].
37. S. M. Koushiappas, Phys. Rev. Lett. **97**, 191301 (2006) [arXiv:astro-ph/0606208];
38. S. Ando, M. Kamionkowski, S. K. Lee and S. M. Koushiappas, arXiv:0809.0886 [astro-ph].
39. S. Profumo, K. Sigurdson and M. Kamionkowski, Phys. Rev. Lett. **97** (2006) 031301 [arXiv:astro-ph/0603373].
40. M. Kamionkowski and S. M. Koushiappas, Phys. Rev. D **77**, 103509 (2008) [arXiv:0801.3269 [astro-ph]].
41. M. Vogelsberger *et al.*, arXiv:0812.0362 [astro-ph].
42. M. Kamionkowski and M. S. Turner, Phys. Rev. D **42** (1990) 3310.
43. S. Profumo and P. Ullio, JCAP **0311**, 006 (2003) [arXiv:hep-ph/0309220].
44. X. l. Chen, M. Kamionkowski and X. m. Zhang, Phys. Rev. D **64**, 021302 (2001) [arXiv:astro-ph/0103452].
45. K. Sigurdson and M. Kamionkowski, Phys. Rev. Lett. **92**, 171302 (2004) [arXiv:astro-ph/0311486].
46. S. Profumo, K. Sigurdson, P. Ullio and M. Kamionkowski, Phys. Rev. D **71**, 023518 (2005) [arXiv:astro-ph/0410714].

47. M. Kaplinghat and A. Rajaraman, Phys. Rev. D **74**, 103004 (2006) [arXiv:astro-ph/0606209]; M. Pospelov, Phys. Rev. Lett. **98**, 231301 (2007) [arXiv:hep-ph/0605215]; K. Kohri and F. Takayama, Phys. Rev. D **76**, 063507 (2007) [arXiv:hep-ph/0605243].
48. M. Kaplinghat, Phys. Rev. D **72**, 063510 (2005) [arXiv:astro-ph/0507300].
49. A. Loeb and M. Zaldarriaga, Phys. Rev. Lett. **92**, 211301 (2004) [arXiv:astro-ph/0312134]; M. Kleban, K. Sigurdson and I. Swanson, JCAP **0708**, 009 (2007) [arXiv:hep-th/0703215].
50. E. Masso, S. Mohanty and S. Rao, arXiv:0906.1979 [hep-ph].
51. B. A. Gradwohl and J. A. Frieman, Astrophys. J. **398** (1992) 407.
52. S. S. Gubser and P. J. E. Peebles, Phys. Rev. D **70**, 123510 (2004) [arXiv:hep-th/0402225].
53. M. Kesden and M. Kamionkowski, Phys. Rev. Lett. **97** (2006) 131303 [arXiv:astro-ph/0606566].
54. M. Kesden and M. Kamionkowski, Phys. Rev. D **74** (2006) 083007 [arXiv:astro-ph/0608095].
55. J. A. Keselman, A. Nusser, and P. J. E. Peebles, arXiv:0902.3452 [astro-ph.GA].
56. M. Kesden, arXiv:0903.4458 [astro-ph.CO].
57. S. M. Carroll, S. Mantry, M. J. Ramsey-Musolf and C. W. Stubbs, arXiv:0807.4363 [hep-ph].
58. S. M. Carroll, S. Mantry and M. J. Ramsey-Musolf, arXiv:0902.4461 [hep-ph].
59. L. Ackerman, M. R. Buckley, S. M. Carroll and M. Kamionkowski, Phys. Rev. D **79**, 023519 (2009) [arXiv:0810.5126 [hep-ph]].
60. J. L. Feng, M. Kaplinghat, H. Tu and H. B. Yu, arXiv:0905.3039 [hep-ph].
61. S. Dodelson and L. M. Widrow, Phys. Rev. Lett. **72**, 17 (1994) [arXiv:hep-ph/9303287].
62. A. Boyarsky, A. Neronov, O. Ruchayskiy, M. Shaposhnikov and I. Tkachev, Phys. Rev. Lett. **97** (2006) 261302 [arXiv:astro-ph/0603660].
63. A. Boyarsky, J. Nevalainen and O. Ruchayskiy, Astron. Astrophys. **471** (2007) 51 [arXiv:astro-ph/0610961].
64. A. Kusenko, Int. J. Mod. Phys. D **13** (2004) 2065 [arXiv:astro-ph/0409521].
65. M. Shaposhnikov, arXiv:astro-ph/0703673.
66. A. Kusenko, arXiv:0906.2968 [hep-ph].
67. R. D. Peccei and H. R. Quinn, Phys. Rev. Lett. **38**, 1440 (1977); F. Wilczek, Phys. Rev. Lett. **40**, 279 (1978); S. Weinberg, Phys. Rev. Lett. **40**, 223 (1978).
68. M. Kamionkowski and J. March-Russell, Phys. Lett. B **282**, 137 (1992) [arXiv:hep-th/9202003]; R. Holman, S. D. H. Hsu, T. W. Kephart, E. W. Kolb, R. Watkins and L. M. Widrow, Phys. Lett. B **282**, 132 (1992) [arXiv:hep-ph/9203206]; S. M. Barr and D. Seckel, Phys. Rev. D **46**, 539 (1992).
69. N. Turok, Phys. Rev. Lett. **76**, 1015 (1996) [arXiv:hep-ph/9511238]; R. Kallosh, A. D. Linde, D. A. Linde and L. Susskind, Phys. Rev. D **52**, 912 (1995) [arXiv:hep-th/9502069]; E. A. Dudas, Phys. Lett. B **325**, 124 (1994) [arXiv:hep-ph/9310260]; K. S. Babu and S. M. Barr, Phys. Lett. B **300**, 367 (1993) [arXiv:hep-ph/9212219].
70. P. Sikivie, Phys. Rev. Lett. **51**, 1415 (1983) [Erratum-ibid. **52**, 695 (1984)].
71. S. Asztalos *et al.*, Phys. Rev. D **64**, 092003 (2001). S. J. Asztalos *et al.*, Astrophys. J. **571**, L27 (2002) [arXiv:astro-ph/0104200].
72. S. Eidelman *et al.* [Particle Data Group], Phys. Lett. B **592**, 1 (2004). See, in particular, p. 394–397 of the review.
73. E. Zavattini *et al.* [PVLAS Collaboration], Phys. Rev. Lett. **96**, 110406 (2006) [arXiv:hep-ex/0507107].
74. R. Rabadan, A. Ringwald and K. Sigurdson, Phys. Rev. Lett. **96**, 110407 (2006) [arXiv:hep-ph/0511103].
75. A. S. .. Chou *et al.* [GammeV (T-969) Collaboration], Phys. Rev. Lett. **100**, 080402 (2008) [arXiv:0710.3783 [hep-ex]].

Chapter 6
Dark Matter: the Particle Physics View

Antonio Masiero

Abstract This is an elementary introduction to some issues of Dark Matter, particularly devoted to readers who are not familiar with the particle physics background needed to address this chapter of astroparticle physics.

6.1 Introduction

The electroweak standard model (SM) is by now more than 40, years old, and it enjoys a full maturity with an extraordinary success in reproducing the many electroweak tests that have been going on since its birth. Not only have its characteristic gauge bosons, W and Z, been discovered and also has the top quark been found in the mass range expected by the electroweak radiative corrections, but the SM has been able to account for an impressively long and very accurate series of measurements. Indeed, in particular at LEP, some of the electroweak observables have been tested with precisions better than the per milli level without finding any discrepancy with the SM predictions. We can safely state that LEP has fully established the validity of the SM as a quantum field theory. At the same time, the SM has successfully passed another very challenging class of exams, namely it has so far accounted for all the very suppressed or forbidden processes where flavor changing neutral currents (FCNC) are present.

Hence, we can firmly state that no matter which physics should lie beyond the SM, necessarily such New Physics, has to reproduce the SM with great accuracy at energies of O (100 GeV). This represents a major breakthrough in our exciting progress toward a unified picture of fundamental interactions.

And, yet, in spite of all this glamorous success of the SM in reproducing the impressive set of experimental electroweak results, we are deeply convinced of the

Physics Department "G. Galilei," University of Padova and INFN, Sez di Padova, ITALY
masiero@infn.pd.it

existence of New Physics beyond this model. We see two main classes of motivations pushing us beyond the SM.

First, we have theoretical reasons to believe that the SM is not the whole story. The SM does not truly unify the elementary interactions (if nothing else, gravity is left out of the game), it does not provide a rationale for the structured pattern of fermion masses and mixings (flavor problem), and it exhibits the gauge hierarchy problem in the scalar sector (namely, the scalar Higgs mass is not protected by any symmetry and, hence, it would tend to acquire large values of the order of the energy scale at which New Physics sets in). Together with these theoretical complaints comes a set of "observational" reasons pushing us beyond the SM: neutrino masses, dark matter (DM), dark energy (DE), the cosmic matter-antimatter asymmetry (baryogenesis), and the need for an inflationary epoch.

Therefore, in particle physics, the discovery and subsequent understanding of the New Physics (NP) beyond the SM is a priority. Most of us are convinced that such NP should be linked to the electroweak symmetry breaking, hence it should set in at an energy scale close to the 100–1000 GeV benchmark we mentioned above as the border of our actual knowledge of fundamental interactions thus far. The reason of this conviction rests upon the crucial point (technically known as the "gauge hierarchy problem") that we need a mechanism to stabilize the energy scale at which the electroweak symmetry breaking occurs, i.e., the TeV scale. The LHC, which is now operating at CERN is the accelerator machine having all the potentialities to get access to such low-energy NP.

The status of our present understanding of cosmology can play an important role in driving the investigation beyond the SM hand understanding the nature of NP. Undoubtedly, the major breakthrough in physics in these last two decades has been represented by the amazing changes we went through in the way we see the Universe and its major, fundamental constituents. In a few years, thanks to a breathtaking progress in observational cosmology, we have been driven to a new Standard Model of Cosmology, the so-called "ΛCDM" model. This is characterized by a critically dense universe where ordinary matter (baryons) constitute only 5% (at most) of the entire energy density. The remaining 95% is in exotic forms, which still demand to be theoretically understood and experimentally revealed. This overwhelming part of the Universe is it is usually divided into two components: dark matter (DM) and dark energy (DE) with a ratio roughly one to three between DM and DE. Moreover, even considering only the baryonic component, a new puzzle emerges, namely that of the overwhelming dominance of baryons over anti baryons in the present Universe.

The (rather strangely looking) picture of the Universe summarized above addresses profound and severe questions to particle physics. What are the sources of DM and DE? Are they related? How can we experimentally reveal them? Can present observations find an explanation within General Relativity (GR) or do they ask for an extension of it? How was the asymmetry between matter and anti matter produced?

Finding an answer to (some of) these questions in the particle physics context implies that we envisage NP beyond the SM of particle physics. As a matter of fact, within the SM there is neither an adequate DM candidate nor any theoretical clue to understand the nature of DE. Moreover, there is no way to generate a cosmic matter-antimatter asymmetry starting from a symmetric initial condition. In other words, today we witness a serious clash between the SM of particle physics and the "ΛCDM" model, whose solution calls for NP and represents one of the greatest challenges in modern science.

It would be a stunning breakthrough if at the LHC, together with the first signals of a (possibly unexpected) TeV NP, we could also collect the first hints to a deeper comprehension of the most fundamental constituents of the whole Universe.

From a cosmological point of view, the most promising candidate for DM was recognized to be a weakly interacting massive particle (WIMP) with a mass in the tens or hundreds of GeV. On the other hand, theoretical research in particle physics leads to the discovery that extensions of the SM introducing New Physics at the electroweak scale were offering, as a "bonus," interesting WIMP candidates (for instance, the lightest supersymmetric (SUSY) particle (LSP) in SUSY extensions of the SM or the lightest Kaluza–Klein mode in theories with extra dimensions). In view of the amazing coincidence between particle physics and cosmology parameters making the WIMP so interesting from the DM point of view, we think that it is nowadays compelling to explore all the possible, rich aspects of the interplay between LHC and DM searches from the WIMP perspective.

Differently from DM, DE is unlikely to call for a particle candidate. Indeed, the accelerated expansion of the Universe demands DE to exhibit a negative pressure, while particle fluids lead to a non-negative pressure. The Einstein cosmological constant can presently account for the data. An interesting alternative is that DE, instead of being a quantity which remains constant, has its own evolution in time, i.e., we have a Dynamical DE (DDE). The prototype of such a proposal is provided by a scalar field with a suitable potential whose energy density keeps varying with time. One of the main goals of next generation experiments on DE is just to ascertain a possible dynamics in DE evolution. This means that we should be able to get information on the accelerated expansion of the Universe at different redshifts establishing whether DE has to be attributed to a cosmological constant or to an evolving DDE. Both the constant and the dynamical interpretations of DE share the problem of the extreme tuning of the energy scale associated to DE (of the millielectronvolt order) with respect to common particle physics scales. Together with it, we have a series of maybe equally profound problems, which can be gathered under the name of "coincidence" questions. Why DM and baryons have comparable densities? Why also DE has a comparable density and has it just in the present cosmic epoch? Motivated in part by the attempt to address these questions, the possibility of an interaction between DDE and DM and/or baryons has also attracted great interest. In particular, it was realized that, even in the absence of a direct DM-DE interaction, the presence of DE could have a major impact on the nature and present abundance of DM. Indeed, the scalar field responsible for the DDE could lead to significant departures from the standard cosmological evolution based on GR, in particular at its the early stages. The most remarkable examples are provided by DDE models based on ST theories of gravity, or by the so-called "kination" scenarios, in which the kinetic energy of the DDE scalar field dominates over the radiation energy density for a period in the early Universe. In particular, in the ST case, it was shown that, even taking into account all the increasingly severe constraints on departures from GR (from big-bang nucleosynthesis BBN, CMB, and the solar system tests), the expansion rate in the decisive moment when the WIMP DM decouples from the thermal bath could have been strongly different from the standard one. This can radically change the predicted abundances of the different DM candidates in a particular model, bringing about phenomenological implications of utmost relevance.

Within the SM, both the amount of CP and the nature of the electroweak phase transition preclude any possibility to have an efficient dynamical mechanism to originate a cosmic matter-antimatter asymmetry. When proceeding beyond the SM, a very interesting mechanism has been suggested: baryogenesis through leptogenesis. In this case, use is made by the CP violating out-of-equilibrium decay of the (heavy) right-handed neutrino entering the so-called "see-saw" mechanism, which is based on a large Majorana mass for the right-handed neutrino.

Leptogenesis makes it possible to link two apparently unrelated observations: the matter-antimatter asymmetry of the Universe in cosmology and neutrino masses and mixing in particle physics. It is intriguing that the measured values are in agreement in a nontrivial way, with some interesting bounds on the involved parameters. In this context, it becomes very attractive to try to tie together leptogenesis, violation of lepton number, and lepton flavor violation (LFV).

In this chapter, I will briefly review the following:

- the main features of the SM such as its spectrum, the Lagrangian and its symmetries, the Higgs mechanism, the successes and shortcomings of the SM;
- two major particle physics candidates for DM: massive (light) neutrinos and the lightest SUSY particle in SUSY extensions of the SM with R parity (to be defined later on). Light neutrinos and the lightest particle are "canonical" examples of the hot and cold DM (CDM), respectively. This choice does not mean that these are the only interesting particle physics candidates for DM. For instance, axions are still of great interest as CDM candidates and their experimental search is proceeding at full steam;
- I'll revisit the DM issue in the context of cosmological scenarios where the expansion rate of the Universe can (even drastically) differ from the standard one at temperatures higher than the MeV scale. i.e., before nucleosynthesis starts.

This chapter is meant to be an introduction to some DM issues for readers who are not familiar with the subject and, in particular, who need an introduction to the particle physics aspects of the DM problem. No discussion on the searches for DM will be presented in this chapter.

6.2 The Standard Model of Particle Physics

In particle physics, the fundamental interactions are described by the Glashow–Weinberg–Salam Standard Theory (GWS) for the electroweak interactions [1, 2, 3] (for a review see [4]) and QCD for the strong one. GWS and QCD are gauge theories

Fermions	Generations			$SU(2)_L \otimes U(1)_Y$
	I	II	III	
$E_{bL} \equiv \begin{pmatrix} \nu_b \\ e_b^- \end{pmatrix}_L$	$\begin{pmatrix} \nu_e \\ e^- \end{pmatrix}_L$	$\begin{pmatrix} \nu_\mu \\ \mu^- \end{pmatrix}_L$	$\begin{pmatrix} \nu_\tau \\ \tau^- \end{pmatrix}_L$	$(2,-1)$
e_{bR}	e_R^-	μ_R^-	τ_R^-	$(1,-2)$
$Q_{bL} \equiv \begin{pmatrix} u_b \\ d_b \end{pmatrix}_L$	$\begin{pmatrix} u \\ d \end{pmatrix}_L$	$\begin{pmatrix} c \\ s \end{pmatrix}_L$	$\begin{pmatrix} t \\ b \end{pmatrix}_L$	$(2,1/3)$
u_{bR}	u_R	c_R	t_R	$(1,4/3)$
d_{bR}	d_R	s_R	b_R	$(1,-2/3)$

Table 6.1 The fermionic spectrum of the SM.

based respectively on the gauge groups $SU(2)_L \times U(1)_Y$ and $SU(3)_c$ where L refers to left, Y to hypercharge, and c to colour. We recall that a gauge theory is invariant under a local symmetry and requires the existence of vector gauge fields living in the adjoint representation of the group. Therefore, in our case, we have the following:

1. three gauge fields W_μ^1, W_μ^2, W_μ^3 for $SU(2)_L$;
2. one gauge field B_μ for $U(1)_Y$;
3. eight gauge bosons λ_μ^a for $SU(3)_c$.

The SM fermions live in the irreducible representations of the gauge group and are reported in Table 1: the indices L and R indicate the left and right fields respectively, $b = 1,2,3$ the generation, the colour is not shown.

The Lagrangian of the SM is dictated by the invariance under the Lorentz and the gauge groups and the request of renormalizability. It is given by the sum of the kinetic fermionic part $\mathscr{L}_{\mathrm{K\,mat}}$ and the gauge one $\mathscr{L}_{\mathrm{K\,gau}}$: $\mathscr{L} = \mathscr{L}_{\mathrm{K\,mat}} + \mathscr{L}_{\mathrm{K\,gau}}$. The fermionic part reads for one generation

$$\begin{aligned}\mathscr{L}_{\mathrm{K\,mat}} = \;& i\bar{Q}_L\gamma^\mu\left(\partial_\mu + igW_\mu^a T_a + i\tfrac{g'}{6}B_\mu\right)Q_L + i\bar{d}_R\gamma^\mu\left(\partial_\mu - i\tfrac{g'}{3}B_\mu\right)d_R \\ & + i\bar{u}_R\gamma^\mu\left(\partial_\mu + i\tfrac{2g'}{3}B_\mu\right)u_R + i\bar{E}_L\gamma^\mu\left(\partial_\mu + igW_\mu^a T_a - i\tfrac{g'}{2}B_\mu\right)E_L \\ & + i\bar{e}_R\gamma^\mu\left(\partial_\mu - ig'B_\mu\right)e_R, \end{aligned} \tag{6.1}$$

where the matrices $T_a = \sigma_a/2$, σ_a are the Pauli matrices, g and g' are the coupling constants of the groups $SU(2)_L$ and $U(1)_Y$, respectively. The Dirac matrices γ^μ are defined as usual. The colour and generation indices are not specified. This

Lagrangian $\mathscr{L}_{\mathrm{K\,mat}}$ is invariant under two global accidental symmetries, the leptonic number and the baryonic one: the fermions belonging to the fields E_{bL} and e_{bR} are called leptons and transform under the leptonic symmetry $U(1)_L$ while the ones belonging to Q_{bL}, u_{bR} and d_{bR} baryons and transform under $U(1)_B$.

The Lagrangian for the gauge fields reads as follows:

$$\mathscr{L}_{\mathrm{K\,gau}} = -\tfrac{1}{4}(\partial_\mu W^a_\nu - \partial_\nu W^a_\mu + \varepsilon^{abc} W^b_\mu W^c_\nu)(\partial^\mu W^{\nu a} - \partial^\nu W^{\mu a} + \varepsilon^{ab'c'} W^{b'}_\nu W^{c'}_\mu)$$
$$-\tfrac{1}{4}(\partial_\mu B_\nu - \partial_\nu B_\mu)(\partial^\mu B^\nu - \partial^\nu B^\mu). \tag{6.2}$$

6.2.1 The Higgs Mechanism and Vector Boson Masses

The gauge symmetry protects the gauge bosons from having mass. Unfortunately, the weak interactions require massive gauge bosons in order to explain the experimental behaviour. However, adding a direct mass term for gauge bosons breaks explicitly the gauge symmetry and spoils renormalizability. To preserve such nice feature of gauge theories, it is necessary to break spontaneously the symmetry. This is achieved through the Higgs mechanism. We introduce in the spectrum a scalar field H, which transforms as a doublet under $SU(2)_L$, carries hypercharge while is colourless. The Higgs doublet has got the following potential V_{Higgs}, kinetic terms $\mathscr{L}_{\mathrm{K}H}$, and Yukawa couplings with the fermions $\mathscr{L}_{Hf}$:

$$\begin{aligned}
V_{\mathrm{Higgs}} &= -\mu^2 H^\dagger H + \lambda (H^\dagger H)^2 \\
\mathscr{L}_{\mathrm{K}H} &= -\left(\partial_\mu H + igW^a_\mu T_a H + i\frac{g'}{2}B_\mu H\right)^\dagger \left(\partial_\mu H + igW^a_\mu T_a H + i\frac{g'}{2}B_\mu H\right) \\
\mathscr{L}_{Hf} &= -\sum_{b,c}^{\text{gener.}} (\lambda^d_{bc}\overline{Q}_{Lb} H D_{Rc} + \lambda^u_{bc}\overline{Q}_{Lb}\widetilde{H} U_{Rc} + \lambda^e_{bc}\overline{E}_{Lb} H E_{Rc}) + \text{h.c.}
\end{aligned} \tag{6.3}$$

where the parameters μ e λ are real constants, and λ^d_{bc}, λ^u_{bc} and λ^e_{bc} are 3×3 matrices on the generation space. $\widetilde{H}$ indicates the charge conjugated of H: $\widetilde{H}^a = \varepsilon^{ab} H^\dagger_b$.

Although the Lagrangian is invariant under the gauge symmetry, the vacuum is not and the neutral component of the doublet H develops a vacuum expectation value (vev):

$$< H^0 > = \begin{pmatrix} 0 \\ v \end{pmatrix}. \tag{6.4}$$

This breaks the symmetry $SU(2)_L \otimes U(1)_Y$ down to $U(1)_{EM}$. We recall that when a global symmetry is spontaneously broken, in the theory appears a massless Goldstone boson; if the symmetry is local (gauge), these Goldstone bosons become the longitudinal components of the vector bosons (it is said that they are eaten up by the gauge bosons). The gauge bosons relative to the broken symmetry acquire a mass as shown in $\mathscr{L}_{\text{M gauge}}$:

$$\mathscr{L}_{\text{M gauge}} = -\frac{1}{2}\frac{v^2}{4}\left[g^2(W_\mu^1)^2 + g^2(W_\mu^2)^2 + (-gW_\mu^3 + g'B_\mu)^2\right]. \tag{6.5}$$

Therefore, there are three massive vectors $W_\mu^\pm$ and Z_μ^0:

$$W_\mu^\pm = \tfrac{1}{\sqrt{2}}(W_\mu^1 \mp iW_\mu^2), \tag{6.6}$$

$$Z_\mu^0 = \tfrac{1}{\sqrt{g^2+g'^2}}(gW_\mu^3 - g'B_\mu), \tag{6.7}$$

whose masses are given by

$$m_W = g\,\tfrac{v}{2}, \tag{6.8}$$

$$m_Z = \sqrt{(g^2+g'^2)}\,\tfrac{v}{2}, \tag{6.9}$$

while the gauge boson $A_\mu \equiv \frac{1}{\sqrt{g^2+g'^2}}(gW_\mu^3 + g'B_\mu)$, relative to $U(1)_{EM}$, remains massless as imposed by the gauge symmetry. Such mechanism is called Higgs mechanism and preserves renormalizability.

6.2.2 Fermion Masses

Fermions are spinors with respect to the Lorentz group $SU(2)\otimes SU(2)$. Weyl spinors are two component spinors which transform under the Lorentz group as

$$\chi_L \text{ as } (\tfrac{1}{2},0) \tag{6.10}$$

$$\eta_R \text{ as } (0,\tfrac{1}{2}) \tag{6.11}$$

and, therefore, are said to be left-handed and right-handed, respectively.

A fermion mass term must be invariant under the Lorentz group. We have two possibilities as follows:

1. a Majorana mass term couples just one spinor with itself:

$$\chi^\alpha \chi^\beta \varepsilon_{\alpha\beta} \text{ or } \eta^{\dot\alpha}\eta^{\dot\beta}\varepsilon_{\dot\alpha\dot\beta}. \tag{6.12}$$

It's not invariant under any local or global symmetry under which the field transforms not trivially;

2. a Dirac mass term involves two different spinors χ_L and η_R:

$$\chi^\alpha \bar\eta^\beta \varepsilon_{\alpha\beta} \text{ or } \bar\chi^{\dot\alpha}\eta^{\dot\beta}\varepsilon_{\dot\alpha\dot\beta}. \tag{6.13}$$

It can be present even if the fields carry quantum numbers.

In the SM, Majorana masses are forbidden by the gauge symmetry in fact we have that for example,

$$\begin{aligned} e_L e_L &\Rightarrow Q \neq 0 \\ \nu_L \nu_L &\Rightarrow SU(2)_L \neq, \end{aligned}$$

and $SU(2)_L$ forbids Dirac mass terms:

$$\overline{e_L} e_R \Rightarrow SU(2)_L \neq . \quad (6.14)$$

Therefore, no direct mass term can be present for fermions in the SM.

However, when the gauge symmetry breaks spontaneously the Yukawa couplings provide Dirac mass terms to fermions which read as follows:

$$\mathscr{L}_{\text{M mat}} = +\frac{1}{\sqrt{2}}\lambda^e v \bar{e}_L e_R + \frac{1}{\sqrt{2}}\lambda^u v \bar{u}_L u_R + \frac{1}{\sqrt{2}}\lambda^d v \bar{d}_L d_R + \text{h.c.} \quad (6.15)$$

with masses:

$$\begin{aligned} m_e &= \tfrac{1}{\sqrt{2}}\lambda_e v \\ m_u &= \tfrac{1}{\sqrt{2}}\lambda_u v \\ m_d &= \tfrac{1}{\sqrt{2}}\lambda_d v. \end{aligned} \quad (6.16)$$

We notice that neutrinos are massless and so remain at any order in perturbation theory:

1. lacking of the right component they cannot have a Dirac mass term;
2. belonging to a $SU(2)_L$ doublet, they cannot have a Majorana mass term.

However, from experimental data, we can infer that neutrinos are massive and that their mass is very small compared with the other mass scales in the SM. The SM cannot provide such mass to neutrinos and hence this constitutes a proof of the existence of physics beyond the SM. The problem of ν masses will be addressed in more detail in Sect. 6.4.1.

6.2.3 Successes and Difficulties of the SM

It is remarkable that the relatively simple structure of the SM succeeds to pass the innumerable experimental tests ranging from the high-energy frontier (high-energy accelerator physics) to the high-intensity frontier (high-precision electroweak physics and flavor physics). However, we see good reasons to expect the existence of Physics beyond the SM. From a theoretical point of view, the SM cannot give an explanation of the existence of three families, of the hierarchy present among their masses, of the fine tuning of some of its parameters, of the lacking of unification of the three fundamental interactions (considering the behaviour of the coupling constants, we see that they tend to unify at a scale $M_X \sim 10^{15}\,\mathrm{GeV}$ where a unified simple group might arise) of the hierarchy problem of the scalar masses which tend to become as large as the highest mass scale in the theory. From an experimental point of view, the nin-vanishing neutrino masses are a proof of Physics beyond the SM. Also cosmo-particle physics gives strong hints in favor of Physics beyond the SM: in particular, baryogenesis cannot find a satisfactory explanation in the SM, inflation is not predicted by SM and finally we have the DM problem.

6.3 The DM Problem: Experimental Evidence

Let's define Ω (for a review see [5] and [6]) as the ratio between the density ρ and the critical density $\rho_{crit} = \frac{3H_0^2}{8\pi G} = 1.88\,h_0^2 \times 10^{-29}\,\mathrm{g\,cm^{-3}}$, where H_0 is the Hubble constant, and G is the gravitational constant:

$$\Omega = \frac{\rho}{\rho_{crit}}. \tag{6.17}$$

The Ω_{lum} due to the contribution of the luminous matter (stars, emitting clouds of gases) is given by

$$\Omega_{\mathrm{lum}} \leq 0.01. \tag{6.18}$$

First evidences of DM come from observations of galactic rotation curves (circular orbital velocity vs. radial distance from the galactic center) using stars and clouds of neutral hydrogen. These curves show an increasing profile for little values of the radial distance r while for bigger ones it becomes flat, finally decreasing again. According to Newtonian mechanic, this behaviour can be explained if the enclosed mass rises linearly with galactocentric distance. However, the light falls off more rapidly and, therefore, we are forced to assume that the main part of matter in galaxies is made of nonshining matter or DM, which extends for a much bigger region than the luminous one. The limit on $\Omega_{\mathrm{galactic}}$ which can be inferred from the study of these curves is

$$\Omega_{\mathrm{galactic}} \geq 0.1. \tag{6.19}$$

The simplest idea is to suppose that the DM is due to baryonic objects which do not shine. However, both BBN and very precise determinations of the acoustic peaks in cosmic background radiation (CBR) (WMAP results) point out that Ω_B cannot exceed 5%, hence making it impossible to account for the whole amount of DM.

One-third of the BBN baryon density is given by stars, cold gas, and warm gas present in galaxies. The other two-third are probably in hot intergalactic gas, warm gas in galaxies and dark stars such as low-mass objects which do not shine (brown dwarfs and planets) or the result of stellar evolution (neutron stars, black holes, white dwarfs). These last ones are called MAssive Compact Halo Objects (MACHOS) and can be detected in our Galaxy through microlensing.

From cluster observations, from the evolution of the abundance of clusters and measurements of the power spectrum of large-scale structures and from the WMAP data on CBR, we obtain a very significant and puzzling result: the energy density of DM accounts for roughly one fourth of the critical energy density.

Hence the major part of DM is nonbaryonic. The crucial point is that the SM does not possess any candidate for such nonbaryonic relics of the early Universe. Hence the demand for non baryonic DM implies the existence of New Physics beyond the SM. Nonbaryonic DM divides into two classes ([5] and [6]): CDM (e.g., neutral heavy particles called WIMPS or very light ones as axions) and hot DM (HDM)(example: the light neutrinos of the SM)

6.4 Lepton Number Violation and Neutrinos as HDM Candidates

The first candidate for DM we will review are neutrinos which can account for HDM: particles that were relativistic at their decoupling from the thermal bath when their rate of interaction became smaller then the expansion rate and they freeze out (or, to be more precise, at the time Galaxy formation starts at $T \sim 300$ eV). The SM has no candidate for HDM; however, it is now well established from experimental data that neutrinos are massive and very light. Therefore, they can account for HDM. We briefly discuss their characteristics.

6.4.1 Neutrino Masses in the SM and Beyond

The SM cannot account for neutrino masses: we cannot construct either a Dirac mass term as there's only a left-handed neutrino and no right-handed component, or a Majorana mass term because such mass would violate the lepton number and the gauge symmetry.

To overcome this problem, many possibilities have been suggested:

- Within the SM spectrum, we can form an $SU(2)_L$ singlet with ν_L using a triplet formed by two Higgs field H as $\nu_L \nu_L H H$. When the Higgs field H develops a VEV this term gives raise to a Majorana mass term. However, this term is not renormalizable, breaks the leptonic symmetry, and do not give an explanation of the smallness of neutrino masses;
- We can introduce a new Higgs triplet Δ and produce a Majorana mass term as in the previous case when Δ acquires a vacuum expectation value;
- However, the most economical way to extend the SM is to introduce a right-handed component N_R, singlet under the gauge group, which couples with the

left-handed neutrinos. The lepton number L can be either conserved or violated. In the former option, neutrinos acquire a "regular" Dirac mass like for all the other charged fermions of the SM. The left- and right-handed components of the neutrino combine together to give rise to a massive four-component Dirac fermion. The problem is that the extreme lightness of neutrinos (in particular of the electron neutrino) requires an exceedingly small neutrino Yukawa coupling of $O(10^{-11})$ or so. Although quite economical, we do not consider this option particularly satisfactory.

The other possibility is to link the presence of neutrino masses to the violation of L. In this case, one introduces a new mass scale, in addition to the electroweak Fermi scale, in the problem. Indeed, lepton number can be violated at a very high- or a very low-mass scale. The former choice represents, in our view, the most satisfactory way to have massive neutrinos with a very small mass. The idea (see-saw mechanism) is to introduce a right-handed neutrino in the fermion mass spectrum with a Majorana mass M much larger than M_W. Indeed, being the right-handed neutrino a singlet under the electroweak symmetry group, its mass is not chirally protected. The simultaneous presence of a very large chirally unprotected Majorana mass for the right-handed component together with a "regular" Dirac mass term (which can be at most of $O(100\ \text{GeV})$ gives rise to two Majorana eigenstates with masses very far apart.

The Lagrangian for neutrino masses is given by

$$\mathscr{L}_{\text{mass}} = -\frac{1}{2}(\overline{\nu}_L\ \overline{N^c_L}) \begin{pmatrix} 0 & m_D \\ m_D & M \end{pmatrix} \begin{pmatrix} \nu^c_R \\ N_R \end{pmatrix} + h.c. \tag{6.20}$$

where ν^c_R is the CP-conjugated of ν_L and N^c_L of N_R. It holds that $m_D \ll M$. Diagonalizing the mass matrix, we find two Majorana eigenstates n_1 and n_2 with masses very far apart:

$$m_1 \simeq \frac{m_D^2}{M}, \quad m_2 \simeq M.$$

The light eigenstate n_1 is mainly in the ν_L direction and is the neutrino that we "observe" experimentally while the heavy one n_2 is in the N_R one. The key-point is that the smallness of its mass (in comparison with all the other fermion masses in the SM) finds a "natural" explanation in the appearance of a new, large mass scale where L is violated explicitly (by two units) in the right-handed neutrino mass term.

6.4.2 Thermal History of Neutrinos

Let us consider a stable massive neutrino (of mass less than 1 MeV) (see, e.g., [5]). If its mass is less than 10^{-4} eV, it is still relativistic today and its contribution to Ω_{M} is negligible. In the opposite case, it is nonrelativistic and its contribution to the energy density of the Universe is simply given by its number density times its mass. The number density is determined by the temperature at which the neutrino decouples and, hence, by the strength of the weak interactions. Neutrinos decouple when their mean free path exceeds the horizon size or equivalently $\Gamma < H$. Using natural units ($c = \hbar = 1$), we have that

$$\Gamma \sim \sigma_\nu n_{e^\pm} \sim G_F^2 T^5 \tag{6.21}$$

and
$$H \sim \frac{T^2}{M_{Pl}} \tag{6.22}$$

so that
$$T_{\nu d} \sim M_{Pl}^{-1/3} G_F^{-2/3} \sim 1\ \text{MeV}, \tag{6.23}$$

where G_F is the Fermi constant, T denotes the temperature, and M_{PL} is the Planck mass. Since this decoupling temperature $T_{\nu d}$ is higher than the electron mass, then the relic neutrinos are slightly colder than the relic photons which are "heated" by the energy released in the electron-positron annihilation. The neutrino number density turns out to be linked to the number density of relic photons n_γ by the relation:

$$n_\nu = \frac{3}{22} g_\nu n_\gamma, \tag{6.24}$$

where $g_\nu = 2$ or 4 according to the Majorana or the Dirac nature of the neutrino, respectively.

Then, one readily obtains the ν contribution to Ω_{M}:

$$\Omega_\nu = 0.01 \times m_\nu(\text{eV}) h_0^{-2} \frac{g_\nu}{2} (\frac{T_0}{2.7})^3. \tag{6.25}$$

Imposing $\Omega_\nu h_0^2$ to be less than one (which comes from the lower bound on the lifetime of the Universe), one obtains the famous upper bound of $200(g_\nu)^{-1}$ eV on the sum of the masses of the light and stable neutrinos:

$$\sum_i m_{\nu_i} \leq 200(g_\nu)^{-1}\ \text{eV}. \tag{6.26}$$

Clearly from Eq.(6.25), one easily sees that it is enough to have one neutrino with a mass in the $1-20$ eV range to obtain Ω_ν in the $0.1-1$ range of interest for the DM problem.

However, the data on neutrino oscillations point to neutrino masses definitely smaller than 1 eV; to be more precise, this is certainly true if we consider schemes where neutrinos possess hierarchical masses (with a direct or inverse hierarchy). In the case of depenerate neutrino masses, one could barely consider neutrinos to be in the eV region. However, as we are going to see below, the data on the large-scale structures disfavor neutrinos with masses > 1eV; indeed, such cosmological data provide the best bound we have so far on the sum of the masses of the stable, light neutrinos.

6.4.3 HDM and Structure Formation

Hence massive neutrinos with mass in the eV range are very natural candidates to contribute an $\Omega_{\rm M}$ larger than 0.1. The actual problem for neutrinos as viable DM candidates concerns their role in the process of large-scale structure formation. The crucial feature of HDM is the erasure of small fluctuations by free streaming: neutrinos stream relativistically for quite a long time till their temperature drops to $T \sim m_\nu$. Therefore, a neutrino fluctuation in order to be preserved must be larger than the distance d_ν travelled by neutrinos during such interval. The mass contained in that space volume is of the order of the supercluster masses:

$$M_{J,\nu} \sim d_\nu^3 m_\nu n_\nu (T = m_\nu) \sim 10^{15} M_\odot, \tag{6.27}$$

where n_ν is the number density of the relic neutrinos and $M_\odot$ is the solar mass. Therefore, the first structures to form are superclusters and smaller structures as galaxies arise from fragmentation in a typical top-down scenario. Unfortunately, in these schemes, one obtains too many structures at super large scales. Hence schemes of pure HDM are strongly disfavored by the demand of a viable mechanism for large-structure formation.

As I mentioned above, not only are such cosmological data ruling out the light neutrinos as being the main source of DM but also they constitute the most powerful way we have at our disposal to put an upper bound on their masses. From the WMAP and the large-scale structure data, we infer that the sum of the light, stable neutrinos has to be less than 1 eV; indeed, if one includes all possible restrictions coming from the LSS data, one should conclude that such mass is <0.2 eV [7].

6.5 Low-energy SUSY and DM

Another kind of DM, widely studied, called CDM is made of particles which were nonrelativistic at their decoupling. Natural candidates for such DM are WIMPs, which are very heavy if compared with neutrinos. The SM does not have nonbaryonic neutral particles, which can account for CDM and, therefore, we need to consider extensions of the SM as SUSY SM in which there are heavy neutral particles remnants of annihilations such as neutralinos (for a review, see [10]).

6.5.1 Neutralinos as the LSP in SUSY Models

One of the major shortcomings of the SM concerns is the protection of the scalar masses once the SM is embedded into some underlying theory (and at least at the Planck scale such New Physics should set in to incorporate gravity into the game). Since there is no typical symmetry protecting scalar masses (while for fermions there is the chiral symmetry and for gauge bosons there are gauge symmetries), the clever idea which was introduced in the early 80s to prevent scalar masses to get too large values was to have a SUSY unbroken down to the weak scale. Since fermion masses are chirally protected and as long as SUSY is unbroken, there must be degeneracy between the fermion and scalar components of a SUSY multiplet, then having a low-energy SUSY, it is possible to have an "induced protection" on scalar masses (for a review, see [8, 9]).

However, the mere supersymmetrization of the SM faces an immediate problem. The most general Lagrangian contains terms that violate baryon and lepton numbers producing a too fast proton decay. To prevent this catastrophic result, we have to add some symmetry which forbids all or part of these dangerous terms with L or B violations. The most familiar solution is the imposition of a discrete symmetry, called R matter parity, which forbids all these dangerous terms. It reads over the fields contained in the theory:

$$R = (-1)^{3(B-L)+2s}. \tag{6.28}$$

R is a multiplicative quantum number reading -1 over the SUSY particles and $+1$ over the ordinary particles. Clearly in models with R parity, the lightest SUSY particle can never decay. This is the famous LSP (lightest SUSY particle) candidate for CDM.

Notice that proton decay does not call directly for R parity. Indeed this decay entails the violation of both B and L. Hence, to prevent a fast proton decay one may impose a discrete symmetry that forbids all the B violating terms in the SUSY Lagrangian, while allowing for terms with L violation (the vice versa is also viable). Models with such alternative discrete symmetries are called SUSY model, with broken R parity. In such models, the stability of the LSP is no longer present and the LSP cannot be a candidate for stable CDM. We will comment later on these alternative models in relation to the DM problem, but we turn now to the more "orthodox" situation with R parity. The favorite LSP is the lightest neutralino.

6.5.2 Neutralinos in the Minimal SUSY Standard Model

If we extend the SM in the minimal way, adding for each SM particle a SUSY partner with the same quantum numbers, we obtain the so-called minimal supersymmetric standard model (MSSM). In this context—the neutralinos are the eigenvectors of the mass matrix of the four neutral fermions partners of the W_3, B, H_1^0, and H_2^0 called, respectively, wino $\tilde{W}_3$, bino $\tilde{B}$, higgsinos $\tilde{H}_1^0$ and $\tilde{H}_2^0$. There are four parameters entering the mass matrix, M_1, M_2, μ, and $\tan\beta$:

$$M = \begin{pmatrix} M_2 & 0 & m_Z\cos\theta_W\cos\beta & -m_Z\cos\theta_W\sin\beta \\ 0 & M_1 & -m_Z\sin\theta_W\cos\beta & m_Z\sin\theta_W\sin\beta \\ m_Z\cos\theta_W\cos\beta & -m_Z\sin\theta_W\cos\beta & 0 & -\mu \\ -m_Z\cos\theta_W\sin\beta & m_Z\sin\theta_W\sin\beta & -\mu & 0 \end{pmatrix}, \tag{6.29}$$

where $m_Z = 91.19 \pm 0.002$ GeV is the mass of the Z boson, θ_W is the weak mixing angle, $\tan\beta \equiv v_2/v_1$ with v_1, v_2 vevs of the scalar fields H_1^0 and H_2^0, respectively.

In general, M_1 and M_2 are two independent parameters, but if one assumes that a grand unification scale takes place, then at grand unification $M_1 = M_2 = M_3$, where M_3 is the gluino mass at that scale. Then, at the M_W scale one obtains:

$$M_1 = \tfrac{5}{3}\tan^2\theta_W M_2 \simeq \tfrac{M_2}{2}, \tag{6.30}$$

$$M_2 = \tfrac{g_2^2}{g_3^2} m_{\tilde{g}} \simeq m_{\tilde{g}}/3, \tag{6.31}$$

where g_2 and g_3 are the SU(2) and SU(3) gauge coupling constants, respectively, and $m_{\tilde{g}}$ is the gluino mass.

The relation (6.30) between M_1 and M_2 reduces to three the number of independent parameters that determine the lightest neutralino composition and mass: $\tan\beta, \mu$, and M_2. The neutralino eigenstates are denoted usually by $\tilde{\chi}_i^0$ being $\tilde{\chi}_1^0$ the lightest one.

If $|\mu| > M_1, M_2$, then $\tilde{\chi}_1^0$ is mainly a gaugino and in particular a bino if $M_1 > m_Z$; if $M_1, M_2 > |\mu|$, then $\tilde{\chi}_1^0$ is mainly a higgsino. The corresponding phenomenology is drastically different leading to different predictions for CDM.

For fixed values of $\tan\beta$, one can study the neutralino spectrum in the (μ, M_2) plane. The major experimental inputs to exclude regions in this plane are the request that the lightest chargino be heavier than $m_Z/2$ and the limits on the invisible width of the Z hence limiting the possible decays $Z \to \tilde{\chi}_1^0\tilde{\chi}_1^0$, $\tilde{\chi}_1^0\tilde{\chi}_2^0$. Moreover, if the GUT assumption is made, then the relation (6.30) between M_2 and $m_{\tilde{g}}$ implies a severe bound on M_2 from the experimental lower bound on $m_{\tilde{g}}$ of Tevatron. The theoretical demand that the electroweak symmetry be broken radiatively, i.e., due to the renormalization effects on the Higgs masses when going from the superlarge scale of supergravity breaking down to M_W, further constrains the available (μ, M_2) region. The first important outcome of this analysis is that the lightest neutralino mass exhibits a lower bound of roughly 30 GeV. The actual bound on the mass of the lightest neutralino $\tilde{\chi}_1^0$ from LEP2 is

$$m_{\tilde{\chi}_1^0} \geq 40 \text{ GeV} \tag{6.32}$$

for any value of $\tan\beta$. This bound becomes stronger if we put further constraints on the MSSM, like, for instance, in the Constrained MSSM (CMSSM) where we have only four independent SUSY-parameters plus the sign of the μ parameter.

It should be reminded that all the above bounds on the lightest neutralino take into account a situation where some unification of the gaugino masses occurs, hence making it possible to limit the mass parameter M_1 through the severe experimental bounds on M_2 as derived from LEP physics. If one removes such unification condition of the gaugino masses, then it is possible to have neutralinos as light as few GeVs [12].

6.5.3 Thermal History of Neutralinos and Ω_{CDM}

Let us focus now on the role played by $\tilde{\chi}_1^0$ as a source of CDM. The lightest neutralino $\tilde{\chi}_1^0$ is kept in thermal equilibrium through its electroweak interactions not only for $T > m_{\tilde{\chi}_1^0}$, but even when T is below $m_{\tilde{\chi}_1^0}$. However, for $T < m_{\tilde{\chi}_1^0}$ the number of $\tilde{\chi}_1^0$s rapidly decreases because of the appearance of the typical Boltzmann suppression factor $\exp(-m_{\tilde{\chi}_1^0}/T)$. When T is roughly $m_{\tilde{\chi}_1^0}/20$ the number of $\tilde{\chi}_1^0$ diminished so much that they do not interact any longer, i.e., they decouple. Hence the contribution to Ω_{CDM} of $\tilde{\chi}_1^0$ is determined by two parameters: $m_{\tilde{\chi}_1^0}$ and the temperature at which $\tilde{\chi}_1^0$ decouples ($T_{\chi d}$) which fixes the number of surviving $\tilde{\chi}_1^0$s. As for the determination of $T_{\chi d}$ itself, one has to compute the $\tilde{\chi}_1^0$ annihilation rate and compare it with the cosmic expansion rate.

Several annihilation channels are possible with the exchange of different SUSY or ordinary particles, $\tilde{f}$, H, Z, etc. Obviously, the relative importance of the channels depends on the composition of $\tilde{\chi}_1^0$.

In the MSSM, there are five new parameters in addition to those already present in the non-SUSY case. Imposing the electroweak radiative breaking further reduces this number to four. Finally, in simple supergravity realizations, the soft parameters A and B are related. Hence, we end up with only three new, independent parameters. One can use the constraint that the relic $\tilde{\chi}_1^0$ abundance provides a correct Ω_{CDM} to restrict the allowed area in this three-dimensional space. Or, at least, one can eliminate points of this space which would lead to $\Omega_{\tilde{\chi}_1^0} > 1$, hence overclosing the Universe.

There exists a vast literature on the subject of SUSY WIMPs and accelerator physics. To review such material is beyond the scope of this chapter. I refer the interested reader to the thorough and broad review of Jungman et al., [10] and the original papers therein quoted for a general discussion of SUSY in the MSSM and to the works in [11] for an updated analysis.

Finally a comment on models without R parity. From the point of view of DM, the major implication is that in this context the LSP is no longer a viable CDM candidate since it decays. There are very special circumstances under which this decay may be so slow that the LSP can still constitute a CDM candidate.

6.6 Changing the Expansion Rate in the Past

In a standard flat FRW universe described by GR, the expansion rate of the Universe, $H_{GR} \equiv \dot{a}/a$, is set by the total energy density, $\tilde{\rho}_{\rm tot}$, according to the Friedmann law,

$$H_{GR}^2 = \frac{1}{3M_p^2}\tilde{\rho}_{\rm tot}\,, \tag{6.33}$$

where M_p is the Planck mass, related to the Newton constant by $M_p = (8\pi G)^{-1/2}$. If the total energy density is dominated by relativistic degrees of freedom, the expansion rate is related to the temperature through the relation

$$H_{GR} \simeq 1.66\, g_*^{1/2} \frac{T^2}{M_p}\,, \tag{6.34}$$

with g_* the effective number of relativistic degrees of freedom.

We will modify the above H–T relation by considering a modification of GR in which an effective Planck mass, different from M_p appears in (6.34). This can be realized in a fully covariant way in ST theories . We will consider the class of ST theories, which can be defined by the following action [13],

$$S = S_g + \sum_i S_i, \tag{6.35}$$

where S_g is the gravitational part, given by the sum of the Einstein–Hilbert and the scalar field actions,

$$S_g = \frac{M_*^2}{2} \int d^4x \sqrt{-g} \left[R + g^{\mu\nu} \partial_\mu \varphi \partial_\nu \varphi - \frac{2}{M_*^2} V(\varphi) \right], \tag{6.36}$$

where $V(\varphi)$ can be either a true potential or a (Einstein frame) cosmological constant, $V(\varphi) = V_0$. The S_i's are the actions for separate "matter" sectors

$$S_i = S_i[\Psi_i, A_i^2(\varphi) g_{\mu\nu}]\ , \tag{6.37}$$

with Ψ_i indicating a generic field of the i-th matter sector, coupled to the metric $A_i^2(\varphi) g_{\mu\nu}$. The actions S_i are constructed starting from the Minkowski actions of Quantum Field Theory, for instance the SM or the MSSM ones, by substituting the flat metric $\eta_{\mu\nu}$ everywhere with $A_i^2(\varphi) g_{\mu\nu}$.

The emergence of such a structure, with different conformal factors A_i^2 for the various sectors can be motivated in extra-dimensional models, assuming that the two sectors live in different portions of the extra-dimensional space.

We consider a flat FRW space-time

$$ds^2 = dt^2 - a^2(t)\, dl^2\ ,$$

where the matter energy-momentum tensors, $T^i_{\mu\nu} \equiv 2(-g)^{-1/2} \delta S_i / \delta g^{\mu\nu}$ admit the perfect-fluid representation

$$T^i_{\mu\nu} = (\rho_i + p_i)\, u_\mu u_\nu\ - p_i\, g_{\mu\nu}\ , \tag{6.38}$$

with $g_{\mu\nu}\ u^\mu u^\nu = 1$.

The cosmological equations then take the form

$$\frac{\ddot{a}}{a} = -\frac{1}{6M_*^2}\left[\sum_i(\rho_i + 3\,p_i) + 2M_*^2\dot{\varphi}^2 - 2V\right], \tag{6.39}$$

$$\left(\frac{\dot{a}}{a}\right)^2 = \frac{1}{3M_*^2}\left[\sum_i \rho_i + \frac{M_*^2}{2}\dot{\varphi}^2 + V\right], \tag{6.40}$$

$$\ddot{\varphi} + 3\frac{\dot{a}}{a}\dot{\varphi} = -\frac{1}{M_*^2}\left[\sum_i \alpha_i(\rho_i - 3p_i) + \frac{\partial V}{\partial \varphi}\right], \tag{6.41}$$

where the coupling functions α_i are given by

$$\alpha_i \equiv \frac{d\log A_i}{d\varphi}. \tag{6.42}$$

The Bianchi identity holds for each matter sector separately, and reads,

$$d(\rho_i a^3) + p_i\, da^3 = (\rho_i - 3\,p_i)\, a^3 d\log A_i(\varphi), \tag{6.43}$$

implying that the energy densities scale as

$$\rho_i \sim A_i(\varphi)^{1-3w_i} a^{-3(1+w_i)}, \tag{6.44}$$

with $w_i \equiv p_i/\rho_i$ the equation of state associated to the i-th energy density (assuming w_i is constant).

6.6.1 GR as a Fixed Point

To start, consider the case of a single matter sector, S_M. In order to compare the ST case with the GR one of Eqs. (6.33, 6.34), it is convenient to Weyl-transform to the so-called Jordan Frame (JF), where the energy-momentum tensor is covariantly conserved. The transformation amounts to a rescaling of the metric according to

$$\tilde{g}_{\mu\nu} = A_M^2(\varphi) g_{\mu\nu}, \tag{6.45}$$

keeping the comoving spatial coordinates and the conformal time $d\eta = dt/a$ fixed. The JF matter energy-momentum tensor, $\tilde{T}^M_{\mu\nu} \equiv 2(-\tilde{g})^{-1/2}\delta S_M/\delta\tilde{g}^{\mu\nu}$, is related to that in Eq. (6.38) by $\tilde{T}^M_{\mu\nu} = A_M^{-2}T^M_{\mu\nu}$, so that energy density and pressure transform as

$$\tilde{\rho}_M = A_M^{-4}\rho_M, \qquad \tilde{p}_M = A_M^{-4}p_M, \tag{6.46}$$

while the cosmic time transforms as $d\tilde{t} = A_M dt$. One can easily verify that the above-defined quantities satisfy the usual Bianchi identity, that is Eq. (6.43) with vanishing RHS, and that, as a consequence, $\tilde{\rho}_M \sim \tilde{a}^{-3(1+w_M)}$. The expansion rate, $H_{ST} \equiv d\log\tilde{a}/d\tilde{t}$, is given by

$$H_{ST} = \frac{1 + \alpha_M(\varphi)\,\varphi'}{A_M(\varphi)}\,\frac{\dot{a}}{a}, \tag{6.47}$$

where we have defined α_M according with Eq. (6.42), and $(\cdot)' \equiv d(\cdot)/d\log a$. Using (6.47) and (6.46) in (6.40), we obtain the Friedmann equation in the ST theory,

$$H_{ST}^2 = \frac{A_M^2(\varphi)}{3M_*^2}\,\frac{(1 + \alpha_M(\varphi)\,\varphi')^2}{1 - (\varphi')^2/6}\left[\tilde{\rho}_M + \tilde{V}\right], \tag{6.48}$$

where $\tilde{V} \equiv A_M^{-4} V$. Comparing with Eq.(6.33), we see that apart from the extra contribution to $\tilde{\rho}_{tot}$ from the scalar field potential, the ST Friedmann equation differs from the standard one of GR by the presence of an effective, field-dependent Planck mass,

$$\frac{1}{3M_p^2} \to \frac{A_M^2(\varphi)}{3M_*^2}\,\frac{(1 + \alpha_M(\varphi)\,\varphi')^2}{1 - (\varphi')^2/6} \simeq \frac{A_M^2(\varphi)}{3M_*^2}, \tag{6.49}$$

where the last equality holds with very good approximations for all the choices of A_i functions considered in the present paper.

If the conformal factor $A_M^2(\varphi)$ is constant, then the full action $S_g + S_M$ is just that of GR (with $M_p = M_*/A_M$) plus a minimally coupled scalar field. Therefore, the coupling function α_M, defined according to Eq. (6.42), measures the "distance" from GR of the ST theory, $\alpha_M = 0$ being the GR limit. Changing A_M, and, therefore, changing the effective Planck mass, opens the way to a modification of the standard relation between H and $\tilde{\rho}$, or T. In order to study the evolution of $A_M(\varphi)$, one should come back to Eq. (6.41). Considering an initial epoch deeply inside radiation domination, we can neglect the contribution from the potential on the RHS. The other contribution, the trace of the energy-momentum tensor $(\rho_M - 3\,p_M)$ is zero for fully relativistic components but turns on to positive values each time the temperature drops below the mass threshold of a particle in the thermal bath. Assuming a mass spectrum – e.g., that of the SM or of the MSSM – one finds that this effect is effective enough to drive the scalar field evolution even in the radiation domination era [14].

The key point to notice is that if there is a field value, φ_0, such that $\alpha_M(\varphi_0) = 0$, this is a *fixed point* of the field evolution . Moreover, if α_M' is positive (negative), the fixed point is attractive (repulsive). Since $\alpha_M = 0$ corresponds to the GR limit, we see that GR is a – possibly attractive – fixed point configuration.

The impact on the DM relic abundance of a scenario based on this mechanism of attraction towards GR was considered in [14, 15].

6.7 Implications for DM in the CMSSM

A modification of the Hubble rate at early times has impact on the formation of DM as a thermal relic, if the particle freeze-out occurs during the period of modification of the expansion rate. ST cosmologies with a Hubble rate increased with respect to the GR case have been discussed in Refs. [14, 15, 16], where the effect on the decoupling of a cold relic was discussed and bounds on the amount of increase of the Hubble rate prior to BBN have been derived from the indirect detection signals of DM in our Galaxy. For cosmological models with an enhanced Hubble rate, the decoupling is anticipated, and the required amount of cold DM is obtained for larger annihilation cross-sections: this, in turn, translates into larger indirect detection rates, which depend directly on the annihilation process. In Refs. [15, 16], we discussed how low-energy antiprotons and gamma–rays fluxes from the galactic center can pose limits on the admissible enchancement of the pre-BBN Hubble rate. We showed that these limits may be severe: for DM particles lighter than about a few hundred GeV antiprotons set the most important limits, which are quite strong for DM masses below 100 GeV. For heavier particles, gamma-rays are more instrumental in determining significant bounds.

In the case of the cosmological models which predict a reduced Hubble rate, the situation is opposite: a smaller expansion rate implies that the cold relic particle remains in equilibrium for a longer time in the early Universe, and, as a consequence, its relic abundance turns out to be smaller than the one obtained in GR. In this case, the required amount of DM is obtained for smaller annihilation cross-sections, and therefore, indirect detection signals are depressed as compared with the standard GR case: as a consequence, no relevant bounds on the pre-BBN expansion rate can be set. On the other hand, for those particle physics models which typically predict large values for the relic abundance of the DM candidate, this class of ST cosmologies may have an important impact in the selection of the regions in parameter space which are cosmologically allowed.

A typical and noticeable case where the relic abundance constraint is very strong is offered by minimal SUGRA models. A reduction of the expansion rate will have a crucial impact on the allowed regions in parameter space, which are, therefore, enlarged. The potential reach of accelerators like the LHC or the International Linear Collider (ILC) on the search of supersymmetry may, therefore, be affected by this broadening of the allowed parameter space, especially for the interesting situation of looking for SUSY configurations able to fully explain the DM problem.

Acknowledgments It is a pleasure to thank my collaborators Riccardo Catena, Nicolao Fornengo, Massimo Pietroni, and Mia Schelke.

References

1. A. Salam, in *Elementary Particle Theory*, ed. N. Svartholm, (Stockholm, Sweden) 1967;
2. S. Weinberg, *Phys. Rev. Lett.* **19**, 1264 (1967);

3. S.L. Glashow, *Nucl. Phys.* **22**, 579 (1961);
4. M.E. Peskin, D.V. Schroeder, *An introduction to quantum field theory.* , Addison-Wesley P.C., Reading Mass. (1995);
5. For an introduction to the DM problem, see, for instance:
R. Kolb and S. Turner, in *The Early universe* (Addison-Wesley, New York, N.Y.) 1990;
Dark Matter, ed. by M. Srednicki, (North-Holland, Amsterdam) 1989;
J. Primack, D. Seckel and B. Sadoulet, *Annu. Rev. Nucl. Part. Sci.* **38**(1988) 751;
6. For a recent review, see G. Bertone, D. Hooper and J. Silk, *Phys. Rep.* **405** (2005) 279;
7. See, for instance, G.L. Fogli, E. Lisi, A. Marrone, A. Melchiorri, A. Palazzo, P. Serra, J. Silk, A. Slosar, *Phys. Rev. D* **75** (2007) 053001
8. For a review, see H.P. Nilles, in *Phys. Rep.* **110C** (1984) 1;
H. Haber and G. Kane, in *Phys. Rep.* **117C** (1985) 1;
9. E. Cremmer, S. Ferrara, L. Girardello and A. van Proeyen, in *Phys. Lett.* **B116** (1982) 231; *Nucl. Phys. B* **B212** (1983) 413;
10. G. Jungman, M. Kamionkowski and K. Griest, *Phys. rep.* **267** (1996) 1, and references therein;
11. For instance, see D. Hooper, in *TASI 2008 Lectures on DM*, arXiv:0901.4090 [hep-ph] and references therein; J. Ellis and K. Olive, in *Particle Dark Matter: Observations, Models and Searches' edited by Gianfranco Bertone, Chapter 8, pp. 142-163*, e-Print: arXiv:1001.3651 [astro-ph.CO]
12. N. Fornengo, A. Bottino, F. Donato, S. Scopel, *Nucl. Phys. Proc. Suppl.* **138** (2005) 28
13. R. Catena, M. Pietroni and L. Scarabello, *Phys. Rev. D* **70** (2004) 103526
14. R. Catena, N. Fornengo, A. Masiero, M. Pietroni and F. Rosati, *Phys. Rev. D* **70**(2004) 063519.
15. M. Schelke, R. Catena, N. Fornengo, A. Masiero and M. Pietroni, *Phys. Rev. D* **74** (2006) 083505.
16. F. Donato, N. Fornengo and M. Schelke, *JCAP* **0703** (2007) 021

Chapter 7
Dark Matter Direct and Indirect Detection

Andrea Giuliani

Abstract Cosmological and astrophysical observations show with outstanding evidence that more than 80% of the matter density in the Universe is nonluminous. Attractive candidates for the composition of this dark cosmic component are still undetected, neutral, heavy particles, which were non-relativistic, or "cold," when they decoupled from ordinary matter. This paper will review the direct and indirect detection methods of these hypothetical particles, with a major emphasis on the previous approach. In the direct search, sophisticated instruments look for the scattering of dark matter particles off nuclei in ultra-low background, deep underground experiments. In the indirect search, space-based and ground-based observatories aim to detect secondary particles that could originate from annihilations of dark matter candidates in various locations in the Milky Way or in close galaxies. Emphasis is given to the most recent developments and to the status of close-future projects.

7.1 Introduction

The concept of dark matter was introduced to solve serious discrepancies between two classes of estimates of the masses of astrophysical objects: from one hand, those based on the luminous and visible parts, and from the other hand, those based on the dynamical behaviour of thc components. At almost every cosmic scale, observations infer a larger dynamic mass than a visible mass, implying a significant dark matter contribution to the gravitational potential. The mass-to-light ratio, M/L is used to quantify this discrepancy, with $M/L \sim 1$ for a star like the Sun.

Andrea Giuliani
Department of Physics and Mathematics, University of Insubria, Via Valleggio 11, I-22100 Como, Italy, e-mail: andrea.giuliani@mib.infn.it

7.1.1 Dark Matter at the Various Scales

At the galactic scale, in most cases, it is possible to observe a clear dark matter component. In spiral galaxies, the stellar motions are dominated by rotation within the disk [1]. The luminous component decreases exponentially from the center, with a characteristic radius of a few kiloparsec. One expects, therefore a Keplerian decline of the star rotation velocities, which should scale as $r^{-1/2}$ outside this radius. On the contrary, star rotation curves usually remain flat far from galactic centers, typically beyond $30-40$ kpc. Quantitatively, this behaviour translates into a mass-to-light ratio $M/L \sim 5-10$. In elliptical galaxies, where the dynamical equilibrium is dominated by pressure rather than the rotation motion, observations indicate an even larger dark matter contribution, with $M/L \sim 10-25$. Considering the various galaxy types, two are particularly dominated by dark matter: the low-surface brightness galaxies (LSB) and the dwarf spheroidal galaxies (DSph). These objects are often characterized by a huge mass-to-light ratio $M/L > 100$ at the border of the field of stars. This feature makes them very attractive sources for indirect dark matter searches, looking in particular for the annihilation γ-rays that they are expected to emit.

At the galaxy-cluster scale, it is worth mention the first claim for the existence of dark matter, which traces back to the famous Zwicky's article in 1933 [2]. In this historical paper, the virial theorem was used to reconstruct the gravitational potential of the Coma cluster concluding that a huge amount of matter was invisible. There are today several other different methods that come to the same conclusion independently, such as x-ray measurements of inter-cluster gas temperature, weak/strong lensing, luminous arcs, and multiple images. They all confirm the presence of dark matter at higher fractions than in galaxies, typically with $M/L \sim 200$ [3]. Globally, these observations translate, in terms of matter density with respect to the critical density, into $\Omega_m \sim 0.2-0.3$.

The recent cosmological-scale measurements, like the cosmic microwave background (CMB) [4], large-scale structure, and supernovae surveys (for instance, SNLS) [5], confirm this matter density and favor the so-called "cosmological concordance model," according to which we live in a cold dark matter (CDM) Universe dominated by a cosmological acceleration term. Quantitatively, the model predicts that the Universe is flat and made of 4% baryons, 20% non-baryonic dark matter (topic of this chapter), and 76% dark energy [6].

7.1.2 The Nature of Dark Matter

On the basis of the observed discrepancies between the visible amounts of baryonic matter and the estimates from big bang nucleosynthesis, we know that dark baryons exist. However, the size of these discrepancies is by far not able to account for the dark matter amount required by the concordance model. Therefore, we conclude that the dark matter in the Universe is essentially nonbaryonic, at the extent specified in the previous section. The theories and the simulations about hierarchical structure formation indicate that the nonbaryonic dark matter assumes the form of a gas of cold and weakly interacting massive particles (WIMPs). There is no viable candidate in the standard mode (SM) of particle physics to the composition of this (CDM) gas. There, are however, theories beyond the SM, which have been developed specifically to solve problems inherent to elementary particle physics (such as the unification of the gauge couplings at high energy and the hierarchy and naturalness problems), that comprise very attractive dark matter candidates. (Extensive reviews on this subject can be found in literatures [7, 8]; we shall provide here a concise summary in Sects. 7.1.2.1 and 7.1.2.2.) In particular, the relic density of these candidates can be estimated, and it turns out almost automatically that their present contribution to the expected nonbaryonic dark matter density lies in the right range [9, 10]. This conspiracy is often referred to as "WIMP miracle." Presently, the most popular candidates for WIMPs come from the supersymmetric and extra-dimensional theories, which are briefly outlined below. In addition to WIMPs, another viable candidate is the axion [11], a particle proposed in the 1970s as an extension to the SM in order to solve the so-called strong CP problem, which is posed by the nonobservation of an electric dipole moment of the neutron. The axion has to be very light, $\sim 10^{-5}$ eV. A review of the experimental searches for axions is beyond the scope of this chapter and will not be given here.

7.1.2.1 SUSY Particles as WIMPs

Supersymmetry (SUSY) is a symmetry between bosons and fermions, which is broken at the presently accessible energy scale. The association of known fermions and bosons in super-multiplets requires to add at least one extra bosonic/fermionic superpartner (called sparticle) to each standard fermion/boson. Therefore, a copious spectrum of new particles would appear if SUSY were a fundamental symmetry of the nature. SUSY is expected to show off beyond the electroweak energy scale, in the range $\sim 0.1 - 1$ TeV.

The minimal supersymmetric extension of the standard model (MSSM) provides a coherent frame for the unification of interactions (except gravity) at the high energy scale, breaks the electroweak symmetry dynamically, and solves partly the hierarchy and naturalness problems. With the implementation of a discrete symmetry called R-parity, which ensures the conservation of $B - L$ quantum number and proton stability, SUSY particles are bound to be created in pairs, and the lightest supersymmetric particle (LSP) is stable. In this framework, among the dark matter candidates, two have been widely studied for the last two decades: the lightest of the four neutralinos (which will be referred to as simply the neutralino *chi*) [12] and the gravitino. The neutralino is a linear combination of the Majorana fermionic superpartners of the gauge and Higgs bosons (gauginos—bino and wino—and higgsinos) [13], while the gravitino is the superpartner of the graviton, appearing when one requires the local invariance of SUSY, which involves automatically gravity. From fundamental considerations, these different phenomenologies can be shown to be connected with the SUSY breaking scenarios. The most popular simplified model, called the minimal supergravity model (mSUGRA), assumes universal gaugino masses $m_{1/2}$ and universal scalar masses m_0 at high energy scale. In this scenario, the MSSM can be described with only five parameters at this scale: the unified scalar masses m_0, the unified gaugino masses $m_{1/2}$, the universal trilinear coupling A_0, and the Higgs doublet vacuum expected value ratio, $tan\beta = v_2/v_1$ and the sign of the Higgs mass mixing parameter μ. In most of the parameter space of this model, the neutralino is the LSP. Its mass is related to those of bino, wino, and two higgsino fields (respectively, M_1, M_2, and the Higgs mixing μ). In mSUGRA, the neutralino is mainly bino-like, with its mass related to the unified gaugino mass as $m_\chi < M_1 < 0.43 m_{1/2} < M_2, \mu$.

Another scenario of SUSY breaking invokes conformal anomalies without any additional hidden sector: the anomaly-mediated SUSY breaking scenario (AMSB). We refer to specialized papers for more insight into these models. We just remark that in AMSB models, the neutralino is usually the LSP, but with a strong wino component.

Finally, gauge-mediated supersymmetry breaking (GMSB) models mediate the breaking to the observable sector from a hidden sector through messenger fields that have gauge interactions. In this scenario, the LSP is mostly the gravitino.

7.1.2.2 Theories with Extradimensions

Modern extradimension (ED) theories derive from the historical approach followed by Kaluza in 1919 [14], who tried to explain electromagnetism as a consequence of the curvature of an additional dimension to the classical 4D space-time. He noticed that by extending general relativity to a 5D space-time, the resulting equations split up to the standard 4D gravitation with an extra term equivalent to Maxwell's equation: a sort of unification of electromagnetism and gravity. In 1926, Klein applied this approach to quantum mechanics [15], suggesting that the extra dimension should be compactified to a very small radius, as small as the Planck scale. These ideas revived in modern string theories, which foresee models with 26 (bosonic strings) or 10 (superstrings) dimensions. Basing on these assumptions, it is possible to build phenomenological models providing solutions to the hierarchy and naturalness problems without introducing SUSY. In general, ED theories introduce new physical states beyond the SM as a consequence of compactification of the EDs. These new particles correspond to excitations in the bulk of all fields which propagate in the compactified EDs. These fields are expanded into a complete series of modes, thus building the so-called Kaluza-Klein (KK) tower.

Today, there are several classes of extradimension models [16]. Those assuming that gravity is the only interaction mediated in the extradimensions (called the "bulk") gives no detectable dark matter candidates. Other scenarios assume that the whole field content of the theory may propagate in all dimensions. Such models define the class called universal extradimensions (UED), which contrasts with previous ED theories by allowing translation invariance along the EDs. Without going into the details, the main point here is that in UED, a new discrete symmetry appears, named KK parity. In analogy with SUSY, the consequence of KK parity conservation is the stability of the lightest KK particle (LKP). In these models, the first KK excitation $B^{(1)}$ of the hypercharge gauge boson and of the neutrino $\nu^{(1)}$ are viable WIMP candidates.

7.1.2.3 Detection Techniques and Distribution of WIMPs in the Galaxy

Two basic methods can be used to detect dark matter, either direct or indirect. Direct searches are based on the detection of dark matter particles actually passing through detectors and physically interacting with them. Chardin [17] provides an exhaustive review (even if not very recent) of the direct detection concepts and experiments. Indirect searches look for secondary products originated when dark matter particles annihilate each other elsewhere, typically in the Galaxy. Our discussion about indirect detection is based mainly on a very comprehensive review of this field [18], even if not containing yet the recent important PAMELA observations. The two methods, being very different, are complementary. Positive evidence seen with these distinct approaches would provide convincing confirmation of the discovery of CDM. Furthermore, in some cases, the two methods are sensitive in different regions of the parameter phase space for the nature of the CDM particle. For both approaches, the distribution of the WIMPs in the Galaxy is crucial. The local density, at the Sun position, is important for direct detection. For indirect detection through the observation of γ-rays or cosmic rays, it is directly the distribution of matter in the halo of the Galaxy, which is relevant. For neutrino searches, the relevant density is that concentrated in the centers of the various possible sources, which is itself related to the halo density at the position of the concentrating body.

The estimate of the WIMP density ρ_W involves both observational and theoretical astrophysics. The density profile can be constrained with galaxy star rotation curves, which allow the reconstruction of the mass profile and by disentangling the baryon contribution, constrain the dark matter distribution. This technique first led to empirical laws for the density distribution, referred to as isothermal profiles [19], with constant rotation velocities and spherical r^{-2} density distributions. Nevertheless, it is rather difficult to estimate the dark matter profile at the center of galaxies; from the observational point of view, the centers are often difficult to observe and characterize, and from the theoretical point of view the determination of the profile involves highly nonlinear calculations. The current theoretical understanding of hierarchical cosmologies and related numerical simulations lead to estimates of galactic halo profiles often based on the formula:

$$\rho_W(r) = \rho_0 \left(\frac{r_0}{r}\right)^{\gamma} \left(\frac{r_0^{\alpha}+a^{\alpha}}{r^{\alpha}+a^{\alpha}}\right)^{\varepsilon}, \tag{7.1}$$

where $\rho_0 = \rho(r_0)$ is a convenient normalization (a frequent choice is $r_0 = R_{Sun}$ in our Galaxy), while a fixes the scale radius below which the profile goes as $r^{-\gamma}$.

Normally, two profiles are considered: one [20] profile characterized by the parameter set $[\gamma = 1, \alpha = 1, \varepsilon = 2]$, and the other profile [21] by $[\gamma = 3/2, \alpha = 1, \varepsilon = 3/2]$. These profiles differ mainly in the central region. Nevertheless, the resolution of these early simulations was not good enough to scrutinize the very center of any CDM halo and the logarithmic coefficient γ was extrapolated down to small radii. More recent work shows that the profile seems to become shallower toward the center of galaxies, and this is supported by a number of observations, especially of LSB galaxies and DSph. In particular, the presence of a supermassive central black hole (SBH) at the center of a galaxy could also steepen the central dark matter profile depending on the cross interaction between stars, DM, and the SBH itself. Without forgetting the many uncertainties that affect the shape profile and the mass distribution in the halo (in addition, a substantial component could be of baryonic origin in the form of MACHOs), a reasonable starting point for discussing direct detection experiments are the following standard assumptions for the local halo density [22], the WIMP density and the WIMP velocity distribution:

- $\rho_0 = 0.3$ GeV cm^{-3}, where ρ_0 is the local (Sun position) halo density.
- $\rho_{WIMP} = \xi \rho_0$, with $\xi < 1$ being the WIMP fraction of the local halo density;
- The WIMP velocity distribution is unknown, but the standard assumption is that it is Maxwellian:

$$dn \propto (\pi v_0^2)^{-\frac{3}{2}} \exp\left[-\left(\frac{v}{v_0}\right)^2\right] d^3\mathbf{v}. \tag{7.2}$$

  To be more exact, v^2 should be replaced by $|\mathbf{v} + \mathbf{v}_E|^2$, where $\mathbf{v}_E$ is the Earth velocity with respect to the dark matter distribution. In addition, the Maxwellian should be truncated at $|\mathbf{v} + \mathbf{v}_E| = v_{esc}$, v_{esc} being the galactic escape velocity. Usual assumptions for the Maxwellian parameters are $v_0 = 230$ km/s and $v_{esc} = 600$ km/s.

An important point for dark matter direct detection concerns the motion of the Earth inside the dark matter distribution. This motion is the composition of the Sun motion in the Galaxy and of the orbital terrestrial motion. The velocity of the Sun in the halo affects the WIMP flux as seen by a terrestrial detector (one speaks about a "WIMP wind"); in addition, the terrestrial orbital velocity adds to the Sun velocity in summer and subtracts from it in winter. (Here and in the following, summer and winter refer to the northern terrestrial hemisphere.) This determines an expected seasonal modulation [23] (typically up to 7%, but with an important dependence on the halo models [24]) in the WIMP interaction rate in terrestrial detectors, with a maximum on 2nd June. As discussed later, this modulation may be a signature for dark matter identification.

The rotational motion of the Earth can also be responsible for a diurnal modulation in the average impact direction of the WIMPs. This effect, more difficult to detect but also much more pronounced (the modulation would be of the order of some 10%), can be a precious tool for dark matter detection as well.

7.2 Direct Detection of WIMPs via Scattering off Ordinary Matter

The hypothetical particles composing dark matter and described in Sects. 7.1.2.1 and 7.1.2.2 interact with ordinary matter by scattering off atomic nuclei as a dominant mechanism [25]. Postponing more detailed calculations, it is instructive to anticipate here order-of-magnitude estimates. For WIMPs with masses of approximately 100 GeV, the local density is about 3000 WIMP per cubic meter and a flux of 6×10^4 WIMPs is traversing each cm^2 of our body every second. Another important aspect is that the average kinetic energy of these WIMPs is 20 keV. This energy is much larger than the $\sim$ eV scale binding energy of nuclei in a solid. In direct searches, the collisions are detected by the measurement of the energy of the recoiling nucleus, as its kinetic energy is deposited in the detector medium.

7.2.1 Rate and Features of the WIMP-Nucleus Interactions

In order to estimate the rate of collisions between WIMP and nucleons, one needs to define which elementary force mediates these interactions. Gravitational forces between a single WIMP and a single nucleus are negligible. Electromagnetic interactions are excluded, since it would mean that WIMP could emit or absorb light. Indeed, the "WIMP miracle" quoted in Sect. 7.1.2 takes place only assuming that the behaviour of the dark matter particles throughout the big bang up to now is ruled by their participation to the weak interactions. If this is the case, this leads to an estimate of the probability of a collision with a nucleus.

7.2.1.1 WIMP-Nucleus and WIMP-Nucleon Interaction

In particle and nuclear physics, the probability of an interaction is usually expressed as deriving from a cross-section, with units of surface. If dN/dt is the number of WIMP-nucleus interactions per unit time, Φ is the WIMP flux and N_t is the number of target nuclei per volume, we have:

$$\frac{dN}{dt} = \Phi \sigma_A N_t, \tag{7.3}$$

where σ_A is the cross-section for a WIMP-nucleus collision. A typical cross-section for a collision on a $A \sim 100$ nucleus involving the nuclear force only is of the order of the size of this nucleus: 10^{-24} cm^{-2} = 1 b. If the nuclear weak force is involved, the cross-section is at most 1 pb. Typical weak cross-sections on single nucleon (a proton or a neutron) are even lower than this ($\sim 10^{-7}$ pb). With such cross-sections, the interaction rate with the WIMP flux can be expected to be at most one collision per kilogram of matter per day, possibly as low as one per year and per ton of detector.

The reason why the weak cross-section on a $A = 100$ nucleus is not simply 100 times that on a single nucleon is that the wavelet associated to the momentum transfer corresponding to a $A = 100$ nucleus with 20 keV kinetic energy is approximately 3 fm, about the size of the entire nucleus. In this case, one must evaluate whether the interaction goes through a spin-dependent or scalar (spin-independent) process. In the first case, only the unpaired nucleon will contribute significantly to the inter-action, as the spins of the A nucleons in a nucleus are systematically anti-aligned. In the second case, all nucleon contributions add coherently: the total amplitude scales as A and the total scattering probability as A^2. Another mass-dependence hidden in the scaling from σ_n to σ_A is that interaction probability depends on the density of states in the final state, which in this case implies that $\sigma_A/\sigma_n = \mu_A^2/\mu_n^2$, where μ_A (μ_n) is the invariant mass of the WIMP-nucleus (WIMP-nucleon) system. In summary, the A-dependence of WIMP-nucleus cross-section is:

$$\sigma_A = \frac{\mu_A^2}{\mu_n^2}\sigma_n A^2 \tag{7.4}$$

for the spin-independent case, and

$$\sigma_A = \frac{\mu_A^2}{\mu_n^2}\sigma_n CJ(J+1) \tag{7.5}$$

for the spin-dependent case, where C is a factor that depends on the details of the structure of the nucleus. It cannot be expressed in a simple form, but it is generally less than unity.

As $\mu_A^2/\mu_n^2 \sim A^2$, the interaction rate per kilogram of target mass is proportional to A^3 in the case of spin-independent interactions and only to A in the case of spin-dependent interactions. Direct searches try to benefit from this scaling by using targets with as large A as possible. In any model where some part of the interaction involves spin-independent interactions, this term dominates the cross-section.

It is sometimes stated that the advantageous A^2 scaling of the spin-independent cross-section arises from the fact that the wavelength associated with the momentum transfer is comparable to the size of the nucleus. To be more precise, full coherence is only achieved when the associated wavelength is much larger than the nucleus size. In this case, one has to take into account interference effects that can be calculated rather precisely using the known form factors. Behind these calculations, there are detailed nuclear structure models. Here, it suffices to say that the net effect in most commonly used target material is to reduce the interaction rate by a factor of 2 to 4, which damps the increase due to the A^2 dependence when $A \sim 100$.

7.2.1.2 Effects of WIMP Interactions with Terrestrial Detectors

The interaction of the WIMPs with ordinary matter determines a nuclear recoil rate in a terrestrial detector. In case of elastic scattering, isotropic in the center of mass, the differential energy spectrum of the nuclear recoil dR/dE_R can be easily evaluated [22]:

$$\frac{dR}{dE_R} = \frac{R_0}{E_0 r} \exp\left[-\left(\frac{E_R}{E_0 r}\right)\right], \tag{7.6}$$

where E_R is the recoil energy, R_0 is the total rate, r is a kinematic factor given by

$$r = \frac{4M_{WIMP}M_N}{(M_{WIMP}+M_N)^2}, \tag{7.7}$$

M_{WIMP} being the WIMP mass, M_N the target nucleus mass, and E_0 a characteristic WIMP velocity expressed by

$$E_0 = \frac{1}{2}M_{WIMP}v_0^2. \tag{7.8}$$

When the finite velocity of the Earth in the Galaxy is accounted for, Eq. (7.6) holds no longer and must be replaced by a more complicate expression, which preserves anyway an almost exponential shape. Therefore, the expected energy spectrum is featureless and dangerously similar to any sort of radioactive background, which can often be well represented by an exponential tail at low energies. The typical energies over which the spectrum extends can be estimated from the expected M_{WIMP} and from the nuclear target mass. It is easy to check with Eq. (7.6) that most of the counts are expected below 20 keV in typical situations, for example, with $M_{WIMP} \sim 50$ GeV and $A = \sim 100$. This means that the spectrum must be searched for in a region very close to the physical threshold of most of the conventional nuclear detectors. In the simplified assumptions that $v_E = 0$ and $v_{esc} = \infty$, the total recoil rate is given by:

$$R_0 = \left(\frac{2}{\pi^{\frac{1}{2}}}\right)\left(\frac{N_{av}1000}{A}\right)\left(\frac{\rho_{WIMP}v_0}{M_{WIMP}}\right)\sigma_0 \tag{7.9}$$

where, after a numerical factor, we can identify the number of targets in 1 kg (second factor), the neutralino flow (third factor), and the cross-section for each target (last factor). Equation (7.9) predicts rates so low to represent a formidable challenge for experimentalists. Since WIMPs relevant for the solution of the dark matter problem are expected to have nucleon cross-section lower than 10^{-41} cm^2, total rates lower than 1 events/ (day kilogram) and than 10^{-3} events / (day kilogram) are predicted for SI and SD couplings, respectively. Figure 7.1 shows the energy recoil spectra for three different nuclear targets in case of SI coupling.

7.2.1.3 The Seasonal Modulation

As pointed out in Sect. 7.1.2.3, the seasonal modulation of the WIMP interaction due to the Earth revolution is a possible clue to disentangle WIMP-induced events from the background [26]. A short simplified discussion of this effect is given here

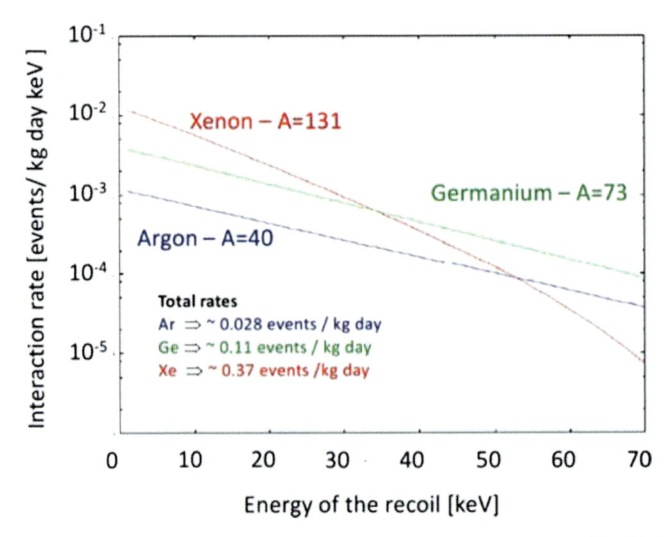

Fig. 7.1 Expected energy recoil spectra induced on different targets by a WIMP with M=100 GeV and $\sigma_n = 2 \times 10^{-7}$ pb (spin-independent interaction), corresponding approximately to the present sensitivities. Total rates are indicated as well. Nuclear form factors are taken into account.

[27]. Details can be found in other reviews. This point is of paramount importance since it is at the center of a controversial experimental observation.

In presence of halo WIMP interactions, a component of the background must present a seasonal modulation with very specific features, hard to mimic with fake effects:

- Modulation present only in a definite energy region.
- Modulation ruled by a cosine function.
- Proper period: T = 1 y.
- Proper phase: 152.5^{th} day in the year (2nd June).
- Proper modulation amplitude: < 7 % in the maximum sensitivity region.

The features of the expected seasonal modulation are illustrated in Fig. 7.2.

In order to have a signal at 1 σ level, we require:

$$S_{sum} + B_{sum} - (S_{win} + B_{win}) > (S_{sum} + B_{sum} + S_{win} + B_{win})^{\frac{1}{2}}, \tag{7.10}$$

where S_{sum} and $_{sum}$ are the signal and background counts in summer, while S_{win} and B_{win} represent the corresponding observables in winter. Equation (7.10) assures that the difference between the summer and winter number of counts is statistically significant. If one assumes that:

$$\begin{aligned} B_{sum} &= B_{win} \\ S_{sum} - S_{win} &= a(dR/dE)M_{det}T\Delta E \\ S_{sum} + S_{win} &= 2(dR/dE)M_{det}T\Delta E \\ B_{sum} + B_{win} &= 2(dR/dE)M_{det}T\Delta E, \end{aligned} \tag{7.11}$$

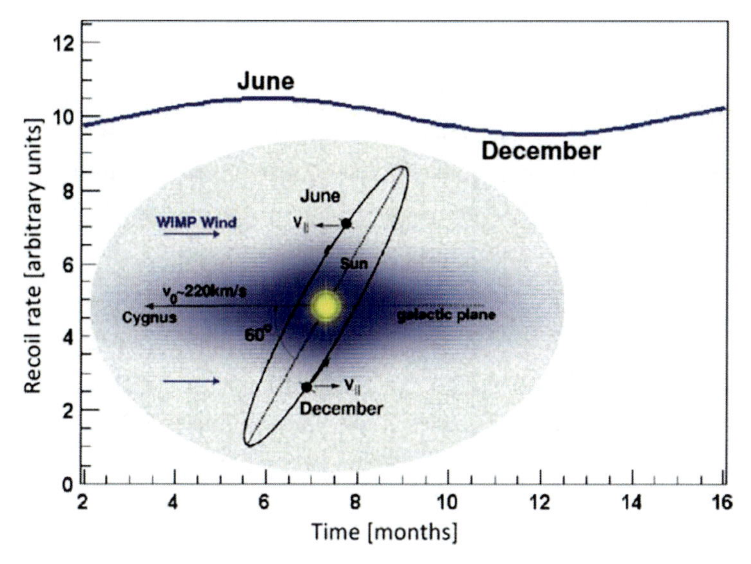

Fig. 7.2 Typical expected behaviour of the recoil rate as a function of time in a 1 year of data taking. The mechanism inducing the seasonal modulation is illustrated in the inset, showing the inclined Earth's orbital plane and the Sun motion in the galactic plane.

where a is the relative modulation amplitude, B is a background coefficient expressed in events/ (day kilogram keV), dR/dE is an average signal rate per unit mass and energy, also expressed in events/ (day kilogram keV), M_{det} is the detector mass, T is the experiment duration, and ΔE is the energy range relevant for the signal expressed in keV. Inserting these observables in (7.10), one gets as a condition on a:

$$a > \left[\frac{2}{(dR/dE)M_{det}T\Delta E}\right]^{\frac{1}{2}} \left[1 + \frac{B}{(dR/dE)M_{det}}\right]^{\frac{1}{2}} \frac{1}{(M_{det}T)}^{\frac{1}{2}}. \qquad (7.12)$$

The second term in the disequality (7.12) represents the lower limit for the modulation amplitude. Therefore, the sensitivity of the experiment scales as $(M_{det}T)^{1/2}$ since the signal, growing as $(M_{det}T)$, is in competition with background fluctuations growing as $(M_{det}T)^{1/2}$.

Unlike experiments aiming at exclusion plot production, searches for a real signal imply large detectors and long exposition time. Of course, the same set up can produce an exclusion plot both from a background measurement and from the non observation of a modulation amplitude. Increasing the detector mass and the exposition time, the second method becomes more stringent than the first, since in the first case, the sensitivity is constant, while in the second case it grows with $(M_{det}T)^{1/2}$. If we take for example $A = 127$, energy threshold 20 keV, $B \sim 1.5$ events/(day kilogram keV), a modulation analysis requires a detector mass around 100 kg to get the same sensitivity as a simple background analysis, assuming $M_{WIMP} \sim 40$ GeV.

7.2.2 Status of the Experimental Search for WIMPs

A large range of techniques and target materials are currently being used to directly search for WIMPs. After a general presentation of the strategies aiming at the measurements of the WIMP interactions, a review will be given of the most recent and significant results.

7.2.2.1 Design of a WIMP Detector

From the above-mentioned discussion, it is immediate to identify the general features of an instrument able to detect the scattering of WIMPs with ordinary matter. The appropriate tool is a low-energy nuclear detector with the following characteristics:

- Very low-energy threshold for nuclear recoils (given the nearly exponential shape of the spectrum, a gain in threshold corresponds to a relevant increase in sensitivity). Thresholds of $\sim$10 keV are reachable with conventional devices, while with phonom-mediated detectors, thresholds down to 300 eV have already been demonstrated.
- Very low raw radioactive background at low energies. In general, it requires a hard work in terms of material selection and cleaning to reduce raw background below 1 events / (day kilogram keV). Backgrounds lower than 10^{-1} events / (day kilogram keV) have already been demonstrated. Furthermore, an underground site is necessary to host high sensitivity experiments, since cosmic rays produce a huge number of counts at low energies.
- Sensitivity to a recoil-specific observable. This allows to reject the ordinary γ and β background for which the energy deposition comes from a primary fast electron. When such an observable is available, the only relevant background source left consists in fast neutrons, which gives rise to slow nuclear recoils as the hypothetical WIMPs.
- Sensitivity to a WIMP-specific observable. It is necessary for an indisputable signature and may consist in the seasonal modulation of the rate, in the diurnal modulation of the nuclear recoil directions (see Sect. 7.1.2.3), and in the correct scaling with A of the target of the candidate WIMP-induced spectrum (see Sect. 7.2.1.1), either using a multi-target detector or comparing experiments with different detectors.

A simple measurement of a background level performed with a low-energy nuclear detector produces information on the WIMPs in the galactic halo. Usually, this information is expressed in the form of an exclusion plot in a (σ_n, M_{WIMP}) plane. The challenge is to test those regions in this plane, which are populated by points corresponding to WIMPs viable for dark matter composition, in the sense explained in Sects. 7.1.2.1 and 7.1.2.2. A simple background measurement cannot prove the existence of neutralinos or KK WIMPs; it can only exclude particles with given features. The parameters that affect the shape of the exclusion plot are the threshold, the background spectrum and the target mass. The exclusion plot is constructed by fixing first a WIMP mass: given the nuclear target mass, this allows to determine the recoil spectrum shape apart from a normalization factor, using the exact version of Eq. (7.6); the value of σ_n, which leads the recoil spectrum to touch the background spectrum at least in one energy bin constitutes the upper limit to σ_n for that WIMP mass. (Higher values of σ_n would produce a recoil spectrum with more counts in that energy bin than those experimentally observed.) The repetition of this procedure over the whole mass range provides the exclusion plot.

The effect on the exclusion plot of the relevant detector parameters can be so summarized:

- reducing the background improves the exclusion plot for any WIMP mass;
- reducing the nuclear target mass, the exclusion plot improves at low WIMP masses, but worsens at high WHVIP masses;
- reducing the threshold improves the exclusion plot mainly at low WIMP masses.

It is not popular nowadays to operate detectors with low target masses (say $A < 50$) since, in this case the region with higher sensitivity is already excluded by accelerator constraints. It is important to point out that the exclusion plot does not improve with longer exposition times, once the background level has settled down to its intrinsic value, or with higher detector masses. Relevant results can, therefore, be achieved even with small detectors and short measurements, provided that the background level is low.

7.2.2.2 Double Read-Out WIMP Detectors

As pointed out in Sect. 7.2.2.1, the first ingredient of an effective WIMP detector is a very low raw background close to the energy threshold, which needs to be around a few keV. The hypothetical WIMP-induced events are in strong competition with the background events determined by particles that interact electromagnetically, such as charged particles and γs originated by residual radioactive contamination or cosmic radiation. In order to extract a nuclear recoil spectrum from this dominant electromagnetic background, it is essential to set up a method enabling to distinguish nuclear recoils from the electron recoils induced by the electromagnetic background. If this operation is successful, the only background source left arises from fast neutrons, which determine slow nuclear recoils as well. The neutron background can be kept under control by combining proper shielding with detector granularity or space resolution. In fact, since the mean free path of fast neutron is normally in the cm / tens of cm range, there is a high chance that a fast neutron, unlike WIMPs, undergoes multiple scattering in a large enough detector.

The event-by-event identification of nuclear recoils can be realized by means of the so-called "double read-out" detectors and can be performed with three different approaches. Without entering the details of the physics of the interaction of radiation with matter, it will be enough to recall here that a fast elementary particle depositing energy in a material produces in all generality three types of elementary excitations in the target medium: electron-hole or electron-ion pairs, phonons, and scintillation photons. The last case requires special properties of the target (an effective scintillation mechanism and acceptable transparency to the emitted light). In order to simplify the terminology, we will refer to the three excitation classes as charge (or ionization), phonons (or heat) and scintillation (or light) in the following.

Normally, a nuclear detector measures the energy deposited by a particle through a signal proportional to the amount of elementary excitations belonging to just one of the three classes. We have for example Ge or Si diodes, which are sensitive to ionization; bolometers, which are sensitive to heat; scintillators, which are sensitive to light. In these single-channel devices, the distinction between an event corresponding to an electron recoil and that corresponding to a nuclear recoil is quite difficult (even if not impossible in particular for scintillators, where different temporal structures of the signal may be used for this purpose). The scenario changes completely if an instrument is able to provide two distinct signals, proportional to two distinct classes of excitations. With these hybrid devices, it is possible to form the ratio between the two signal amplitudes and hence to identify nuclear recoils. The logic of the double-readout approach is represented by the scheme reported in Fig. 7.3.

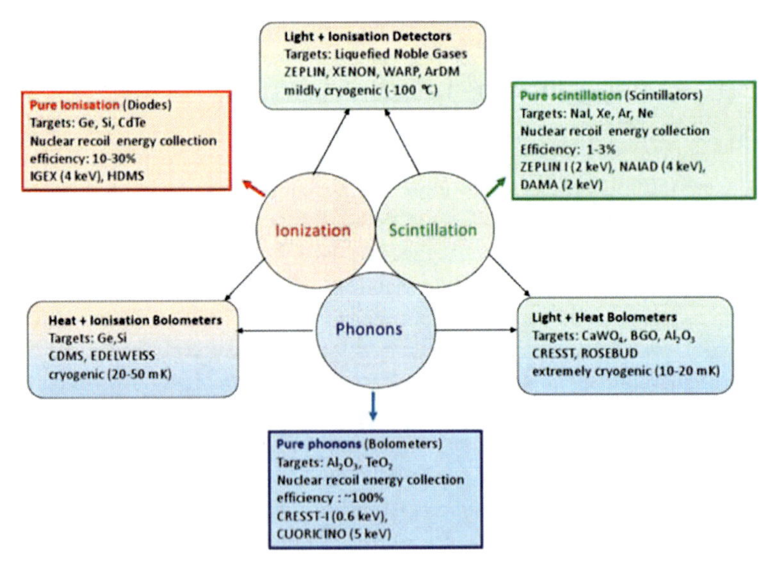

Fig. 7.3 The philosophy of the double read-out approach in a WIMP detector is depicted schematically. Representative experiments for each class are indicated, together with relevant experimental parameters (electron-equivalent energy threshold in keV, nuclear recoil energy collection efficiency, and operation temperatures of the cryogenic devices).

To be more quantitative, let us define S_c, S_h, and S_l are the signal amplitudes related to charge, heat, and ionization, respectively. We can form three ratios, depending on which type of hybrid detector is involved:

$$
\begin{aligned}
R_{c+h} &\equiv \frac{S_c}{S_h} \quad \text{for charge + heat hybrid detectors} \\
R_{l+h} &\equiv \frac{S_l}{S_h} \quad \text{for light + heat hybrid detectors} \\
R_{c+l} &\equiv \frac{S_c}{S_l} \quad \text{for charge + light hybrid detectors,}
\end{aligned}
\tag{7.13}
$$

These ratios are very useful and represent recoil-specific observables in the sense specified in Sect. 7.2.2.1. In fact, it happens that in all the three cases R is different for events, which determine nuclear recoils from those that determine electron recoils. In particular,

$$
\frac{R_{c+h,l+h,c+l}[\text{nuclear recoil}]}{R_{c+h,l+h,c+l}[\text{electron recoil}]} \ll 1. \tag{7.14}
$$

Usually, the ratios R are taken equal to 1 by construction for electron recoils, and therefore, a nuclear recoil can be identified when its corresponding R is less than 1 with statistical significance. The distribution of R for electron and nuclear recoils tend to merge at low energies, when the signal-to-noise ratio is low. If the separation is good at energies of interest for WIMP-induced recoil, the method enables to reject efficiently electromagnetic background.

The reasons why the ratios defined in Eq. (7.13) identify nuclear recoils are readily explained. In general, slow nuclei are very inefficient in producing ionization with respect to fast electrons. A suppression of the light signal is expected too, but to a less extent. This accounts for the low values of R_{c+h} and R_{c+l} for nuclear recoils. On the contrary, slow nuclei produce a high phonon yield since the energy is directly delivered to the medium elastic field in this case. This explains the low value of R_{l+h} and once more that of R_{c+h}.

Now we will describe the general features of three classes of double read-out devices belonging to the three cases discussed above.

1. Charge + Heat. The detectors consist of arrays of large Ge or Si diodes operated as conventional semiconductor devices (providing the charge signal S_c) with an additional phonon sensor for the heat signal $S-h$. The latter element may be either a properly doped Ge crystal, sensitive to thermal phonons through a strong dependence of the resistivity on the temperature, or a set of thin superconducting films sensitive to the out-of-equilibrium phonons produced in first instance by the impinging particle. The detectors must be operated at very low temperatures, typically in the range 10–100 mK, in order to work as bolometers and to be sensitive to the phonon signal. It is useful if the total mass is large enough to make the research competitive in terms of seasonal modulation sensitivity (WIMP-specific observable). Therefore, the array should consist of tens of individual elements,

providing also the granularity necessary to reject/assess residual neutron background. The raw background and the energy threshold must be conveniently low. The double readout provides the recoil-specific observable R_{c+h}. Experiments using this approach are CDMS (and its future expansion SuperCDMS) and EDELWEISS.

2. Light + Heat. The detectors consist of an array of large scintillators with an additional phonon sensor (these devices are often defined "scintillating bolometer"). The phonon sensor provides the charge signal S_c. The same considerations as above repeat for the phonon sensor and for the total mass, threshold, background and operational temperatures. Light detector converts the scintillation photon in a signal S_l. A remarkable technical difficulty consists of the necessity to operate this light detector at very low temperatures, with a very low threshold, in the few-photon range. Currently, these light detectors have been implemented through auxiliary bolometers consisting of thin wafers of low specific heat materials and provided with proper phonon sensors. The light signal is, therefore, converted into a temperature signal. In this case, the double readout provides the recoil-specific observable R_{l+h}. Experiments based on this technology are CRESST, and ROSEBUD. EDELWEISS, CRESST, and ROSEBUD are joining their efforts in view of a unified large European cryogenic experiment named EURECA.
3. Charge + Light. The detectors use liquid noble elements as dark matter target. These materials are ideal to build large, homogeneous, and position-sensitive devices. Liquified noble gases are intrinsic good scintillators and have high ionization yields. If a high electric field ($\sim$ 1 kV/cm) is applied, ionization electrons can also be detected, either directly or through the secondary process of proportional scintillation. In the latter case, the ionization electrons are drifted out of the liquid in a double phase arrangement, and they are detected through a scintillation signal produced in the vapor. In this application, a prompt scintillation signal gives S_l and corresponds to the interaction of the impinging particle in the liquid, while a delayed scintillation signal provides S_c and is proportionally related to the amount of ionization electrons drifting in the vapor. The recoil-specific observable R_{c+l} allows to distinguish between electron and nuclear recoils. Several experiments use noble elements for WIMP search: CLEAN, ArDM, WARP, DEAP, XENON10 and XENON100, ZEPLIN-II and ZEPLIN-III, LUX, and XMASS.

7.2.2.3 Review of Experiments for Direct Dark Matter Detection

In this section, we will provide a short review of the experiments searching for the direct interaction of WIMPs and presenting the current more stringent limits. Then, we will discuss the prospects for the future. Given the large number of projects searching for dark matter, this review will be incomplete. We will focus on the most sensitive and relevant searches; however, aiming, at covering the full spectrum of the technical approaches.

When renouncing a recoil-specific observable, it is possible to realize detectors with a simpler structure with respect to those described in Sect. 7.2.2.2. In order to be competitive, this single-readout devices need to exhibit a very low raw background spectrum and, at least in prospect, a large total mass, to be sensitive to the WIMP-specific observable represented by annual modulation (see Sect. 7.2.1.3) and to compensate in this way the lack of a method for nuclear recoil identification. Scintillating crystals like sodium iodine (NaI) are a convenient solution to accumulate large masses on detector material. However it is difficult to achieve radiopurity comparable to Ge. NaI-based searches, such as DAMA-LIBRA [28], ELEGANT [29], or NAIAD [30], originally attempted to use pulse shape discrimination to statistically identify a WIMP component in their observed rate. It was found that the low number of detected scintillation photon per keV of incident energy restricts the usefulness of this method at low energies.

The limitation of pulse shape analysis at low energy enticed the **DAMA** collaboration to turn to a WIMP discrimination based on annual modulation. With a data set of 290 kg·y recorded with a 87.3 kg array of NaI scintillators over 8 years in the Laboratori Sotterranei del Gran Sasso (Italy), DAMA reported the observation of a modulation originally interpreted as a WIMP with a mass of 52 GeV and $\sigma_n = 7.2 \times 10^{-42}$ cm^2. Such a result corresponds to a total rate of approximately 1 nuclear recoil per kg·day above a threshold of 2 keV (over an energy scale calibrated with electrons). The effective threshold for a iodine nucleus recoil is 22 keV since, in this case, it holds for the so-called quenching factor Q:

$$Q = \frac{S_l[nuclear\ \ recoil]}{S_l[electron\ \ recoil]} = 0.09. \tag{7.15}$$

This result was confirmed by the subsequent expansion of the DAMA experiment, named **LIBRA**, comprising 232.8 kg of NaI detectors (see Fig. 7.4). With a much higher statistics corresponding to 530 kg·y, the modulation effect was confirmed at a 8.3 σ confidence level [31]. The period of the modulation ($T = 0.998 \pm 0.003$ y), its phase ($t_0 = 144 \pm 8$ days), and the involved energy region (2-6 keV with electron

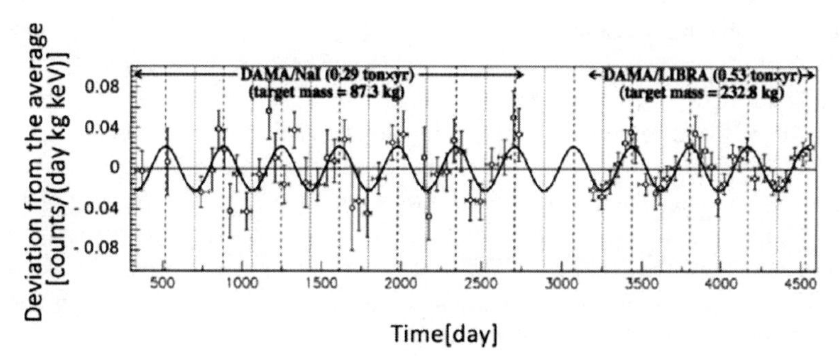

Fig. 7.4 The controversial annual modulation of the counting rate observed by the DAMA/LIBRA experiment in the Gran Sasso Laboratory, first with 83.7 kg (larger error bars), later with 232.8 kg detector mass (small error bars). (Figure adapted from e-Print arXiv:0804.2741v1.)

calibration) correspond exactly to what expected by the interaction with particles composing the galactic halo, as pointed out in Sect. 7.2.1.3. The rate can be fitted by the relationship:

$$R(E,t) = S_0(E) + S_m(E)\cos[\omega(t - t_0)] \tag{7.16}$$

with $S_m = (0.0215 \pm 0.0026)$ counts/(day kg keV). In spite of these remarkable correspondences between the experimental results and the expected signature for WIMP detection, reconciliating the reported modulation effect with the published exclusion limits based on other direct and indirect searches requires very strong excursions from the usual supersymmetric neutralino scenario depicted in Sect. 7.1.2.1. The DAMA-LIBRA collaboration has put forward several possible explanations of this effect, pointing at more exotic models for dark matter particles, that could interact electromagnetically and result, therefore, invisible for searches that reject electromagnetic background through the double-readout approach. An independent cross-check with NaI will be provided by ANAIS [32] (Laboratorio Subterràneo de Canfranc, Spain) with more than 100 kg of NaI detectors. The DAMA-LIBRA result will be also scrutinized with a similar technology by the Korean group KIMS [33], which is developing low background CsI crystal detectors with low threshold for the direct WIMP search.

Among the searches based on a single-readout approach, it is worth mention experiments based on pure ionization Ge detectors with a very low threshold and an impressively low raw background (TEXONO [34] and CoGeNT [35]). In spite of their inability to identify nuclear recoils, these searches reach excellent sensitivities at low WIMP masses, both for the spin independent and spin dependent channels. A recent result achieved by CoGeNT [36] is particularly relevant. An irreducible exponential background at low energy, close to the record threshold of 400 eV (electron-calibrated energy scale), is compatible with the signal expected by a light WIMP (7-11 GeV) with spin independent coupling. It has been proposed [37] that a population of relic neutralinos can fit these results, the DAMA/LIBRA data on the annual modulation effect, and a tiny effect observed by the CDMS collaboration, discussed later. This coincidence needs to be further investigated in searches sensitive to low mass region.

We shall review now experiments based on bolometers. As already pointed out in Sect. 7.2.2.2, cryogenic calorimeters [38] are meeting crucial characteristics of a successful WIMP detector: low-energy threshold (< 5 keV), excellent energy resolution ($< 1\%$ at 10 keV), and the ability to differentiate nuclear from electron recoils on an event-by-event basis. Their development was driven by the exciting possibility of doing a calorimetric energy measurement down to very low energies with unsurpassed energy resolution. Because of the T^3 dependence of the heat capacity of a dielectric crystal, at low temperatures, a small energy deposition can significantly change the temperature of the absorber. The signal is recorded either after the phonons reach equilibrium, or thermalize, or when they are still out of equilibrium, or athermal, the latter providing additional information about the location of an event.

CDMS [39] and **SuperCDMS** [40] belong to class 1 "charge + heat" of the three introduced in Sect. 7.2.2.2. The Cold Dark Matter Search experiment operates 30 low-temperature Ge and Si detectors at the Soudan Underground Laboratory in Minnesota (USA). The dark matter targets are high-purity Ge and Si crystals with a flat cylindrical shape, 1 cm thick, and 7.6 cm in diameter. The single crystal mass is 250 g and 100 g, respectively. Charge electrodes are used for the ionization measurement. They are divided into an inner disk, covering 85% of the cylinder base, and an outer ring, which is used to reject events near the edges of the crystal, where background interactions are more likely to occur. Superconducting transition edge sensors photolitographically patterned onto one of the crystal base detect the athermal phonons from particle interactions, providing $\sim$1-mm space resolution in the x-y plane. If an event occurs close to the detector's base, the phonon signal is faster than for events far from the surface, because of phonon interactions in the thin metallic films. The risetime of the phonon pulses and the time difference between the charge and phonon signals, allow to reject surface events caused by electron recoils. This spacial resolution in the z coordinate is crucial, since close-to-surface electron recoils may give rise to incomplete charge collection and mimic nuclear recoils (see Fig. 7.5). Presently, together with XENON, CDMS has set the most stringent limit on WIMPs, publishing a spin-independent WIMP-nucleon cross-section of $(3.8 \times 10^{-44}$ cm^2 at 70 GeV mass. This limit does not come from a zero-background

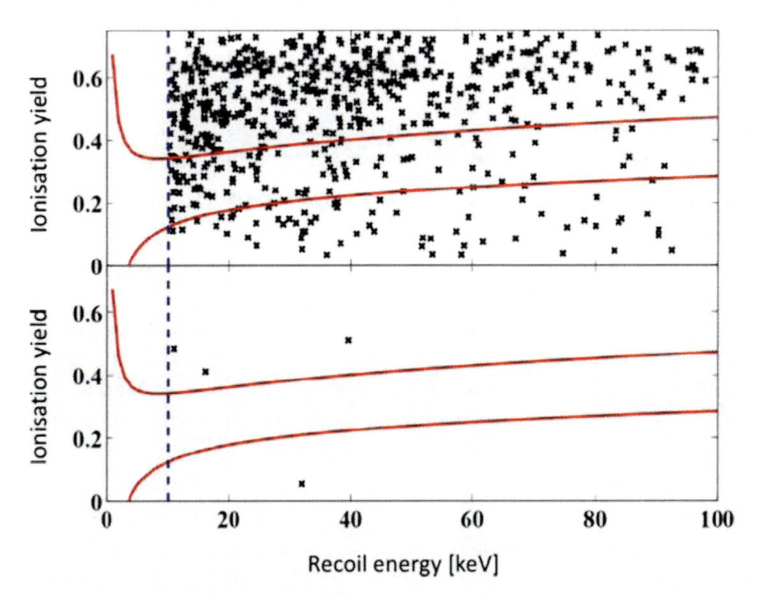

Fig. 7.5 Rejection power for electron recoils of the CDMS detector. The top graph shows the ionization yield (a quantity proportional to R_{c+h} defined in the text) as a function of the recoil energy without cuts on the pulse risetime. The red lines define the low R_{c+h} region corresponding to nuclear recoils. The bottom graph shows the same parameter space after application of the time cuts described in the text, which reject surface events mimicking nuclear recoils. No counts in the signal band is registered above 10 keV threshold (dashed blue line) in an exposure of 121.3 kg·day. (Figure adapted from e-Print arXiv:0802.3530v2.)

exposure. Two events in the WIMP acceptance region at recoil energies of 12.3 keV and 15.5 keV were observed. The probability to have observed two or more surface events in this exposure is 20.

The **EDELWEISS** experiment operates Ge bolometers at ~20 mK in the Laboratoire Souterrain de Modane (France) [41]. This search uses the "charge + heat" approach as well. The bolometers simultaneously detect the phonon and the ionization signals, allowing a discrimination against bulk electron recoils of better than 99.9% above 15 keV recoil energy. The charge signal is measured by Al electrodes sputtered on each side of the crystals, while the phonon signal is recorded by a Ge neutron transmutation-doped (NTD) heat sensor glued onto one of the charge collection electrodes. The NTD sensors read out the thermal phonon signal on a time scale of about 100 ms, and cannot provide space information as in the CDMS case. Therefore, EDELWEISS has the problem to reject close-to-surface events. In the past, much effort was dedicated to a design based on NbSi thin-film sensors. These films, being sensitive to the athermal phonon component of the signal, show a strong difference in the pulse shape, depending on the interaction depth of an event. Now, a very promising new design of the charge collection electrodes, based on an interdigit pattern [42], has allowed to reject with high-efficiency surface events using the ionization channel. Thanks to these additional tools, the present sensitivity of EDELWEISS is expected to be very high and competitive with XENON and CDMS. This expectation has been confirmed by the initial results of EDELWEISS-II [43], which has brought this experiment into the leading group with a limit of $\sim 10^{-43}$ cm^2 at !0 GeV mass. The EDELWEISS-II set-up uses a specially developed 50 liter low-radioactivity dilution refrigerator, able to house up to 120 detectors. The collaboration seems to have in its hands the technology and the means to achieve a sensitivity below 10^{-44} cm^2 in the next years.

The **CRESST** collaboration has developed cryogenic detectors based on $CaWO_4$ crystals [44], which show a higher light yield at low temperatures compared with other scintillating materials. The detectors are also equipped with a separate, cryogenic light detector, presently consisting of a sapphire wafer of 40 mm diameter and 0.4 mm thickness, with an epitaxially grown silicon layer on one side for photon absorption. The light detector is mounted close to a flat surface of the scintillating crystal. The temperature rise in both $CaWO_4$ and light detector is measured with tungsten superconducting phase transition thermometers, kept around 10 mK, across their transition between the superconducting and normal conducting state. A nuclear recoil in the 300 g $CaWO_4$ detector has a different scintillation light yield than an electron recoil of the same energy, allowing to discriminate between the two type of events according to the principles exposed in Sect. 7.2.2.2 for the second "light + heat" detector class. The advantage of the CRESST approach is the low-energy threshold in the phonon signal, and the fact that no light yield degradation for surface events has been detected so far. The limitation is in that only a few tens of photons are emitted per keV electron recoil, a number that is further diminished for nuclear recoils, because of the involved quenching factors. The experiment, located in the Laboratori Nazionali del Gran Sasso (Italy), was recently upgraded with a 66-channels SQUID readout system, with a neutron moderator made out of polyethylene, and with a scintillator muon veto. The whole structure, cooled down to 10 mK, can accommodate 33 detector modules, for a total target mass of about 10 kg. Data obtained with two detector modules for a total exposure of 48 kg·days show only three events in the tungsten recoils acceptance region, corresponding to a rate of 0.063 per kg·day. Standard assumptions on the dark matter flux, coherent or spin-independent interactions, yield a limit for WIMP-nucleon scattering of 4.8×10^{-43} cm^2, at $M_{WIMP} \sim 50$ GeV. The few events observed in the nuclear recoil region are not compatible with the neutron background estimated with Monte Carlo simulations. There is at the moment no conclusive explanation for these few candidate events from conventional radioactive or particle sources.

The expertize developed by EDELWEISS, CRESST, and ROSEBUD [45] (another class 2 experiment at the Laboratorio Subterràneo de Canfranc) represents the basis for the proposed design study **EURECA** (European Underground Rare Event search with Calorimeter Array), aiming at the development of a 100 kg-1 ton cryogenic, multi-target experiment located in the Laboratoire Souterrain de Modane (France) [46].

We shall move now to experiments belonging to class 3 "charge + light", discussed in Sect. 7.2.2.2. The **XENON** collaboration has operated a 15 kg (active mass) dual phase detector time projection chamber in the Laboratori Nazionali del Gran Sasso, named XENON10 [47]. It uses two arrays of UV-sensitive photomultipliers (PMTs) to detect the prompt and proportional light signals induced by particles interacting in the sensitive xenon volume. The bottom array of 41 PMTs is located below the cathode, fully immersed in LXe, and mainly detects the prompt light signal. The 48 PMTs of the top array are located in the cold gas above the liquid, detecting the proportional light. XENON10 has full 3D position sensitivity: the time separation between the two pulses of direct and proportional light (with a of maximum 75 μs) provides the event depth of an interaction (< 1 mm resolution), the hit pattern in the top PMT array providing the x-y position (few mm resolution). In addition to the recoil-specific observable R_{c+l}, the position sensitivity, along with the self-shielding of liquid xenon, serves as an important background rejection feature. Based on calibration and on a period of nonblind WIMP search data, the WIMP search region was defined between 4.5 and 29.6 keV nuclear recoil energy, 3σ below the mean of the nuclear-recoil band (thus, at 50% nuclear recoil acceptance). The rejection power can be appreciated in Fig. 7.6. From a total of 1800 events in the 58.6 live days of blind WIMP search data, 10 events were observed in the WIMP search region, with $7.0^{+1.4}_{-1.0}$ events expected based on statistical (Gaussian) leakage alone. Given the uncertainty in the number of estimated leakage events from electron recoils (no neutron = induced recoil events were expected for above exposure), conservative limits with no background subtraction were calculated for spin-independent WIMP cross-sections. The 90% C.L. upper limit for $M_{WIMP} \sim$ 100 GeV is 8.8×10^{-44} cm^2, the most stringent ever obtained together with CDMS. The XENON collaboration has now installed in Gran Sasso XENON100 [48] with an active target of 62 kg of ultrapure liquid xenon. The global detector configuration is substantially a scaling up of the XENON10 structure. However, the raw background is about two orders of magnitude better, because of the selection

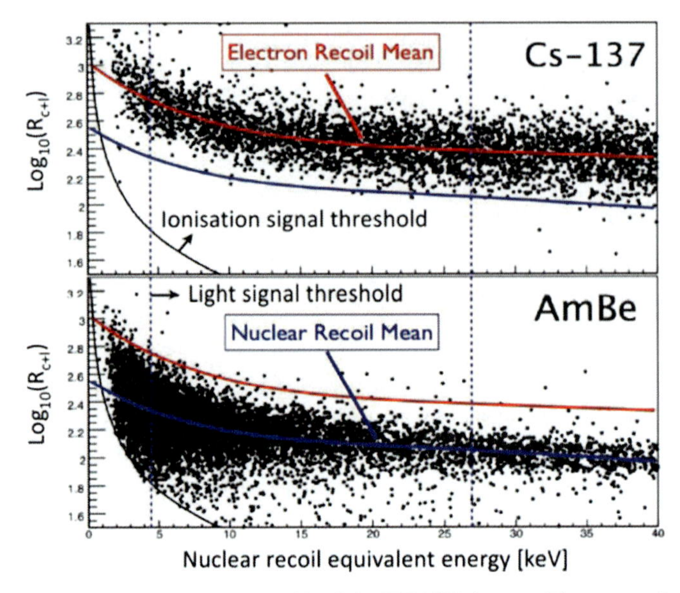

Fig. 7.6 Rejection power for electron recoils of the XENON detector. The top graph shows R_{c+l} in log scale as a function of the nuclear recoil equivalent energy, with events generated by a γ calibration source. The bottom graph shows the same parameter space with nuclear recoil events generated by a neutron source. The separation between the two event classes is apparent, even if not so complete as in the phonon + ionization detectors. A high rejection power ($>$ 99.5%) is obtained only by defining an acceptance box at low R_{c+l} with 50% efficiency for nuclear recoils.(Figure adapted from e-Print arXiv:0706.0039v2.)

of low-activity materials and the location of the cryocooler (a possible source of radioactive background) outside the shields and the veto. The dark matter search has started at the end of 2009, and the first preliminary results are very promising, with a limit of 3.4×10^{-44} cm^2 for 55 GeV WIMP mass. Its expected sensitivity extends down to cross-sections of the order of 2×10^{-45} cm^2 for standard WIMPs inducing nuclear recoils. In addition, the preliminary runs exhibit a very low raw background, of the order of 0.02 counts/(kev kg day) below 10 keV with 50 kg fiducial volume. This performance could enable XENON100 to switch off the recoil identification mode and to look at the raw spectrum in order to check if a seasonal modulation appears, as in the DAMA experiment, when searching for particles with electromagnetic interactions. On the contrary, the capability of XENON100 to constrain low WIMP masses, as those to which CoGeNT is sensitive, is controversial, due to a confuse experimental situation about the light yield of slow nuclear recoils in liquid xenon. For recent measurements of this parameter, see [49] and references therein.

Other searches, here only mentioned, exploit the charge + light mode to reject nuclear recoils. Besides XENON, LUX [50] and ZEPLIN-III use xenon as sensitive material as well. The recent results of ZEPLIN-III are competitive with the best ever obtained with double-readout experiments [51]. WARP [52] and ArDM [53] are based on a double-phase argon target. With respect to xenon, argon has an additional tool able to perform nuclear vs. electron recoil discrimination, that is pulse shape analysis in the primary scintillation signal. However, the lower A makes cross-section for spin-independent interaction significantly lower. Other experiments use liquid target without recoil discrimination, aiming at a very low raw background and a large mass, like the Japanese XMASS [54] (xenon as target), CLEAN [55] (neon as target) and DEAP [56] (argon as target).

It is worth mention also a more exotic approach, which might lead to an unexpected breakthrough in the direct search for dark matter. The idea is to build large WIMP detectors using superheated liquids. An energy deposition can destroy the metastable state, leading to the formation of bubbles, which can be detected and recorded both acoustically and optically. Since a minimal energy deposition is required to induce a phase-transition, these detectors are so-called threshold devices. The operating temperatures and pressure can be adjusted such that only nuclear recoils (large stopping powers dE/dx) lead to the formation of bubbles. COUPP [57], PICASSO [58], and SIMPLE [59] belong to this class of experiments.

Finally, directional experiments deserve a special mention. As pointed out in Sect. 7.1.2.3, a diurnal modulation is expected, since all the nuclear recoils induced by WIMP events should globally point at the Cygnus constellation direction due to the Earth motion in the Galaxy. A strong forward/backward asymmetry is then expected, once taking into account the Earth rotation. This is a powerful signature, which cannot be mimicked by any conceivable background source. The challenge is to measure 3D tracks of low-energy nuclear recoils with directional detectors based on low-pressure gas chambers. Low pressures (of the order of few tens of torr) are required for the slow nuclear recoil to have an appreciable range. This makes quite difficult to increase the sensitive mass. However, order of 10 events would be sufficient to provide a sgnificant signal. The projects belonging to this class are DRIFT-II [60], DM-TPC [61], NEWAGE [62], and MIMAC [63].

Summarizing, the experiments with the current higher sensitivities to WIMPs are CDMS and XENON. They represent the two most promising appoaches to this search in terms of discovery potential (charge + heat and charge + light). The future will tell us which method is more effective to reach sensitivities in the 10^{-46} cm^2 range, necessary to explore most of the parameter space for SUSY and KK WIMPs. The competition between these two collaborations and technologies can be appreciated in Fig. 7.7, showing the SI cross-section experimental bounds. The limits shown there will be improved soon, since both collaborations are performing runs with higher statistics and improved apparatus. In addition, other searches have the potential to get similar or better performance, e.g., EDELWEISS-II [43], whose new results are expected at the end of 2009 or beginning of 2010.

7.3 Indirect Detection via Annihilation of Dark Matter Particles

The self-annihilation rate of WIMPs is proportional to the square of the particle density. Therefore, the rates and experimental sensitivities for all indirect searches depend heavily on the dark matter distribution. The most obvious annihilation

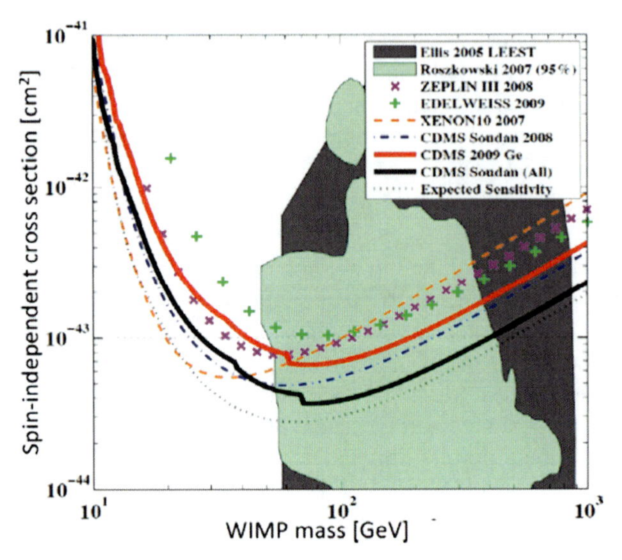

Fig. 7.7 Spin-independent WIMP-nucleon cross-section σ_n upper limits (90% C.L.) as a function of the WIMP mass, with CDMS results compared with the present most sensitive searches. The CDMS data are reported in three curves: the lower solid curve (black) is the combined analysis of all the data; the upper solid curve (red) refers to the last set of data, released in 2009; the upper curve (dash dot) is the result of a re-analysis of CDMS data taken in 2008. The calculated sensitivity of the CDMS experiment is reported as well. Parameter ranges expected from supersymmetric models described in two different theoretical papers are also reported.(Figure adapted from e-Print arXiv:0912.3592v1.)

source is the center of the Galaxy, where the dark matter density is expected to rise substantially, even though the shape of the density profile is controversial (see Sect. 7.1.2.3). The hypothesized accumulation of dark matter around the central black hole of the Galaxy is rather contested. A model predicts a dark matter spike with a distribution $\rho(r) \propto r^{-A}$, with $2.25 < A < 2.5$ [64], which could provide large fluxes in particular in neutrino searches. There has been an extensive discussion about the existence of this dark matter spike and the situation remains unclear, at the point that a null search for annihilation radiation from the galactic center could not be interpreted as evidence against WIMP dark matter, due to the uncertainty on this matter distribution.

Other sources are considered besides the galactic centers. Enhancements of the dark matter annihilation rate are often discussed, arising from the existence of dark matter substructures within halos of galaxies, named *clumps*. In terms of the observable cosmic-ray flux, the presence of such over-densities are often translated into a so-called *boost factor* applied to the whole primary spectrum [65]. However, the situation is rather unclear and controversial concerning the possible values of this boost factor.

The accumulation of dark matter in massive objects is less controversial and is proposed for the Sun and the Earth for any WIMP-like particle. After the encounter of a WIMP with a massive celestial object, it may happen that the WIMP velocity becomes less than the escape velocity and the particle gets gravitationally trapped. The particles then migrate and accumulate at the core of the massive object. At the core, the density of WIMPs is such that annihilations take place and the balance between accumulation and annihilation can lead to equilibrium in very massive bodies such as the Sun. In the case of less massive bodies, such as the Earth, this equilibrium is not yet attained and the lower density continues to increase.

7.3.1 Introduction to Annihilation Mechanisms and Products

In order to be able to self-annihilate, any dark matter candidate must either be a Majorana particle or a Dirac particle with no matter-antimatter asymmetry. The weak interaction cross-section required by the cosmology constraints (the so-called "WIMP miracle") implies that a high local density of the particles is essential to have an experimentally observable rate, the annihilation rate being related quadratically to the density. In all the possible annihilation sites, the relative velocity of the WIMPs is low and usually annihilation rates are calculated in the null velocity limit. In this limit, the annihilation products at the leading order in perturbation theory are mostly pairs of standard model fermions/antifermions and neutral pair combinations of gauge or Higgs bosons of all types. Any annihilation product $\chi + \chi \to A + B$ is potentially created if kinematically allowed, i.e., $2m_\chi > m_A + m_B$. The particle species observed in the experiments are the results of decays and hadronization of the tree level final state particles. Further, mainly for the charged particle species, it is crucial to take into account propagation and interaction effects between the source and the Earth.

7.3.1.1 Annihilation Channel in SUSY

In supersymmetric WIMPs, the features of the annihilation mechanism are strictly related to the gaugino/higgsino content of the neutralino. Only the higgsino and wino parts of the neutralino allow couplings to the gauge bosons, so that only mixed, higgsino- or wino-like neutralinos can annihilate into massive gauge and Higgs bosons (Z^0Z^0, W^+W^-, $Z^0h(H)$, Z^0A, $W^{+/-}H^{-/+}$, H^+H^-, and pair combinations of A, h, H). We remind that H^+, H^-, A, h, and H are the five Higgs bosons expected in MSSM (see Sect. 7.1.2.1). These configurations occur in the general MSSM when the Higgs mass parameter $\mu \leq M_1, M_2$, as well as in the mSUGRA when the unified scalar mass, m_0, is in the multi-TeV range (the so-called *focus point* regions).

For annihilation in fermion/antifermion pairs and in the low velocity limit, annihilation at rest prevents neutralinos from having their spins parallel, due to Pauli blocking. This results at the end into the helicity suppression of low mass and massless fermion/antifermion pairs in the final state. For massive fermions, this favors the pair creation of down-type quarks, with an enhancement factor of $\tan\beta$, whereas there is a $(\tan\beta)^{-1}$ suppression factor for up-type quark pair production. Hence, the bottom dominates over the top quark final states and the τ is the only charged lepton created.

Since mSUGRA neutralinos are bino-like in most of the parameter space, they will mainly annihilate into $b\bar{b}$ pairs with pair production of τ also often relevant. When massive enough and when carrying a higgsino or wino component, neutralinos will annihilate into massive gauge bosons, for instance, in the general MSSM, in focus point regions of mSUGRA, in nonunified gaugino mass models and in AMSB models.

7.3.1.2 Annihilation Channel in Theories with Extradimensions

As already pointed out in Sect. 7.1.2.2, the LKP in UED is generally the $B^{(1)}$, the first KK excitation of the hypercharge boson and, therefore, a vector particle. As such, it does not suffer the constraint of helicity suppression of low mass final fermion states in the annihilation rates. Moreover, the annihilation amplitudes are simply proportional to the square of the hypercharge of the generated particles, leading to a neat phenomenology if compared with SUSY. The annihilation gives rise mainly to pairs of charged leptons ($\sim 20\%$ per generation), in up-type quarks ($\sim 11\%$ per generation), Higgs bosons ($\sim 2.3\%$), neutrinos ($\sim 1.2\%$ per generation), and down-type quarks ($\sim 0.7\%$ per generation). Since the preferred mass range is above 400 GeV for LKPs, all standard model particles are created through annihilation.

In other ED models, the annihilation products depend on the WIMP mass. Once again, for low masses, these products are mainly quarks, neutrinos, and charged leptons, while the dominant channels are $t\bar{t}$, W^+W^- and Zh for high masses.

In a sentence, the most significant difference in the annihilation channels in ED theories is the possibility of creating charged lepton and neutrino pairs that are suppressed in SUSY phenomenologies.

7.3.1.3 Energy Spectrum of Final Annihilation Products

Indirect detection is based on precise signatures, which derive from a detailed knowledge of the injected cosmic-ray spectrum. In the case of γ-rays and neutrinos, the propagation is direct, and the detected spectrum can only be affected by absorption and also by oscillations for neutrinos. On the contrary, charged particles are sensitive to diffusion in magnetic turbulence and details of their shapes are likely to be deformed during their travel to the Earth. Spectral information is in any case essential to extract a primary signature of dark matter annihilation from a conventional cosmic-ray background.

The clearest signatures for a detection would be single energy lines in γ-ray, neutrino or charged lepton cosmic ray spectra, which would give unambiguous information about the WIMP mass as well. However, γ-ray lines are suppressed in all models considered here, making them a weak signal. In ED theories, due to lepton/antilepton pair production, neutrino lines could emerge from the continuum contribution. Electron/positron pairs can also be produced with significant branching ratios up to 20%, so distorting clearly the primary spectrum despite propagation effects.

All annihilation mechanisms resulting into quark/antiquark pairs or massive bosons are followed by parton fragmentation, which gives rise to charged and neutral pions sharing equally the energy budget. These hadronic cascades give rise to a continuum spectrum for γ-rays, neutrinos, and for antimatter cosmic rays (positrons, antiprotons, antideuterons). The spectrum details can be estimated either with Monte Carlo simulations or directly from parameterized fragmentation functions with simple phase space arguments. In both cases, the resulting spectra generally exhibit a simple dependence on the WIMP mass.

The W^+W^- decays are recognizable; thanks to particular spectral features for positron and neutrino production due to the large branching ratios of $W \to \nu l$.

7.3.2 Indirect Search Exploiting the Antimatter Component in Cosmic Rays

The indirect dark matter searches aim at detecting rare species of particles in the cosmic-ray extra-atmospheric flux, generally antimatter components, which are not expected to be involved in the cosmic acceleration mechanisms. Of course, secondary interactions take place during the propagation of the primary cosmic rays and generate antiparticles. Therefore, in order to observe an indirect dark matter signal in the data, it is essential to have accurate background estimates of the antimatter component. The efficacy of the indirect searches relies on the confidence of the flux and energy spectra used in these background calculations. The data on charged particle spectra in cosmic rays come from experiments in the high atmosphere flying in balloons or above it in satellites.

Positrons are a rare component in cosmic rays. Electrons themselves represent only 1–2% of the total of the cosmic-ray flux reaching the Earth, and the positron component is only $\sim$ 10% of the electron flux at $\sim$1 GeV. As above mentioned it is expected that the bulk of electrons are primary particles injected and accelerated, in cosmic accelerators, while the positrons are secondary interaction products originated by nuclear collisions in interstellar matter. Antiprotons are an even rarer component of the cosmic-ray flux than positrons: the antiproton/proton flux ratio is only 10^{-5} at 1 GeV. As for the positron case, antiprotons are mainly produced in collisions of primary cosmic-rays with the interstellar medium.

A historical balloon experiment looking for positrons in the cosmic-ray flux is the high-energy antimatter telescope [66], which reported a longly disputed excess in the positron data with respect to the estimated background. Presently, the leader apparatus in this search is the space-based PAMELA [67] (which can provide also antiproton measurements), whose results will be discussed in Sect. 7.3.3.1, together with those about the global cosmic ray electron spectrum achieved by ATIC [68] and PPB-BETS [69] (balloon), and now Fermi-LAT (space) [70]. At very high energy, the ground-based experiment H.E.S.S. [71] can measure the electron component through the identification of the initial electromagnetic nature of the atmospheric cascades. Precise antiproton data were copiously collected previously by the balloon-borne experiment with a superconducting spectrometer (BESS) detector [72], while other data came from AMS-01 [73] and CAPRICE [74] experiments. In the future, AMS-02 will provide a formidable wealth of information given its complex and complete structure [75]. It is an actual high-energy particle physics experiment in space, scheduled to be launched in 2010 for a 3-year mission. The AMS-02 detector is a complete particle spectrometer with different elements designed to measure particle type and kinematic parameters.

7.3.3 Indirect Search with γ-rays and Neutrinos

The great advantage in searching for dark matter through annihilation to γ rays is that this channel retains the information of the source location, in contrast to charged cosmic rays that are diffused by the galactic magnetic turbulence. Hence, γ-rays provide an additional experimental signature in the indirect dark matter searches, and further, γ-rays allow searches beyond the Milky Way into other galaxies in the local cluster. The experiments performing dark matter searches with γ-rays are space-based for low energies and ground-based for high energies. Of course, even in this case, the signal has to be disantangled from a background represented by a continuum in the energy spectrum.

The main γ-ray production mechanism that contributes to this diffuse flux, at energies from 100 MeV to tens of GeV, is the interaction of charged cosmic rays with the interstellar matter, which produces π_0 and consequently γ-rays via their decay: $\pi_0 \rightarrow \gamma\,\gamma$. Other sources are the inverse Compton scattering of cosmic-ray electrons off the interstellar photons and the electron bremsstrahlung in the interstellar medium. The complexity of the involved phenomena explain why the γ-ray background calculations contain significant uncertainties, which complicate the interpretation of the experiments searching for dark matter in this channel.

A historical survey of the γ-ray sky was made by the space-based instrument EGRET [76] in the 1990s. Even in this case, an excess with respect to background prediction was interpreted as a possible dark matter signal, even if not conclusively. At very high energies, the γ-ray telescopes CANGAROO [77], Whipple [78], and the already quoted HESS have observed γs coming from the galactic center regions. In the low-energy region ($\sim$ MeV), most of the information in the past was collected by the experiment INTEGRAL [79]. Presently, the γ-ray sky is observed by the satellite AGILE [80] (up to 50 GeV) and by the already mentioned Fermi-LAT (prepared and launched with the name GLAST), which has extended the energy range up to 300 GeV.

As γ-rays, neutrinos are not deviated by magnetic fields and so point back to their source. In addition, their weakly interacting nature allows them to exit from annihilation sources embedded inside dense matter distributions such as the centers of the Sun and the Earth. Unfortunately, this same nature implies that very massive detectors are needed to detect neutrinos. With the present and projected generation of neutrino telescopes (ANTARES [81], NESTOR [82], NEMO [83], Baikal [84], AMANDA [85], ICECUBE [86], KM3NET [87]), only dark matter concentrations in massive bodies could give detectable annihilation fluxes and those from the galactic halo are negligible.

7.3.3.1 The PAMELA Effect

The PAMELA telescope [88] is a complete particle detector containing a magnetic spectrometer, a silicon tracker, a time-of-flight system, an electromagnetic calorimeter and also a neutron detector. The geometrical acceptance of the detector is $\sim$20 cm^2 sr. The detector was launched from Baikonour on June 15, 2006 on the Russian Resurs DK1 satellite.

Recently, the data collected by PAMELA generated discussions and expectations inside the cosmic-ray and dark-matter scientific communities [90, 91]. The reason is that the PAMELA positron fraction data exhibit a very clear excess above $\sim$10 GeV, as reported in Fig. 7.8. This feature cannot be explained by secondary production.

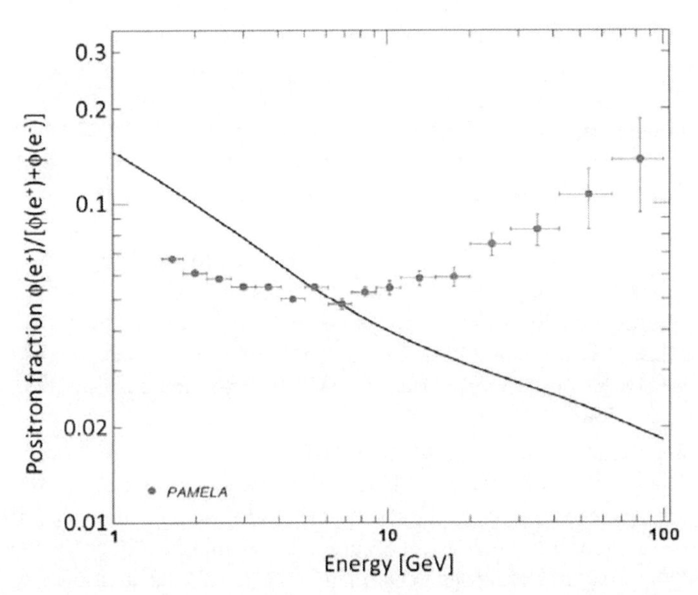

Fig. 7.8 The positron fraction observed by PAMELA [67] compared with a theoretical model [89]. The solid line shows a calculation for pure secondary production of positrons induced by the propagation of cosmic-rays in the Galaxy. Between 5–10 GeV, the PAMELA positron fraction is compatible with other measurements, not shown here. The PAMELA data at low energies (below 10 GeV) differ from the theoretical model because of solar modulation effects. (Figure adapted from e-Print arXiv:0810.4995.)

The data show also a change in the positron fraction data below $\sim$10 GeV (when compared with previous searches), which is probably due to the solar modulation and change in the polarity of the solar magnetic field with respect to the previous cycle. These results were and are interpreted in terms of annihilation of dark matter particles in our Galaxy. However, this is not the only interpretation, and further, it is not the most convincing. Therefore, the PAMELA collaboration has resisted to the temptation to claim the discovery of dark matter. There are actually competing astrophysical sources, such as pulsars, that can originate an additional relevant flux of primary positrons and electrons.

It is worth mention that an independent confirmation about an anomalous behaviour of the lepton component in cosmic rays comes from measurements of high-energy electrons. The cosmic-ray electron flux was not measured very precisely in the past, in particular at high energies because of the very steep spectrum, that requires a high rejection power and long exposure.

Simulations of the electron propagation from local sources has shown that features in the electron spectrum may be expected in the TeV range where the flux of Galactic cosmic-ray electrons gradually steepens. On the other hand, annihilation of KK particles may produce spectral features in sub-TeV range. The first indication of a feature (or excess) in the electron spectrum at a few hundred GeV came from PPB-BETS flight a couple of years ago, while a recent confirmation of the excess by ATIC gives more confidence that we are not in presence of an instrumental artifact.

The anomalies registered by PAMELA and complemented by ATIC / PPB-BETS observations (an unexpected bump in the total electron + positron flux in the 300–600 GeV energy range) need to be integrated with the very recent results of Fermi-LAT about the electron–positron component in cosmic rays. This experiment has reported high precision measurements of the energy spectrum of these particles between 20 GeV and 1 TeV. The spectrum shows no prominent spectral features, and is significantly harder than that inferred from several previous experiments.

The combined analysis of the available recent results show that, while the reported Fermi-LAT data alone can be interpreted in terms of a single component scenario, when considering other complementary experimental results several combinations of parameters, involving both the pulsar and dark matter scenarios, allow a consistent description of the observations. These complementary results are, in particular, the positron fraction reported by PAMELA and the cosmic ray electron spectrum measured by H.E.S.S.

A logic path to the interpretation of the observations consists of a first step which takes into account only the Fermi-LAT, H.E.S.S., and low-energy electron data [91]. The spectra can be reasonably explained by a conventional cosmic ray diffusion model. Thc agreement between data and observations improve either by following a "statistical" approach, which tries to estimate the effect of the stochasticity of the sources (assumed as active supernova remnants), or by trying to model the contribution of actually observed nearby sources.

The second step involves the consideration of the positron PAMELA data. Here, a serious problem occurs: the positron fraction foreseen by the conventional diffusion model, corrected with any possible source effect, is totally inconsistent with

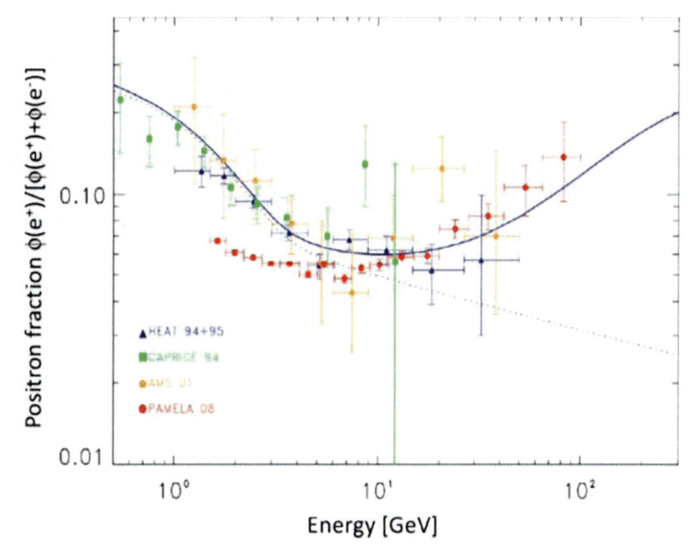

Fig. 7.9 The positron fraction observed by PAMELA [67] and other experiments compared with a calculation (black dotted line) for pure secondary production of positrons induced by the propagation of cosmic-rays in the Galaxy (large-scale galactic component) and with a calculation taking into account observed pulsars with distance <1 kpc (blue continuous line) [91]. Two close pulsars (Monogem and Geminga) give a dominant contribution. The PAMELA anomaly can be accounted for considering the nearby pulsar contributions. (Figure adapted from e-Print arXiv:0905.0636v3.)

the PAMELA results. In particular, the precise measurements provided by Fermi with its hard $\sim E^{-3}$ observed electron spectrum aggravate the inconsistency with respect to the pre-Fermi situation. The models which explain the Fermi data do not account for the rise in the positron fraction seen by PAMELA. Some additional sources of positrons are required. However, no new physics is needed to conciliate data and models. In particular, pulsars are undisputed sources of relativistic electrons and positrons, believed to be produced in the magnetosphere and possibly reaccelerated by the pulsar winds or the supernova remnant shocks. At energies between 100 GeV and 1 TeV, the electron flux reaching the Earth may be the sum of an almost homogeneous and isotropic cosmic-ray component produced by galactic supernova remnants plus the local contribution of a few pulsars, with the latter expected to contribute more and more significantly as the energy increases [91]. The quantitative predictions based on this assumption are in remarkable agreement with the whole set of data (see Fig. 7.9).

Of course, a dark matter annihilation contribution to the observed positron effect cannot be totally excluded. A dark matter interpretation of the Fermi-LAT and of the PAMELA data is an open possibility. However, this interpretation is disfavored for at least the three following reasons [91]:

1. Astrophysical sources can explain both the observed spectral features and the positron ratio measurement: no exotic mechanism is actually required to fit the data.

2. Dark matter annihilation produces antiprotons and protons in addition to electrons and positrons. However, the antiproton data collected by PAMELA and BESS are consistent with each other and with the antiproton flux predicted by the standard secondary production. This sets very stringent constraints on the dominant dark matter annihilation modes. In particular, for ordinary particle dark matter models, such as neutralino dark matter (see Sect. 7.1.2.1) or the lightest KK particle of UED (see Sect. 7.1.2.2), the antiproton bound rules out most of the parameter space.
3. Assuming that dark matter particles are weakly interacting and that their production in the early Universe was due to an ordinary freeze-out process involving the same annihilation mechanism that dark matter would undergo in today's cold Universe (see Sect. 7.1.2), the annihilation rate in the Galaxy would be roughly two orders of magnitude too small to explain the anomalous electron–positron observation. On the other hand, a highly clumpy Galactic dark matter density profile, or the presence of a nearby concentrated clump, can provide sufficient enhancements to the rate of dark matter annihilation (see Sect. 7.1.2.3).

In spite of these caveats, we conclude remarking that there is a copious literature on possible dark matter interpretation of the excess high-energy electrons and positrons (e.g., see references in [91]). Although the pulsar explanation looks perfectly in line with the Fermi-LAT data, a clear discrimination between the pulsar and the dark matter scenarios is not possible on the basis of the currently available data and requires to consider additional complementary observations.

7.4 Conclusions

The candidates to the composition of dark matter are actively pursued by physicists using three different complementary methods. First, a vast international program to detect the tiny energy deposited by WIMP scattering in an ultra-low background detector is underway. After decades of technological developments, experiments operating deep underground have now reached the sensitivities to test realistic particle physics models. In parallel, an arsenal of different instruments, space-based and ground-based, are analyzing with increasing precision the composition of the extra-atmospheric cosmic rays in a wide energy range, with the chance to detect above the background the secondary particles emitted by dark matter annihilation. Finally, in the LHC era, the possibility is open to produce directly in the laboratory the elusive particles that hundreds of scientists are chasing in the Cosmos or in underground installations. The concurrence of these three approaches, separate in terms of involved technologies and science communities but unified by a common fundamental goal, represents an extraordinary intellectual adventure that could soon culminate into the discovery of the dominant form of matter in our Universe.

References

1. Sofue Y. and Rubin V., Ann. Rev. Astron. Astrophys. **39**, 137 (2001)
2. Zwicky F, Helvetica Phys. Acta **6**, 110 (1933)
3. Bahcall N. et al, Astrophys. J. Suppl.**148**, 243 (2003)
4. Bennett C. L et al, (WMAP Collaboration) 2003 Astrophys. J. Suppl. **148**, 1 (2003); Spergel D N et al, Astrophys. J. Suppl. **148**, 175 (2003)
5. Astier P. et al, Astron. Astrophys. **447**, 31 (2006)
6. Turner M. S., Science **315**, 59 (2007)
7. Bertone G., Hooper D and Silk J, Phys. Rep. **405**, 279 (2005)
8. Bergstrom L, Rep. Prog. Phys. **63**, 793 (2000)
9. Ellis J. et al, Nucl. Phys. B **238**, 453 (1984)
10. Feng J., J. Phys. G **32**, R1 (2006)
11. Duffy L. D. and van Bibber K., New J. Phys. **11**, 105008 (2009)
12. Jungman G, Kamionkowski M.and Griest K., Phys. Rep. **267**, 195 (1996)
13. Goldberg H., Phys. Lett. **50**, 1419 (1983)
14. Kaluza T. F. E., Sitzungsberichte Preussische Akademie der Wissenschaften **96**, 69 (1921)
15. Klein O., Z. Phys. **37**, 895 (1926)
16. Hewett J. and March-Russell J., Phys. Lett. B **592**, 1 (2004)
17. Chardin G., e-Print arXiv:astro-ph/0411503v3
18. Carr J., Lamanna G. and Lavalle J., Rep. Prog. Phys. **69**, 2475 (2006)
19. Michie R. W., Mon. Not. R. Astron. Soc. **125**, 127 (1963)
20. Navarro J. F. et al, Astrophys. J. **462**, 563 (1996)
21. Moore B. et al, Phys. Rev. D **64**, 063508 (2001)
22. Lewin J. D. and Smith P. F., Astropart. Phys. **6**, 87 (1996)
23. Drukier A., Freese K. and Spergel D. N., Phys. Rev. D **30**, 3495 (1986)
24. Copi C. J. and Krauss L. M., Phys. Rev. D **67**, 103507 (2003)
25. Goodman M. W. and Witten E., Phys. Rev. D **31**, 3059 (1985)
26. Freese K. et al, Phys. Rev. D **37**, 3388 (1988)
27. Bernabei R., Riv. N. Cim. **18**, (1995)
28. Bernabei R. et al, Riv. N. Cim. **26**, 1 (2003)
29. Fushimi K. et al, Astropart. Phys. **12**, 185 (2000)
30. Alner G. J. et al, Phys Lett B **616**, 17 (2005)
31. Bernabei R. et al, Eur. Phys. J. C **56**, 333 (2008)
32. Amaré J. et al, J. Phys. Conf. Ser.**39**, 123 (2006)
33. Lee H. S. et al, Phys. Rev. Lett. **99**, 091301 (2007)
34. Lin S. T. et al., Phys.Rev.D **79**, 061101 (2009)
35. Aalseth C.E. et al, Phys. Rev. Lett. **101**, 251301 (2008)
36. Aalseth C.E. et al, e-Print arXiv:1002.4703v2[astro-ph.CO]
37. A. Bottino et al., e-Print arXiv:0912.4025v2[hep-ph]
38. Giuliani A., Physica B **280**, 501 (2000)
39. Ahmed Z. et al, Phys. Rev. Lett. **102**, 011301 (2009); Ahmed, Z. et al., e-Print arXiv:0912.3592v1[astro-ph.CO].
40. Akerib D. S. et al, J. Low Temp. Phys. **151**, 818 (2008)
41. Sanglard V. et al, Phys. Rev. D **71**, 122002 (2005)
42. Broniatowski A. et al, Phys. Lett. B **681**, 305 (2009)
43. Armengaud E. et al., Phys. Lett. B **687** 294 (2010)
44. Angloher G. et al, Astropart. Phys. **31**, 270 (2009)
45. Cebrian S. et al, Physics Letters B **563**, 48 (2003)
46. Kraus H. et al, Nucl. Phys. B (Proc. Suppl.) **173**, 168 (2007)
47. Angle J. et al, Phys. Rev. Lett. **100**, 021303 (2008)
48. Aprile E. et al., e-Print arXiv:1005.0380v2[astro-ph.CO]
49. A. Manzur et al., e-Print arXiv:0909.1063v4[physics.ins-det]

50. Gaitskell R.: LUX - Large Underground Xenon Dark Matter Experiment - Report on Design, Construction and Detector Testing, to be published in the Proceedings of IDM08, Stockholm, August 19, (2008)
51. Akimov D. Yu. et al., e-Print arXiv:1003.5626v2[hep-ex]
52. Benetti P. et al, Astropart. Phys. **28**, 495 (2008)
53. Lanfranchi M. and Rubbia A., J. Phys. Conf. Ser. **65**, 012014 (2007)
54. Suzuki, Y: XMASS experiment, to be published in the Proceedings of IDM08, Stockholm, August 19, (2008)
55. Horowitz C. J et al, Phys. Rev.D **68**, 023005 (2003)
56. Boulay M. G. and Hime A., Astropart. Phys. **25**, 179 (2006)
57. Behnke E. et al, Science **319**, 933 (2008)
58. Archambault S. et al, Phys.Lett. B **682**, 185 (2009)
59. Girard T. A. et al, Phys. Lett. B **621**, 233 (2005)
60. Alner G. J. et al, Nucl. Instr. Meth. Phys. Res. A **555**, 173 (2005)
61. Sciolla G. et al, e-Print arXiv: 0903.3895 [astro-ph-IM], to be published in the Journal of Physics: Conference Series
62. Nishimura H. et al, Astropart. Phys. **31**, 185 (2009)
63. Moulin E. et al, Phys. Lett. B **614**, 143 (2005)
64. Gondolo P. and Silk J., Phys. Rev. Lett. **83**, 1719 (1999)
65. Silk J. and Stebbins A., Astrophys. J. **411**, 439 (1993)
66. Barwick S. W. et al, Astrophys. J. **482**, L191 (1997)
67. Adriani O. et al, Nature **458**, 607 (2009)
68. Chang J. et al, Nature **456**, 362 (2008)
69. Torii S. et al, e-Print arXiv:0809.0760 [astro-ph]
70. Abdo A. A. et al, Phys. Rev. Lett. **102**, 181101 (2009)
71. Aharonian F. et al, Phys. Rev. Lett. **101**, 261104 (2008)
72. K. Abe et al, Phys.Lett.B **670**, 103 (2008)
73. Aguilar M. et al., Phys. Reports **366**, 331 (2002)
74. Boezio M. et al, Astrophys. J.**561**, 787 (2001)
75. Battiston R., J. Phys. Conf. Ser. **116**, 012001 (2008)
76. Mayer-Hasselwander H. A. et al, Astron. Astrophys. **335**, 161 (1998)
77. Tsuchiya K. et al, Astrophys. J. **606**, L115 (2004)
78. Kosack K. et al, Astrophys. J. **608**, L97 (2004)
79. Jean P. et al, Astron. Astrophys. **407**, L55 (2003)
80. Tavani M. et al, Astron.Astrophys. **502**, 995 (2009)
81. Ageron M. et al, Astropart. Phys., **31**, 277 (2009)
82. Aggouras G. et al, Nucl. Instrum. Meth. Phys. Res. A **552**, 420 (2005)
83. Capone A., Nucl. Instrum. Meth. Phys. Res. A **602**, 47 (2009)
84. Aynutdinov V. et al, Astropart.Phys.**25**, 140 (2006)
85. Ackermann M. et al, Astropart.Phys. **22**, 127 (2004)
86. Abbasi R. et al, Astrophys.J.**701**, L47 (2009)
87. Lyons K.: The KM3NeT Project, XXth Rencontres de Blois, Blois France May 18-23, (2008)
88. Adriani O. et al, Nucl. Instrum. Meth. Phys. Res. A **478**, 114 (2002)
89. Strong A. W. and Moskalenko I. V., Astrophys. J. **509**, 212 (1998)
90. Morselli A. and Moskalenko I. V., e-Print arXiv:0811.3526 [astro-ph]
91. Grasso D. et al, e-Print arXiv:0905.0636v3 [astro-ph.HE]

Part III
Dark Energy

Chapter 8
Dark Energy: Investigation and Modeling

Shinji Tsujikawa

Abstract Constantly accumulating observational data continue to confirm that about 70% of the energy density today consists of dark energy responsible for the accelerated expansion of the Universe. We present recent observational bounds on dark energy constrained by the type Ia supernovae, cosmic microwave background, and baryon acoustic oscillations. We review a number of theoretical approaches that have been adopted thus far to explain the origin of dark energy. This includes the cosmological constant, modified matter models (such as quintessence, k-essence, coupled dark energy, unified models of dark energy and dark matter), modified gravity models (such as $f(R)$ gravity, scalar-tensor theories, braneworlds), and inhomogeneous models. We also discuss observational and experimental constraints on those models and clarify which models are favored or ruled out in current observations.

8.1 Introduction

The discovery of the late-time cosmic acceleration reported in 1998 [1, 2] based on the type Ia supernovae (SN Ia) observations opened up a new field of research in cosmology. The source for this acceleration, dubbed dark energy [3], has been still a mystery in spite of tremendous efforts to understand its origin over the last decade [4–12]. Dark energy is distinguished from ordinary matter in that it has a negative pressure whose equation of state $w_{\rm DE}$ is close to -1. Independent observational data such as SN Ia [13–18], Cosmic Microwave Background (CMB) [19–22], and Baryon Acoustic Oscillations (BAO) [23–25] have continued to confirm that about 70% of the energy density of the present Universe consists of dark energy.

The simplest candidate for dark energy is the so-called cosmological constant Λ whose equation of state is $w_{\rm DE} = -1$. If the cosmological constant originates from

Department of Physics, Faculty of Science, Tokyo University of Science, 1-3 Kagurazaka, Shinjuku-ku, Tokyo, 162-8601, Japan e-mail: shinji@rs.kagu.tus.ac.jp a vacuum energy of particle physics, its energy scale is significantly larger than the dark energy density today [26] ($\rho_{\rm DE}^{(0)} \simeq 10^{-47}$ GeV^4). Hence we need to find a mechanism to obtain the tiny value of Λ consistent with observations. Several efforts have been made in this direction under the framework of particle physics. For example, the recent development of string theory shows that it is possible to construct de Sitter vacua by compactifying extra dimensions in the presence of fluxes with an account of non-perturbative corrections [27].

The first step toward understanding the property of dark energy is to clarify whether it is a simple cosmological constant or it originates from other sources that dynamically change in time. The dynamical dark energy models can be distinguished from the cosmological constant by considering the evolution of $w_{\rm DE}$. The scalar field models of DE such as quintessence [28–38] and k-essence [39–41] predict a wide variety of variations of $w_{\rm DE}$, but still the current observational data are not sufficient to provide some preference of such models over the Λ-Cold-Dark-Matter (ΛCDM) model. Moreover, the field potentials need to be sufficiently flat such that the field evolves slowly enough to drive the present cosmic acceleration. This demands that the field mass is extremely small ($m_\phi \simeq 10^{-33}$ eV) relative to typical mass scales appearing in particle physics [42, 43]. However, it is not entirely hopeless to construct viable scalar-field dark energy models in the framework of particle physics. We note that there is another class of modified matter models based on perfect fluids–so-called (generalized) the Chaplygin gas model [44, 45]. If these models are responsible for explaining the origin of dark matter as well as dark energy, then they are severely constrained from the matter power spectrum in galaxy clustering [46].

There exists another class of dynamical dark energy models that modify General Relativity (GR). The models that belong to this class are $f(R)$ gravity [47, 49–52] (f is a function of the Ricci scalar R), scalar-tensor theories [53–57], and Dvali, Gabadadze, and Porrati (DGP) braneworld model [58]. The attractive feature of these models is that the cosmic acceleration can be realized without recourse to a dark energy component. If we modify gravity from General Relativity, however, there are stringent constraints coming from local gravity tests as well as a number of observational constraints such as large-scale structure (LSS) and CMB. Hence the restriction on modified gravity models is in general very tight compared with modified matter models. We shall construct viable modified gravity models and discuss their observational and experimental signatures.

In addition to the above-mentioned models, there are attempts to explain the cosmic acceleration without dark energy. One example is the void model in which an apparent accelerated expansion is induced by a large spatial inhomogeneity [59–63]. Another example is the so-called backreaction model in which the backreaction of spatial inhomogeneities on the Friedmann–Lemaître–Robertson–Walker (FLRW) background is responsible for the real acceleration [64–66]. We shall discuss these models as well.

This review is organized as follows. In Section 8.2, we provide recent observational constraints on dark energy obtained by SN Ia, CMB, and BAO data. In Section 8.3, we review theoretical attempts to explain the origin of the cosmological constant consistent with the low-energy scale of dark energy. In Section 8.4, we discuss modified gravity models of dark energy—including quintessence, k-essence, coupled dark energy, and unified models of dark energy and dark matter. In Section 8.5, we review modified gravity models and provide a number of ways to distinguish those models observationally from the ΛCDM model. Section 8.6, is devoted to the discussion about the cosmic acceleration without using dark energy. We conclude in Section 8.7.

We use units such that $c = \hbar = 1$, where c is the speed of light and $\hbar$ is the reduced Planck's constant. The gravitational constant G is related to the Planck mass $m_{\rm pl} = 1.2211 \times 10^{19}\,\text{GeV}$ via $G = 1/m_{\rm pl}^2$ and the reduced Planck mass $M_{\rm pl} = 2.4357 \times 10^{18}\,\text{GeV}$ via $\kappa^2 \equiv 8\pi G = 1/M_{\rm pl}^2$, respectively. We write the Hubble constant today as $H_0 = 100\,h\,\text{km}\,\text{sec}^{-1}\,\text{Mpc}^{-1}$, where h describes the uncertainty on the value H_0. We use the metric signature $(-,+,+,+)$.

8.2 Observational Constraints on Dark Energy

The late-time cosmic acceleration is supported by a number of independent observations, such as (i) supernovae observations, (ii) Cosmic Microwave Background (CMB), and (iii) Baryon acoustic oscillations (BAO). In this section, we discuss observational constraints on the property of dark energy.

8.2.1 Supernovae Ia Observations

In 1998, Riess et al. [1] and Perlmutter et al., [2] independently reported the late-time cosmic acceleration by observing distant supernovae of type Ia (SN Ia). The line-element describing a 4-dimensional homogeneous and isotropic Universe, which is called the FLRW space-time, is given by [67]

$$\mathrm{d}s^2 = g_{\mu\nu}\mathrm{d}x^\mu \mathrm{d}x^\nu = -\mathrm{d}t^2 + a^2(t)\left[\frac{\mathrm{d}r^2}{1-Kr^2} + r^2(\mathrm{d}\theta^2 + \sin^2\theta\,\mathrm{d}\phi^2)\right], \tag{8.1}$$

where $a(t)$ is the scale factor with cosmic time t, and $K = +1, -1, 0$ corresponds to closed, open, and flat geometries, respectively. The redshift z is defined by $z = a_0/a - 1$, where $a_0 = 1$ is the scale factor today.

In order to discuss the cosmological evolution in the low-redshift regime ($z < \mathcal{O}(1)$), let us consider nonrelativistic matter with energy density ρ_m and dark energy with energy density $\rho_{\rm DE}$ and pressure $P_{\rm DE}$, satisfying the continuity equations

$$\dot{\rho}_m + 3H\rho_m = 0, \tag{8.2}$$

$$\dot{\rho}_{\rm DE} + 3H(\rho_{\rm DE} + P_{\rm DE}) = 0, \tag{8.3}$$

which correspond to the conservation of the energy-momentum tensor $T_{\mu\nu}$ for each component ($\nabla_\mu T^{\mu\nu}=0$, where ∇ represents a covariant derivative). Note that a dot represents a derivative with respect to t. The cosmological dynamics is known by solving the Einstein equations

$$G_{\mu\nu} = 8\pi G T_{\mu\nu}\,, \tag{8.4}$$

where $G_{\mu\nu}$ is the Einstein tensor. For the metric (8.1), the (00) component of the Einstein equations gives [67]

$$H^2 = \frac{8\pi G}{3}(\rho_m + \rho_{\rm DE}) - \frac{K}{a^2}\,, \tag{8.5}$$

where $H \equiv \dot{a}/a$ is the Hubble parameter. We define the density parameters

$$\Omega_m \equiv \frac{8\pi G\rho_m}{3H^2}\,, \qquad \Omega_{\rm DE} \equiv \frac{8\pi G\rho_{\rm DE}}{3H^2}\,, \qquad \Omega_K \equiv -\frac{K}{(aH)^2}\,, \tag{8.6}$$

which satisfy the relation $\Omega_m + \Omega_{\rm DE} + \Omega_K = 1$ from Eq. (8.5). Integrating Eqs. (8.2) and (8.3), we obtain

$$\rho_m = \rho_m^{(0)}(1+z)^3\,, \qquad \rho_{\rm DE} = \rho_{\rm DE}^{(0)} \exp\left[\int_0^z \frac{3(1+w_{\rm DE})}{1+\tilde{z}}{\rm d}\tilde{z}\right]\,, \tag{8.7}$$

where "0" represents the values today and $w_{\rm DE} = P_{\rm DE}/\rho_{\rm DE}$ is the equation of state of dark energy. Plugging these relations into Eq. (8.5), it follows that

$$H^2(z) = H_0^2\left[\Omega_m^{(0)}(1+z)^3 + \Omega_{\rm DE}^{(0)}\exp\left\{\int_0^z \frac{3(1+w_{\rm DE})}{1+\tilde{z}}{\rm d}\tilde{z}\right\} + \Omega_K^{(0)}(1+z)^2\right]. \tag{8.8}$$

The expansion rate $H(z)$ can be known observationally by measuring the luminosity distance $d_L(z)$ of SN Ia. The luminosity distance is defined by $d_L^2 \equiv L_s/(4\pi\mathscr{F})$, where L_s is the absolute luminosity of a source and $\mathscr{F}$ is an observed flux. It is a textbook exercise [9, 12, 67] to derive $d_L(z)$ for the FLRW metric (8.1):

$$d_L(z) = \frac{1+z}{H_0\sqrt{\Omega_K^{(0)}}}\sinh\left(\sqrt{\Omega_K^{(0)}}\int_0^z \frac{{\rm d}\tilde{z}}{E(\tilde{z})}\right)\,, \tag{8.9}$$

where $E(z) \equiv H(z)/H_0$. The function $f_K(\chi) \equiv 1/\sqrt{\Omega_K^{(0)}}\sinh(\sqrt{\Omega_K^{(0)}}\chi)$ can be understood as $f_K(\chi) = \sin\chi$ (for $K=+1$), $f_K(\chi) = \chi$ (for $K=0$), and $f_K(\chi) = \sinh\chi$ (for $K=-1$). For the flat case ($K=0$), Eq. (8.9) reduces to $d_L(z) = (1+z)\int_0^z {\rm d}\tilde{z}/H(\tilde{z})$, i.e.,

$$H(z) = \left[\frac{\rm d}{{\rm d}z}\left(\frac{d_L(z)}{1+z}\right)\right]^{-1}. \tag{8.10}$$

Hence the measurement of the luminosity distance $d_L(z)$ of SN Ia allows us to find the expansion history of the Universe for $z < \mathscr{O}(1)$.

The luminosity distance d_L is expressed in terms of an apparent magnitude m and an absolute magnitude M of an object, as

$$m - M = 5\log_{10}\left(\frac{d_L}{10\ \mathrm{pc}}\right). \tag{8.11}$$

The absolute magnitude M at the peak of brightness is the same for any SN Ia under the assumption of standard candles, which is around $M \simeq -19$ [1, 2]. The luminosity distance $d_L(z)$ is known from Eq. (8.11) by observing the apparent magnitude m. The redshift z of an object is known by measuring the wavelength λ_0 of light relative to its wavelength λ in the rest frame, i.e., $z = \lambda_0/\lambda - 1$. The observations of many SN Ia provide the dependence of the luminosity distance d_L in terms of z.

Expanding the function (8.9) around $z = 0$, it follows that

$$\begin{aligned} d_L(z) &= \frac{1}{H_0}\left[z + \left\{1 - \frac{E'(0)}{2}\right\}z^2 + \mathscr{O}(z^3)\right] \\ &= \frac{1}{H_0}\left[z + \frac{1}{4}\left(1 - 3w_{\mathrm{DE}}\Omega_{\mathrm{DE}}^{(0)} + \Omega_K^{(0)}\right)z^2 + \mathscr{O}(z^3)\right], \end{aligned} \tag{8.12}$$

where a prime represents a derivative with respect to z. Note that, in the second line, we have used Eq. (8.8). In the presence of dark energy ($w_{\mathrm{DE}} < 0$ and $\Omega_{\mathrm{DE}}^{(0)} > 0$), the luminosity distance gets larger than that in the flat Universe without dark energy. For smaller (negative) w_{DE} and for larger $\Omega_{\mathrm{DE}}^{(0)}$, this tendency becomes more significant. The open Universe without dark energy can also give rise to a larger value of $d_L(z)$, but the density parameter $\Omega_K^{(0)}$ is constrained to be close to 0 from the WMAP data (more precisely, $-0.0175 < \Omega_K^{(0)} < 0.0085$ [21]). Hence, in the low redshift regime ($z \lesssim 1$), the luminosity distance in the open Universe is hardly different from that in the flat Universe without dark energy.

As we see from Eq. (8.12), the observational data in the high redshift regime ($z > 0.5$) allow us to confirm the presence of dark energy. The SN Ia data released by Riess et al. [1] and Perlmutter et al. [2] in 1998 in the redshift regime $0.2 < z < 0.8$ showed that the luminosity distances of observed SN Ia tend to be larger than those predicted in the flat Universe without dark energy. Assuming a flat Universe with a dark energy equation of state $w_{\mathrm{DE}} = -1$ (i.e., the cosmological constant), Perlmutter et al. [2] found that the cosmological constant is present at the 99% confidence level. According to their analysis, the density parameter of non-relativistic matter today was constrained to be $\Omega_m^{(0)} = 0.28^{+0.09}_{-0.08}$ (68% confidence level) in the flat Universe with the cosmological constant.

Over the past decade, more SN Ia data have been collected by a number of high-redshift surveys such as SuperNova Legacy Survey (SNLS) [13], Hubble Space Telescope (HST) [14, 15], and "Equation of State: SupErNovae trace Cosmic Expansion" (ESSENCE) [16, 17] survey. These data also confirmed that the

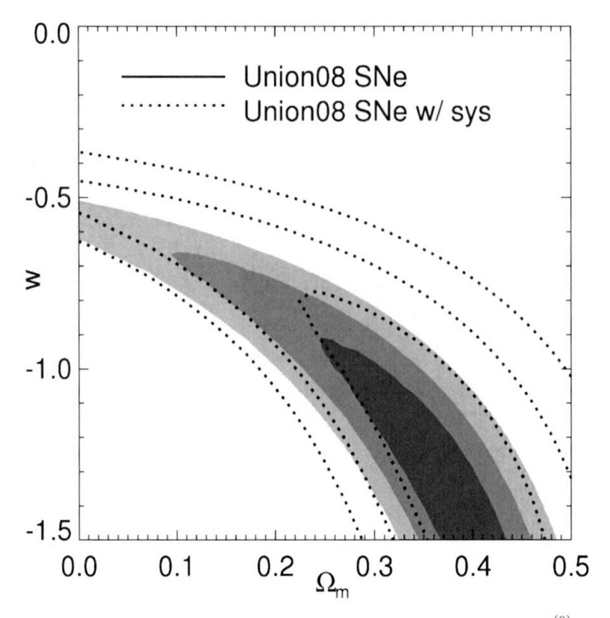

Fig. 8.1 68.3%, 95.4%, and 99.7% confidence level contours on $w_{\rm DE}$ and $\Omega_m^{(0)}$ (denoted as w and Ω_m in the figure) constrained by the Union08 SN Ia data set. The equation of state $w_{\rm DE}$ is assumed to be constant. From Ref. [18].

Universe entered the epoch of cosmic acceleration after the matter-dominated epoch. If we allow the case in which dark energy is different from the cosmological constant (i.e., $w_{\rm DE} \neq -1$), then observational constraints on $w_{\rm DE}$ and $\Omega_{\rm DE}^{(0)}$ (or $\Omega_m^{(0)}$) are not so stringent. In Fig. 8.1, we show the observational contours on $(w_{\rm DE}, \Omega_m^{(0)})$ for constant $w_{\rm DE}$ obtained from the "Union08" SN Ia data by Kowalski et al. [18]. Clearly, the SN Ia data alone are not yet sufficient to place tight bounds on $w_{\rm DE}$.

In the flat Universe dominated by dark energy with constant $w_{\rm DE}$, it follows from Eq. (8.8) that $H^2 \simeq H_0^2 \Omega_{\rm DE}^{(0)}(1+z)^{3(1+w_{\rm DE})} \propto a^{-3(1+w_{\rm DE})}$. Integrating this equation, we find that the scale factor evolves as $a \propto t^{2/(3(1+w_{\rm DE}))}$ for $w_{\rm DE} > -1$ and $a \propto e^{Ht}$ for $w_{\rm DE} = -1$. The cosmic acceleration occurs for $-1 \leq w_{\rm DE} < -1/3$. In fact, Fig. 8.1 shows that $w_{\rm DE}$ is constrained to be smaller than $-1/3$. If $w_{\rm DE} < -1$, which is called phantoms or ghosts [68], the solution corresponding to the expanding Universe is given by $a \propto (t_s - t)^{2/(3(1+w_{\rm DE}))}$, where t_s is a constant. In this case, the Universe ends at $t = t_s$ with a so-called big rip singularity [69, 70] at which the curvature grows toward infinity.[1] The current observations allow the possibility of the phantom equation of state. We note, however, that the dark energy equation of state smaller than -1 does not necessarily imply the appearance of the big rip

[1] There are other classes of finite-time singularities studied in Refs. [71–79]. In some cases, quantum effects can moderate such singularities [75, 80–82]

singularity. In fact, in some of modified gravity models such as $f(R)$ gravity, it is possible to realize $w_{\rm DE} < -1$ without having a future big rip singularity [83].

If the dark equation of state is not constant, we need to parametrize $w_{\rm DE}$ as a function of the redshift z. This smoothing process is required because the actual observational data have discrete values of redshifts with systematic and statistical errors. There are several ways of parametrizations proposed so far. In general, one can write such parametrizations in the form

$$w_{\rm DE}(z) = \sum_{n=0} w_n x_n(z)\,, \tag{8.13}$$

where n's are integers. We show a number of examples for the expansions:

$$\text{(i) Redshift}: \qquad x_n(z) = z^n\,, \tag{8.14}$$

$$\text{(ii) Scale factor}: \qquad x_n(z) = (1-a)^n = \left(\frac{z}{1+z}\right)^n\,, \tag{8.15}$$

$$\text{(iii) Logarithmic}: \qquad x_n(z) = [\ln(1+z)]^n\,. \tag{8.16}$$

The parametrization (i) was introduced by Huterer and Turner [3] and Weller and Albrecht [84] with $n \le 1$, i.e., $w_{\rm DE} = w_0 + w_1 z$. Chevalier and Polarski [85] and Linder [86] proposed the parametrization (ii) with $n \le 1$, i.e.,

$$w_{\rm DE}(z) = w_0 + w_1(1-a) = w_0 + w_1\frac{z}{1+z}\,. \tag{8.17}$$

This has a dependence $w_{\rm DE}(z) = w_0 + w_1$ for $z \to \infty$ and $w_{\rm DE}(z) \to w_0$ for $z \to 0$. A more general form, $w_{\rm DE}(z) = w_0 + w_1 z/(1+z)^p$, was proposed by Jassal et al. [87]. The parametrization (iii) with $n \le 1$ was introduced by Efstathiou [88]. A functional form that can be used for a fast transition of $w_{\rm DE}(z)$ was also proposed [89, 90]. In addition to the parametrization of $w_{\rm DE}$, a number of authors assumed parametric forms of $d_L(z)$ [91], or $H(z)$ [92–94]. Many works placed observational constraints on the property of dark energy by using such parametrizations [95–109].

In Fig. 8.2, we show the SN Ia constraints combined with other measurements such as the WMAP 7-year [22] and the BAO data [25]. The parametrization (8.17) is used in this analysis. The Gaussian prior on the present-day Hubble constant [110], $H_0 = 74.2 \pm 3.6$ km sec^{-1} Mpc^{-1} (68% confidence level), is also included in the analysis (obtained from the magnitude-redshift relation of 240 low-z SN Ia at $z < 0.1$). In Fig. 8.2, "$D_{\Delta t}$" means a constraint coming from the measurement of gravitational lensing time delays [111]. The joint constraint from WMAP + BAO + H_0 + $D_{\Delta t}$ + SN gives the bound

$$w_0 = -0.93 \pm 0.13\,, \qquad w_1 = -0.41^{+0.72}_{-0.71}\,, \tag{8.18}$$

at the 68% confidence level. Hence the current observational data are consistent with the flat Universe in the presence of the cosmological constant ($w_0 = -1, w_1 = 0$).

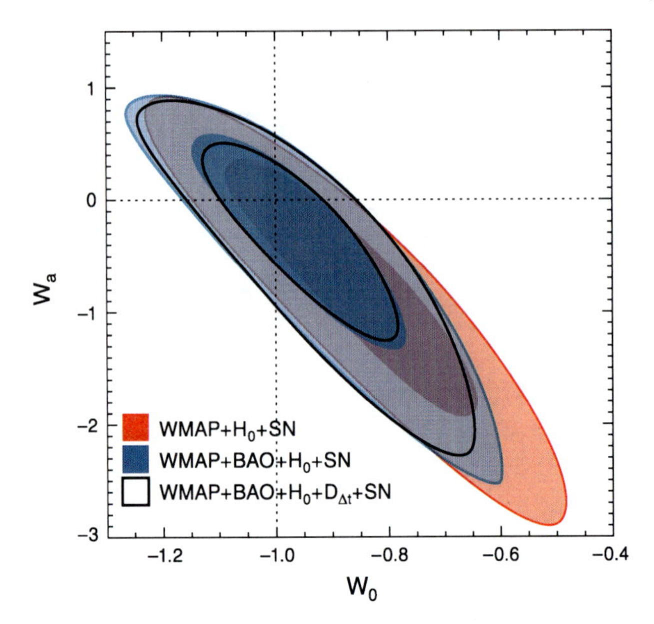

Fig. 8.2 Observational constraints on the parameters w_0 and w_1 (denoted as w_a in the figure) for the parametrization (8.17). The contours show the 68% and 95% confidence level from WMAP + H_0 + SN (red), WMAP + BAO + H_0 + SN (blue), and WMAP + BAO + H_0 + $D_{\Delta t}$ + SN (black), for a flat Universe. From Ref. [22].

8.2.2 CMB

The temperature anisotropies in CMB are affected by the presence of dark energy. The position of the acoustic peaks in CMB anisotropies depends on the expansion history from the decoupling epoch to the present. Hence the presence of dark energy leads to the shift for the positions of acoustic peaks. There is also another effect called the Integrated Sachs-Wolfe (ISW) effect [112] induced by the variation of the gravitational potential during the epoch of the cosmic acceleration. Since the ISW effect is limited to large-scale perturbations, the former effect is typically more important.

The cosmic inflation in the early Universe [113–116] predicts nearly scale-invariant spectra of density perturbations through the quantum fluctuation of a scalar field. This is consistent with the CMB temperature anisotropies observed by COBE [117] and WMAP [19]. The perturbations are "frozen" after the scale $\lambda = (2\pi/k)a$ (k is a comoving wave number) leaves the Hubble radius H^{-1} during inflation ($\lambda > H^{-1}$) [118, 119]. After inflation, the perturbations cross inside the Hubble radius again ($\lambda < H^{-1}$) and they start to oscillate as sound waves. This second horizon crossing occurs earlier for larger k (i.e., for smaller scale perturbations).

We define the sound horizon as $r_s(\eta) = \int_0^\eta d\tilde{\eta}\, c_s(\tilde{\eta})$, where c_s is the sound speed and $d\eta = a^{-1}dt$. The sound speed squared is given by

$$c_s^2 = 1/[3(1+R_s)]\,, \qquad R_s = 3\rho_b/(4\rho_\gamma)\,, \tag{8.19}$$

where ρ_b and ρ_γ are the energy densities of baryons and photons, respectively. The characteristic angle for the location of CMB acoustic peaks is [120]

$$\theta_A \equiv \frac{r_s(z_{\rm dec})}{d_A^{(c)}(z_{\rm dec})}\,, \tag{8.20}$$

where $d_A^{(c)}$ is the comoving angular diameter distance related with the luminosity distance d_L via the duality relation $d_A^{(c)} = d_L/(1+z)$ [12, 67], and $z_{\rm dec} \simeq 1090$ is the redshift at the decoupling epoch. The CMB multipole ℓ_A that corresponds to the angle (8.20) is

$$\ell_A = \frac{\pi}{\theta_A} = \pi \frac{d_A^{(c)}(z_{\rm dec})}{r_s(z_{\rm dec})}\,. \tag{8.21}$$

Using Eq. (8.9) and the background equation $3H^2 = 8\pi G(\rho_m + \rho_r)$ for the redshift $z > z_{\rm dec}$ (where ρ_m and ρ_r are the energy density of non-relativistic matter and radiation, respectively), we obtain [121, 122]

$$\ell_A = \frac{3\pi}{4}\sqrt{\frac{\omega_b}{\omega_\gamma}}\left[\ln\left(\frac{\sqrt{R_s(a_{\rm dec})+R_s(a_{\rm eq})}+\sqrt{1+R_s(a_{\rm dec})}}{1+\sqrt{R_s(a_{\rm eq})}}\right)\right]^{-1}\mathscr{R}\,, \tag{8.22}$$

where $\omega_b \equiv \Omega_b^{(0)} h^2$ and $\omega_\gamma \equiv \Omega_\gamma^{(0)} h^2$, and $\mathscr{R}$ is the so-called CMB shift parameter defined by [123]

$$\mathscr{R} \equiv \sqrt{\frac{\Omega_m^{(0)}}{\Omega_K^{(0)}}}\,\sinh\left(\sqrt{\Omega_K^{(0)}}\int_0^{z_{\rm dec}}\frac{{\rm d}z}{E(z)}\right)\,. \tag{8.23}$$

The quantity $R_s = 3\rho_b/(4\rho_\gamma)$ can be expressed as

$$R_s(a) = (3\omega_b/4\omega_\gamma)a\,. \tag{8.24}$$

In Eq. (8.22), $a_{\rm dec}$ and $a_{\rm eq}$ correspond to the scale factor at the decoupling epoch and at the radiation-matter equality, respectively.

The change of cosmic expansion history from the decoupling epoch to the present affects the CMB shift parameter, which gives rise to the shift for the multipole ℓ_A. The general relation for all peaks and troughs of observed CMB anisotropies is given by [124]

$$\ell_m = \ell_A(m - \phi_m)\,, \tag{8.25}$$

where m represents peak numbers ($m = 1$ for the first peak, $m = 1.5$ for the first trough, ...) and ϕ_m is the shift of multipoles. For a given cosmic curvature $\Omega_K^{(0)}$, the quantity ϕ_m depends weakly on ω_b and $\omega_m \equiv \Omega_m^{(0)} h^2$. The shift of the first peak can be fitted as $\phi_1 = 0.265$ [124]. The WMAP 5-year bound on the CMB shift parameter is given by [21]

$$\mathscr{R} = 1.710 \pm 0.019\,, \tag{8.26}$$

at the 68% confidence level. Taking $\mathscr{R} = 1.710$ together with other values $\omega_b = 0.02265$, $\omega_m = 0.1369$, and $\omega_\gamma = 2.469 \times 10^{-5}$ constrained by the WMAP 5-year data, we obtain $\ell_A \simeq 300$ from Eq. (8.22). Using the relation (8.25) with $\phi_1 = 0.265$, we find that the first acoustic peak corresponds to $\ell_1 \simeq 220$, as observed in CMB anisotropies.

In the flat Universe ($K = 0$), the CMB shift parameter is simply given by $\mathscr{R} = \sqrt{\Omega_m^{(0)}} \int_0^{z_{\rm dec}} {\rm d}z/E(z)$. For smaller $\Omega_m^{(0)}$ (i.e., for larger $\Omega_{\rm DE}^{(0)}$), $\mathscr{R}$ tends to be smaller. For the cosmological constant ($w_{\rm DE} = -1$), the normalized Hubble expansion rate is given by $E(z) = [\Omega_m^{(0)}(1+z)^3 + \Omega_{\rm DE}^{(0)}]^{1/2}$. Under the bound (8.26), the density parameter is constrained to be $0.72 < \Omega_{\rm DE}^{(0)} < 0.77$. This is consistent with the bound coming from the SN Ia data. One can also show that, for increasing $w_{\rm DE}$, the observationally allowed values of $\Omega_m^{(0)}$ get larger. However, $\mathscr{R}$ depends weakly on the $w_{\rm DE}$. Hence the CMB data alone do not provide a tight constraint on $w_{\rm DE}$. In Fig. 8.3, we show the joint observational constraints on

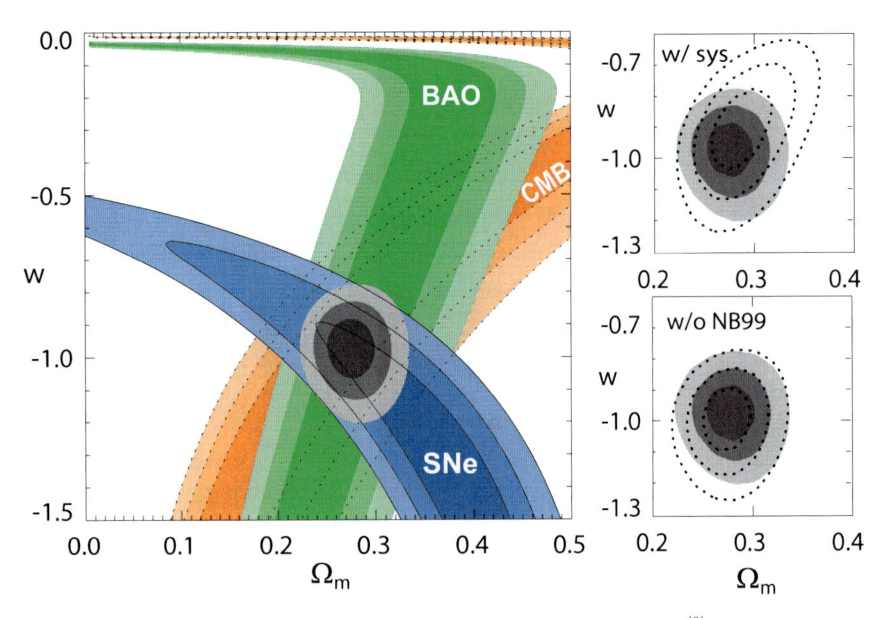

Fig. 8.3 68.3%, 95.4% and 99.7% confidence level contours on $w_{\rm DE}$ and $\Omega_m^{(0)}$ (denoted as w and Ω_m in the figure, respectively) for a flat Universe. The left panel illustrates the individual constraints from SN Ia, CMB, and BAO, as well as the combined constraints (filled gray contours, statistical errors only). The upper right panel shows the effect of including systematic errors. The lower right panel illustrates the impact of the Supernova Cosmology Project (SCP) Nearby 1999 data. From Ref. [18].

$w_{\rm DE}$ and $\Omega_m^{(0)}$ (for constant $w_{\rm DE}$) obtained from the WMAP 5-year data and the Union08 SN Ia data [18]. The joint observational constraints provide much tighter bounds compared with the individual constraint from CMB and SN Ia. For the flat Universe, Kowalski et al. [18] obtained the bounds $w_{\rm DE} = -0.955^{+0.060+0.059}_{-0.066-0.060}$ and $\Omega_m^{(0)} = 0.265^{+0.022+0.018}_{-0.021-0.016}$ (with statistical and systematic errors) from the combined data analysis of CMB and SN Ia. See also Refs. [95, 125–133] for related observational constraints.

8.2.3 BAO

The detection of BAO first reported in 2005 by Eisenstein et. al. [23] in a spectroscopic sample of 46,748 luminous red galaxies observed by the Sloan Digital Sky Survey (SDSS) has provided another test for probing the property of dark energy. Since baryons are strongly coupled to photons prior to the decoupling epoch, the oscillation of sound waves is imprinted in baryon perturbations as well as CMB anisotropies.

The sound horizon at which baryons were released from the Compton drag of photons determines the location of BAO. This epoch, called the drag epoch, occurs at the redshift z_d. The sound horizon at $z = z_d$ is given by $r_s(z_d) = \int_0^{\eta_d} {\rm d}\eta\, c_s(\eta)$, where c_s is the sound speed. According to the fitting formula of z_d by Eisenstein and Hu [134], z_d and $r_s(z_d)$ are constrained to be around $z_d \approx 1020$ and $r_s(z_d) \approx 150$ Mpc.

We observe the angular and redshift distributions of galaxies as a power spectrum $P(k_\perp, k_\parallel)$ in the redshift space, where $k_\perp$ and $k_\parallel$ are the wave numbers perpendicular and parallel to the direction of light, respectively. In principle, we can measure the following two ratios [135]

$$\theta_s(z) = \frac{r_s(z_d)}{d_A^{(c)}(z)}, \qquad \delta z_s(z) = \frac{r_s(z_d)H(z)}{c}, \tag{8.27}$$

where the speed of light c is recovered for clarity. In the first equation, $d_A^{(c)}$ is the comoving angular diameter distance related with the proper angular diameter distance d_A via the relation $d_A^{(c)} = d_A/a = d_A(1+z)$. The quantity $\theta_s(z)$ characterizes the angle orthogonal to the line of sight, whereas the quantity δz_s corresponds to the oscillations along the line of sight.

The current BAO observations are not sufficient to measure both $\theta_s(z)$ and $\delta z_s(z)$ independently. From the spherically averaged spectrum, one can find a combined distance scale ratio given by [135]

$$\left[\theta_s(z)^2 \delta z_s(z)\right]^{1/3} \equiv \frac{r_s(z_d)}{[(1+z)^2 d_A^2(z) c/H(z)]^{1/3}}, \tag{8.28}$$

or, alternatively, the effective distance ratio [23]

$$D_V(z) \equiv \left[(1+z)^2 d_A^2(z) cz/H(z)\right]^{1/3}. \tag{8.29}$$

In 2005, Eisenstein et al. [23] obtained the constraint $D_V(z) = 1370 \pm 64$ Mpc at the redshift $z = 0.35$. In 2007, Percival et al. [24] measured the effective distance ratio defined by

$$r_{\rm BAO}(z) \equiv r_s(z_d)/D_V(z), \tag{8.30}$$

at the two redshifts: $r_{\rm BAO}(z=0.2) = 0.1980 \pm 0.0058$ and $r_{\rm BAO}(z=0.35) = 0.1094$ ± 0.0033. This is based on the data from the 2-degree Field (2dF) Galaxy Redshift Survey. These data provide the observational contour of BAO plotted in Fig. 8.3. From the joint data analysis of SN Ia [18], WMAP 5-year [21], and BAO data [24], Kowalski et al. [18] placed the constraints $w_{\rm DE} = -0.969^{+0.059}_{-0.063}({\rm stat})^{+0.063}_{-0.066}({\rm sys})$ and $\Omega_m^{(0)} = 0.274^{+0.016}_{-0.016}({\rm stat})^{+0.013}_{-0.012}({\rm sys})$ for the constant equation of state of dark energy.

The recent measurement of the 2dF as well as the SDSS data provided the effective distance ratio to be $r_{\rm BAO}(z=0.2) = 0.1905 \pm 0.0061$ and $r_{\rm BAO}(z=0.35) = 0.1097 \pm 0.0036$ [25]. Using these data together with the WMAP 7-year data [22] and the Gaussian prior on the Hubble constant $H_0 = 74.2 \pm 3.6$ km sec^{-1} Mpc^{-1} [110], Komatsu et al. [22] derived the constraint $w_{\rm DE} = -1.10 \pm 0.14$ (68% confidence level) for the constant equation of state in the flat Universe. Adding the high-z SN Ia in their analysis they found the most stringent bound: $w_{\rm DE} = -0.980 \pm 0.053$ (68% confidence level). Hence the ΛCDM model is well consistent with a number of independent observational data.

Finally, we should mention that there are other constraints coming from the cosmic age [136], large-scale clustering [137–139], gamma ray bursts [140–144], and weak lensing [145–150]. So far, we have not found strong evidence for supporting dynamical dark energy models over the ΛCDM model, but future high-precision observations may break this degeneracy.

8.3 Cosmological Constant

The cosmological constant Λ is one of the simplest candidates of dark energy, and as we have seen in the previous section, it is favored by a number of observations. However, if the origin of the cosmological constant is a vacuum energy, it suffers from a serious problem of its energy scale relative to the dark energy density today [26]. The zero-point energy of some field of mass m with momentum k and frequency ω is given by $E = \omega/2 = \sqrt{k^2+m^2}/2$. Summing over the zero-point energies of this field up to a cut-off scale $k_{\rm max}$ ($\gg m$), we obtain the vacuum energy density

$$\rho_{\rm vac} = \int_0^{k_{\rm max}} \frac{{\rm d}^3 k}{(2\pi)^3} \frac{1}{2} \sqrt{k^2+m^2}. \tag{8.31}$$

Since the integral is dominated by the mode with large k ($\gg m$), we find that

$$\rho_{\rm vac} \approx \int_0^{k_{\rm max}} \frac{4\pi k^2 {\rm d}k}{(2\pi)^3} \frac{1}{2} k = \frac{k_{\rm max}^4}{16\pi^2} \,. \tag{8.32}$$

Taking the cut-off scale $k_{\rm max}$ to be the Planck mass $m_{\rm pl}$, the vacuum energy density can be estimated as $\rho_{\rm vac} \simeq 10^{74}$ GeV^4. This is about 10^{121} times larger than the observed value $\rho_{\rm DE}^{(0)} \simeq 10^{-47}\,{\rm GeV}^4$.

Before the observational discovery of dark energy in 1998, most people believed that the cosmological constant is exactly zero and tried to explain why it is so. The vanishing of a constant may imply the existence of some symmetry. In supersymmetric theories, the bosonic degree of freedom has its Fermi counter part, which contributes to the zero point energy with an opposite sign[2]. If supersymmetry is unbroken, an equal number of bosonic and fermionic degrees of freedom is present such that the total vacuum energy vanishes. However, it is known that supersymmetry is broken at sufficient high energies (for the typical scale $M_{\rm SUSY} \approx 10^3$ GeV). Therefore, the vacuum energy is generally nonzero in the world of broken supersymmetry.

Even if supersymmetry is broken, there is a hope to obtain a vanishing Λ or a tiny amount of Λ. In supergravity theory, the effective cosmological constant is given by an expectation value of the potential V for chiral scalar fields φ^i [151]:

$$V(\varphi,\varphi^*) = e^{\kappa^2 K}\left[D_i W (K^{ij^*})(D_j W)^* - 3\kappa^2 |W|^2\right], \tag{8.33}$$

where K and W are the so-called Kähler potential and the superpotential, respectively, which are the functions of φ^i and its complex conjugate φ^{i*}. The quantity K^{ij^*} is an inverse of the derivative $K_{ij^*} \equiv \partial^2 K/\partial\varphi^i \partial\varphi^{j^*}$, whereas the derivative $D_i W$ is defined by $D_i W \equiv \partial W/\partial\varphi^i + \kappa^2 W(\partial K/\partial\varphi^i)$.

The condition $D_i W \neq 0$ corresponds to the breaking of supersymmetry. In this case, it is possible to find scalar field values leading to the vanishing potential ($V = 0$), but this is not in general an equilibrium point of the potential V. Nevertheless, there is a class of Kähler potentials and superpotentials giving a stationary scalar-field configuration at $V = 0$. The gluino condensation model in $E_8 \times E_8$ superstring theory proposed by Dine [153] belongs to this class. The reduction of the 10-dimensional action to the 4-dimensional action gives rise to a so-called modulus field T. This field characterizes the scale of the compactified 6-dimensional manifold. Generally one has another complex scalar field S corresponding to 4-dimensional dilaton/axion fields. The fields T and S are governed by the Kähler potential

$$K(T,S) = -(3/\kappa^2)\ln(T+T^*) - (1/\kappa^2)\ln(S+S^*)\,, \tag{8.34}$$

---

[2] The readers who are not familiar with supersymmetric theories may consult the books [151, 152].

where $(T+T^*)$ and $(S+S^*)$ are positive definite. The field S couples to the gauge fields, while T does not. An effective superpotential for S can be obtained by integrating out the gauge fields under the use of the R-invariance [154]:

$$W(S) = M_{\rm pl}^3 \left[c_1 + c_2 \exp(-3S/2c_3)\right] , \tag{8.35}$$

where c_1, c_2, and c_3 are constants.

Substituting Eqs. (8.34) and (8.35) into Eq. (8.33), we obtain the field potential

$$\begin{aligned} V &= \frac{1}{(T+T^*)^3(S+S^*)}(D_S W)K^{SS^*}(D_S W)^* \\ &= \frac{M_{\rm pl}^4}{(T+T^*)^3(S+S^*)}\left| c_1 + c_2 \exp(-3S/2c_3)\left\{1+\frac{3}{2c_3}(S+S^*)\right\}\right|^2 , \end{aligned} \tag{8.36}$$

where, in the first line, we have used the property $(D_T W)K^{TT^*}(D_T W)^* = 3\kappa^2|W|^2$ for the modulus term. This potential is positive because of the cancellation of the last term in Eq. (8.33). The stationary field configuration with $V=0$ is realized under the condition $D_S W = \partial W/\partial S - W/(S+S^*) = 0$. The derivative, $D_T W = \kappa^2 W \partial K/\partial T = -3W/(T+T^*)$, does not necessarily vanish. When $D_T W \neq 0$, the supersymmetry is broken with a vanishing potential energy. Therefore, it is possible to obtain a stationary field configuration with $V=0$ even if supersymmetry is broken.

The above discussion is based on the lowest-order perturbation theory. This picture is not necessarily valid to all finite orders of perturbation theory because the nonsupersymmetric field configuration is not protected by any symmetry. Moreover, some nonperturbative effect can provide a large contribution to the effective cosmological constant [43]. The so-called flux compactification in type IIB string theory allows us to realize a metastable dS vacuum by taking into account a nonperturbative correction to the superpotential (coming from brane instantons) as well as a number of anti D3branes in a warped geometry [27]. Hence it is not hopeless to obtain a small value of Λ or a vanishing Λ even in the presence of some nonperturbative corrections.

Kachru, Kallosh, Linde, and Trivedi (KKLT) [27] constructed dS solutions in type II string theory compactified on a Calabi–Yau manifold in the presence of flux. The construction of the dS vacua in the KKLT scenario consists of two steps. The first step is to freeze all moduli fields in the flux compactification at a supersymmetric Anti de Sitter (AdS) vacuum. Then, a small number of the anti-D3-brane is added in a warped geometry with a throat so that the AdS minimum is uplifted to yield a dS vacuum with broken supersymmetry. If we want to use the KKLT dS minimum derived above for the present cosmic acceleration, we require that the potential energy $V_{\rm dS}$ at the minimum is of the order of $V_{\rm dS} \simeq 10^{-47}\,{\rm GeV}^4$. Depending on the number of fluxes, there are a vast of dS vacua, which opened up a notion called string landscape [155].

The question why the vacuum we live in has a very small energy density among many possible vacua has been sometimes answered with the anthropic principle

[156, 157]. Using the anthropic arguments, Weinberg put the bound on the vacuum energy density [158]

$$-10^{-123} m_{\rm pl}^4 \lesssim \rho_\Lambda \lesssim 3 \times 10^{-121} m_{\rm pl}^4 \,. \tag{8.37}$$

The upper bound comes from the requirement that the vacuum energy does not dominate over the matter density for the redshift $z \gtrsim 1$. Meanwhile, the lower bound comes from the condition that ρ_Λ does not cancel the present cosmological density. Some people have studied landscape statistics by considering the relative abundance of long-lived, low-energy vacua satisfying the bound (8.37) [159–162]. These statistical approaches are still under study, but it will be interesting to pursue the possibility to obtain high probabilities for the appearance of low-energy vacua.

Even in 1980s, there were some pioneering works for finding a mechanism to make the effective cosmological constant small. For example, let us consider a 4-form field $F^{\mu\nu\lambda\sigma}$ expressed by a unit totally antisymmetric tensor $\varepsilon^{\mu\nu\lambda\sigma}$, as $F^{\mu\nu\lambda\sigma} = c\varepsilon^{\mu\nu\lambda\sigma}$ (c is a constant). Then, the energy density of the 4-form field is given by $F^{\mu\nu\lambda\sigma}F_{\mu\nu\lambda\sigma}/(2\cdot 4!) = c^2/2$. Taking into account a scalar field ϕ with a potential energy $V(\phi)$, the total energy density is $\Lambda = V(\phi) + c^2/2$. In 1984, Linde [163] considered the quantum creation of the Universe and claimed that the final value of Λ can appear with approximately the same probability because $V(\phi)$ can take any initial value such that $\Lambda \approx m_{\rm pl}^4$.

In 1987–1988, Brown and Teltelboim [164, 165] studied the quantum creation of closed membranes by totally antisymmetric tensor and gravitational fields to neutralize the effective cosmological constant with small values. The constant c appearing in the energy density of the 4-form field can be quantized in integer multiples of the membrane charge q, i.e., $c = nq$. If we consider a negative bare cosmological constant $-\Lambda_b$ (as in the KKLT model) in the presence of the flux energy density $n^2q^2/2$, then the effective gravitational constant is given by $\Lambda = -\Lambda_b + n^2q^2/2$. The field strength of the 4-form field is slowly discharged by a quantum Schwinger pair creation of field sources $[nq \to (n-1)q]$. However, in order to get a tiny value of Λ consistent with the dark energy density today, the membrane change q is constrained to be very small (for natural choices of Λ_b) [166].

In 2000, Bousso and Polchinski [167] considered multiple 4-form fields that arise in M-theory compactifications and showed that the small value of Λ can be explained for natural choices of q. More precisely, if we consider $J (> 1)$ 4-form fields as well as J membrane species with charges $q_1, q_2, \cdots, q_J$ and the quantized flux $F_i^{\mu\nu\lambda\sigma} = n_i q_i \varepsilon^{\mu\nu\lambda\sigma}$, the effective cosmological constant is given by

$$\Lambda = -\Lambda_b + \sum_{i=1}^{J} n_i^2 q_i^2/2 \,. \tag{8.38}$$

Bousso and Polchinski [167] showed that, for natural values of charges ($q_i < \mathcal{O}(0.1)$), there exists integers n_i such that $2\Lambda_b < \sum_{i=1}^{J} n_i^2 q_i^2 < 2(\Lambda_b + \Delta\Lambda)$ with $\Delta\Lambda \approx 10^{-47}\,{\rm GeV}^4$. This can be realized for $J > 100$ and $\Lambda_b \approx m_{\rm pl}^4$.

There are some interesting works for decoupling Λ from gravity. In the cascading gravity scenario proposed in Ref. [168], the cosmological constant can be made gravitationally inactive by shutting off large-scale density perturbations. In Ref. [169], an incompressible gravitational Aether fluid was introduced to degraviate the vacuum. In Refs. [170, 171], Padmanabhan showed an example to gauge away the cosmological constant from gravity according to the variational principle different from the standard method. See also Refs. [172–184] for other possibilities to solve the cosmological constant problem. If the cosmological constant is completely decoupled from gravity, it is required to find alternative models of dark energy consistent with observations.

In the subsequent sections, we shall consider alternative models of dark energy, under the assumption that the cosmological constant problem is solved in such a way that it vanishes completely.

8.4 Modified Matter Models

In this section, we will discuss "modified matter models" in which the energy-momentum tensor $T_{\mu\nu}$ on the r.h.s. of the Einstein equations contains an exotic matter source with a negative pressure. The models that belong to this class are quintessence, k-essence, coupled dark energy, and generalized Chaplygin gas.

8.4.1 Quintessence

A canonical scalar field ϕ responsible for dark energy is dubbed quintessence [36, 37] (see also Refs. [28–34] for earlier works). The action of quintessence is described by

$$S = \int \mathrm{d}^4x\sqrt{-g}\left[\frac{1}{2\kappa^2}R - \frac{1}{2}g^{\mu\nu}\partial_\mu\phi\partial_\nu\phi - V(\phi)\right] + S_M\,, \qquad (8.39)$$

where R is a Ricci scalar, and ϕ is a scalar field with a potential $V(\phi)$. As a matter action S_M, we consider perfect fluids of radiation (energy density ρ_r, equation of state $w_r = 1/3$) and nonrelativistic matter (energy density ρ_m, equation of state $w_m = 0$).

In the flat FLRW, background radiation and non-relativistic matter satisfy the continuity equations $\dot{\rho}_r + 4H\rho_r = 0$ and $\dot{\rho}_m + 3H\rho_m = 0$, respectively. The energy density ρ_ϕ and the pressure P_ϕ of the field are $\rho_\phi = \dot{\phi}^2/2 + V(\phi)$ and $P_\phi = \dot{\phi}^2/2 - V(\phi)$, respectively. The continuity equation, $\dot{\rho}_\phi + 3H(\rho_\phi + P_\phi) = 0$, translates to

$$\ddot{\phi} + 3H\dot{\phi} + V_{,\phi} = 0\,, \qquad (8.40)$$

where $V_{,\phi} \equiv dV/d\phi$. The field equation of state is given by

$$w_\phi \equiv \frac{P_\phi}{\rho_\phi} = \frac{\dot{\phi}^2 - 2V(\phi)}{\dot{\phi}^2 + 2V(\phi)}. \tag{8.41}$$

From the Einstein equations (8.4), we obtain the following equations

$$H^2 = \frac{\kappa^2}{3}\left[\frac{1}{2}\dot{\phi}^2 + V(\phi) + \rho_m + \rho_r\right], \tag{8.42}$$

$$\dot{H} = -\frac{\kappa^2}{2}\left(\dot{\phi}^2 + \rho_m + \frac{4}{3}\rho_r\right). \tag{8.43}$$

Although $\{\rho_r, \rho_m\} \gg \rho_\phi$ during radiation and matter eras, the field energy density needs to dominate at late times to be responsible for dark energy. The condition to realize the late-time cosmic acceleration corresponds to $w_\phi < -1/3$, i.e., $\dot{\phi}^2 < V(\phi)$ from Eq. (8.41). This means that the scalar potential needs to be flat enough for the field to evolve slowly. If the dominant contribution to the energy density of the Universe is the slowly rolling scalar field satisfying the condition $\dot{\phi}^2 \ll V(\phi)$, we obtain the approximate relations $3H\dot{\phi} + V_{,\phi} \simeq 0$ and $3H^2 \simeq \kappa^2 V(\phi)$ from Eqs. (8.40) and (8.42), respectively. Hence the field equation of state in Eq. (8.41) is approximately given by

$$w_\phi \simeq -1 + 2\varepsilon_s/3, \tag{8.44}$$

where $\varepsilon_s \equiv \left(V_{,\phi}/V\right)^2/(2\kappa^2)$ is the so-called slow-roll parameter [118]. During the accelerated expansion of the Universe, ε_s is much smaller than 1 because the potential is sufficiently flat. Unlike the cosmological constant, the field equation of state deviates from -1 ($w_\phi > -1$).

Introducing the dimensionless variables $x_1 \equiv \kappa\dot{\phi}/(\sqrt{6}H)$, $x_2 \equiv \kappa\sqrt{V}/(\sqrt{3}H)$, and $x_3 \equiv \kappa\sqrt{\rho_r}/(\sqrt{3}H)$, we obtain the following equations from Eqs. (8.40)–(8.43) [9, 35, 185, 186]:

$$x_1' = -3x_1 + \frac{\sqrt{6}}{2}\lambda x_2^2 + \frac{1}{2}x_1(3 + 3x_1^2 - 3x_2^2 + x_3^2), \tag{8.45}$$

$$x_2' = -\frac{\sqrt{6}}{2}\lambda x_1 x_2 + \frac{1}{2}x_2(3 + 3x_1^2 - 3x_2^2 + x_3^2), \tag{8.46}$$

$$x_3' = -2x_3 + \frac{1}{2}x_3(3 + 3x_1^2 - 3x_2^2 + x_3^2), \tag{8.47}$$

where a prime represents a derivative with respect to $N = \ln a$, and λ is defined by $\lambda \equiv -V_{,\phi}/(\kappa V)$. The density parameters of the field, radiation, and non-relativistic matter are given by $\Omega_\phi = x_1^2 + x_2^2$, $\Omega_r = x_3^2$ and $\Omega_m = 1 - x_1^2 - x_2^2 - x_3^2$, respectively. One has constant λ for the exponential potential [35]

$$V(\phi) = V_0 e^{-\kappa\lambda\phi}, \tag{8.48}$$

in which case the fixed points of the system (8.45) (8.47) can be derived by setting $x_i' = 0$ ($i = 1,2,3$). The fixed point that can be used for dark energy is given by

$$(x_1, x_2, x_3) = \left(\lambda/\sqrt{6}, \sqrt{1-\lambda^2/6}, 0\right), \quad w_\phi = -1+\lambda^2/3, \quad \Omega_\phi = 1. \quad (8.49)$$

The cosmic acceleration can be realized for $w_\phi < -1/3$, i.e., $\lambda^2 < 2$. One can show that in this case the accelerated fixed point is a stable attractor [35]. Hence the solutions finally approach the fixed point (8.49) after the matter era [characterized by the fixed point $(x_1, x_2, x_3) = (0,0,0)$].

If λ varies with time, we have the following relation

$$\lambda' = -\sqrt{6}\lambda^2(\Gamma - 1)x_1, \quad (8.50)$$

where $\Gamma \equiv VV_{,\phi\phi}/V_{,\phi}^2$. For monotonically decreasing potentials, one has $\lambda > 0$ and $x_1 > 0$ for $V_{,\phi} < 0$ and $\lambda < 0$ and $x_1 < 0$ for $V_{,\phi} > 0$. If the condition

$$\Gamma = \frac{VV_{,\phi\phi}}{V_{,\phi}^2} > 1, \quad (8.51)$$

is satisfied, the absolute value of λ decreases toward 0 irrespective of the signs of $V_{,\phi}$ [38]. Then, the solutions finally approach the accelerated "instantaneous" fixed point (8.49) even if λ^2 is larger than 2 during radiation and matter eras [185, 186]. In this case, the field equation of state gradually decreases to -1, so the models showing this behavior are called "freezing" models [187]. The condition (8.51) is the so-called tracking condition under which the field density eventually catches up that of the background fluid.

A representative potential of the freezing model is the inverse power-law potential $V(\phi) = M^{4+n}\phi^{-n}$ ($n > 0$) [31, 38], which can appear in the fermion condensate model as a dynamical supersymmetry breaking [188]. In this case, one has $\Gamma = (n+1)/n > 1$ and hence the tracking condition is satisfied. Unlike the cosmological constant, even if the field energy density is not negligible relative to the background fluid density around the beginning of the radiation era, the field eventually enters the tracking regime to lead to the late-time cosmic acceleration [38]. Another example of freezing models is $V(\phi) = M^{4+n}\phi^{-n}\exp(\alpha\phi^2/m_{\rm pl}^2)$, which has a minimum with a positive energy density at which the field is eventually trapped. This potential is motivated in the framework of supergravity [189].

There is another class of quintessence potentials called "thawing" models [187]. In thawing models, the field with mass m_ϕ has been frozen by the Hubble friction (i.e., the term $H\dot{\phi}$) until recently and then it begins to evolve after H drops below m_ϕ. At early times, the equation of state of dark energy is $w_\phi \simeq -1$, but it begins to grow for $H < m_\phi$. The representative potentials that belong to this class are (i) $V(\phi) = V_0 + M^{4-n}\phi^n$ ($n > 0$) and (ii) $V(\phi) = M^4\cos^2(\phi/f)$. The potential (i) with $n = 1$ was originally proposed by Linde [190] to replace the cosmological constant by a slowly evolving scalar field. In Ref. [191], this was revised to allow for negative values of $V(\phi)$. The Universe will collapse in the future if the system enters the region with $V(\phi) < 0$. The potential (ii) is motivated by the Pseudo-Nambu-Goldstone Boson (PNGB), which was introduced in Ref. [192] in response to the first tentative suggestions for the existence of the cosmological constant. The small mass of the PNGB model required for dark energy is protected against radiative corrections, so this model is favored theoretically. In fact, there are a number of interesting works to explain the small energy scale $M \approx 10^{-3}$ eV required for the PNGB quintessence in supersymmetric theories [193–196]. See Refs. [197–208] for the construction of quintessence potentials in the framework of supersymmetric theories.

The freezing models and the thawing models are characterized by the conditions $w'_\phi \equiv \mathrm{d}w_\phi/\mathrm{d}N < 0$ and $w'_\phi > 0$, respectively. More precisely, the allowed regions for the freezing and thawing models are given by $3w_\phi(1+w_\phi) \lesssim w'_\phi \lesssim 0.2w_\phi(1+w_\phi)$ and $1+w_\phi \lesssim w'_\phi \lesssim 3(1+w_\phi)$, respectively [187] (see Ref. [209] for details). These regions are illustrated in Fig. 8.4. Although the observational data available till now are not sufficient to distinguish freezing and thawing models by the variation of w_ϕ, we may be able to do so with the next decade high-precision observations.

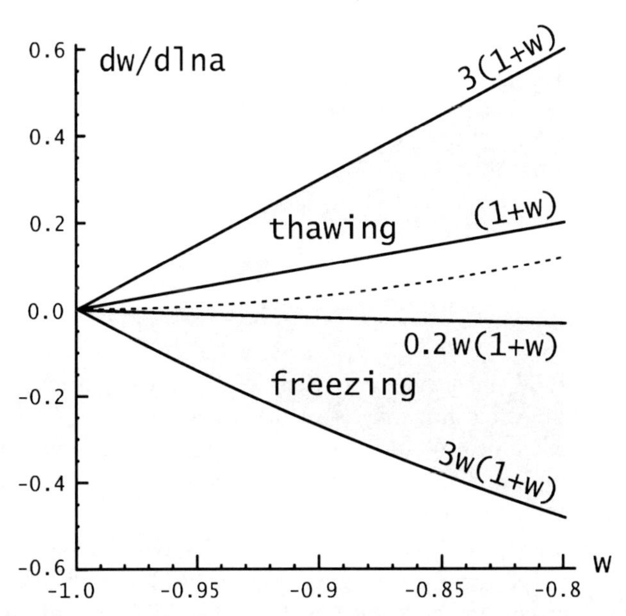

Fig. 8.4 The allowed region in the (w_ϕ, w'_ϕ) plane for thawing and freezing models of quintessence (w_ϕ is denoted as w in the figure). The thawing models correspond to the region between two curves: (a) $w'_\phi = 3(1+w_\phi)$ and (b) $w'_\phi = 1+w_\phi$, whereas the freezing models are characterized by the region between two curves: (c) $w'_\phi = 0.2w_\phi(1+w_\phi)$ and (d) $w'_\phi = 3w_\phi(1+w_\phi)$. The dotted line shows the border between the acceleration and deceleration of the field ($\ddot{\phi} = 0$), which corresponds to $w'_\phi = 3(1+w_\phi)^2$. From Ref. [187].

There is also another useful measure called "statefinder" [210] by which the quintessence models can be distinguished from the ΛCDM model (see also Ref. [211]). The statefinder parameters are defined by

$$r \equiv \frac{1}{aH^3}\frac{\mathrm{d}^3 a}{\mathrm{d}t^3}\,, \qquad s \equiv \frac{r-1}{3(q-1/2)}\,, \tag{8.52}$$

where $q = -\ddot{a}/(aH^2)$ is the deceleration parameter. Let us consider the case in which the dark energy density $\rho_{\rm DE}$ satisfies the continuity equation $\dot{\rho}_{\rm DE} + 3H(1+w_{\rm DE})\rho_{\rm DE} = 0$. In Einstein gravity in which the Friedmann equation $3H^2 = \kappa^2(\rho_{\rm DE}+\rho_m)$ holds, we obtain

$$r = 1 + \frac{9w_{\rm DE}\Omega_{\rm DE}}{2}s\,, \qquad s = 1 + w_{\rm DE} - \frac{\dot{w}_{\rm DE}}{3Hw_{\rm DE}}\,, \tag{8.53}$$

where $\Omega_{\rm DE} \equiv \kappa^2\rho_{\rm DE}/(3H^2)$. The ΛCDM model ($w_{\rm DE} = -1$) corresponds to a point $(r,s) = (1,0)$ in the (r,s) plane, but the quintessence models are characterized by the curves in the region $r < 1$ and $s > 0$ [93]. The Chaplygin gas model we discuss in Sec. 8.4.4 gives rise to the curves in the region $r > 1$ and $s < 0$. Hence one can distinguish between dark energy models by using the statefinders defined in Eq. (8.52).

It is possible to reconstruct quintessence potentials from the observational data of SN Ia. Neglecting the contribution of radiation and using the relation $\rho_m = \rho_m^{(0)}(1+z)^3$ for nonrelativistic matter, we get the following equations from Eqs. (8.42) and (8.43) [91, 92, 212, 213]:

$$\frac{\kappa^2}{2}\left(\frac{\mathrm{d}\phi}{\mathrm{d}z}\right)^2 = \frac{1}{1+z}\frac{\mathrm{d}\ln E(z)}{\mathrm{d}z} - \frac{3\Omega_m^{(0)}}{2}\frac{1+z}{E(z)^2}\,, \tag{8.54}$$

$$\frac{\kappa^2 V}{3H_0^2} = E(z) - \frac{1+z}{6}\frac{\mathrm{d}E(z)^2}{\mathrm{d}z} - \frac{1}{2}\Omega_m^{(0)}(1+z)^3\,, \tag{8.55}$$

where $E(z) = H(z)/H_0$. Note that we have changed the time derivative to the derivative with respect to the redshift z, by using the relation $\mathrm{d}t = -\mathrm{d}z/[H(1+z)]$. Integrating Eq. (8.54), the field ϕ is known as a function of z. Inverting $\phi(z)$ to $z(\phi)$ and plugging it into Eq. (8.55), one can reconstruct the potential V with respect to ϕ by using the information of the observationally known values of $H(z)$ and $H'(z)$ as well as $\Omega_m^{(0)}$. We caution, however, that the actual observational data (such as the luminosity distance) are obtained at discrete values of redshifts. Hence we need some smoothing process for reconstructing the potential $V(\phi)$ and the field equation of state $w_\phi(z)$. This smoothing was already discussed in Sec. 8.2.1.

From the viewpoint of particle physics, the quintessence energy density can be comparable to the background fluid density in the early Universe. It is possible to construct quintessence models in which the field energy density is proportional to the fluid density during radiation and matter eras. For the exponential potential (8.48), there is a fixed point giving a constant field density parameter $\Omega_\phi = 3(1+w_M)/\lambda^2$ with $w_\phi = w_M$, where w_M is the fluid equation of state [35]. This is called a scaling solution, which is stable for $\lambda^2 > 3(1+w_M)$. During radiation and matter eras one has $\Omega_\phi = 4/\lambda^2$ and $\Omega_\phi = 3/\lambda^2$, respectively. The big bang nucleosynthesis (BBN) places the bound $\Omega_\phi < 0.045$ (95% confidence level) around the temperature $T = 1$ MeV [214], which gives the bound $\lambda > 9.4$. In this case, however, the scalar field does not exit to the accelerated fixed point given by Eq. (8.49).

There are several ways to allow a transition from the scaling regime to the epoch of cosmic acceleration. One of them is to introduce a field potential that becomes shallow at late times, $V(\phi) = c_1 e^{-\kappa\lambda\phi} + c_2 e^{-\kappa\mu\phi}$ with $\lambda^2 > 3(1+w_M)$ and $\mu^2 < 2$ [215] (see Refs. [214, 216–221] for related works). For this double exponential potential, the field equation of state of the final attractor is given by $w_\phi = -1+\mu^2/3$. Another way is to consider multiple scalar fields with exponential potentials, $V(\phi_1,\phi_2) = c_1 e^{-\kappa\lambda_1\phi_1} + c_2 e^{-\kappa\lambda_2\phi_2}$ [222, 223]. In this case, the phenomenon called assisted inflation [224] occurs for the multi-field exponential potential, even if the individual field has too steep a potential to lead to cosmic acceleration. The scalar field equation of state finally approaches the value $w_\phi = -1+\lambda_{\rm eff}^2/3$, where $\lambda_{\rm eff} = (1/\lambda_1^2 + 1/\lambda_2^2)^{-1/2}$ is smaller than each λ_i $(i=1,2)$. In the presence of three assisting scalar fields, it is possible to realize the observational bound $w_\phi(z=0) < -0.8$ today [225].

There is a class of models dubbed quintessential inflation [226] in which a single scalar field ϕ is responsible for both inflation and dark energy. One example of quintessential inflation is given by $V(\phi) = \lambda(\phi^4+M^4)$ for $\phi < 0$ and $V(\phi) = \lambda M^4/[1+(\phi/M)^n]$ $(n>0)$ for $\phi \geq 0$. In the regime $\phi < 0$ with $|\phi| \gg M$, the potential behaves as $V(\phi) \simeq \lambda\phi^4$, which leads to inflation. In the regime $\phi > 0$ with $\phi \gg M$, one has $V(\phi) \simeq \lambda M^{4+n}\phi^{-n}$, which leads the late-time cosmic acceleration. Since the potential does not have a minimum, the reheating after inflation proceeds with a gravitational particle production. Although this process is not efficient in general, it may be possible to make the reheating more efficient under the instant preheating mechanism proposed in Ref. [227]. See Refs. [228–239] about related works on quintessential inflation.

8.4.2 k-Essence

Scalar fields, with non-canonical kinetic terms often appear in particle physics. In general the action for such theories can be expressed as

$$S = \int d^4x\sqrt{-g}\left[\frac{1}{2\kappa^2}R + P(\phi,X)\right] + S_M\,, \tag{8.56}$$

where $P(\phi,X)$ is a function in terms of a scalar field ϕ and its kinetic energy $X = -(1/2)g^{\mu\nu}\partial_\mu\phi\partial_\nu\phi$, and S_M is a matter action. Even in the absence of the field potential $V(\phi)$, it is possible to realize the cosmic acceleration due to the kinetic energy X [240]. The application of these theories to dark energy was first carried out by Chiba et al. [39]. In Ref. [40], this was extended to more general cases and the models based on the action (8.56) were named "k-essence." The action (8.56) includes a wide variety of theories listed below.

- (i) Low-energy effective string theory
  The action of low-energy effective string theory in the presence of a higher-order derivative term $(\tilde{\nabla}\phi)^4$ is given by [241, 242]

$$S = \frac{1}{2\kappa^2}\int \mathrm{d}^4\tilde{x}\sqrt{-\tilde{g}}\left[F(\phi)\tilde{R} + \omega(\phi)(\tilde{\nabla}\phi)^2 + \alpha' B(\phi)(\tilde{\nabla}\phi)^4 + \mathcal{O}(\alpha'^2)\right], \quad (8.57)$$

  which is derived by the expansion in terms of the Regge slope parameter α' (this is related to the string mass scale M_s via the relation $M_s = \sqrt{2/\alpha'}$). The scalar field ϕ, dubbed a dilaton field, is coupled to the Ricci scalar R with the strength $F(\phi)$. This frame is called the Jordan frame, in which the tilde is used in the action (8.57). Under a conformal transformation, $g_{\mu\nu} = F(\phi)\tilde{g}_{\mu\nu}$, we obtain the action in the Einstein frame [240]

$$S_E = \int \mathrm{d}^4x\sqrt{-g}\left[\frac{1}{2\kappa^2}R + K(\phi)X + L(\phi)X^2 + \cdots\right], \quad (8.58)$$

  where $K(\phi) = 3(F_{,\phi}/F)^2 - 2\omega/F$ and $L(\phi) = 2\alpha' B(\phi)/\kappa^2$.

- (ii) Ghost condensate
  The theories with a negative kinetic energy $-X$ generally suffers from the vacuum instability [243, 244], but the presence of the quadratic term X^2 can evade this problem. The model constructed in this vein is the ghost condensate model characterized by the Lagrangian [245]

$$P = -X + X^2/M^4, \quad (8.59)$$

  where M is a constant. A more general version of this model, called the dilatonic ghost condensate [246], is

$$P = -X + e^{\kappa\lambda\phi}X^2/M^4, \quad (8.60)$$

  which is motivated by a dilatonic higher-order correction to the tree-level action [as we have discussed in the case (i)].

- (iii) Tachyon
  A tachyon field appears as an unstable mode of D-branes non-Bogomol'nyi-Prasad–Sommerfield [non-BPS] branes. The effective 4-dimensional Lagrangian is given by [247–249]

$$P = -V(\phi)\sqrt{1-2X}, \quad (8.61)$$

where $V(\phi)$ is a potential of the tachyon field ϕ. Although the tachyon model is difficult to be compatible with inflation in the early Universe because of the problem for ending inflation [250–252], one can use it for dark energy provided that the potential is shallower than $V(\phi) = V_0\phi^{-2}$ [253–258].

- (iv) Dirac-Born-Infeld (DBI) theory
In the so-called "D-cceleration" mechanism in which a scalar field ϕ parametrizes a direction on the approximate Coulomb branch of the system in $\mathcal{N} = 4$ supersymmetric Yang Mills theory, the field dynamics can be described by the DBI action for a probe D3-brane moving in a radial direction of the AdS space-time [259, 260]. The Lagrangian density with the field potential $V(\phi)$ is given by

$$P = -f(\phi)^{-1}\sqrt{1-2f(\phi)X} + f(\phi)^{-1} - V(\phi), \tag{8.62}$$

where $f(\phi)$ is a warped factor of the AdS throat. In this theory, one can realize the acceleration of the Universe even in the regime where $2f(\phi)X$ is close to 1. The application of this theory to dark energy has been carried out in Refs. [261–263].

For the theories with the action (8.56), the pressure P_ϕ and the energy density ρ_ϕ of the field are $P_\phi = P$ and $\rho_\phi = 2XP_{,X} - P$, respectively. The equation of state of k-essence is given by

$$w_\phi = \frac{P_\phi}{\rho_\phi} = \frac{P}{2XP_{,X} - P}. \tag{8.63}$$

As long as the condition $|2XP_{,X}| \ll |P|$ is satisfied, w_ϕ can be close to -1. In the ghost condensate model (8.59), we have

$$w_\phi = \frac{1 - X/M^4}{1 - 3X/M^4}, \tag{8.64}$$

which gives $-1 < w_\phi < -1/3$ for $1/2 < X/M^4 < 2/3$. In particular, the de Sitter solution ($w_\phi = -1$) is realized at $X/M^4 = 1/2$. Since the field energy density is $\rho_\phi = M^4/4$ at the de Sitter point, it is possible to explain the cosmic acceleration today for $M \simeq 10^{-3}$ eV.

In order to discuss stability conditions of k-essence in the ultraviolet (UV) regime, we decompose the field into the homogenous and perturbed parts as $\phi(t, \mathbf{x}) = \phi_0(t) + \delta\phi(t, \mathbf{x})$ in the Minkowski background and derive the Lagrangian and the Hamiltonian for perturbations. The resulting second-order Hamiltonian reads [246]

$$\delta H = (P_{,X} + 2XP_{,XX})\frac{(\delta\dot{\phi})^2}{2} + P_{,X}\frac{(\nabla\delta\phi)^2}{2} - P_{,\phi\phi}\frac{(\delta\phi)^2}{2}. \tag{8.65}$$

The term $P_{,\phi\phi}$ is related with the effective mass of the field, which is unimportant in the UV regime as long as the field is responsible for dark energy. The positivity of the first two terms in Eq. (8.65) leads to the following stability conditions

$$P_{,X} + 2XP_{,XX} \geq 0, \qquad P_{,X} \geq 0. \tag{8.66}$$

The phantom model with a negative kinetic energy $-X$ with a potential $V(\phi)$, i.e., $P = -X - V(\phi)$, do not satisfy the above conditions. Although the phantom model with $P = -X - V(\phi)$ can lead to the background cosmological dynamics allowed by SN Ia observations ($w_\phi < -1$) [243, 264, 265], it suffers from a catastrophic particle production of ghosts and normal fields because of the instability of the vacuum [243, 244] (see Refs. [136, 266–280] for related works). This problem is overcome in the ghost condensate model (8.59) in which the conditions (8.66) are satisfied for $X/M^4 > 1/2$ [3]. Thus, the successful k-essence models need to be constructed to be consistent with the conditions (8.66), while the field is responsible for dark energy under the condition $|2XP_{,X}| \ll |P|$.

The propagation speed c_s of the field is given by [282]

$$c_s^2 = \frac{P_{\phi,X}}{\rho_{\phi,X}} = \frac{P_{,X}}{P_{,X} + 2XP_{,XX}}, \tag{8.67}$$

which is positive under the conditions (8.66). The speed c_s remains sub-luminal provided that

$$P_{,XX} > 0. \tag{8.68}$$

This condition is ensured for the models (8.59–8.62).

There are some k-essence models proposed to solve the coincidence problem of dark energy. One example is [40, 41]

$$P = \frac{1}{\phi^2}\left(-2.01 + 2\sqrt{1+X} + 3 \cdot 10^{-17}X^3 - 10^{-24}X^4\right). \tag{8.69}$$

In these models, the solutions can finally approach the accelerating phase even if they start from relatively large values of the k-essence energy density Ω_ϕ in the radiation era. In such cases, however, there is a period in which the sound speed becomes superluminal before reaching the accelerated attractor [283]. Moreover, it was shown that the basins of attraction of a radiation scaling solution in such models are restricted to be very small [284]. We stress that these problems arise only for the k-essence models constructed to solve the coincidence problem.

8.4.3 Coupled Dark Energy

Since the energy density of dark energy is the same order as that of dark matter in the present Universe, this implies that dark energy may have some relation with dark matter. In this section, we discuss the cosmological viability of coupled dark energy models and related topics such as scaling solutions, the chameleon mechanism, and varying α.

---

[3] It is possible that the dilatonic ghost condensate model crosses the cosmological constant boundary $w_\phi = -1$, but the quantum instability problem is present in the region $w_\phi < -1$. This crossing does not occur for the single-field k-essence Lagrangian with a linear function of X [281].

8.4.3.1 The Coupling between Dark Energy and Dark Matter

In the flat FLRW cosmological background, a general coupling between dark energy (with energy density $\rho_{\rm DE}$ and equation of state $w_{\rm DE}$) and dark matter (with energy density ρ_m) may be described by the following equations

$$\dot{\rho}_{\rm DE} + 3H(1 + w_{\rm DE})\rho_{\rm DE} = -\beta\,, \tag{8.70}$$

$$\dot{\rho}_m + 3H\rho_m = +\beta\,, \tag{8.71}$$

where β is the rate of the energy exchange in the dark sector.

There are several forms of couplings proposed so far. Two simple examples are given by

$$\text{(A)}\ \ \beta = \kappa Q \rho_m \dot{\phi}\,, \tag{8.72}$$

$$\text{(B)}\ \ \beta = \alpha H \rho_m\,, \tag{8.73}$$

where Q and α are dimensionless constants. The coupling (A) arises in scalar-tensor theories after the conformal transformation to the Einstein frame [285–288]. In general, the coupling Q is field-dependent [289, 290], but Brans–Dicke (BD) theory [291] (including $f(R)$ gravity) gives rise to a constant coupling [292]. The coupling (B) is more phenomenological, but this form is useful to place observational bounds from the cosmic expansion history. Several authors studied other couplings of the forms $\beta = (\alpha_m \rho_m + \alpha_{\rm DE}\rho_{\rm DE})H$ [293–295], $\beta = \alpha \Omega_{\rm DE} H$ [296–298], and $\beta = \Gamma \rho_m$ [299–301]. See also Refs. [302–318] for related works.

(A) The Coupling (A)

Let us consider the coupling (A) in the presence of a coupled quintessence field with the exponential potential (8.48). We assume that the coupling Q is constant. Taking into account radiation uncoupled to dark energy ($\rho_r \propto a^{-4}$) the Friedmann equation is given by $3H^2 = \kappa^2(\rho_{\rm DE} + \rho_m + \rho_r)$, where $\rho_{\rm DE} = \dot{\phi}^2/2 + V(\phi)$. Introducing the dimensionless variables $x_1 = \kappa\dot{\phi}/(\sqrt{6}H)$, $x_2 = \kappa\sqrt{V}/(\sqrt{3}H)$, and $x_3 = \kappa\sqrt{\rho_r}/(\sqrt{3}H)$ as in Sec. 8.4.1, we obtain

$$x_1' = -3x_1 + \frac{\sqrt{6}}{2}\lambda x_2^2 + \frac{1}{2}x_1(3 + 3x_1^2 - 3x_2^2 + x_3^2) - \frac{\sqrt{6}}{2}Q(1 - x_1^2 - x_2^2 - x_3^2)\,, \tag{8.74}$$

and the same differential equations for x_2 and x_3 as given in Eqs. (8.46) and (8.47). For this dynamical system, there is a scalar-field dominated fixed point given in Eq. (8.49) as well as the radiation point $(x_1, x_2, x_3) = (0, 0, 1)$. In the presence of the coupling Q, the standard matter era is replaced by a "ϕ-matter-dominated epoch (ϕMDE)" [287] characterized by

$$(x_1, x_2, x_3) = (-\sqrt{6}Q/3, 0, 0)\,, \qquad \Omega_\phi = 2Q^2/3\,, \qquad w_\phi = 1\,. \tag{8.75}$$

Defining the effective equation of state

$$w_{\rm eff} = -1 - 2\dot{H}/(3H^2)\,, \tag{8.76}$$

one has $w_{\rm eff} = 2Q^2/3$ for the ϕMDE, which is different from 0 in the uncoupled case. Provided that $2Q^2/3 < 1$, the ϕMDE is a saddle followed by the accelerated point (8.49) [287].

The evolution of the scale factor during the ϕMDE is given by $a \propto t^{2/(3+2Q^2)}$, which is different from that in the uncoupled quintessence. This leads to a change to the CMB shift parameter defined in Eq. (8.23). From the CMB likelihood analysis, the strength of the coupling is constrained to be $|Q| < 0.1$ [287]. The evolution of matter density perturbations is also subject to change by the effect of the coupling. Under a quasi-static approximation on sub-horizon scales, the matter perturbation δ_m obeys the following equation [319, 320]

$$\ddot{\delta}_m + (2H + Q\dot{\phi})\dot{\delta}_m - 4\pi G_{\rm eff}\rho_m\delta_m \simeq 0\,, \tag{8.77}$$

where the effective gravitational coupling is given by $G_{\rm eff} = (1+2Q^2)G$. During the ϕMDE, one can obtain the analytic solution to Eq. (8.77), as $\delta_m \propto a^{1+2Q^2}$. Hence the presence of the coupling Q leads to a larger growth rate relative to the uncoupled quintessence. We can parameterize the growth rate of matter perturbations, as [321]

$$f \equiv \frac{\dot{\delta}_m}{H\delta_m} = (\Omega_m)^\gamma\,, \tag{8.78}$$

where $\Omega_m \equiv \kappa^2\rho_m/(3H^2)$ is the density parameter of nonrelativistic matter. In the ΛCDM model the growth index γ can be approximately given by $\gamma \simeq 0.55$ [322, 323]. In the coupled quintessence, the growth rate can be fitted to the numerical solution by the formula $f = (\Omega_m)^\gamma(1+cQ^2)$, where $c = 2.1$ and $\gamma = 0.56$ are the best-fit values [324]. Using the galaxy and Lyman-α power spectra, the growth index γ and the coupling Q are constrained to be $\gamma = 0.6^{+0.4}_{-0.3}$ and $|Q| < 0.52$ (95% confidence level), respectively. This is weaker than the bound coming from the CMB constraint [324]. We also note that the equation for matter perturbations has been derived for the coupled k-essence scenario with a field-dependent coupling $Q(\phi)$ [325, 326]. In principle, it is possible to reconstruct the coupling from observations if the evolution of matter perturbations is known accurately [327].

There is another interesting coupled dark energy scenario called mass-varying neutrino [328–332] in which the mass m_ν of the neutrino depends on the quintessence field ϕ. The energy density ρ_ν and the pressure P_ν of neutrinos can be determined by assuming a Fermi-Dirac distribution with the neglect of the chemical potential. It then follows that the field ϕ obeys the equation of motion [331]

$$\ddot{\phi} + 3H\dot{\phi} + V_{,\phi} = -\kappa Q(\phi)(\rho_\nu - 3P_\nu)\,, \tag{8.79}$$

where $Q(\phi) \equiv \frac{{\rm d}\ln m_\nu(\phi)}{{\rm d}\phi}$. In the relativistic regime in which the neutrino mass m_ν is much smaller than the neutrino temperature T_ν, the r.h.s. of Eq. (8.79) is suppressed because of the relation $\rho_\nu \simeq 3P_\nu$. In the nonrelativistic regime with $m_\nu \gg T_\nu$, the pressure P_ν is much smaller than the energy density ρ_ν, in which case the field equation (8.79) mimics Eq. (8.70) with the coupling (A). In the mass-varying neutrino scenario, the field-dependent mass of neutrinos determines the strength of the coupling $Q(\phi)$. In the nonrelativistic regime, the neutrino energy density is approximately given by $\rho_\nu \simeq n_\nu m_\nu(\phi)$, where n_ν is the number density of neutrinos. Then, the effective potential of the field is given by $V_{\rm eff}(\phi) \simeq V(\phi) + n_\nu m_\nu(\phi)$, which gives rise to a minimum for a runaway quintessence potential $V(\phi)$. Since the field equation of state in this regime is $w_\phi \simeq -V(\phi)/[V(\phi)+n_\nu m_\nu(\phi)]$, one has $w_\phi \simeq -1$ for $n_\nu m_\nu \ll V(\phi)$. See Refs. [333–338] for a number of cosmological consequences of the mass-varying neutrino scenario.

(B) The Coupling (B)

Let us consider the coupling (B) given in Eq. (8.73). For constant δ, one can integrate Eq. (8.71) as $\rho_m = \rho_m^{(0)}(1+z)^{3-\alpha}$, where z is a redshift. If the equation of state $w_{\rm DE}$ is constant, one obtains the following integrated solution for Eq. (8.70):

$$\rho_{\rm DE} = \rho_{\rm DE}^{(0)}(1+z)^{3(1+w_{\rm DE})} + \rho_m^{(0)}\frac{\alpha}{\alpha+3w_{\rm DE}}\left[(1+z)^{3(1+w_{\rm DE})} - (1+z)^{3-\alpha}\right]. \quad (8.80)$$

The Friedmann equation, $3H^2 = \kappa^2(\rho_{\rm DE}+\rho_m)$, gives the parametrization for the normalized Hubble parameter $E(z) = H(z)/H_0$, as

$$E^2(z) = \Omega_{\rm DE}^{(0)}(1+z)^{3(1+w_{\rm DE})} + \frac{1-\Omega_{\rm DE}^{(0)}}{\alpha+3w_{\rm DE}}\left[\alpha(1+z)^{3(1+w_{\rm DE})} + 3w_{\rm DE}(1+z)^{3-\alpha}\right]. \quad (8.81)$$

This parametrization can be used to place observational constraints on the coupling α. The combined data analysis using the observational data of the 5-year Supernova Legacy Survey (SNLS) [13], the CMB shift parameter from the 3-year WMAP [20], and the BAO [23] shows that α and $w_{\rm DE}$ are constrained to be $-0.08 < \alpha < 0.03$ and $-1.16 < w_{\rm DE} < -0.91$ (95% confidence level) [298]. In Ref. [300], it was shown that, for constant $w_{\rm DE}$, cosmological perturbations are subject to nonadiabatic instabilities in the early radiation era. This problem can be alleviated by considering the time-dependent $w_{\rm DE}$ satisfying the condition $w_{\rm DE} > -4/5$, at early times [317].

It is also possible to extend the analysis to the case in which the coupling α varies in time. Dalal et al. [296] assumed the scaling relation, $\rho_{\rm DE}/\rho_m = (\rho_{\rm DE}^{(0)}/\rho_m^{(0)})a^\xi$, where ξ is a constant. For constant $w_{\rm DE}$, the coupling α is expressed in the form

$$\alpha(z) = \frac{\alpha_0}{\Omega_{\rm DE}^{(0)} + (1-\Omega_{\rm DE}^{(0)})(1+z)^\xi}, \quad (8.82)$$

where $\alpha_0 = -(\xi + 3w_{\rm DE})\Omega_{\rm DE}^{(0)}$. If $\xi > 0$, $\alpha(z)$ decreases for higher z. In this case, the Hubble parameter can be parametrized as

$$E^2(z) = (1+z)^3\left[1-\Omega_{\rm DE}^{(0)} + \Omega_{\rm DE}^{(0)}(1+z)^{-\xi}\right]^{-3w_{\rm DE}/\xi}. \quad (8.83)$$

The combined data analysis using the SNLS, WMAP 3-year, and the BAO gives the bounds $-0.4 < \alpha_0 < 0.1$ and $-1.18 < w_{\rm DE} < -0.91$ (95% confidence level) [298]. The ΛCDM model ($\alpha = 0$, $w_{\rm DE} = -1$) remains a good fit to the data.

8.4.3.2 Coupled Dark Energy and Coincidence Problem

In the coupled quintessence with an exponential potential ($V(\phi) = V_0 e^{-\kappa\lambda\phi}$ with $\lambda > 0$), the ϕMDE scaling solution with $\Omega_\phi = 2Q^2/3 =$ constant replaces the standard matter era. In this model, there is another scaling solution given by [287, 306]

$$(x_1, x_2, x_3) = \left(\frac{\sqrt{6}}{2(Q+\lambda)}, \sqrt{\frac{2Q(Q+\lambda)+3}{2(Q+\lambda)^2}}, 0\right), \qquad \Omega_\phi = \frac{Q(Q+\lambda)+3}{(Q+\lambda)^2},$$

$$w_\phi = -\frac{Q(Q+\lambda)}{Q(Q+\lambda)+3}, \qquad w_{\rm eff} = -\frac{Q}{Q+\lambda}. \tag{8.84}$$

In the presence of the coupling Q, the condition for the late-time cosmic acceleration, $w_{\rm eff} < -1/3$, is satisfied for $Q > \lambda/2 > 0$ or $Q < -\lambda < 0$. Then, the scaling solution (8.84) can give rise to the global attractor with $\Omega_\phi \simeq 0.7$. The ϕMDE solution (8.75) followed by the accelerated scaling solution (8.84) may be used for alleviating the coincidence problem because dark energy and dark matter follow the same scaling relation from the end of the radiation era. In Ref. [287] it was shown, however, that the coupled quintessence with an exponential potential does not allow for such cosmological evolution for the viable parameter space in the (Q, λ) plane consistent with observational constraints.

It is possible to extend the analysis to coupled k-essence models described by the action (8.56) in the presence of the coupling (8.72). From the requirement that the density parameter Ω_ϕ ($\neq 0$) and the equation of state w_ϕ are constants to realize scaling solutions, one can show that the Lagrangian density takes the following form [246, 339]

$$P = X\, g(Xe^{\kappa\lambda\phi}), \tag{8.85}$$

where λ is a constant, and g is an arbitrary function in terms of $Y \equiv Xe^{\kappa\lambda\phi}$. The result (8.85) is valid for constant Q, but it can be generalized to a field-dependent coupling $Q(\phi)$ [340]. The quintessence with an exponential potential ($P = X - ce^{-\kappa\lambda\phi}$) corresponds to $g(Y) = 1 - c/Y$. Since the dilatonic ghost condensation model (8.60) corresponds to $g(Y) = -1 + Y/M^4$, this model has a scaling solution. We can also show that the tachyon field with the Lagrangian density $P = -V(\phi)\sqrt{1-2X}$ also has a scaling solution for the potential $V(\phi) \propto \phi^{-2}$ [257]. Even when the Hubble parameter squared H^2 is proportional to the energy density ρ^n, it is possible to obtain the Lagrangian density having scaling solutions in the form $P = X^{1/n} g(Xe^{n\kappa\lambda\phi})$ [339]. The cosmological dynamics of scaling solutions in such cases (including the high-energy regime [341, 342] in the Randall-Sundrum scenario [343, 344]) have been discussed by a number of authors [345–347].

For the Lagrangian density (8.85), there are two fixed points relevant to dark energy. Defining the dimensionless variables $x \equiv \dot{\phi}/(\sqrt{6}H)$ and $y \equiv e^{-\lambda\phi/2}/(\sqrt{3}H)$ (in the unit of $\kappa^2 = 1$), they are given by [348]

- (A) Scalar-field dominated solution

$$x_A = \frac{\lambda}{\sqrt{6}P_{,X}}, \quad \Omega_\phi = 1, \quad w_{\rm eff} = w_\phi = -1 + \frac{\lambda^2}{3P_{,X}}. \tag{8.86}$$

- (B) Scaling solution

$$x_B = \frac{\sqrt{6}}{2(Q+\lambda)}, \quad \Omega_\phi = \frac{Q(Q+\lambda)+3P_{,X}}{(Q+\lambda)^2}.$$

$$w_{\rm eff} = -\frac{Q}{Q+\lambda}, \quad w_\phi = -\frac{Q(Q+\lambda)}{Q(Q+\lambda)+3P_{,X}}. \tag{8.87}$$

The points (A) and (B) are responsible for the cosmic acceleration for (A) $\lambda^2/P_{,X} < 2$ and (B) $Q > \lambda/2 > 0$ or $Q < -\lambda < 0$, respectively. From the stability analysis about the fixed points it follows that, when the point (A) is stable, the point (B) is not stable, and vice versa [348].

The ϕMDE corresponds to a fixed point at which the kinetic energy of the field dominates over the potential energy, i.e., $x \neq 0$ and $y = 0$. Since the quantity $Y = Xe^{\kappa\lambda\phi}$ can be expressed as $Y = x^2/y^2$, the function $g(Y)$ cannot be singular at $y = 0$ for the existence of the ϕMDE. Then, the function $g(Y)$ should be expanded in negative powers of Y, i.e.,

$$g(Y) = c_0 + \sum_{n>0} c_n Y^{-n} = c_0 + \sum_{n>0} c_n (y^2/x^2)^n, \tag{8.88}$$

which includes the quintessence with an exponential potential. For this form of $g(Y)$, there is the following ϕMDE point (C):

$$(x_C, y_C) = \left(-\frac{\sqrt{6}Q}{3c_0}, 0\right), \quad \Omega_\phi = w_{\rm eff} = \frac{2Q^2}{3c_0}, \quad w_\phi = 1, \tag{8.89}$$

together with the purely kinetic point $(x,y) = (\pm 1/\sqrt{c_0}, 0)$ and $\Omega_\phi = 1$ for $c_0 > 0$.

An ideal cosmological trajectory that alleviates the coincidence problem of dark energy should be the ϕMDE (C) followed by the point (B). However, it was shown in Ref. [340] that such a trajectory is not allowed because the solutions cannot cross the singularity at $x = 0$ as well as another singularity associated with the sound speed. For example, when $c_0 > 0$, we find that $x_B > 0$ and $x_C < 0$ for $Q > \lambda/2 > 0$, whereas $x_B < 0$ and $x_C > 0$ for $Q < -\lambda < 0$. These points are separated between the line $x = 0$ at which the function (8.88) diverges. The ϕMDE solution chooses the accelerated point (A) as a final attractor. The above discussion shows that the coincidence problem is difficult to be solved even for the general Lagrangian density (8.85) that has scaling solutions. This problem mainly comes from the fact that a large coupling Q required for the existence of a viable scaling solution (B) is not compatible with a small coupling Q required for the existence of the ϕMDE. We need a rapidly growing coupling to realize such a transition [349].

8.4.3.3 Chameleon Mechanism

If a scalar field ϕ is coupled to baryons as well as dark matter, this gives rise to a fifth force interaction that can be constrained experimentally. A large coupling of the order of unity arises in modified gravity theories as well as superstring theories. In such cases, we need to suppress such a strong interaction with baryons for the compatibility with local gravity experiments. There is a way to screen the fifth force under the so-called chameleon mechanism [350, 351] in which the field mass is different depending on the matter density in the surrounding environment. If the field is sufficiently heavy in the regions of high density, a spherically symmetric body can have a "thin shell" around its surface such that the effective coupling between the field and matter is suppressed outside the body.

The action of a chameleon scalar field ϕ with a potential $V(\phi)$ is given by

$$S = \int \mathrm{d}^4 x \sqrt{-g} \left[\frac{1}{2\kappa^2} R - \frac{1}{2} g^{\mu\nu} \partial_\mu \phi \partial_\nu \phi - V(\phi) \right] + \int \mathrm{d}^4 x \mathscr{L}_M (g_{\mu\nu}^{(i)}, \Psi_M^{(i)}), \quad (8.90)$$

where g is the determinant of the metric $g_{\mu\nu}$ (in the Einstein frame) and $\mathscr{L}_M$ is a matter Lagrangian with matter fields $\Psi_M^{(i)}$ coupled to a metric $g_{\mu\nu}^{(i)}$. The metric $g_{\mu\nu}^{(i)}$ is related to the Einstein frame metric $g_{\mu\nu}$ via $g_{\mu\nu}^{(i)} = e^{2\kappa Q_i \phi} g_{\mu\nu}$, where Q_i are the strengths of the couplings for each matter component with the field ϕ. The typical field potential is chosen to be of the runaway type (such as $V(\phi) = M^{4+n}\phi^{-n}$). We also restrict the form of the potential such that $|V_{,\phi}| \to \infty$ as $\phi \to 0$.

Varying the action (8.90) with respect to ϕ, we obtain the field equation

$$\Box \phi - V_{,\phi} = -\sum_i \kappa Q_i e^{4\kappa Q_i \phi} g_{(i)}^{\mu\nu} T_{\mu\nu}^{(i)}, \quad (8.91)$$

where $T_{\mu\nu}^{(i)} = -(2/\sqrt{-g^{(i)}}) \delta \mathscr{L}_M / \delta g_{(i)}^{\mu\nu}$ is the stress-energy tensor for the i-th form of matter. For non-relativistic matter, we have $g_{(i)}^{\mu\nu} T_{\mu\nu}^{(i)} = -\tilde{\rho}_i$, where $\tilde{\rho}_i$ is an energy density. It is convenient to introduce the energy density $\rho_i \equiv \tilde{\rho}_i e^{3\kappa Q_i \phi}$, which is conserved in the Einstein frame. In the following, let us consider the case in which the couplings Q_i are the same for all species, i.e., $Q_i = Q$. In a spherically symmetric space-time under the weak gravitational background (i.e., neglecting the backreaction of gravitational potentials), Eq. (8.91) reads

$$\frac{\mathrm{d}^2 \phi}{\mathrm{d}r^2} + \frac{2}{r} \frac{\mathrm{d}\phi}{\mathrm{d}r} = \frac{\mathrm{d}V_{\mathrm{eff}}}{\mathrm{d}\phi}, \quad (8.92)$$

where r is a distance from the center of symmetry, and $V_{\rm eff}$ is the effective potential given by

$$V_{\rm eff}(\phi) = V(\phi) + e^{\kappa Q \phi} \rho \tag{8.93}$$

and $\rho \equiv \sum_i \rho_i$. For the runaway potential with $V_{,\phi} < 0$, the positive coupling Q leads to a minimum of the effective potential. In $f(R)$ gravity, the negative coupling ($Q = -1/\sqrt{6}$) gives rise to a minimum for the potential with $V_{,\phi} > 0$ (as we will see in Sec. 8.5.1.3).

We assume that a spherically symmetric body has a constant density $\rho = \rho_A$ inside the body ($r < r_c$) and that the energy density outside the body ($r > r_c$) is $\rho = \rho_B$. The mass M_c of the body and the gravitational potential Φ_c at the radius r_c are given by $M_c = (4\pi/3) r_c^3 \rho_A$ and $\Phi_c = GM_c/r_c$, respectively. The effective potential $V_{\rm eff}(\phi)$ has two minima at the field values ϕ_A and ϕ_B satisfying $V'_{\rm eff}(\phi_A) = 0$ and $V'_{\rm eff}(\phi_B) = 0$, respectively. The former corresponds to the region with a high density that gives rise to a heavy mass squared $m_A^2 \equiv V''_{\rm eff}(\phi_A)$, whereas the latter to the lower density region with a lighter mass squared $m_B^2 \equiv V''_{\rm eff}(\phi_B)$. When we consider the "dynamics" of the field ϕ according to Eq. (8.92), we need to consider the inverted effective potential $(-V_{\rm eff})$ having two *maxima* at $\phi = \phi_A$ and $\phi = \phi_B$.

The boundary conditions for the field are given by $\frac{{\rm d}\phi}{{\rm d}r}(r = 0) = 0$ and $\phi(r \to \infty) = \phi_B$. The field ϕ is at rest at $r = 0$ and begins to roll down the potential when the matter-coupling term $\kappa Q \rho_A e^{\kappa Q \phi}$ becomes important at a radius r_1 in Eq. (8.92). As long as r_1 is close to r_c so that $\Delta r_c \equiv r_c - r_1 \ll r_c$, the body has a thin shell inside the body. Since the field acquires a sufficient kinetic energy in the thin shell regime ($r_1 < r < r_c$), it climbs up the potential hill outside the body ($r > r_c$). The field profile can be obtained by matching the solutions of Eq. (8.92) at the radius $r = r_1$ and $r = r_c$. Neglecting the mass term m_B, we obtain the thin shell field profile outside the body [350–352]

$$\phi(r) \simeq \phi_B - \frac{2Q_{\rm eff}}{\kappa} \frac{GM_c}{r}, \tag{8.94}$$

where

$$Q_{\rm eff} = 3Q\varepsilon_{\rm th}, \qquad \varepsilon_{\rm th} \equiv \frac{\kappa(\phi_B - \phi_A)}{6Q\Phi_c}. \tag{8.95}$$

Here, $\varepsilon_{\rm th}$ is called the thin shell parameter. Under the conditions $\Delta r_c/r_c \ll 1$ and $1/(m_A r_c) \ll 1$, the thin shell parameter is approximately given by $\varepsilon_{\rm th} \simeq \Delta r_c/r_c + 1/(m_A r_c)$ [352]. As long as $\varepsilon_{\rm th} \ll 1$, the amplitude of the effective coupling $Q_{\rm eff}$ can be much smaller than 1. Hence it is possible for the large coupling models ($|Q| = \mathcal{O}(1)$) to be consistent with local gravity experiments if the body has a thin shell.

Let us study the constraint on the thin shell parameter from the possible violation of the equivalence principle (EP). The tightest bound comes from the solar system tests of weak EP using the free-fall acceleration of Moon ($a_{\rm Moon}$) and Earth ($a_\oplus$)

toward Sun [351]. The experimental bound on the difference of two accelerations is given by [353, 354]

$$\frac{|a_{\rm Moon}-a_{\oplus}|}{(a_{\rm Moon}+a_{\oplus})/2} < 10^{-13}. \tag{8.96}$$

If Earth, Sun, and Moon have thin shells, the field profiles outside the bodies are given by Eq. (8.94) with the replacement of corresponding quantities. The acceleration induced by a fifth force with the field profile $\phi(r)$ and the effective coupling $Q_{\rm eff}$ is $a^{\rm fifth} = |Q_{\rm eff}\nabla\phi(r)|$. Using the thin shell parameter $\varepsilon_{{\rm th},\oplus}$ for Earth, the accelerations $a_{\oplus}$ and $a_{\rm Moon}$ toward Sun (mass $M_{\odot}$) are [351]

$$a_{\oplus} \simeq \frac{GM_{\odot}}{r^2}\left[1+18Q^2\varepsilon_{{\rm th},\oplus}^2\frac{\Phi_{\oplus}}{\Phi_{\odot}}\right], \tag{8.97}$$

$$a_{\rm Moon} \simeq \frac{GM_{\odot}}{r^2}\left[1+18Q^2\varepsilon_{{\rm th},\oplus}^2\frac{\Phi_{\oplus}^2}{\Phi_{\odot}\Phi_{\rm Moon}}\right], \tag{8.98}$$

where $\Phi_{\odot} \simeq 2.1\times10^{-6}$, $\Phi_{\oplus} \simeq 7.0\times10^{-10}$, and $\Phi_{\rm Moon} \simeq 3.1\times10^{-11}$ are the gravitational potentials of Sun, Earth, and Moon, respectively. Hence the condition (8.96) translates into

$$\varepsilon_{{\rm th},\oplus} < 8.8\times10^{-7}/|Q|. \tag{8.99}$$

Since the condition $|\phi_B| \gg |\phi_A|$ is satisfied for the field potentials under consideration, one has $\varepsilon_{{\rm th},\oplus} \simeq \kappa\phi_B/(6Q\Phi_{\oplus})$ from Eq. (8.95). Then, the condition (8.99) translates into

$$|\kappa\phi_{B,\oplus}| < 3.7\times10^{-15}. \tag{8.100}$$

For example, let us consider the inverse power-law potential $V(\phi) = M^{4+n}\phi^{-n}$. In this case, we have $\phi_{B,\oplus} = [(n/Q)(M_{\rm pl}^4/\rho_B)(M/M_{\rm pl})^{n+4}]^{1/(n+1)}M_{\rm pl}$, where we recovered the reduced Planck mass $M_{\rm pl} = 1/\kappa$. For n and Q of the order of unity, the constraint (8.100) gives $M < 10^{-(15n+130)/(n+4)}M_{\rm pl}$. When $n=1$, e.g., one has $M < 10^{-2}$ eV. If the same potential is responsible for dark energy, the mass M is constrained to be larger than this value [355]. For the potential $V(\phi) = M^4\exp(M^n/\phi^n)$, however, we have that $V(\phi) \approx M^4 + M^{4+n}\phi^{-n}$ for $\phi > M$, which is responsible for dark energy for $M \approx 10^{-3}$ eV. This can be compatible with the mass scale M constrained by (8.100) [355]. See Refs. [356–363] for a number of cosmological and experimental aspects of the chameleon field.

8.4.3.4 Varying α

So far we have discussed the coupling between dark energy and non-relativistic matter. In this section, we discuss the case in which dark energy is coupled to an electromagnetic field. In fact, a temporal variation of the effective fine structure "constant" α has been reported by a number of authors. The variation of α constrained by the Oklo natural fission reactor is given by $-0.9 \times 10^{-7} < \Delta\alpha/\alpha \equiv (\alpha - \alpha_0)/\alpha_0 < 1.2 \times 10^{-7}$ at the redshift $z \approx 0.16$, where α_0 is the value of α today [364]. From the absorption line spectra of distant quasars, we have the constraints $\Delta\alpha/\alpha = (-0.574 \pm 0.102) \times 10^{-5}$ over the redshift range $0.2 < z < 3.7$ [365, 366] and $\Delta\alpha/\alpha = (-0.06 \pm 0.06) \times 10^{-5}$ for $0.4 < z < 2.3$ [367]. Although the possibility of systematic errors still remains [368], this may provide important implications for the existence of a light scalar field related with dark energy.

The Lagrangian density describing such a coupling between the field ϕ and the electromagnetic field $F_{\mu\nu}$ is given by

$$\mathscr{L}_F(\phi) = -\frac{1}{4} B_F(\phi) F_{\mu\nu} F^{\mu\nu}. \tag{8.101}$$

The coupling of the form $B_F(\phi) = e^{-\zeta\kappa(\phi-\phi_0)}$ was originally introduced by Bekenstein [369], where ζ is a coupling constant and ϕ_0 is the field value today. There are also other choices of the coupling, see e.g., Refs. [370–377]. The fine structure "constant" α is inversely proportional to $B_F(\phi)$ so that this can be expressed as $\alpha = \alpha_0/B_F(\phi)$, where α_0 is the present value. The exponential coupling $B_F(\phi) = e^{-\zeta\kappa(\phi-\phi_0)}$ has a linear dependence $B_F(\phi) \simeq 1 - \zeta\kappa(\phi - \phi_0)$ in the regime $|\zeta\kappa(\phi - \phi_0)| \ll 1$ so that the variation of α is given by

$$\frac{\Delta\alpha}{\alpha} = \frac{\alpha - \alpha_0}{\alpha_0} \simeq \zeta\kappa(\phi - \phi_0). \tag{8.102}$$

Using the constraint $\Delta\alpha/\alpha \simeq -10^{-5}$ around $z = 3$ [366] obtained from quasar absorption lines, the coupling ζ can be expressed as

$$\zeta \simeq -\frac{10^{-5}}{\kappa\phi(z=3) - \kappa\phi(z=0)}. \tag{8.103}$$

Let us consider the case in which the scalar field has a power-law dependence in terms of the scale factor a, i.e.,

$$\phi = \phi_0 a^p = \phi_0 (1+z)^{-p}. \tag{8.104}$$

In fact, in the so-called tracking regime of the matter-dominated era [38], the inverse power-law potential $V(\phi) = M^{4+n}\phi^{-n}$ gives rise to a constant field equation of state: $w_\phi = -2/(n+2)$, which corresponds to the field evolution (8.104) with $p = 3/(n+2)$. Using Eq. (8.104), the coupling ζ in Eq. (8.103) reads

$$\zeta \simeq \frac{10^{-5}}{1 - 4^{-p}} (\kappa\phi_0)^{-1}. \tag{8.105}$$

Since $\kappa\phi_0$ is of the order of unity in order to realize the present cosmic acceleration [38], the coupling ζ is constrained to be $\zeta \approx 10^{-5}$ for p of the order of 1.

The above discussion is valid for the potentials having the solution (8.104) in the tracking regime. The variation of α for other quintessence potentials was discussed in Ref. [376]. There is also a k-essence model in which a tachyon field is coupled to electromagnetic fields [378].

8.4.4 Unified Models of Dark Energy and Dark Matter

There are a number of works to explain the origin of dark energy and dark matter using a single fluid or a single scalar field. Let us first discuss the generalized Chaplygin gas (GCG) model as an example of a single fluid model [44, 45]. In this model, the pressure P of the perfect fluid is related to its energy density ρ via

$$P = -A\rho^{-\alpha}, \tag{8.106}$$

where A is a positive constant. The original Chaplygin gas model corresponds to $\alpha = 1$ [44].

Plugging the relation (8.106) into the continuity equation $\dot{\rho} + 3H(\rho + P) = 0$, we obtain the integrated solution

$$\rho(t) = \left[A + \frac{B}{a^{3(1+\alpha)}}\right]^{1/(1+\alpha)}, \tag{8.107}$$

where B is an integration constant. In the early epoch ($a \ll 1$), the energy density evolves as $\rho \propto a^{-3}$, which means that the fluid behaves as dark matter. In the late epoch ($a \gg 1$), the energy density approaches a constant value $A^{1/(a+\alpha)}$ and hence the fluid behaves as dark energy. A fluid with the generalized Chaplygin gas therefore interpolates between dark matter and dark energy.

Although this model is attractive to provide unified description of two dark components, it is severely constrained by the matter power spectrum in large-scale structure. The gauge-invariant matter perturbation δ_m with a comoving wave number k obeys the following equation of motion [46]

$$\ddot{\delta}_m + \left(2 + 3c_s^2 - 6w\right) H\dot{\delta}_m - \left[\frac{3}{2}H^2(1 - 6c_s^2 - 3w^2 + 8w) - \left(\frac{c_s k}{a}\right)^2\right]\delta_m = 0, \tag{8.108}$$

where $w = P/\rho$ is the fluid equation of state, and c_s is the sound speed given by

$$c_s^2 = \frac{\mathrm{d}P}{\mathrm{d}\rho} = -\alpha w. \tag{8.109}$$

Since $w \to 0$ and $c_s^2 \to 0$ in the limit $z \gg 1$, the sound speed is much smaller than unity in the deep matter era and starts to grow around the end of it. Since w is negative, c_s^2 is positive for $\alpha > 0$ and negative for $\alpha < 0$.

From Eq. (8.108), the perturbations satisfying the following condition grow via the gravitational instability

$$|c_s^2| < \frac{3}{2}\left(\frac{aH}{k}\right)^2. \tag{8.110}$$

When $|c_s^2| > (3/2)(aH/k)^2$, the perturbations exhibit either rapid growth or damped oscillations depending on the sign of c_s^2. The violation of the condition (8.110) mainly occurs around the present epoch in which $|w|$ is of the order of unity and hence $|c_s^2| \sim |\alpha|$. The smallest scale relevant to the galaxy matter power spectrum in the linear regime corresponds to the wave number around $k = 0.1\,h\,\mathrm{Mpc}^{-1}$. Then the constraint (8.110) gives the upper bound on the values of $|\alpha|$ [46]:

$$|\alpha| \lesssim 10^{-5}. \tag{8.111}$$

Hence the generalized Chaplygin gas model is hardly distinguishable from the ΛCDM model. In particular, the original Chaplygin gas model ($\alpha = 1$) is excluded from the observations of large-scale structure. Although nonlinear clustering may change the evolution of perturbations in this model [379, 380], it is unlikely that the constraint (8.111) is relaxed significantly.

The above conclusion comes from the fact that in the Chaplygin gas model, the sound speed is too large to match with observations. There is a way to avoid this problem by adding a non-adiabatic contribution to Eq. (8.108) to make c_s vanish [381]. It is also possible to construct unified models of dark energy and dark matter using a purely kinetic scalar field [382]. Let us consider k-essence models in which the Lagrangian density $P(X)$ has an extremum at some value $X = X_0$, e.g., [382]

$$P = P_0 + P_2(X - X_0)^2. \tag{8.112}$$

The pressure $P_\phi = P$ and the energy density $\rho_\phi = 2XP_{,X} - P$ satisfy the continuity equation $\dot{\rho}_\phi + 3H(\rho_\phi + P_\phi) = 0$, i.e.,

$$(P_{,X} + 2XP_{,XX})\dot{X} + 6HP_{,X}X = 0. \tag{8.113}$$

The solution around $X = X_0$ can be derived by introducing a small parameter $\varepsilon = (X - X_0)/X_0$. Plugging Eq. (8.112) into Eq. (8.113), we find that ε satisfies the equation $\dot{\varepsilon} = -3H\varepsilon$ at linear order. Hence we obtain the solution $X = X_0\left[1 + \varepsilon_1(a/a_1)^{-3}\right]$, where ε_1 and a_1 are constants. The validity of the above approximation demands that $\varepsilon_1(a/a_1)^{-3} \ll 1$. Since $P_\phi \simeq P_0$ and $\rho_\phi \simeq -P_0 + 4P_2X_0^2\varepsilon_1(a/a_1)^{-3}$ in the regime where X is close to X_0, the field equation of state is given by

$$w_\phi \simeq -\left[1 - \frac{4P_2}{P_0}X_0^2\varepsilon_1\left(\frac{a}{a_1}\right)^{-3}\right]^{-1}. \tag{8.114}$$

Since $w_\phi \to -1$ at late times, it is possible to give rise to the cosmic acceleration. One can also realize $w_\phi \approx 0$ during the matter era, provided that the condition $4P_2X_0^2/|P_0| \gg 1$ is satisfied. The sound speed squared defined in Eq. (8.67) is approximately given by

$$c_s^2 \simeq \frac{1}{2}\varepsilon_1 \left(\frac{a}{a_1}\right)^{-3}, \tag{8.115}$$

which is much smaller than unity. Hence the large sound speed problem can be evaded in the model (8.112). In Ref. [383], it was shown that the above purely k-essence model is equivalent to a fluid with a closed-form barotropic equation of state plus a constant term that works as a cosmological constant to all orders in structure formation. See Refs. [384–388] for generalized versions of the above model.

8.5 Modified Gravity Models

There is another class of dark energy models in which gravity is modified from GR. We review a number of cosmological and gravitational aspects of $f(R)$ gravity, Gauss–Bonnet (GB) gravity, scalar-tensor theories, and a braneworld model. We also discuss observational signatures of those models to distinguish them from other dark energy models.

8.5.1 *f(R) Gravity*

The simplest modification to GR is $f(R)$ gravity with the action

$$S = \frac{1}{2\kappa^2}\int d^4x\sqrt{-g}f(R) + \int d^4x\,\mathscr{L}_M(g_{\mu\nu},\Psi_M), \tag{8.116}$$

where f is a function of the Ricci scalar R, and $\mathscr{L}_M$ is a matter Lagrangian for perfect fluids. The Lagrangian $\mathscr{L}_M$ depends on the metric $g_{\mu\nu}$ and the matter fields Ψ_M. We do not consider a direct coupling between the Ricci scalar and matter (such as $f_1(R)\mathscr{L}_M$ studied in Refs. [389–391]).

8.5.1.1 Viable *f(R)* Dark Energy Models

In the standard variational approach called the metric formalism, the affine connections $\Gamma^\lambda_{\mu\nu}$ are related with the metric $g_{\mu\nu}$ [67]. In this formalism, the field equation can be derived by varying the action (8.116) with respect to $g_{\mu\nu}$:

$$F(R)R_{\mu\nu}(g) - \frac{1}{2}f(R)g_{\mu\nu} - \nabla_\mu\nabla_\nu F(R) + g_{\mu\nu}\Box F(R) = \kappa^2 T_{\mu\nu}, \tag{8.117}$$

where $F(R) \equiv \partial f/\partial R$, and $T_{\mu\nu} = -(2/\sqrt{-g})\delta \mathscr{L}_M/\delta g^{\mu\nu}$ is the energy-momentum tensor of matter. Note that there is another way for the variation of the action called the Palatini formalism in which the metric and the connections are treated as independent variables. In Sec. 8.5.1.4, we shall briefly mention the application of Palatini $f(R)$ gravity to dark energy. The trace of Eq. (8.117) is given by

$$3\Box F(R) + F(R)R - 2f(R) = \kappa^2 T\,, \tag{8.118}$$

where $T = g^{\mu\nu}T_{\mu\nu} = -\rho_M + 3P_M$. Here ρ_M and P_M are the energy density and the pressure of matter, respectively.

The de Sitter point corresponds to a vacuum solution at which the Ricci scalar is constant. Since $\Box F(R) = 0$ at this point, we obtain

$$F(R)R - 2f(R) = 0\,. \tag{8.119}$$

The model $f(R) = \alpha R^2$ satisfies this condition and hence it gives rise to an exact de Sitter solution. In fact the first model of inflation proposed by Starobinsky [113] corresponds to $f(R) = R + \alpha R^2$, in which the cosmic acceleration ends when the term αR^2 becomes smaller than R. Dark energy models based on $f(R)$ theories can be also constructed to realize the late-time de Sitter solution satisfying the condition (8.119).

The possibility of the late-time cosmic acceleration in $f(R)$ gravity was first suggested by Capozziello [47] in 2002. An $f(R)$ dark energy model of the form $f(R) = R - \mu^{2(n+1)}/R^n$ $(n > 0)$ was proposed in Refs. [49–52] (see also Refs. [392–397]), but it became clear that this model suffers from a number of problems such as the matter instability [398], absence of the matter era [399, 400], and inability to satisfy local gravity constraints [401–406]. This problem arises from the fact that $f_{,RR} < 0$ in this model.

In order to see why the models with negative values of $f_{,RR}$ are excluded, let us consider local fluctuations on a background characterized by a curvature R_0 and a density ρ_0. We expand Eq. (8.118) in powers of fluctuations under a weak field approximation. We decompose the quantities $F(R)$, $g_{\mu\nu}$, and $T_{\mu\nu}$ into the background part and the perturbed part: $R = R^{(0)} + \delta R$, $F = F^{(0)}(1+\delta_F)$, $g_{\mu\nu} = \eta_{\mu\nu} + h_{\mu\nu}$, and $T_{\mu\nu} = T^{(0)}_{\mu\nu} + \delta T_{\mu\nu}$, where we have used the approximation that $g^{(0)}_{\mu\nu}$ corresponds the metric $\eta_{\mu\nu}$ in the Minkowski space-time. Then, the trace equation (8.118) reads [403, 404]

$$\left(\frac{\partial^2}{\partial t^2} - \nabla^2\right)\delta_F + M^2\,\delta_F = -\frac{\kappa^2}{3F^{(0)}}\delta T\,, \tag{8.120}$$

where $\delta T \equiv \eta^{\mu\nu}\delta T_{\mu\nu}$, and

$$M^2 \equiv \frac{1}{3}\left[\frac{f_{,R}(R^{(0)})}{f_{,RR}(R^{(0)})} - R^{(0)}\right] = \frac{R^{(0)}}{3}\left[\frac{1}{m(R^{(0)})} - 1\right]. \tag{8.121}$$

Here the quantity $m = Rf_{,RR}/f_{,R}$ characterizes the deviation from the ΛCDM model ($f(R) = R - 2\Lambda$). In the homogeneous and isotropic cosmological background (without a Hubble friction), δ_F is a function of the cosmic time t only and Eq. (8.120) reduces to

$$\ddot{\delta}_F + M^2\,\delta_F = \frac{\kappa^2}{3F^{(0)}}\rho\,, \tag{8.122}$$

where $\rho \equiv -\delta T$. For the models where the deviation from the ΛCDM model is small, we have $m(R^{(0)}) \ll 1$ so that $|M^2|$ is much larger than $R^{(0)}$. If $M^2 < 0$, the perturbation δ_F exhibits a violent instability. Then, the condition $M^2 \simeq f_{,R}(R^{(0)})/(3f_{,RR}(R^{(0)})) > 0$ is needed for the stability of cosmological perturbations. We also require that $f_{,R}(R^{(0)}) > 0$ to avoid antigravity (i.e., to avoid that the graviton becomes a ghost). Hence the condition $f_{,RR}(R^{(0)}) > 0$ needs to hold for avoiding a tachyonic instability associated with the negative mass squared [407–411].

For the consistency with local gravity constraints in solar system, the function $f(R)$ needs to be close to that in the ΛCDM model in the region of high density (in the region where the Ricci scalar R is much larger than the cosmological Ricci scalar R_0 today). We also require the existence of a stable late-time de Sitter point given in Eq. (8.119). From the stability analysis about the de Sitter point, one can show that it is stable for $0 < m = Rf_{,RR}/f_{,R} < 1$ [412–414]. Then, we can summarize the conditions for the viability of $f(R)$ dark energy models:

- (i) $f_{,R} > 0$ for $R \geq R_0$.
- (ii) $f_{,RR} > 0$ for $R \geq R_0$.
- (iii) $f(R) \to R - 2\Lambda$ for $R \gg R_0$.
- (iv) $0 < Rf_{,RR}/f_{,R} < 1$ at the de Sitter point satisfying $Rf_{,R} = 2f$.

The examples of viable models satisfying all these requirements are [415–417]

$$\text{(A)}\ f(R) = R - \mu R_c\frac{(R/R_c)^{2n}}{(R/R_c)^{2n}+1} \qquad \text{with}\ \ n,\mu,R_c > 0\,, \tag{8.123}$$

$$\text{(B)}\ f(R) = R - \mu R_c\left[1 - \left(1 + R^2/R_c^2\right)^{-n}\right] \qquad \text{with}\ \ n,\mu,R_c > 0\,, \tag{8.124}$$

$$\text{(C)}\ f(R) = R - \mu R_c \tanh\left(R/R_c\right) \qquad \text{with}\ \ \mu,R_c > 0\,, \tag{8.125}$$

where μ, R_c, and n are constants. Models similar to (C) were proposed in Refs. [418, 419]. Note that R_c is roughly of the order of the present cosmological Ricci scalar R_0. If $R \gg R_c$, the models are close to the ΛCDM model ($f(R) \simeq R - \mu R_c$), so that GR is recovered in the region of high density. The models (A) and (B) have the following asymptotic behavior

$$f(R) \simeq R - \mu R_c\left[1 - (R/R_c)^{-2n}\right]\,, \qquad (R \gg R_c)\,, \tag{8.126}$$

which rapidly approaches the ΛCDM model for $n \gtrsim 1$. The model (C) shows an even faster decrease of m in the region $R \gg R_c$. The model $f(R) = R - \mu R_c(R/R_c)^n$ ($0 < n < 1$) proposed in Refs. [414, 420] is also viable, but it does not allow the rapid decrease of m in the region of high density required for the consistency with local gravity tests.

For example, let us consider the model (B). The de Sitter point given by the condition (8.119) satisfies

$$\mu = \frac{x_1(1+x_1^2)^{n+1}}{2[(1+x_1^2)^{n+1} - 1 - (n+1)x_1^2]}, \tag{8.127}$$

where $x_1 \equiv R_1/R_c$ and R_1 is the Ricci scalar at the de Sitter point. The stability condition ($0 < m < 1$) at this point gives [416]

$$(1+x_1^2)^{n+2} > 1 + (n+2)x_1^2 + (n+1)(2n+1)x_1^4. \tag{8.128}$$

The condition (8.128) gives the lower bound on the parameter μ. When $n = 1$, one has $x_1 > \sqrt{3}$ and $\mu > 8\sqrt{3}/9$. Under Eq. (8.128), one can show that the conditions $f_{,R} > 0$ and $f_{,RR} > 0$ are also satisfied for $R \geq R_1$.

8.5.1.2 Observational Signatures of $f(R)$ Dark Energy Models

In the flat FLRW space-time, we obtain the following equations of motion from Eqs. (8.117) and (8.118):

$$3FH^2 = \kappa^2\rho_m + (FR - f)/2 - 3H\dot{F}, \tag{8.129}$$

$$2F\dot{H} = -\kappa^2\rho_m - \ddot{F} + H\dot{F}, \tag{8.130}$$

where, for the perfect fluid, we have taken into account only the nonrelativistic matter with energy density ρ_m. In order to confront $f(R)$ dark energy models with SN Ia observations, we rewrite Eqs. (8.129) and (8.130) as follows:

$$3AH^2 = \kappa^2(\rho_m + \rho_{\rm DE}), \tag{8.131}$$

$$-2A\dot{H} = \kappa^2(\rho_m + \rho_{\rm DE} + P_{\rm DE}), \tag{8.132}$$

where A is some constant and

$$\kappa^2\rho_{\rm DE} \equiv (1/2)(FR - f) - 3H\dot{F} + 3H^2(A - F), \tag{8.133}$$

$$\kappa^2 P_{\rm DE} \equiv \ddot{F} + 2H\dot{F} - (1/2)(FR - f) - (3H^2 + 2\dot{H})(A - F). \tag{8.134}$$

By defining $\rho_{\rm DE}$ and $P_{\rm DE}$ in this way, one can easily show that the following continuity equation holds

$$\dot{\rho}_{\rm DE} + 3H(\rho_{\rm DE} + P_{\rm DE}) = 0. \tag{8.135}$$

We define the dark energy equation of state $w_{\rm DE} \equiv P_{\rm DE}/\rho_{\rm DE}$, which is directly related to the one used in SN Ia observations. From Eqs. (8.131) and (8.132), it is given by [83, 421]

$$w_{\rm DE} = -\frac{2A\dot{H} + 3AH^2}{3AH^2 - \kappa^2 \rho_m} = \frac{w_{\rm eff}}{1 - (F/A)\tilde{\Omega}_m}, \qquad (8.136)$$

where $\tilde{\Omega}_m \equiv \kappa^2 \rho_m/(3FH^2)$. The viable $f(R)$ models approach the ΛCDM model in the past, i.e., $F \to 1$ as $R \to \infty$. In order to reproduce the standard matter era in the high-redshift regime we can choose $A = 1$ in Eqs. (8.131) and (8.132). Another possible choice is $A = F_0$, where F_0 is the present value of F. This choice is suitable if the deviation of F_0 from 1 is small (as in scalar-tensor theory with a massless scalar field [422, 423]). In both cases, the equation of state $w_{\rm DE}$ can be smaller than -1 before reaching the de Sitter attractor [83, 415, 417, 419, 424]. This originates from the fact that the presence of non-relativistic matter makes the denominator in Eq. (8.136) smaller than 1 (unlike Refs. [425, 426] in which the authors did not take into account the contribution of non-relativistic matter). Thus, $f(R)$ dark energy models give rise to a phantom equation of state without violating stability conditions of the system. The models (A) and (B) are allowed from the SN Ia observations provided that n is larger than the order of unity [427–430].

The modification of gravity manifests itself in the effective gravitational coupling that appears in the equation of cosmological perturbations. The full perturbation

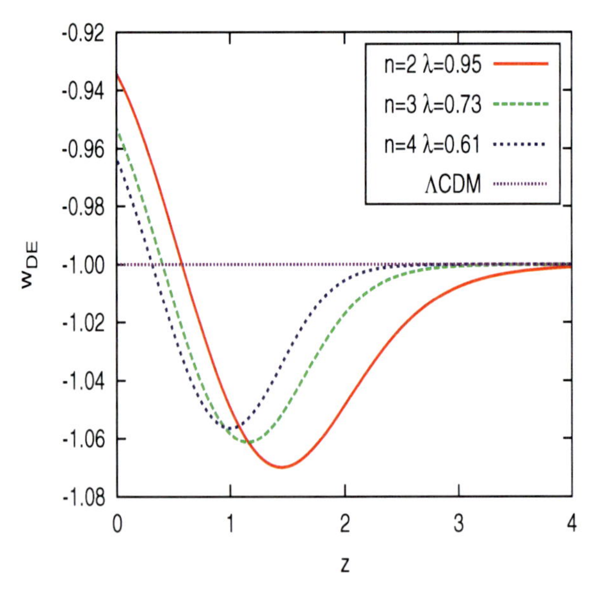

Fig. 8.5 Evolution of the dark energy equation of state $w_{\rm DE}$ for the model (B) with $A = 1$ in Eqs. (8.131) and (8.132). The phantom equation of state and the cosmological constant boundary crossing are realized. From Ref. [424].

equations in $f(R)$ gravity are presented in Refs. [431–433]. When we confront $f(R)$ models with the observations of large-scale structure, the wave numbers k of interest are sub-horizon modes with $k/a \gg H$. We can employ a so-called quasi-static approximation under which the dominant terms in perturbation equations correspond to those including k^2/a^2, $\delta\rho_m$, and M^2 [9, 91, 423, 434]. Then, the matter density perturbation δ_m approximately satisfies the following equation [434, 435]

$$\ddot{\delta}_m + 2H\dot{\delta}_m - 4\pi G_{\rm eff}\rho_m\delta_m \simeq 0\,, \tag{8.137}$$

where ρ_m is the energy density of nonrelativistic matter, and

$$G_{\rm eff} = \frac{G}{f_{,R}}\frac{1+4mk^2/(a^2R)}{1+3mk^2/(a^2R)}\,, \tag{8.138}$$

where $m \equiv Rf_{,RR}/f_{,R}$. This approximation is accurate for viable $f(R)$ dark energy models as long as an oscillating mode of the scalar-field degree of freedom is suppressed relative to the matter-induced mode [416, 417, 436–439].

In the regime where the deviation from the ΛCDM model is small such that $mk^2/(a^2R) \ll 1$, the effective gravitational coupling $G_{\rm eff}$ is very close to the gravitational constant G. Then, the matter perturbation evolves as $\delta_m \propto t^{2/3}$ during the matter dominance. Meanwhile, in the regime $mk^2/(a^2R) \gg 1$, one has $G_{\rm eff} \simeq 4G/(3f_{,R})$, so that the evolution of δ_m during the matter era is given by $\delta_m \propto t^{(\sqrt{33}-1)/6}$ [416, 417]. The transition from the former regime to the latter regime occurs at the critical redshift [440]

$$z_k \simeq \left[\left(\frac{k}{a_0H_0}\right)^2 \frac{2n(2n+1)}{\mu^{2n}}\frac{(2(1-\Omega_m^{(0)}))^{2n+1}}{(\Omega_m^{(0)})^{2(n+1)}}\right]^{1/(6n+4)} - 1\,, \tag{8.139}$$

where "0" represents the values today. The time t_k at the transition has a scale-dependence $t_k \propto k^{-3/(6n+4)}$, which means that the transition occurs earlier for larger k. The matter power spectrum $P_{\delta_m} = |\delta_m|^2$ at the onset of cosmic acceleration (at time t_Λ) shows a difference compared with the case of the ΛCDM model [416]:

$$\frac{P_{\delta_m}}{P_{\delta_m}{}^{\Lambda\rm CDM}} = \left(\frac{t_\Lambda}{t_k}\right)^{2\left(\frac{\sqrt{33}-1}{6}-\frac{2}{3}\right)} \propto k^{\frac{\sqrt{33}-5}{6n+4}}\,. \tag{8.140}$$

The ratio of the two power spectra today, i.e., $P_{\delta_m}(t_0)/P_{\delta_m}{}^{\Lambda\rm CDM}(t_0)$, is in general different from Eq. (8.140), but the difference is small for n of the order of unity [417].

The modified evolution of perturbations for the redshift $z < z_k$ gives rise to the integrated Sachs-Wolfe (ISW) effect in CMB anisotropies [420, 441–443], but this is limited to very large scales (low multipoles). Since the CMB spectrum on the scales relevant to the large-scale structure ($k \gtrsim 0.01\,h\,{\rm Mpc}^{-1}$) is hardly affected by this modification, there is a difference between the spectral indices of the CMB spectrum and the galaxy power spectrum: $\Delta n_s = (\sqrt{33}-5)/(6n+4)$. Observationally, we do not find any strong signature for the difference of slopes of the two spectra. If we take the mild bound $\Delta n_s < 0.05$, we obtain the constraint $n > 2$.

The growth index γ defined in Eq. (8.78) can be as close as 0.4 today in viable $f(R)$ models given in Eqs. (8.123)–(8.125) [421]. Depending on the epoch at which the perturbations cross the transition redshift z_k, the spatial dispersion of the growth index $\gamma_0 \equiv \gamma(z=0)$ appears in the range of values $0.40 < \gamma < 0.55$. There are also regions in parameter space for which γ_0 converges to values around $0.40 < \gamma < 0.43$ [440]. These unusual dispersed or converged spectra will be useful to distinguish between $f(R)$ gravity models and the ΛCDM model in future high-precision observations. Since the modified evolution of matter perturbations directly affects the shear power spectrum in weak lensing, this is another important test for probing $f(R)$ gravity observationally [444–448]. We also note that the nonlinear evolution of matter perturbations in $f(R)$ gravity (corresponding to the scales $k \gtrsim 0.1\,h\,\mathrm{Mpc}^{-1}$) has been studied in Refs. [449–457].

8.5.1.3 Local Gravity Constraints

Let us discuss local gravity constraints on $f(R)$ dark energy models. In the region of high density where gravitational experiments are carried out, the linear expansion of R in terms of the cosmological value $R^{(0)}$ and the perturbation δR is no longer valid because of the violation of the condition $\delta R \ll R^{(0)}$. In such a nonlinear regime, the chameleon mechanism [350, 351] can be at work to suppress the effective coupling between dark energy and nonrelativistic matter. In order to study how the chameleon mechanism works in $f(R)$ gravity, we transform the action (8.116) to the Einstein frame action under the conformal transformation $\tilde{g}_{\mu\nu} = F g_{\mu\nu}$: [458]

$$S_E = \int \mathrm{d}^4x \sqrt{-\tilde{g}} \left[\frac{1}{2\kappa^2}\tilde{R} - \frac{1}{2}\tilde{g}^{\mu\nu}\partial_\mu\phi\partial_\nu\phi - V(\phi) \right] + \int \mathrm{d}^4x\, \mathscr{L}_M(g_{\mu\nu}, \Psi_m), \quad (8.141)$$

where $\kappa\phi \equiv \sqrt{3/2}\ln F$, $V(\phi) = (RF - f)/(2\kappa^2 F^2)$, and a tilde represents quantities in the Einstein frame.

The action (8.141) is the same as (8.90) with the correspondence that $g_{\mu\nu}$ in the Jordan frame is equivalent to $g^{(i)}_{\mu\nu}$ in the action (8.90). Since the quantity F is given by $F = e^{-2\kappa Q\phi}$ with $Q = -1/\sqrt{6}$ in metric $f(R)$ gravity, the field ϕ is coupled to non-relativistic matter (including baryons and dark matter) with a universal coupling $Q = -1/\sqrt{6}$. Let us consider the models (8.123) and (8.124), which behave as Eq. (8.126), in the region of high density ($R \gg R_c$). For the functional form (8.126), the effective potential defined in Eq. (8.93) is

$$V_{\mathrm{eff}}(\phi) \simeq \frac{\mu R_c}{2\kappa^2} e^{-4\kappa\phi/\sqrt{6}} \left[1 - (2n+1)\left(\frac{-\kappa\phi}{\sqrt{6}n\mu} \right)^{2n/(2n+1)} \right] + \rho e^{-\kappa\phi/\sqrt{6}}, \quad (8.142)$$

where

$$F = e^{2\kappa\phi/\sqrt{6}} = 1 - 2n\mu(R/R_c)^{-(2n+1)}. \tag{8.143}$$

Inside and outside a spherically symmetric body, the effective potential (8.142) has the following minima given, respectively, by

$$\kappa\phi_A \simeq -\sqrt{6}n\mu\left(\frac{R_c}{\kappa^2\rho_A}\right)^{2n+1}, \quad \kappa\phi_B \simeq -\sqrt{6}n\mu\left(\frac{R_c}{\kappa^2\rho_B}\right)^{2n+1}. \tag{8.144}$$

One has $|\phi_B| \gg |\phi_A|$ provided that $\rho_A \gg \rho_B$.

The bound (8.100) translates into

$$\frac{n\mu}{x_1^{2n+1}}\left(\frac{R_1}{\rho_B}\right)^{2n+1} < 1.5 \times 10^{-15}, \tag{8.145}$$

where $x_1 = R_1/R_c$ and R_1 is the Ricci scalar at the de Sitter point. Let us consider the case in which the Lagrangian density is given by (8.126) for $R \geq R_1$. In the original models of Hu and Sawicki [415] and Starobinsky [416], there are some modification to the estimation of R_1, but this change is not significant when we place constraints on model parameters. The de Sitter point for the model (8.126) corresponds to $\mu = x_1^{2n+1}/[2(x_1^{2n} - n - 1)]$. Substituting this relation into Eq. (8.145), we find

$$\frac{n}{2(x_1^{2n} - n - 1)}\left(\frac{R_1}{\rho_B}\right)^{2n+1} < 1.5 \times 10^{-15}. \tag{8.146}$$

The stability of the de Sitter point requires that $m(R_1) < 1$, which translates into the condition $x_1^{2n} > 2n^2 + 3n + 1$. Then, the term $n/[2(x_1^{2n} - n - 1)]$ is smaller than 0.25 for $n > 0$. Using the approximation that R_1 and ρ_B are of the orders of the present cosmological density 10^{-29} g/cm^3 and the baryonic/dark matter density 10^{-24} g/cm^3 in our galaxy, respectively, we obtain the following constraint from (8.146): [459]

$$n > 0.9. \tag{8.147}$$

Thus, n does not need to be much larger than unity. Under the condition (8.147), the deviation from the ΛCDM becomes important as R decreases to the order of R_c.

From (8.143), we find that there is a curvature singularity with $R \to \infty$ (and $M^2 \to \infty$) at $\phi = 0$ for the models (8.123) and (8.124). At this singularity, the field potential is finite, while its derivative goes to infinity. This singularity can be accessible as we go back to the past [460], unless the oscillating mode of the scalar-field degree of freedom is suppressed. This amounts to the fine-tuning of initial conditions for the field perturbation [416]. This past singularity can be cured by taking into account the R^2 term [461]. The model of the type $f(R) = R - \alpha R_c \ln(1 + R/R_c)$ was also proposed to address this problem [462], but it satisfies neither local gravity constraints [463] nor observational constraints of large-scale structure [464].

Frolov [460] anticipated that the curvature singularity may be accessed in a strong gravitational background such as neutron stars. Kobayashi and Maeda [465, 466] showed the difficulty of obtaining static spherically symmetric solutions because of the presence of the singularity. On the other hand, the choice of accurate boundary conditions confirms the existence of static solutions in a strong gravitational background with $\Phi_c \lesssim 0.3$ [467–470].

8.5.1.4 Palatini *f(R)* Gravity

In the so-called Palatini formalism of $f(R)$ gravity, the connections $\Gamma^{\lambda}_{\mu\nu}$ are treated as independent variables when we vary the action (8.116) [471–477]. Variation of the action (8.116) with respect to $g_{\mu\nu}$ gives

$$F(R)R_{\mu\nu}(\Gamma) - \frac{1}{2}f(R)g_{\mu\nu} = \kappa^2 T_{\mu\nu}\,, \tag{8.148}$$

where $F(R) = \partial f/\partial R$, $R_{\mu\nu}(\Gamma)$ is the Ricci tensor corresponding to the connections $\Gamma^{\lambda}_{\mu\nu}$, and $T_{\mu\nu}$ is the energy-momentum tensor of matter[4]. $R_{\mu\nu}(\Gamma)$ is in general different from the Ricci tensor calculated in terms of metric connections $R_{\mu\nu}(g)$. Taking the trace of Eq. (8.148), we find

$$F(R)R - 2f(R) = \kappa^2 T\,, \tag{8.149}$$

where $T = g^{\mu\nu}T_{\mu\nu}$. The trace T directly determines the Ricci scalar $R(T)$, which is related with the Ricci scalar $R(g) = g^{\mu\nu}R_{\mu\nu}(g)$ in the metric formalism via [482]

$$\begin{aligned} R(T) = R(g) &+ \frac{3}{2(f'(R(T)))^2}(\nabla_\mu f'(R(T)))(\nabla^\mu f'(R(T))) \\ &+ \frac{3}{f'(R(T))}\Box f'(R(T))\,, \end{aligned} \tag{8.150}$$

where a prime represents a derivative in terms of $R(T)$. Variation of the action (8.116) with respect to the connection leads to the following equation

$$\begin{aligned} R_{\mu\nu}(g) - \frac{1}{2}g_{\mu\nu}R(g) = & \frac{\kappa^2 T_{\mu\nu}}{F} - \frac{FR(T) - f}{2F}g_{\mu\nu} + \frac{1}{F}(\nabla_\mu\nabla_\nu F - g_{\mu\nu}\Box F) \\ & - \frac{3}{2F^2}\left[\partial_\mu F \partial_\nu F - \frac{1}{2}g_{\mu\nu}(\nabla F)^2\right]. \end{aligned} \tag{8.151}$$

Unlike the trace equation (8.118) in the metric formalism, the kinetic term $\Box F$ is not present in the corresponding Eq. (8.149) in the Palatini formalism. Since the time derivatives of the scalar-field degree of freedom do not appear in Palatini $f(R)$

---

[4] There is another way for the variation of the action, known as the metric-affine formalism [478], in which the matter Lagrangian $\mathcal{L}_M$ depends not only on the metric $g_{\mu\nu}$ but also on the connection $\Gamma^{\lambda}_{\mu\nu}$. See Refs. [479–481] for the detail of such an approach.

gravity, cosmological solutions are not plagued by the dominance of the oscillating mode in the past. In fact, the sequence of radiation, matter, and de Sitter epochs can be realized even for the model $f(R) = R - \alpha/R^n$ $(n > 0)$ [483–489]. The combined data analysis of SN Ia, BAO, and the CMB shift parameter places the bound $n \in [-0.23, 0.42]$ on the model $f(R) = R - \alpha/R^n$ $(n > -1)$ [489].

Although the background cosmology is well behaved in Palatini $f(R)$ gravity, the evolution of non-relativistic matter perturbations exhibits a distinguished feature relative to that in the ΛCDM model [435, 490–492]. Under the quasi-static approximation on sub-horizon scales, the equation of matter perturbations is given by [435]

$$\ddot{\delta}_m + 2H\dot{\delta}_m - \frac{\kappa^2 \rho_m}{2F}\left[1 + \frac{mk^2/(a^2R)}{1-m}\right]\delta_m \simeq 0 . \tag{8.152}$$

Although the matter perturbation evolves as $\delta_m \propto t^{2/3}$ in the regime $|m|k^2/(a^2R) \ll 1$, the evolution of δ_m in the regime $|m|k^2/(a^2R) \gg 1$ is completely different from that in GR. After the perturbations enter the regime $|m|k^2/(a^2R) \gtrsim 1$, they exhibit violent growth or damped oscillations depending on the signs of m [435]. The $f(R)$ models are consistent with observations of LSS if the perturbations do not enter the regime $|m|k^2/(a^2R) \gtrsim 1$ by today. This translates into the condition

$$|m(z=0)| \lesssim (a_0 H_0/k)^2 . \tag{8.153}$$

If we take the maximum wave number $k \approx 0.2\,h\,\mathrm{Mpc}^{-1}$ (i.e., $k \approx 600 a_0 H_0$), Eq. (8.153) gives the bound $|m(z=0)| \lesssim 3 \times 10^{-6}$. Hence the $f(R)$ models in the Palatini formalism are hardly distinguishable from the ΛCDM model.

There are also a number of problems in Palatini $f(R)$ dark energy models associated with the non-dynamical nature of the scalar-field degree of freedom. The dark energy model $f(R) = R - \mu^4/R$ is in conflict with the Standard Model of particle physics [475–477, 493–497] because of large non-perturbative corrections to the matter Lagrangian. If we consider the models $f(R) = R - \mu^{2(n+1)}/R^n$, the only way to make such corrections small is to choose n very close to 0 [433]. Hence the deviation from the ΛCDM model needs to be very small. It was also shown that, for $f(R)$ dark energy models, a divergent behavior arises for the Ricci scalar at the surface of a static spherically symmetric star with a polytropic equation of state $P = c\rho_0^{\Gamma}$ $(3/2 < \Gamma < 2)$, where P is the pressure and ρ_0 is the rest-mass density [497, 498]. These results show that Palatini $f(R)$ dark energy models are difficult to be compatible with observational and experimental constraints, although this may not be the case for $f(R)$ models close to the Planck scale [499–501].

8.5.2 Gauss–Bonnet Dark Energy Models

It is possible to extend $f(R)$ gravity to more general theories in which the Lagrangian density f is an arbitrary function of R, $P \equiv R_{\mu\nu}R^{\mu\nu}$, and $Q \equiv R_{\mu\nu\alpha\beta}R^{\mu\nu\alpha\beta}$, where $R_{\mu\nu}$ and $R_{\mu\nu\alpha\beta}$ are Ricci tensor and Riemann tensor respectively [502, 503]. The appearance of spurious spin-2 ghosts can be avoided by taking a Gauss–Bonnet (GB) combination [504–507]

$$\mathcal{G} = R^2 - 4R_{\mu\nu}R^{\mu\nu} + R_{\mu\nu\alpha\beta}R^{\mu\nu\alpha\beta} . \tag{8.154}$$

A simple dark energy model motivated from low-energy effective string theory [242] is given by [508]

$$S = \int d^4x\sqrt{-g}\left[\frac{1}{2\kappa^2}R - \frac{1}{2}g^{\mu\nu}\partial_\mu\phi\partial_\nu\phi - V(\phi) - f(\phi)\mathcal{G}\right] + S_M , \tag{8.155}$$

where $V(\phi)$ and $f(\phi)$ are functions of a scalar field ϕ, and S_M is a matter action. The coupling $f(\phi)\mathcal{G}$ allows the presence of a de Sitter solution even for a runaway field potential $V(\phi)$ [509–512]. For the exponential potential $V(\phi) = V_0 e^{-\kappa\lambda\phi}$ and the coupling $f(\phi) = (f_0/\mu)e^{\mu\kappa\phi}$, it was shown in Refs. [509, 510] that a scaling matter era can be followed by the late-time de Sitter solution for $\mu > \lambda$.

Koivisto and Mota [509] found that the parameter λ in this model is constrained to be $3.5 < \lambda < 4.5$ (95% confidence level) from the observational data of SN Ia and WMAP 3-year. The parameter λ is constrained to be $3.5 < \lambda < 4.5$ at the 95% confidence level. In the second paper [511], they showed that the model is strongly disfavored from the combined data analysis including the constraints coming from BBN, LSS, BAO, and solar system data. It was shown in Refs. [510, 513] that, when the GB term dominates the dynamics, tensor perturbations are subject to negative instabilities. Amendola et al. [514] studied local gravity constraints on the above model and showed that the energy contribution coming from the GB term needs to be strongly suppressed for the consistency with solar system experiments. The above results imply that the GB term with the scalar-field coupling $f(\phi)\mathcal{G}$ can hardly be the source for dark energy.

The dark energy models based on the Lagrangian density $\mathcal{L} = R/(2\kappa^2) + f(\mathcal{G})$, have been studied by a number of authors [515–524]. In the presence of a perfect fluid with an energy density ρ_M, the Friedmann equation is given by [515, 516]

$$3H^2 = \mathcal{G}f_{,\mathcal{G}} - f - 24H^2 f_{,\mathcal{G}\mathcal{G}}\dot{\mathcal{G}} + \rho_M . \tag{8.156}$$

The Hubble parameter $H = H_1$ at the de Sitter point satisfies $3H_1^2 = \mathcal{G}_1 f_{,\mathcal{G}}(\mathcal{G}_1) - f(\mathcal{G}_1)$, where $\mathcal{G}_1 = 24H_1^4$. The stability of the de Sitter point requires the condition $0 < H_1^6 f_{,\mathcal{G}\mathcal{G}}(H_1) < 1/384$ [520]. In order to avoid the instability of solutions during radiation and matter eras, we also need the condition $f_{,\mathcal{G}\mathcal{G}} > 0$. In Ref. [520], the authors presented a number of $f(\mathcal{G})$ models that are cosmologically viable at the background level. One of such viable models is given by

$$f(\mathcal{G}) = \lambda\frac{\mathcal{G}}{\sqrt{\mathcal{G}_*}}\arctan\left(\frac{\mathcal{G}}{\mathcal{G}_*}\right) - \alpha\lambda\sqrt{\mathcal{G}_*} , \tag{8.157}$$

where α, λ, and $\mathscr{G}_*$ are constants. This model can satisfy solar system constraints for a wide range of parameter space [523].

If we consider cosmological perturbations, however, there is a UV instability in $f(\mathscr{G})$ models associated with a negative propagation speed squared of a scalar-field degree of freedom [516, 524]. This growth of perturbations gets stronger on smaller scales, which is difficult to be compatible with the observed galaxy spectrum unless the deviation from GR is very small. Thus, $f(\mathscr{G})$ dark energy models are effectively ruled out as an alternative to the ΛCDM model.

8.5.3 Scalar-Tensor Theories

There is another class of modified gravity called scalar-tensor theories in which the Ricci scalar R is coupled to a scalar field φ. One of the simplest examples is Brans–Dicke (BD) theory [291] with the action

$$S = \int \mathrm{d}^4x\sqrt{-g}\left[\frac{1}{2}\varphi R - \frac{\omega_{\mathrm{BD}}}{2\varphi}(\nabla\varphi)^2 - U(\varphi)\right] + S_M(g_{\mu\nu}, \Psi_M), \qquad (8.158)$$

where ω_{BD} is the BD parameter, $U(\varphi)$ is the field potential, and S_M is a matter action that depends on the metric $g_{\mu\nu}$ and matter fields Ψ_m. The original BD theory [291] does not have the field potential $U(\varphi)$.

The general action for scalar-tensor theories can be written as

$$S = \int \mathrm{d}^4x\sqrt{-g}\left[\frac{1}{2}f(\varphi, R) - \frac{1}{2}\omega(\varphi)(\nabla\varphi)^2\right] + S_M(g_{\mu\nu}, \Psi_M), \qquad (8.159)$$

where f is a general function of the scalar field φ and the Ricci scalar R, ω is a function of φ. We choose the unit $\kappa^2 = 1$. We consider theories of the type

$$f(\varphi, R) = F(\varphi)R - 2U(\varphi). \qquad (8.160)$$

Under the conformal transformation $\tilde{g}_{\mu\nu} = F g_{\mu\nu}$, the action in the Einstein frame is given by [458]

$$S_E = \int \mathrm{d}^4x\sqrt{-\tilde{g}}\left[\frac{1}{2}\tilde{R} - \frac{1}{2}(\tilde{\nabla}\phi)^2 - V(\phi)\right] + S_M(g_{\mu\nu}, \Psi_M), \qquad (8.161)$$

where $V = U/F^2$. We have introduced a new scalar field ϕ in order to make the field kinetic term canonical:

$$\phi \equiv \int \mathrm{d}\varphi\sqrt{\frac{3}{2}\left(\frac{F_{,\varphi}}{F}\right)^2 + \frac{\omega}{F}}. \qquad (8.162)$$

We define the coupling between dark energy and non-relativistic matter, as

$$Q \equiv -\frac{F_{,\phi}}{2F} = -\frac{F_{,\varphi}}{F}\left[\frac{3}{2}\left(\frac{F_{,\varphi}}{F}\right)^2 + \frac{\omega}{F}\right]^{-1/2}. \tag{8.163}$$

In $f(R)$ gravity, we have $\omega = 0$ and hence $F = \exp(\sqrt{2/3}\phi)$ from Eq. (8.162). Then, the coupling is given by $Q = -1/\sqrt{6}$ from Eq. (8.163). If Q is constant as in $f(R)$ gravity, the following relations hold from Eqs. (8.162) and (8.163):

$$F = e^{-2Q\phi}, \quad \omega = (1-6Q^2)F\left(\frac{\mathrm{d}\phi}{\mathrm{d}\varphi}\right)^2. \tag{8.164}$$

In this case, the action (8.159) in the Jordan frame reads [292]

$$S = \int \mathrm{d}^4x\sqrt{-g}\left[\frac{1}{2}F(\phi)R - \frac{1}{2}(1-6Q^2)F(\phi)(\nabla\phi)^2 - U(\phi)\right] + S_M(g_{\mu\nu}, \Psi_m). \tag{8.165}$$

In the limit that $Q \to 0$, the action (8.165) reduces to the one for a minimally coupled scalar field ϕ with the potential $U(\phi)$. The transformation of the Jordan frame action (8.165) under the conformal transformation $\tilde{g}_{\mu\nu} = e^{-2Q\phi}g_{\mu\nu}$ gives rise to the Einstein frame action (8.161) with a constant coupling Q.

One can compare (8.165) with the action (8.158) in BD theory. Setting $\varphi = F = e^{-2Q\phi}$, one finds that two actions are equivalent if the parameter $\omega_{\rm BD}$ is related to Q via the relation [351, 292]

$$3 + 2\omega_{\rm BD} = \frac{1}{2Q^2}. \tag{8.166}$$

Using this relation, we find that the General Relativistic limit ($\omega_{\rm BD} \to \infty$) corresponds to the vanishing coupling ($Q \to 0$). Since $Q = -1/\sqrt{6}$ in $f(R)$ gravity, this corresponds to the BD parameter $\omega_{\rm BD} = 0$ [525, 526, 401]. One can show that the field equation (8.151) in Palatini $f(R)$ gravity is equivalent to the one derived in BD theory with $\omega_{\rm BD} = -3/2$ [482, 433]. Hence Palatini $f(R)$ gravity corresponds to the infinite coupling ($Q^2 \to \infty$).

There are also other scalar-tensor theories that give rise to field-dependent couplings $Q(\phi)$. For a nonminimally coupled scalar field with $F(\varphi) = 1 - \xi\varphi^2$ and $\omega(\varphi) = 1$ in the action (8.159) with (8.160), the coupling is field dependent, i.e., $Q(\varphi) = \xi\varphi/[1-\xi\varphi^2(1-6\xi)]^{1/2}$. The cosmological dynamics of dark energy models based on such theories have been studied by a number of authors [286, 527–532].

Let us consider BD theory with the action (8.165). In the absence of the potential $U(\phi)$ the BD parameter $\omega_{\rm BD}$ is constrained to be $\omega_{\rm BD} > 4.0 \times 10^4$ from solar-system experiments [354]. This bound also applies to the case of a nearly massless field with the potential $U(\phi)$, in which the Yukawa correction e^{-Mr} is close to unity (where M is the scalar field mass and r is an interaction length). Using the bound $\omega_{\rm BD} > 4.0 \times 10^4$ in Eq. (8.166), we find

$$|Q| < 2.5 \times 10^{-3}. \tag{8.167}$$

In this case, the cosmological evolution for such theories is hardly distinguishable from the $Q = 0$ case. Even for scalar-tensor theories with such small couplings, it was shown that the phantom equation state of dark energy can be realized without the appearance of a ghost state [533–536].

In the presence of the field potential, it is possible for large coupling models ($|Q| = \mathcal{O}(1)$) to satisfy local gravity constraints under the chameleon mechanism, provided that the mass M of the field ϕ is sufficiently large in the region of high density. In metric $f(R)$ gravity ($Q = -1/\sqrt{6}$), the field potential $U(\phi)$ in Eq. (8.165) corresponds to $U = (FR - f)/2$ with $\phi = \sqrt{3/2}\ln F$. The viable $f(R)$ dark energy models (8.123) and (8.124) have the asymptotic form (8.126), in which case the field potential is given by

$$U(\phi) = \frac{\mu R_c}{2}\left[1 - \frac{2n+1}{(2n\mu)^{2n/(2n+1)}}\left(1 - e^{2\phi/\sqrt{6}}\right)^{2n/(2n+1)}\right]. \tag{8.168}$$

For BD theories with the constant coupling Q, one can generalize the potential (8.168) to the form

$$U(\phi) = U_0\left[1 - C(1 - e^{-2Q\phi})^p\right] \qquad (U_0 > 0,\ C > 0,\ 0 < p < 1). \tag{8.169}$$

As $\phi \to 0$, the potential (8.169) approaches the finite value U_0 with a divergence of the field mass squared $M^2 = U_{,\phi\phi} \to \infty$. This model has a curvature singularity at $\phi = 0$ as in the case of the $f(R)$ models (8.123) and (8.124). The mass M decreases as the field evolves away from $\phi = 0$. The late-time cosmic acceleration can be realized by the potential (8.169) provided that U_0 is of the order of H_0^2.

Since the action (8.161) in the Einstein frame is equivalent to the action (8.90), the chameleon mechanism can be at work even for BD theories with large couplings ($|Q| = \mathcal{O}(1)$). Considering a spherically symmetric body with homogenous densities ρ_A and ρ_B inside and outside bodies respectively, the effective potential $V_{\rm eff} = V(\phi) + e^{Q\phi}\rho$ in the Einstein frame (where $V(\phi) = U(\phi)/F^2$) has two minima characterized by

$$\phi_A \simeq \frac{1}{2Q}\left(\frac{2U_0 pC}{\rho_A}\right)^{1/(1-p)}, \qquad \phi_B \simeq \frac{1}{2Q}\left(\frac{2U_0 pC}{\rho_B}\right)^{1/(1-p)}. \tag{8.170}$$

Using the experimental bound (8.100) coming from the violation of equivalence principle together with the condition for realizing the cosmic acceleration today, we obtain the constraint [292]

$$p > 1 - \frac{5}{13.8 - \log_{10}|Q|}. \tag{8.171}$$

When $|Q| = 10^{-2}$ and $|Q| = 10^{-1}$, we have $p > 0.68$ and $p > 0.66$, respectively. In $f(R)$ gravity, the above bound corresponds to $p > 0.65$, which translates into $n > 0.9$ for the model (8.126).

The evolution of cosmological perturbations in scalar-tensor theories has been discussed in Refs. [292, 423, 434, 537, 538]. Under the quasi-static approximation on sub-horizon scales, the matter perturbation δ_m for the theory (8.165) obeys the following equation of motion [292, 538]

$$\ddot{\delta}_m + 2H\dot{\delta}_m - 4\pi G_{\rm eff}\rho_m\delta_m \simeq 0\,, \tag{8.172}$$

where the effective (cosmological) gravitational coupling is

$$G_{\rm eff} = \frac{G}{F}\frac{(k^2/a^2)(1+2Q^2)F + M^2}{(k^2/a^2)F + M^2}\,. \tag{8.173}$$

Here, $M^2 \equiv U_{,\phi\phi}$ is the field mass squared. In the "General Relativistic" regime characterized by $M^2/F \gg k^2/a^2$, one has $G_{\rm eff} \simeq G/F$ and $\delta_m \propto t^{2/3}$. In the "scalar-tensor" regime characterized by $M^2/F \ll k^2/a^2$, it follows that $G_{\rm eff} \simeq (1+2Q^2)G/F$ and $\delta_m \propto t^{(\sqrt{25+48Q^2}-1)/6}$. If the transition from the former regime to the latter regime occurs during the matter era, this gives rise to a difference between the spectral indices of the matter power spectrum and of the CMB spectrum on the scales $0.01\,h\,{\rm Mpc}^{-1} \lesssim k \lesssim 0.2\,h\,{\rm Mpc}^{-1}$ [292]:

$$\Delta n_s = \frac{(1-p)(\sqrt{25+48Q^2}-5)}{4-p}\,. \tag{8.174}$$

Under the criterion $\Delta n_s < 0.05$, we obtain the bounds $p > 0.957$ for $Q = 1$ and $p > 0.855$ for $Q = 0.5$. As long as p is close to 1, the model can be consistent with both cosmological and local gravity constraints.

For the perturbed metric $ds^2 = -(1+2\Psi)dt^2 + a^2(t)(1-2\Phi)\delta_{ij}dx^i dx^j$, the gravitational potentials obey the following equations under a quasi-static approximation on sub-horizon scales [292]

$$\frac{k^2}{a^2}\Psi \simeq -\frac{4\pi G}{F}\frac{(k^2/a^2)(1+2Q^2)F + M^2}{(k^2/a^2)F + M^2}\rho_m\delta_m\,, \tag{8.175}$$

$$\frac{k^2}{a^2}\Phi \simeq -\frac{4\pi G}{F}\frac{(k^2/a^2)(1-2Q^2)F + M^2}{(k^2/a^2)F + M^2}\rho_m\delta_m\,, \tag{8.176}$$

where we have recovered the gravitational constant G. The results (8.175) and (8.176) include those in $f(R)$ gravity by setting $Q = -1/\sqrt{6}$. In the regime $M^2/F \ll k^2/a^2$, the evolution of Ψ and Φ is subject to change compared with that in the GR regime characterized by $M^2/F \gg k^2/a^2$. In general, the difference from GR may be quantified by the parameters q and ζ [444]:

$$\frac{k^2}{a^2}\Phi = -4\pi G q\rho_m\delta_m\,, \qquad \frac{\Phi-\Psi}{\Phi} = \zeta\,. \tag{8.177}$$

In the regime $M^2/F \ll k^2/a^2$ of scalar-tensor theory (8.165), it follows that $q \simeq (1-2Q^2)/F$ and $\zeta \simeq -4Q^2/(1-2Q^2)$.

In order to confront dark energy models with the observations of weak lensing, it may be convenient to introduce the following quantity [444]

$$\Sigma \equiv q(1-\zeta/2)\,. \tag{8.178}$$

From the definition (8.177), we find that the weak lensing potential $\psi = \Phi + \Psi$ can be expressed as

$$\psi = -8\pi G \frac{a^2}{k^2} \rho_m \delta_m \Sigma\,. \tag{8.179}$$

In scalar-tensor theory (8.165), one has $\Sigma = 1/F$. The effect of modified gravity theories manifests itself in weak lensing observations in at least two ways. One is the multiplication of the term Σ on the r.h.s. of Eq. (8.179). Another is the modification of the evolution of δ_m. The latter depends on two parameters q and ζ, or equivalently, Σ and ζ. Thus, two parameters (Σ, ζ) will be useful to detect signatures of modified gravity theories from future surveys of weak lensing. See Refs. [539–551] for related works about testing gravitational theories in weak lensing observations.

8.5.4 DGP Model

In the so-called Dvali, Gabadadze, and Porrati (DGP) [58] braneworld, it is possible to realize a "self-accelerating Universe" even in the absence of dark energy. In braneworlds standard model, particles are confined on a 3-dimensional (3D) brane embedded in the 5-dimensional bulk space-time with large extra dimensions. In the DGP braneworld model [58], the 3-brane is embedded in a Minkowski bulk space-time with infinitely large extra dimensions. Newton gravity can be recovered by adding a 4D Einstein DGP Hilbert action sourced by the brane curvature to the 5D action. Such a 4D term may be induced by quantum corrections coming from the bulk gravity and its coupling with matter on the brane. In the DGP model, the standard 4D gravity is recovered for small distances, whereas the effect from the 5D gravity manifests itself for large distances. The late-time cosmic acceleration can be realized without introducing a dark energy component [552, 553] (see also Ref. [554] for a generalized version of the DGP model).

The action for the DGP model is given by

$$S = \frac{1}{2\kappa_{(5)}^2} \int \mathrm{d}^5X \sqrt{-\tilde{g}}\,\tilde{R} + \frac{1}{2\kappa_{(4)}^2} \int \mathrm{d}^4X \sqrt{-g}R - \int \mathrm{d}^5X \sqrt{-\tilde{g}}\,\mathscr{L}_M\,, \tag{8.180}$$

where $\tilde{g}_{AB}$ is the metric in the 5D bulk and $g_{\mu\nu} = \partial_\mu X^A \partial_\nu X^B \tilde{g}_{AB}$ is the induced metric on the brane with $X^A(x^c)$ being the coordinates of an event on the brane labelled by x^c. The 5D and 4D (reduced) gravitational constants, $\kappa_{(5)}^2$ and $\kappa_{(4)}^2$, are related with the 5D and 4D Planck masses, $M_{(5)}$ and $M_{(4)}$, via $\kappa_{(5)}^2 = 1/M_{(5)}^3$ and $\kappa_{(4)}^2 = 1/M_{(4)}^2$. The first and second terms in Eq. (8.180) correspond to Einstein–Hilbert actions in the 5D bulk and on the brane, respectively. The matter action consists of a brane-localized matter whose action is given by $\int d^4x\sqrt{-g}\,(\sigma + \mathscr{L}_M^{\rm brane})$, where σ is the 3-brane tension and $\mathscr{L}_M^{\rm brane}$ is the Lagrangian density on the brane. Since the tension is not related to the Ricci scalar R, it can be adjusted to be zero.

The Einstein equation in the 5D bulk is given by $G_{AB}^{(5)} = 0$, where $G_{AB}^{(5)}$ is the 5D Einstein tensor. Imposing the Israel junction conditions on the brane with a Z_2 symmetry, we obtain the 4D Einstein equation [555]

$$G_{\mu\nu} - \frac{1}{r_c}(K_{\mu\nu} - Kg_{\mu\nu}) = \kappa_{(4)}^2 T_{\mu\nu}\,, \qquad (8.181)$$

where $K_{\mu\nu}$ is the extrinsic curvature on the brane and $T_{\mu\nu}$ is the energy-momentum tensor of localized matter. The cross-over scale r_c is defined by $r_c \equiv \kappa_{(5)}^2/(2\kappa_{(4)}^2)$. The Friedmann equation on the flat FLRW brane takes a simple form [552, 553]

$$H^2 - \frac{\varepsilon}{r_c}H = \frac{\kappa_{(4)}^2}{3}\rho_M\,, \qquad (8.182)$$

where $\varepsilon = \pm 1$, and ρ_M is the energy density of matter on the brane (with pressure P_M) satisfying the continuity equation

$$\dot{\rho}_M + 3H(\rho_M + P_M) = 0\,. \qquad (8.183)$$

If r_c is much larger than the Hubble radius H^{-1}, the first term in Eq. (8.182) dominates over the second one. In this case, the standard Friedmann equation, $H^2 = \kappa_{(4)}^2\rho_M/3$, is recovered. Meanwhile, in the regime $r_c < H^{-1}$, the presence of the second term in Eq. (8.182) leads to a modification to the standard Friedmann equation. In the Universe dominated by non-relativistic matter ($\rho_M \propto a^{-3}$), the Universe approaches a de Sitter solution for $\varepsilon = +1$: $H \to H_{\rm dS} = 1/r_c$. Hence it is possible to realize the present cosmic acceleration provided that r_c is of the order of the present Hubble radius H_0^{-1}.

Although the DGP braneworld is an attractive model allowing a self-acceleration, the joint constraints from SNLS, BAO, and CMB data shows that this model is disfavored observationally [556–561]. There is a modified version of the DGP model characterized by the Friedmann equation $H^2 - H^\alpha/r_c^{2-\alpha} = \kappa_{(4)}^2\rho_m/3$, where ρ_m is the energy density of non-relativistic matter [562]. In Fig. 8.6, we show 1σ and 2σ contours in the $(\Omega_m^{(0)}, \alpha)$ plane constrained from the joint data analysis of SN Ia, BAO, CMB, gamma ray bursts, and the linear growth factor of matter perturbations [561]. The parameter α is constrained to be $\alpha = 0.254 \pm 0.153$ (68% confidence level) and hence the flat DGP model ($\alpha = 1$) is incompatible with observations.

The evolution of density perturbations in the DGP model has been studied in Refs. [563–571]. Under the quasi-static approximation on sub-horizon scales, the

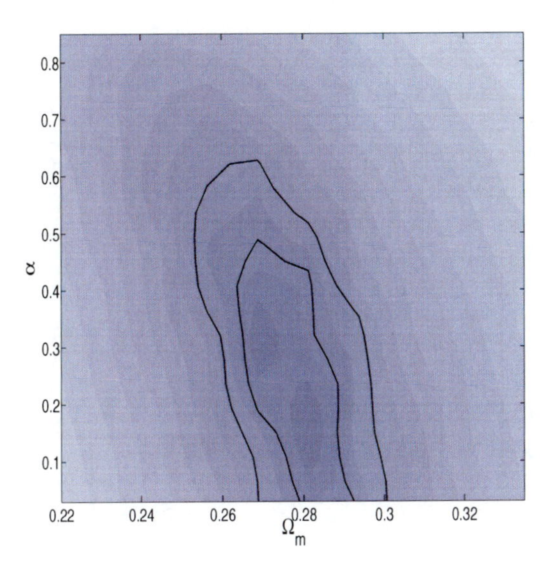

Fig. 8.6 Combined observational constraints on the modified DGP model characterized by the Friedmann equation $H^2 - H^\alpha/r_c^{2-\alpha} = \kappa_{(4)}^2\rho_m/3$. The two curves show 1σ and 2σ contours in the $(\Omega_m^{(0)}, \alpha)$ plane. The original DGP model ($\alpha = 1$) is incompatible with observations. From Ref. [561].

linear matter perturbation δ_m (with a homogenous density ρ_m) obeys the following equation

$$\ddot{\delta}_m + 2H\dot{\delta}_m - 4\pi G_{\rm eff}\rho_m\delta_m = 0\,, \qquad G_{\rm eff} = [1 + 1/(3\beta)]G\,, \tag{8.184}$$

where $\beta \equiv 1 - 2Hr_c[1 + \dot{H}/(3H^2)]$. In the deep matter era ($Hr_c \gg 1$), β is largely negative ($|\beta| \gg 1$). In this regime, the matter perturbation evolves as $\delta_m \propto t^{2/3}$ as in GR. Around the late-time de Sitter solution, one has $\beta \simeq 1 - 2Hr_c \simeq -1$ and $1 + 1/(3\beta) \simeq 2/3$, so that the growth rate of δ_m gets smaller than that in the ΛCDM model. The growth index γ defined in Eq. (8.78) is given by $\gamma \approx 0.68$ [323].

In the massless regime ($M^2/F \ll k^2/a^2$) in BD theory, the effective gravitational coupling (8.173) is given by $G_{\rm eff} = (1 + 2Q^2)G/F = (4 + 2\omega_{\rm BD})/(3 + 2\omega_{\rm BD})G/F$, where we used the relation (8.166). Comparing this with the effective coupling (8.184), we find that the DGP model is related to BD theory via $\omega_{\rm BD} = (3/2)(\beta - 1)$ with $F = 1$. Since $\beta < 0$ for the self-accelerating DGP solution, this implies that $\omega_{\rm BD} < -3/2$ and hence the DGP model contains a ghost mode. It is, however, possible to construct a generalized DGP model free from the ghost problem by embedding our visible 3-brane with a 4-brane in a flat 6D bulk [168].

In the DGP model, a brane bending mode ϕ in the bulk corresponds to a scalar-field degree of freedom. In general, such a field can mediate a long-range fifth force incompatible with local gravity experiments, but the presence of a self-interaction of ϕ allows the so-called Vainshtein mechanism [572] to work within a radius

$r_* = (r_g r_c^2)^{1/3}$ (r_g is the Schwarzschild radius of a source). The DGP model can be consistent with local gravity constraints under some range of conditions on the energy-momentum tensor [573–575].

The DGP model stimulated other approaches for constructing ghost-free theories in the presence of nonlinear self-interactions of a scalar field ϕ. It is important to keep the field equations at second order in time derivatives to avoid that an extra degree of freedom gives rise to a ghost state. In particular, Nicolis et al. [577] imposed a constant gradient-shift symmetry ("Galilean" symmetry), $\partial_\mu \phi \to \partial_\mu \phi + b_\mu$, to restrict the equations of motion at second order, while keeping a universal gravitational coupling with matter. In the 4-dimensional Minkowski space-time, they found five terms $\mathcal{L}_i$ ($i = 1, \cdots 5$) giving rise to equations of motion satisfying the Galilean symmetry. The first three terms are given by $\mathcal{L}_1 = \phi$, $\mathcal{L}_2 = \nabla_\mu \phi \nabla^\mu \phi$, and $\mathcal{L}_3 = \Box\phi \nabla_\mu \phi \nabla^\mu \phi$. The term $\mathcal{L}_3$ is the nonlinear field derivative that appears in the DGP model, which allows the possibility for the consistency with solar system experiments through the Vainshtein mechanism. Deffayet et al. [578, 579] derived the covariant expression of the terms $\mathcal{L}_i$ ($i = 1, \cdots 5$) by extending the analysis to the curved space time.

Silva and Koyama [580] considered BD theory (without a field potential) in the presence of a nonlinear derivative term $\xi(\phi)\Box\phi\nabla_\mu\phi\nabla^\mu\phi$, i.e.,

$$S = \int \mathrm{d}^4x\sqrt{-g}\left[\frac{1}{2}\phi R - \frac{\omega_{\rm BD}}{2\phi}(\nabla\phi)^2 + \xi(\phi)\Box\phi\nabla_\mu\phi\nabla^\mu\phi\right] + S_M\,, \tag{8.185}$$

where S_M is the matter action. Although the nonlinear derivative term does not satisfy the Galilean invariance in the FLRW cosmological background, the field equation of motion remains at second order (see also Ref. [581]). This term also arises as one of higher-order derivative corrections to low-energy effective string theory. Interestingly, for the function $\xi(\phi) = 1/(M^2\phi^2)$, there exists a dS solution responsible for dark energy (provided that $M \approx H_0$). Moreover, because of the presence of the nonlinear interaction, the problems of the appearance of ghosts and instabilities can be avoided for the BD parameter $\omega_{\rm BD}$ smaller than -2 [580]. At early times, General Relativity can be recovered by the cosmological Vainshtein mechanism. A number of interesting observational signatures, such as modified growth of matter perturbations as well as a distinguished ISW effect, have been studied in Refs. [580, 582, 583].

8.6 Cosmic Acceleration without Dark Energy

There are attempts to explain the apparent cosmic acceleration by inhomogeneities in the distribution of matter without recourse to a dark energy component (see Ref. [584] for review).

One of such approaches is the void model in which the presence of underdense bubbles leads to the faster expansion of the Universe compared with the outside. In other words, we live in the middle of a huge spherical region and we interpret the evolution of this underdense region as an apparent cosmic acceleration. Originally, Tomita [59, 60] introduced a local homogenous void separated from the outside described by a homogenous FLRW space with a singular mass shell. The analysis was extended to the models with a continuous transition between the inside and outside the void [63]. This can be described by a class of the Lemaître–Tolman–Bondi (LTB) spherically symmetric models. Theoretical and observational aspects of the LTB model have been extensively studied as an alternative to dark energy [61–63, 585–610].

The second approach is based on the backreaction of cosmological perturbations arising from perturbing the homogeneous Universe [64–66]. Unlike the void model, this tries to explain a *real* cosmic acceleration by arranging inhomogeneities that come from the deviation of the FLRW metric.

There is another approach for explaining the apparent cosmic acceleration based on the "Ultra Strong" version of equivalence principle [611, 612]. In this model, the standard geometric description of space-time as a metric manifold holds as a small distance approximation and hence General Relativity can be modified on large scales by a curvature-dependent subleading effect. Although the original model proposed in Ref. [613] do not explain the observational data of SN Ia very well, its modified version can be consistent with the SN Ia data with a rather low value of the Hubble constant, $H_0 \approx 50\,\mathrm{km\,sec^{-1}\,Mpc^{-1}}$.

In the following, we shall briefly review the first two approaches.

8.6.1 Inhomogeneous LTB Model

In order to discuss a spherical inhomogeneity in local regions, we take the LTB metric given by

$$\mathrm{d}s^2 = -\mathrm{d}t^2 + X^2(t,r)\,\mathrm{d}r^2 + R^2(t,r)\,\mathrm{d}\Omega^2\,, \tag{8.186}$$

where the expansion factor along the radial coordinate r is different relative to the surface line element $\mathrm{d}\Omega^2 = \mathrm{d}\theta^2 + \sin^2\theta\,\mathrm{d}\phi^2$. Solving the (0,1) component of the Einstein equation $G_{01} = 0$ for the fluid at rest, it follows that $X(t,r)$ is separable as $X(t,r) = R'(t,r)/\sqrt{1+\beta(r)}$. Here a prime represents a partial derivative with respect to r and $\beta(r)$ is a function of r. Then, the metric (8.186) is given by

$$\mathrm{d}s^2 = -\mathrm{d}t^2 + \frac{[R'(t,r)]^2}{1+\beta(r)}\mathrm{d}r^2 + R^2(t,r)\mathrm{d}\Omega^2\,. \tag{8.187}$$

The metric (8.187) recovers the one in the FLRW space-time by the choice $R = a(t)r$ and $\beta = -Kr^2$, where K is a cosmic curvature. In other cases, the metric (8.187) describes a spherical inhomogeneity centered on the origin. We define the transverse Hubble function $H_\perp$ and the radial Hubble function $H_{||}$, as $H_\perp \equiv \dot{R}'/R'$

and $H_\perp = \dot{R}/R$. The Einstein equations in the presence of nonrelativistic matter (energy density ρ_m) give [63]

$$H_\perp^2 + 2H_{||}H_\perp - \frac{\beta}{R^2} - \frac{\beta'}{RR'} = 8\pi G\rho_m\,, \tag{8.188}$$

$$6\frac{\ddot{R}}{R} + 2H_\perp^2 - 2\frac{\beta}{R^2} - 2H_{||}H_\perp + \frac{\beta'}{RR'} = -8\pi G\rho_m\,. \tag{8.189}$$

Eliminating the term ρ_m from Eqs. (8.188) and (8.189), we obtain the relation $2R\ddot{R}+\dot{R}^2 = \beta(r)$. Integrating this equation, it follows that

$$H_\perp^2 = \frac{\alpha(r)}{R^3} + \frac{\beta(r)}{R^2}\,, \tag{8.190}$$

where $\alpha(r)$ is an arbitrary function of r. From this, we can introduce the today's effective density parameters of matter and the spatial curvature, respectively, as

$$\Omega_m^{(0)}(r) \equiv \frac{\alpha(r)}{R_0^3 H_{\perp 0}^2}\,, \qquad \Omega_K^{(0)}(r) = 1 - \Omega_m^{(0)}(r) = \frac{\beta(r)}{R_0^2 H_{\perp 0}^2}\,. \tag{8.191}$$

We define the time $t = 0$ at the decoupling epoch (the redshift $z \simeq 1090$) with $R(r,t=0) = 0$. Introducing the conformal time η as $\mathrm{d}\eta = (\sqrt{\beta}/R)\,\mathrm{d}t$, we obtain the following parametric solutions of Eqs. (8.188) and (8.189) for $\beta > 0$ [63]:

$$R = \frac{\alpha(r)}{2\beta(r)}(\cosh\eta - 1) = \frac{R_0\Omega_m^{(0)}(r)}{2[1-\Omega_m^{(0)}(r)]}(\cosh\eta - 1)\,, \tag{8.192}$$

$$t = \frac{\alpha(r)}{2\beta^{3/2}(r)}(\sinh\eta - \eta) = \frac{\Omega_m^{(0)}(r)}{2[1-\Omega_m^{(0)}(r)]^{3/2}H_{\perp 0}}(\sinh\eta - \eta)\,. \tag{8.193}$$

The structure of the void with an under density can be accommodated by choosing $\Omega_m^{(0)}(r)$ and $h \equiv H_{\perp 0}/(100\,\mathrm{km\,sec^{-1}\,Mpc^{-1}})$ in the following form [598] (see also Ref. [63] for another choice)

$$\Omega_m^{(0)}(r) = \Omega_{\rm out} + (\Omega_{\rm in} - \Omega_{\rm out})f(r,r_0,\Delta)\,, \tag{8.194}$$

$$h(r) = h_{\rm out} + (h_{\rm in} - h_{\rm out})f(r,r_0,\Delta)\,, \tag{8.195}$$

where the function $f(r,r_0,\Delta) = [1-\tanh((r-r_0)/2\Delta)]/[1+\tanh(r_0/2\Delta)]$ describes a transition of a shell of radius r_0 and thickness Δ ("in" and "out" represent quantities inside and outside the void, respectively).

The trajectory of photons arriving at $r = 0$ today is characterized by a path $t = \hat{t}(r)$ satisfying [590]

$$\frac{\mathrm{d}\hat{t}}{\mathrm{d}r} = -\frac{R'(r,\hat{t})}{\sqrt{1+\beta}}\,. \tag{8.196}$$

Then, one can show that the redshift $z = z(r)$ of photons obeys the differential equation [62]

$$\frac{dz}{dr} = (z+1)\frac{\dot{R}'(r,\hat{t})}{\sqrt{1+\beta}}, \tag{8.197}$$

with $z(r=0) = 0$. The luminosity distance $d_L(z)$ is related to the diameter distance $d_A(z) = R(r,\hat{t})$ according to the usual duality relation

$$d_L(z) = (1+z)^2 R(r,\hat{t}). \tag{8.198}$$

Now, we are ready to confront the inhomogeneous LTB model with the SN Ia observations. From the requirement that the CMB acoustic peak is not spoiled, we require that local density parameter $\Omega_{\rm in}$ is in the range 0.1–0.3, whereas $\Omega_{\rm out} = 1$ as predicted by inflation. The observed local value of H is around $h_{\rm in} \approx 0.7$, whereas outside the void one requires $h \approx 0.5$ to be consistent with $\Omega_{\rm out} = 1$. Then, the two parameters r_0 and Δ are constrained by the SN Ia data. In Ref. [598], it was shown that the inhomogeneous LTB model can be consistent with the SN Ia data for $r_0 = 2.3 \pm 0.9$ Gpc and $\Delta/r_0 > 0.2$.

We note, however, that one can place other constraints on the void model. If we do not live around the center of the void, the observed CMB dipole becomes much larger than that allowed by observations. The maximum distance r_c to the center is constrained to be smaller than 10–20 Mpc [590, 605]. Even if we happen to live very close to the center of the void, we observe distant off-centered galaxy clusters. Such off-centered clusters should see a large CMB dipole in their reference frame. For us, this manifests itself observationally as a kinematic Sunyaev–Zeldovich effect. Using the observational data of only nine clusters, the inhomogeneous LTB model with void sizes greater than 1.5 Gpc can be ruled out [599, 604]. This is already in mild conflict with the constraint derived by the SN Ia data. It remains to see whether the void model can be ruled out or not in future observations.

8.6.2 *Backreaction of Cosmological Perturbations*

Let us finally discuss the possibility of realizing a real cosmic acceleration by the backreaction of inhomogeneities to the FLRW space–time. In general, averaging the inhomogeneities and then solving the Einstein equations (the standard approach) might not be the same as solving the full inhomogeneous Einstein equations first and then averaging them. In other words, the expected value of a nonlinear function of x is not the same as the nonlinear function of the expected value of x. The argument is complicated and controversial, so we mention the basic ideas only briefly. The readers who are interested in the detail of this line of research may have a look at the original papers [64–66, 614–628].

Let us decompose the Einstein tensor $G_{\mu\nu}$ and the energy-momentum tensor $T_{\mu\nu}$ into the background (0-th order FLRW Universe) and the perturbed parts, as

$G_{\mu\nu} = G^{(0)}_{\mu\nu} + G^{(1)}_{\mu\nu}$ and $T_{\mu\nu} = T^{(0)}_{\mu\nu} + T^{(1)}_{\mu\nu}$, respectively. Then, the (00) component of the Einstein equation (8.4) gives

$$G^{(0)}_{00} = 8\pi G\,(T^{(0)}_{00} + T^{(1)}_{00}) - G^{(1)}_{00}\,. \tag{8.199}$$

Identifying the average matter density at this order as $\langle\rho\rangle = T^{(0)}_{00} + T^{(1)}_{00}$ and averaging over Eq. (8.199), we obtain

$$\langle G^{(0)}_{00}\rangle = 8\pi G\langle\rho\rangle - \langle G^{(1)}_{00}\rangle\,, \tag{8.200}$$

where $G^{(0)}_{00} = 3H^2$ in the FLRW background. This shows that $3H^2 \neq 8\pi G\langle\rho\rangle$ in general because of the presence of the term $\langle G^{(1)}_{00}\rangle$. If we first average the metric as $\langle g^{(1)}_{\mu\nu}\rangle = 0$, it then follows that $G^{(1)}_{00} = G_{00}(\langle g^{(1)}_{\mu\nu}\rangle) = 0$ and hence $3H^2 = 8\pi G\langle\rho\rangle$. The above argument can be extended to the second order, i.e.,

$$\langle G^{(0)}_{00}\rangle = 8\pi G\langle\rho\rangle - \langle G^{(1)}_{00} + G^{(2)}_{00}\rangle\,. \tag{8.201}$$

The cosmological evolution depends on the averaging procedure. For example, if we take the average of the function $f(t,x)$ as [65]

$$\langle f\rangle(t) = \frac{\int \mathrm{d}^3x\sqrt{\gamma(t,x)}\,f(t,x)}{\int \mathrm{d}^3x\sqrt{\gamma(t,x)}}\,, \tag{8.202}$$

where γ corresponds to the determinant of the perturbed metric of spatial constant-time hypersurfaces, then the second-order term $G^{(2)}_{00}$ contributes to the expansion rate of the Universe (which is typically of the order of 10^{-5}). If we use other ways of averaging, the amplitude of such a term is subject to change [616]. Also the results are affected by adding the contributions higher than the second order. In fact, Ref. [614] pointed out the danger of arbitrarily stopping at some order by showing several examples in which many contributions cancel each other.

The backreaction scenario is very attractive if it really works because it is the most economical way of explaining the cosmic acceleration without using dark energy. We hope that further progress will be expected in this direction.

8.7 Conclusions

We summarize the results presented in this review.

- The cosmological constant ($w_{\rm DE} = -1$) is favored by a number of observations, but theoretically, it is still challenging to explain why its energy scale is very small.
- Quintessence leads to the variation of the field equation of state in the region $w_\phi > -1$, but the current observations are not sufficient to distinguish between quintessence potentials.

- In k-essence, it is possible to realize the cosmic acceleration by a field kinetic energy while avoiding the instability problem associated with a phantom field. The k-essence models that aim to solve the coincidence problem inevitably leads to the superluminal propagation of the sound speed.
- In coupled dark energy models, there is an upper bound on the strength of the coupling from the observations of CMB, large-scale structure and SN Ia.
- The generalized Chaplygin gas model allows the unified description of dark energy and dark matter, but it needs to be very close to the ΛCDM model to explain the observed matter power spectrum. There is a class of viable unified models of dark energy and dark matter using a purely k-essence field.
- In $f(R)$ gravity and scalar-tensor theories, it is possible to construct viable models that satisfy both cosmological and local gravity constraints. These models leave several interesting observational signatures such as the modifications to the matter power spectrum and to the weak lensing spectrum.
- The dark energy models based on the Gauss-Bonnet term are in conflict with a number of observations and experiments in general and hence they are excluded as an alternative to the ΛCDM model.
- The DGP model allows the self-acceleration of the Universe, but it is effectively ruled out from observational constraints and the ghost problem. However, some of the extension of works such as Galileon gravity allow the possibility for avoiding the ghost problem while satisfying cosmological and local gravity constraints.
- The models based on the inhomogeneities in the distribution of matter allow the possibility for explaining the apparent accelerated expansion of the Universe. The void model can be consistent with the SN Ia data, but it is still challenging to satisfy all other constraints coming from the CMB and the kinematic Sunyaev–Zeldovich effect.

When the author submitted a review article [9] on dark energy to *International Journal of Modern Physics D* in March 2006, we wrote in concluding section that "over 900 papers with the words 'dark energy' in the title have appeared on the archives since 1998, and nearly 800 with the words 'cosmological constant' have appeared." Now in April 2010, I need to change the sentence to "over 2250 papers with the words 'dark energy' in the title have appeared on the archives since 1998, and nearly 1750 with the words 'cosmological constant' have appeared." This means that over 4000 papers about dark energy and cosmological constant have been already written, with more than 2300 papers over the past 4 years. Many cosmologists, astrophysicists, and particle physicists have extensively worked on this new field of research after the first discovery of the cosmic acceleration in 1998. We hope that the future progress of both theory and observations will provide some exciting clue to reveal the origin of dark energy.

Acknowledgments I thank Sabino Matarrese to invite me to write this article in a chapter "dark energy: investigation and modeling" of a book published in Springer. I am also grateful to all my collaborators and colleagues with whom I discussed about dark energy. This work was supported by Grant-in-Aid for Scientific Research Fund of the JSPS (No. 30318802) and Grant-in-Aid for Scientific Research on Innovative Areas (No. 21111006).

References

1. A. G. Riess et al. [Supernova Search Team Collaboration], Astron. J. **116**, 1009 (1998).
2. S. Perlmutter et al. [Supernova Cosmology Project Collaboration], Astrophys. J. **517**, 565 (1999).
3. D. Huterer and M. S. Turner, Phys. Rev. D **60**, 081301 (1999).
4. V. Sahni and A. A. Starobinsky, Int. J. Mod. Phys. D **9**, 373 (2000).
5. S. M. Carroll, Living Rev. Rel. **4**, 1 (2001).
6. P. J. E. Peebles and B. Ratra, Rev. Mod. Phys. **75**, 559 (2003).
7. T. Padmanabhan, Phys. Rept. **380**, 235 (2003).
8. V. Sahni, Lect. Notes Phys. **653**, 141 (2004)
9. E. J. Copeland, M. Sami and S. Tsujikawa, Int. J. Mod. Phys. D **15**, 1753 (2006).
10. R. Durrer and R. Maartens, Gen. Rel. Grav. **40**, 301 (2008).
11. R. R. Caldwell and M. Kamionkowski, Ann. Rev. Nucl. Part. Sci. **59**, 397 (2009).
12. L. Amendola and S. Tsujikawa, *Dark energy–Theory and observations*, Cambridge University Press (2010).
13. P. Astier et al. [The SNLS Collaboration], Astron. Astrophys. **447**, 31 (2006).
14. A. G. Riess et al. [Supernova Search Team Collaboration], Astrophys. J. **607**, 665 (2004).
15. A. G. Riess et al., Astrophys. J. **659**, 98 (2007).
16. W. M. Wood-Vasey et al. [ESSENCE Collaboration], Astrophys. J. **666**, 694 (2007).
17. T. M. Davis et al., Astrophys. J. **666**, 716 (2007).
18. M. Kowalski et al. [Supernova Cosmology Project Collaboration], Astrophys. J. **686**, 749 (2008).
19. D. N. Spergel et al. [WMAP Collaboration], Astrophys. J. Suppl. **148**, 175 (2003).
20. D. N. Spergel et al. [WMAP Collaboration], Astrophys. J. Suppl. **170**, 377 (2007).
21. E. Komatsu et al. [WMAP Collaboration], Astrophys. J. Suppl. **180**, 330 (2009).
22. E. Komatsu et al., arXiv:1001.4538 [astro-ph.CO].
23. D. J. Eisenstein et al. [SDSS Collaboration], Astrophys. J. **633**, 560 (2005).
24. W. J. Percival, S. Cole, D. J. Eisenstein, R. C. Nichol, J. A. Peacock, A. C. Pope and A. S. Szalay, Mon. Not. Roy. Astron. Soc. **381**, 1053 (2007).
25. W. J. Percival et al., Mon. Not. Roy. Astron. Soc. **401**, 2148 (2010).
26. S. Weinberg, Rev. Mod. Phys. **61**, 1 (1989).
27. S. Kachru, R. Kallosh, A. D. Linde and S. P. Trivedi, Phys. Rev. D **68**, 046005 (2003).
28. Y. Fujii, Phys. Rev. D **26**, 2580 (1982).
29. L. H. Ford, Phys. Rev. D **35**, 2339 (1987).
30. C. Wetterich, Nucl. Phys B. **302**, 668 (1988).
31. B. Ratra and J. Peebles, Phys. Rev D **37**, 321 (1988).
32. T. Chiba, N. Sugiyama and T. Nakamura, Mon. Not. Roy. Astron. Soc. **289**, L5 (1997).
33. P. G. Ferreira and M. Joyce, Phys. Rev. Lett. **79**, 4740 (1997).
34. P. G. Ferreira and M. Joyce, Phys. Rev. D **58**, 023503 (1998).
35. E. J. Copeland, A. R. Liddle and D. Wands, Phys. Rev. D **57**, 4686 (1998).
36. R. R. Caldwell, R. Dave and P. J. Steinhardt, Phys. Rev. Lett. **80**, 1582 (1998).
37. I. Zlatev, L. M. Wang and P. J. Steinhardt, Phys. Rev. Lett. **82**, 896 (1999).
38. P. J. Steinhardt, L. M. Wang and I. Zlatev, Phys. Rev. D **59**, 123504 (1999).
39. T. Chiba, T. Okabe and M. Yamaguchi, Phys. Rev. D **62**, 023511 (2000).
40. C. Armendariz-Picon, V. F. Mukhanov and P. J. Steinhardt, Phys. Rev. Lett. **85**, 4438 (2000).
41. C. Armendariz-Picon, V. F. Mukhanov and P. J. Steinhardt, Phys. Rev. D **63**, 103510 (2001).
42. S. M. Carroll, Phys. Rev. Lett. **81**, 3067 (1998).
43. C. F. Kolda and D. H. Lyth, Phys. Lett. B **458**, 197 (1999).
44. A. Y. Kamenshchik, U. Moschella and V. Pasquier, Phys. Lett. B **511**, 265 (2001).
45. M. C. Bento, O. Bertolami and A. A. Sen, Phys. Rev. D **66**, 043507 (2002).
46. H. Sandvik, M. Tegmark, M. Zaldarriaga and I. Waga, Phys. Rev. D **69**, 123524 (2004).
47. S. Capozziello, Int. J. Mod. Phys. D **11**, 483 (2002).
48. S. Capozziello, S. Carloni and A. Troisi, Recent Res. Dev. Astron. Astrophys. **1**, 625 (2003).

49. S. Capozziello, S. Carloni and A. Troisi, Recent Res. Dev. Astron. Astrophys. **1**, 625 (2003).
50. S. Capozziello, V. F. Cardone, S. Carloni and A. Troisi, Int. J. Mod. Phys. D **12**, 1969 (2003).
51. S. M. Carroll, V. Duvvuri, M. Trodden and M. S. Turner, Phys. Rev. D **70**, 043528 (2004).
52. S. Nojiri and S. D. Odintsov, Phys. Rev. D **68**, 123512 (2003).
53. L. Amendola, Phys. Rev. D **60**, 043501 (1999).
54. J. P. Uzan, Phys. Rev. D **59**, 123510 (1999).
55. T. Chiba, Phys. Rev. D **60**, 083508 (1999).
56. N. Bartolo and M. Pietroni, Phys. Rev. D **61** 023518 (2000).
57. F. Perrotta, C. Baccigalupi and S. Matarrese, Phys. Rev. D **61**, 023507 (2000).
58. G. R. Dvali, G. Gabadadze and M. Porrati, Phys. Lett. B **485**, 208 (2000).
59. K. Tomita, Astrophys. J. **529**, 38 (2000).
60. K. Tomita, Mon. Not. Roy. Astron. Soc. **326**, 287 (2001).
61. M. N. Celerier, Astron. Astrophys. **353**, 63 (2000).
62. H. Iguchi, T. Nakamura and K. i. Nakao, Prog. Theor. Phys. **108**, 809 (2002).
63. H. Alnes, M. Amarzguioui and O. Gron, Phys. Rev. D **73**, 083519 (2006).
64. S. Rasanen, JCAP **0402**, 003 (2004).
65. E. W. Kolb, S. Matarrese, A. Notari and A. Riotto, Phys. Rev. D **71**, 023524 (2005).
66. E. W. Kolb, S. Matarrese and A. Riotto, New J. Phys. **8**, 322 (2006).
67. S. Weinberg, *Gravitation and Cosmology*, Wiley and Sons, New York (1972).
68. R. R. Caldwell, Phys. Lett. B **545**, 23 (2002).
69. A. A. Starobinsky, Grav. Cosmol. **6**, 157 (2000).
70. R. R. Caldwell, M. Kamionkowski and N. N. Weinberg, Phys. Rev. Lett. **91**, 071301 (2003).
71. J. D. Barrow, G. Galloway and F. Tipler, Mon. Not. Roy. astr. Soc. **223**, 835 (1986).
72. J. D. Barrow, Class. Quant. Grav. **21**, L79 (2004).
73. J. D. Barrow, Class. Quant. Grav. **21**, 5619 (2004).
74. H. Stefancic, Phys. Rev. D **71**, 084024 (2005).
75. S. Nojiri, S. D. Odintsov and S. Tsujikawa, Phys. Rev. D **71**, 063004 (2005).
76. I. H. Brevik and O. Gorbunova, Gen. Rel. Grav. **37**, 2039 (2005).
77. M. P. Dabrowski, Phys. Lett. B **625**, 184 (2005).
78. M. Sami, A. Toporensky, P. V. Tretjakov and S. Tsujikawa, Phys. Lett. B **619**, 193 (2005).
79. M. Bouhmadi-Lopez and J. A. Jimenez Madrid, JCAP **0505**, 005 (2005).
80. S. Nojiri and S. D. Odintsov, Phys. Lett. B **595**, 1 (2004).
81. M. Sami, P. Singh and S. Tsujikawa, Phys. Rev. D **74**, 043514 (2006).
82. D. Samart and B. Gumjudpai, Phys. Rev. D **76**, 043514 (2007).
83. L. Amendola and S. Tsujikawa, Phys. Lett. B **660**, 125 (2008).
84. J. Weller and A. J. Albrecht, Phys. Rev. D **65**, 103512 (2002).
85. M. Chevallier and D. Polarski, Int. J. Mod. Phys. D **10**, 213 (2001).
86. E. V. Linder, Phys. Rev. Lett. **90**, 091301 (2003).
87. H. K. Jassal, J. S. Bagla and T. Padmanabhan, Mon. Not. Roy. Astron. Soc. **356**, L11 (2005).
88. G. Efstathiou, Mon. Not. R. Astron. Soc. **342**, 810 (2000).
89. B. A. Bassett, M. Kunz, J. Silk and C. Ungarelli, Mon. Not. Roy. Astron. Soc. **336**, 1217 (2002).
90. B. A. Bassett, P. S. Corasaniti and M. Kunz, Astrophys. J. **617**, L1 (2004).
91. A. A. Starobinsky, JETP Lett. **68**, 757 (1998).
92. T. D. Saini, S. Raychaudhury, V. Sahni and A. A. Starobinsky, Phys. Rev. Lett. **85**, 1162 (2000).
93. U. Alam, V. Sahni, T. D. Saini and A. A. Starobinsky, Mon. Not. Roy. Astron. Soc. **344**, 1057 (2003).
94. U. Alam, V. Sahni, T. D. Saini and A. A. Starobinsky, Mon. Not. Roy. Astron. Soc. **354**, 275 (2004).
95. I. Maor, R. Brustein, J. McMahon and P. J. Steinhardt, Phys. Rev. D **65**, 123003 (2002).
96. P. S. Corasaniti and E. J. Copeland, Phys. Rev. D **67**, 063521 (2003).
97. Y. Wang and P. Mukherjee, Astrophys. J. **606**, 654 (2004).
98. S. Nesseris and L. Perivolaropoulos, Phys. Rev. D **70**, 043531 (2004).
99. R. Lazkoz, S. Nesseris and L. Perivolaropoulos, JCAP **0511**, 010 (2005).

100. A. Upadhye, M. Ishak and P. J. Steinhardt, Phys. Rev. D **72**, 063501 (2005).
101. Z. K. Guo, N. Ohta and Y. Z. Zhang, Phys. Rev. D **72**, 023504 (2005).
102. Y. Wang and P. Mukherjee, Astrophys. J. **650**, 1 (2006).
103. S. Nesseris and L. Perivolaropoulos, JCAP **0702**, 025 (2007).
104. R. Crittenden, E. Majerotto and F. Piazza, Phys. Rev. Lett. **98**, 251301 (2007).
105. K. Ichikawa and T. Takahashi, JCAP **0702**, 001 (2007).
106. G. B. Zhao, J. Q. Xia, B. Feng and X. Zhang, Int. J. Mod. Phys. D **16**, 1229 (2007).
107. G. B. Zhao, J. Q. Xia, H. Li, C. Tao, J. M. Virey, Z. H. Zhu and X. Zhang, Phys. Lett. B **648**, 8 (2007).
108. B. A. Bassett et al., JCAP **0807**, 007 (2008).
109. K. Ichikawa and T. Takahashi, JCAP **0804**, 027 (2008).
110. A. G. Riess et al., Astrophys. J. **699**, 539 (2009).
111. S. H. Suyu et al., Astrophys. J. **711**, 201 (2010).
112. R. K. Sachs and A. M. Wolfe, Astrophys. J. 147, 73 (1967).
113. A. A. Starobinsky, Phys. Lett. B **91** (1980) 99.
114. D. Kazanas, Astrophys. J. **241** L59 (1980).
115. K. Sato, Mon. Not. R. Astron. Soc. **195**, 467 (1981).
116. A. H. Guth, Phys. Rev. D **23**, 347 (1981).
117. G. F. Smoot et al., Astrophys. J. **396**, L1 (1992).
118. A. R. Liddle and D. H. Lyth, *Cosmological inflation and large-scale structure*, Cambridge University Press (2000).
119. B. A. Bassett, S. Tsujikawa and D. Wands, Rev. Mod. Phys. **78**, 537 (2006).
120. L. Page et al. [WMAP Collaboration], Astrophys. J. Suppl. **148**, 233 (2003).
121. W. Hu and N. Sugiyama, Astrophys. J. **444**, 489 (1995).
122. W. Hu and N. Sugiyama, Astrophys. J. **471**, 542 (1996).
123. G. Efstathiou and J. R. Bond, Mon. Not. Roy. Astron. Soc. **304**, 75 (1999).
124. M. Doran and M. Lilley, Mon. Not. Roy. Astron. Soc. **330**, 965 (2002).
125. A. Melchiorri, L. Mersini, C. J. Odman and M. Trodden, Phys. Rev. D **68**, 043509 (2003).
126. S. Hannestad and E. Mortsell, Phys. Rev. D **66**, 063508 (2002).
127. J. Weller and A. M. Lewis, Mon. Not. Roy. Astron. Soc. **346**, 987 (2003).
128. P. S. Corasaniti, T. Giannantonio and A. Melchiorri, Phys. Rev. D **71**, 123521 (2005).
129. W. Lee and K. W. Ng, Phys. Rev. D **67**, 107302 (2003).
130. Y. Wang and M. Tegmark, Phys. Rev. Lett. **92**, 241302 (2004).
131. D. Rapetti, S. W. Allen and J. Weller, Mon. Not. Roy. Astron. Soc. **360**, 555 (2005).
132. L. Pogosian, P. S. Corasaniti, C. Stephan-Otto, R. Crittenden and R. Nichol, Phys. Rev. D **72**, 103519 (2005).
133. J. Q. Xia, G. B. Zhao, B. Feng, H. Li and X. Zhang, Phys. Rev. D **73**, 063521 (2006).
134. D. J. Eisenstein and W. Hu, Astrophys. J. **496**, 605 (1998).
135. M. Shoji, D. Jeong and E. Komatsu, Astrophys. J. **693**, 1404 (2009).
136. B. Feng, X. L. Wang and X. M. Zhang, Phys. Lett. B **607**, 35 (2005).
137. M. Tegmark et al. [SDSS Collaboration], Phys. Rev. D **69**, 103501 (2004).
138. U. Seljak et al., Phys. Rev. D **71**, 103515 (2005).
139. M. Tegmark et al. [SDSS Collaboration], Phys. Rev. D **74**, 123507 (2006).
140. D. Hooper and S. Dodelson, Astropart. Phys. **27**, 113 (2007).
141. M. Oguri and K. Takahashi, Phys. Rev. D **73**, 123002 (2006).
142. S. Basilakos and L. Perivolaropoulos, Mon. Not. Roy. Astron. Soc. **391**, 411 (2008).
143. Y. Wang, Phys. Rev. D **78**, 123532 (2008).
144. R. Tsutsui, T. Nakamura, D. Yonetoku, T. Murakami, S. Tanabe, Y. Kodama and K. Takahashi, Mon. Not. Roy. Astron. Soc. **394**, L31 (2009).
145. B. Jain and A. Taylor, Phys. Rev. Lett. **91**, 141302 (2003).
146. M. Takada and B. Jain, Mon. Not. Roy. Astron. Soc. **348**, 897 (2004).
147. M. Ishak, Mon. Not. Roy. Astron. Soc. **363**, 469 (2005).
148. C. Schimd et al., Astron. Astrophys. **463**, 405 (2007).
149. M. Takada and S. Bridle, New J. Phys. **9**, 446 (2007).
150. L. Hollenstein, D. Sapone, R. Crittenden and B. M. Schaefer, JCAP **0904**, 012 (2009).

151. D. Bailin and A. Love, *Supersymmetric gauge field theory and string theory,* Institute of Physics Publishing (1994).
152. M. Green, J. H. Schwarz and E. Witten, *Superstring theory,* Cambridge University Press, Cambridge (1987).
153. M. Dine, R. Rohm, N. Seiberg and E. Witten, Phys. Lett. B **156**, 55 (1985).
154. I. Affleck, M. Dine and N. Seiberg, Nucl. Phys. B **256**, 557 (1985).
155. L. Susskind, arXiv:hep-th/0302219.
156. B. Carter, *Large number coincidences and the anthropic principle in cosmology*, IAU Symposium 63: Con- frontation of cosmological theories with observational data, 291 (1974).
157. J. Barrow and F. Tipler, *The cosmological anthropic principle*, Oxford University Press, Oxford (1988).
158. S. Weinberg, Phys. Rev. Lett. **59**, 2607 (1987).
159. J. Garriga and A. Vilenkin, Phys. Rev. D **61**, 083502 (2000).
160. F. Denef and M. R. Douglas, JHEP **0405**, 072 (2004).
161. J. Garriga, A. D. Linde and A. Vilenkin, Phys. Rev. D **69**, 063521 (2004).
162. R. Blumenhagen et al., Nucl. Phys. B **713**, 83 (2005).
163. A. D. Linde, Rept. Prog. Phys. **47**, 925 (1984).
164. J. D. Brown and C. Teitelboim, Phys. Lett. B **195**, 177 (1987).
165. J. D. Brown and C. Teitelboim, Nucl. Phys. B **297**, 787 (1988).
166. R. Bousso, Gen. Rel. Grav. **40**, 607 (2008).
167. R. Bousso and J. Polchinski, JHEP **0006**, 006 (2000).
168. C. de Rham, G. Dvali, S. Hofmann, J. Khoury, O. Pujolas, M. Redi and A. J. Tolley, Phys. Rev. Lett. **100**, 251603 (2008).
169. N. Afshordi, arXiv:0807.2639 [astro-ph].
170. T. Padmanabhan, Gen. Rel. Grav. **40**, 529 (2008).
171. T. Padmanabhan, arXiv:0807.2356 [gr-qc].
172. S. W. Hawking, Phys. Lett. B **134**, 403 (1984).
173. S. Kachru, M. B. Schulz and E. Silverstein, Phys. Rev. D **62**, 045021 (2000).
174. N. Arkani-Hamed, S. Dimopoulos, N. Kaloper and R. Sundrum, Phys. Lett. B **480**, 193 (2000).
175. J. L. Feng, J. March-Russell, S. Sethi and F. Wilczek, Nucl. Phys. B **602**, 307 (2001).
176. S. H. H. Tye and I. Wasserman, Phys. Rev. Lett. **86**, 1682 (2001).
177. J. Garriga and A. Vilenkin, Phys. Rev. D **64**, 023517 (2001).
178. J. Yokoyama, Phys. Rev. Lett. **88**, 151302 (2002).
179. C. P. Burgess, R. Kallosh and F. Quevedo, JHEP **0310**, 056 (2003).
180. Y. Aghababaie, C. P. Burgess, S. L. Parameswaran and F. Quevedo, Nucl. Phys. B **680**, 389 (2004).
181. S. Mukohyama and L. Randall, Phys. Rev. Lett. **92**, 211302 (2004).
182. M. Ahmed, S. Dodelson, P. B. Greene and R. Sorkin, Phys. Rev. D **69**, 103523 (2004).
183. G. L. Kane, M. J. Perry and A. N. Zytkow, Phys. Lett. B **609**, 7 (2005).
184. A. D. Dolgov and F. R. Urban, Phys. Rev. D **77**, 083503 (2008).
185. A. de la Macorra and G. Piccinelli, Phys. Rev. D **61**, 123503 (2000).
186. S. C. C. Ng, N. J. Nunes and F. Rosati, Phys. Rev. D **64**, 083510 (2001).
187. R. R. Caldwell and E. V. Linder, Phys. Rev. Lett. **95**, 141301 (2005).
188. P. Binetruy, Phys. Rev. D **60**, 063502 (1999).
189. P. Brax and J. Martin, Phys. Lett. B **468**, 40 (1999).
190. A. D. Linde, “Inflation and quantum cosmology”, in: Three hundred years of gravitation, Eds.: S. W. Hawking and W. Israel, Cambridge Univeristy Press, 604 (1987).
191. R. Kallosh et al., JCAP **0310**, 015 (2003).
192. J. A. Frieman, C. T. Hill, A. Stebbins and I. Waga, Phys. Rev. Lett. **75**, 2077 (1995).
193. Y. Nomura, T. Watari and T. Yanagida, Phys. Lett. B **484**, 103 (2000).
194. K. Choi, Phys. Rev. D **62**, 043509 (2000).
195. J. E. Kim and H. P. Nilles, Phys. Lett. B **553**, 1 (2003).
196. L. J. Hall, Y. Nomura and S. J. Oliver, Phys. Rev. Lett. **95**, 141302 (2005).

197. E. J. Copeland, N. J. Nunes and F. Rosati, Phys. Rev. D **62**, 123503 (2000).
198. P. K. Townsend, JHEP **0111**, 042 (2001).
199. S. Hellerman, N. Kaloper and L. Susskind, JHEP **0106**, 003 (2001).
200. R. Kallosh, A. D. Linde, S. Prokushkin and M. Shmakova, Phys. Rev. D **65**, 105016 (2002).
201. R. Kallosh, A. D. Linde, S. Prokushkin and M. Shmakova, Phys. Rev. D **66**, 123503 (2002).
202. P. Fre, M. Trigiante and A. Van Proeyen, Class. Quant. Grav. **19**, 4167 (2002).
203. L. A. Boyle, R. R. Caldwell and M. Kamionkowski, Phys. Lett. B **545**, 17 (2002).
204. M. Gasperini, F. Piazza and G. Veneziano, Phys. Rev. D **65**, 023508 (2002).
205. T. Damour, F. Piazza and G. Veneziano, Phys. Rev. Lett. **89**, 081601 (2002).
206. A. J. Albrecht, C. P. Burgess, F. Ravndal and C. Skordis, Phys. Rev. D **65**, 123507 (2002).
207. C. P. Burgess, C. Nunez, F. Quevedo, G. Tasinato and I. Zavala, JHEP **0308**, 056 (2003).
208. C. P. Burgess, Annals Phys. **313**, 283 (2004).
209. E. V. Linder, Phys. Rev. D **73**, 063010 (2006).
210. V. Sahni, T. D. Saini, A. A. Starobinsky and U. Alam, JETP Lett. **77**, 201 (2003).
211. V. Sahni, A. Shafieloo and A. A. Starobinsky, Phys. Rev. D **78**, 103502 (2008).
212. T. Nakamura and T. Chiba, Mon. Not. Roy. Astron. Soc. **306**, 696 (1999).
213. T. Chiba and T. Nakamura, Phys. Rev. D **62**, 121301 (2000).
214. R. Bean, S. H. Hansen and A. Melchiorri, Phys. Rev. D **64**, 103508 (2001).
215. T. Barreiro, E. J. Copeland and N. J. Nunes, Phys. Rev. D **61**, 127301 (2000).
216. V. Sahni and L. M. Wang, Phys. Rev. D **62**, 103517 (2000).
217. A. J. Albrecht and C. Skordis, Phys. Rev. Lett. **84**, 2076 (2000).
218. S. Dodelson, M. Kaplinghat and E. Stewart, Phys. Rev. Lett. **85**, 5276 (2000).
219. L. A. Urena-Lopez and T. Matos, Phys. Rev. D **62**, 081302 (2000).
220. A. A. Sen and S. Sethi, Phys. Lett. B **532**, 159 (2002).
221. D. Blais and D. Polarski, Phys. Rev. D **70**, 084008 (2004).
222. A. A. Coley and R. J. van den Hoogen, Phys. Rev. D **62**, 023517 (2000).
223. S. A. Kim, A. R. Liddle and S. Tsujikawa, Phys. Rev. D **72**, 043506 (2005).
224. A. R. Liddle, A. Mazumdar and F. E. Schunck, Phys. Rev. D **58**, 061301 (1998).
225. J. Ohashi and S. Tsujikawa, Phys. Rev. D **80**, 103513 (2009).
226. P. J. E. Peebles and A. Vilenkin, Phys. Rev. D **59**, 063505 (1999).
227. G. N. Felder, L. Kofman and A. D. Linde, Phys. Rev. D **59**, 123523 (1999).
228. M. Giovannini, Class. Quant. Grav. **16**, 2905 (1999).
229. M. Peloso and F. Rosati, JHEP **9912**, 026 (1999).
230. A. B. Kaganovich, Phys. Rev. D **63**, 025022 (2001).
231. M. Yahiro, G. J. Mathews, K. Ichiki, T. Kajino and M. Orito, Phys. Rev. D **65**, 063502 (2002).
232. K. Dimopoulos and J. W. F. Valle, Astropart. Phys. **18**, 287 (2002).
233. N. J. Nunes and E. J. Copeland, Phys. Rev. D **66**, 043524 (2002).
234. K. Dimopoulos, Phys. Rev. D **68**, 123506 (2003).
235. H. Tashiro, T. Chiba and M. Sasaki, Class. Quant. Grav. **21**, 1761 (2004).
236. M. Sami and V. Sahni, Phys. Rev. D **70**, 083513 (2004).
237. R. Rosenfeld and J. A. Frieman, JCAP **0509**, 003 (2005).
238. A. R. Liddle and L. A. Urena-Lopez, Phys. Rev. Lett. **97**, 161301 (2006).
239. I. P. Neupane, Class. Quant. Grav. **25**, 125013 (2008).
240. C. Armendariz-Picon, T. Damour and V. F. Mukhanov, Phys. Lett. B **458**, 209 (1999).
241. M. Gasperini and G. Veneziano, Astropart. Phys. **1**, 317 (1993).
242. M. Gasperini and G. Veneziano, Phys. Rept. **373**, 1 (2003).
243. S. M. Carroll, M. Hoffman and M. Trodden, Phys. Rev. D **68**, 023509 (2003).
244. J. M. Cline, S. Jeon and G. D. Moore, Phys. Rev. D **70**, 043543 (2004).
245. N. Arkani-Hamed, H. C. Cheng, M. A. Luty and S. Mukohyama, JHEP **0405**, 074 (2004).
246. F. Piazza and S. Tsujikawa, JCAP **0407**, 004 (2004).
247. M. R. Garousi, Nucl. Phys. B **584**, 284 (2000).
248. A. Sen, JHEP **0204**, 048 (2002).
249. G. W. Gibbons, Phys. Lett. B **537**, 1 (2002).
250. M. Fairbairn and M. H. G. Tytgat, Phys. Lett. B **546**, 1 (2002).
251. A. Feinstein, Phys. Rev. D **66**, 063511 (2002).

252. L. Kofman and A. D. Linde, JHEP **0207**, 004 (2002).
253. T. Padmanabhan, Phys. Rev. D **66**, 021301 (2002).
254. L. R. W. Abramo and F. Finelli, Phys. Lett. B **575**, 165 (2003).
255. J. M. Aguirregabiria and R. Lazkoz, Phys. Rev. D **69**, 123502 (2004).
256. M. R. Garousi, M. Sami and S. Tsujikawa, Phys. Rev. D **70**, 043536 (2004).
257. E. J. Copeland, M. R. Garousi, M. Sami and S. Tsujikawa, Phys. Rev. D **71**, 043003 (2005).
258. G. Calcagni and A. R. Liddle, Phys. Rev. D **74**, 043528 (2006).
259. E. Silverstein and D. Tong, Phys. Rev. D **70**, 103505 (2004).
260. M. Alishahiha, E. Silverstein and D. Tong, Phys. Rev. D **70**, 123505 (2004).
261. J. Martin and M. Yamaguchi, Phys. Rev. D **77**, 123508 (2008).
262. Z. K. Guo and N. Ohta, JCAP **0804**, 035 (2008).
263. E. J. Copeland, S. Mizuno and M. Shaeri, arXiv:1003.2881 [hep-th].
264. P. Singh, M. Sami and N. Dadhich, Phys. Rev. D **68**, 023522 (2003).
265. M. Sami and A. Toporensky, Mod. Phys. Lett. A **19**, 1509 (2004).
266. M. P. Dabrowski, T. Stachowiak and M. Szydlowski, Phys. Rev. D **68**, 103519 (2003).
267. E. Elizalde, S. Nojiri and S. D. Odintsov, Phys. Rev. D **70**, 043539 (2004).
268. Z. K. Guo, Y. S. Piao, X. M. Zhang and Y. Z. Zhang, Phys. Lett. B **608**, 177 (2005).
269. M. z. Li, B. Feng and X. m. Zhang, JCAP **0512**, 002 (2005).
270. H. Wei, R. G. Cai and D. F. Zeng, Class. Quant. Grav. **22**, 3189 (2005).
271. I. Y. Aref'eva, A. S. Koshelev and S. Y. Vernov, Phys. Rev. D **72**, 064017 (2005).
272. H. Wei and R. G. Cai, Phys. Rev. D **72**, 123507 (2005).
273. R. G. Cai and A. Wang, JCAP **0503**, 002 (2005).
274. G. B. Zhao, J. Q. Xia, M. Li, B. Feng and X. Zhang, Phys. Rev. D **72**, 123515 (2005).
275. X. F. Zhang, H. Li, Y. S. Piao and X. M. Zhang, Mod. Phys. Lett. A **21**, 231 (2006).
276. W. Zhao, Phys. Rev. D **73**, 123509 (2006).
277. Z. K. Guo, Y. S. Piao, X. Zhang and Y. Z. Zhang, Phys. Rev. D **74**, 127304 (2006).
278. S. Nojiri and S. D. Odintsov, Gen. Rel. Grav. **38**, 1285 (2006).
279. Y. f. Cai, M. z. Li, J. X. Lu, Y. S. Piao, T. t. Qiu and X. m. Zhang, Phys. Lett. B **651**, 1 (2007).
280. Y. F. Cai, E. N. Saridakis, M. R. Setare and J. Q. Xia, arXiv:0909.2776 [hep-th].
281. A. Vikman, Phys. Rev. D **71**, 023515 (2005).
282. J. Garriga and V. F. Mukhanov, Phys. Lett. B **458**, 219 (1999).
283. C. Bonvin, C. Caprini and R. Durrer, Phys. Rev. Lett. **97**, 081303 (2006).
284. M. Malquarti, E. J. Copeland, A. R. Liddle and M. Trodden, Phys. Rev. D **67**, 123503 (2003).
285. C. Wetterich, Astron. Astrophys. **301**, 321 (1995).
286. L. Amendola, Phys. Rev. D **60**, 043501 (1999).
287. L. Amendola, Phys. Rev. D **62**, 043511 (2000).
288. D. J. Holden and D. Wands, Phys. Rev. D **61**, 043506 (2000).
289. G. Huey and B. D. Wandelt, Phys. Rev. D **74**, 023519 (2006).
290. S. Das, P. S. Corasaniti and J. Khoury, Phys. Rev. D **73**, 083509 (2006).
291. C. Brans and R. H. Dicke, Phys. Rev. **124**, 925 (1961).
292. S. Tsujikawa, K. Uddin, S. Mizuno, R. Tavakol and J. Yokoyama, Phys. Rev. D **77**, 103009 (2008).
293. W. Zimdahl, D. Pavon and L. P. Chimento, Phys. Lett. **B521**, 133 (2001).
294. W. Zimdahl and D. Pavon, Gen. Rel. Grav. **35**, 413 (2003).
295. L. P. Chimento, A. S. Jakubi, D. Pavon and W. Zimdahl, Phys. Rev. D **67**, 083513 (2003).
296. N. Dalal, K. Abazajian, E. E. Jenkins and A. V. Manohar, Phys. Rev. Lett. **87**, 141302 (2001).
297. L. Amendola, G. Camargo Campos and R. Rosenfeld, Phys. Rev. D **75**, 083506 (2007).
298. Z. K. Guo, N. Ohta and S. Tsujikawa, Phys. Rev. D **76**, 023508 (2007).
299. C. G. Boehmer, G. Caldera-Cabral, R. Lazkoz and R. Maartens, Phys. Rev. D **78**, 023505 (2008).
300. J. Valiviita, E. Majerotto and R. Maartens, JCAP **0807**, 020 (2008).
301. C. G. Boehmer, G. Caldera-Cabral, N. Chan, R. Lazkoz and R. Maartens, arXiv:0911.3089 [gr-qc].
302. D. Pavon, S. Sen and W. Zimdahl, JCAP **0405**, 009 (2004).
303. R. G. Cai and A. Wang, JCAP **0503**, 002 (2005).

304. Z. K. Guo, R. G. Cai and Y. Z. Zhang, JCAP **0505**, 002 (2005).
305. M. Nishiyama, M. a. Morita and M. Morikawa, arXiv:astro-ph/0403571.
306. B. Gumjudpai, T. Naskar, M. Sami and S. Tsujikawa, JCAP **0506**, 007 (2005).
307. T. Koivisto, Phys. Rev. D **72**, 043516 (2005).
308. S. del Campo, R. Herrera, G. Olivares and D. Pavon, Phys. Rev. D **74**, 023501 (2006).
309. S. Lee, G. C. Liu and K. W. Ng, Phys. Rev. D **73**, 083516 (2006).
310. H. M. Sadjadi and M. Alimohammadi, Phys. Rev. D **74**, 103007 (2006).
311. O. Bertolami, F. Gil Pedro and M. Le Delliou, Phys. Lett. B **654**, 165 (2007).
312. R. Mainini and S. Bonometto, JCAP **0706**, 020 (2007).
313. H. Wei and S. N. Zhang, Phys. Lett. B **644**, 7 (2007).
314. T. Fukuyama, M. Morikawa and T. Tatekawa, JCAP **0806**, 033 (2008).
315. G. Caldera-Cabral, R. Maartens and L. A. Urena-Lopez, Phys. Rev. D **79**, 063518 (2009).
316. G. Caldera-Cabral, R. Maartens and B. M. Schaefer, JCAP **0907**, 027 (2009).
317. E. Majerotto, J. Valiviita and R. Maartens, Mon. Not. Roy. Astron. Soc. **402**, 2344 (2010).
318. H. Wei, arXiv:1004.0492 [gr-qc].
319. L. Amendola and D. Tocchini-Valentini, Phys. Rev. D **66**, 043528 (2002).
320. L. Amendola, Phys. Rev. D **69**, 103524 (2004).
321. J. Peebles, *The large-scale structure of the Universe*, Princeton University Press (1980).
322. L. M. Wang and P. J. Steinhardt, Astrophys. J. **508**, 483 (1998).
323. E. V. Linder, Phys. Rev. D **72**, 043529 (2005).
324. C. Di Porto and L. Amendola, Phys. Rev. D **77**, 083508 (2008).
325. L. Amendola, Phys. Rev. Lett. **93**, 181102 (2004).
326. L. Amendola, S. Tsujikawa and M. Sami, Phys. Lett. B **632**, 155 (2006).
327. S. Tsujikawa, Phys. Rev. D **72**, 083512 (2005).
328. P. Q. Hung, arXiv:hep-ph/0010126.
329. P. Gu, X. Wang and X. Zhang, Phys. Rev. D **68**, 087301 (2003).
330. R. Fardon, A. E. Nelson and N. Weiner, JCAP **0410**, 005 (2004).
331. R. D. Peccei, Phys. Rev. D **71**, 023527 (2005).
332. R. Takahashi and M. Tanimoto, Phys. Lett. B **633**, 675 (2006).
333. N. Afshordi, M. Zaldarriaga and K. Kohri, Phys. Rev. D **72**, 065024 (2005).
334. A. W. Brookfield, C. van de Bruck, D. F. Mota and D. Tocchini-Valentini, Phys. Rev. Lett. **96**, 061301 (2006).
335. A. W. Brookfield, C. van de Bruck, D. F. Mota and D. Tocchini-Valentini, Phys. Rev. D **73**, 083515 (2006).
336. L. Amendola, M. Baldi and C. Wetterich, Phys. Rev. D **78**, 023015 (2008).
337. K. Ichiki and Y. Y. Keum, JHEP **0806**, 058 (2008).
338. K. Bamba, C. Q. Geng and S. H. Ho, JCAP **0809**, 001 (2008).
339. S. Tsujikawa and M. Sami, Phys. Lett. B **603**, 113 (2004).
340. L. Amendola, M. Quartin, S. Tsujikawa and I. Waga, Phys. Rev. D **74**, 023525 (2006).
341. P. Binetruy, C. Deffayet and D. Langlois, Nucl. Phys. B **565**, 269 (2000).
342. T. Shiromizu, K. i. Maeda and M. Sasaki, Phys. Rev. D **62**, 024012 (2000).
343. L. Randall and R. Sundrum, Phys. Rev. Lett. **83**, 3370 (1999).
344. L. Randall and R. Sundrum, Phys. Rev. Lett. **83**, 4690 (1999).
345. K. i. Maeda, Phys. Rev. D **64**, 123525 (2001).
346. S. Mizuno and K. i. Maeda, Phys. Rev. D **64**, 123521 (2001).
347. E. J. Copeland, S. J. Lee, J. E. Lidsey and S. Mizuno, Phys. Rev. D **71**, 023526 (2005).
348. S. Tsujikawa, Phys. Rev. D **73**, 103504 (2006).
349. L. Amendola and D. Tocchini-Valentini, Phys. Rev. D **64**, 043509 (2001).
350. J. Khoury and A. Weltman, Phys. Rev. Lett. **93**, 171104 (2004).
351. J. Khoury and A. Weltman, Phys. Rev. D **69**, 044026 (2004).
352. T. Tamaki and S. Tsujikawa, Phys. Rev. D **78**, 084028 (2008).
353. C. M. Will, Living Rev. Rel. **4**, 4 (2001).
354. C. M. Will, Living Rev. Rel. **9**, 3 (2005).
355. P. Brax, C. van de Bruck, A. C. Davis, J. Khoury and A. Weltman, Phys. Rev. D **70**, 123518 (2004).

356. S. S. Gubser and J. Khoury, Phys. Rev. D **70**, 104001 (2004).
357. H. Wei and R. G. Cai, Phys. Rev. D **71**, 043504 (2005).
358. D. F. Mota and D. J. Shaw, Phys. Rev. Lett. **97**, 151102 (2006).
359. P. Brax, C. van de Bruck, A. C. Davis, D. F. Mota and D. J. Shaw, Phys. Rev. D **76**, 124034 (2007).
360. P. Brax, C. van de Bruck and A. C. Davis, Phys. Rev. Lett. **99**, 121103 (2007).
361. D. F. Mota and D. J. Shaw, Phys. Rev. D **75**, 063501 (2007).
362. A. C. Davis, C. A. O. Schelpe and D. J. Shaw, Phys. Rev. D **80**, 064016 (2009).
363. L. Hui, A. Nicolis and C. Stubbs, Phys. Rev. D **80**, 104002 (2009).
364. Y. Fujii et al., Nucl. Phys. B **573**, 377 (2000).
365. M. T. Murphy et al., Mon. Not. Roy. Astron. Soc. **327**, 1208 (2001).
366. J. K. Webb et al., Phys. Rev. Lett. **87**, 091301 (2001).
367. H. Chand, R. Srianand, P. Petitjean and B. Aracil, Astron. Astrophys. **417**, 853 (2004).
368. M. T. Murphy, J. K. Webb, V. V. Flambaum and S. J. Curran, Astrophys. Space Sci. **283**, 577 (2003).
369. J. D. Bekenstein, Phys. Rev. D **25**, 1527 (1982).
370. H. B. Sandvik, J. D. Barrow and J. Magueijo, Phys. Rev. Lett. **88**, 031302 (2002).
371. G. R. Dvali and M. Zaldarriaga, Phys. Rev. Lett. **88**, 091303 (2002).
372. T. Chiba and K. Kohri, Prog. Theor. Phys. **107**, 631 (2002).
373. J. P. Uzan, Rev. Mod. Phys. **75**, 403 (2003).
374. D. Parkinson, B. A. Bassett and J. D. Barrow, Phys. Lett. B **578**, 235 (2004).
375. S. Lee, K. A. Olive and M. Pospelov, Phys. Rev. D **70**, 083503 (2004).
376. E. J. Copeland, N. J. Nunes and M. Pospelov, Phys. Rev. D **69**, 023501 (2004).
377. D. F. Mota and J. D. Barrow, Phys. Lett. B **581**, 141 (2004).
378. M. R. Garousi, M. Sami and S. Tsujikawa, Phys. Rev. D **71**, 083005 (2005).
379. P. P. Avelino, L. M. G. Beca, J. P. M. de Carvalho, C. J. A. Martins and E. J. Copeland, Phys. Rev. D **69**, 041301 (2004).
380. L. Amendola, F. Finelli, C. Burigana and D. Carturan, JCAP **0307**, 005 (2003).
381. L. Amendola, I. Waga and F. Finelli, JCAP **0511**, 009 (2005).
382. R. J. Scherrer, Phys. Rev. Lett. **93**, 011301 (2004).
383. D. Giannakis and W. Hu, Phys. Rev. D **72**, 063502 (2005).
384. C. Armendariz-Picon and E. A. Lim, JCAP **0508**, 007 (2005).
385. D. Bertacca, S. Matarrese and M. Pietroni, Mod. Phys. Lett. A **22**, 2893 (2007).
386. D. Bertacca, N. Bartolo, A. Diaferio and S. Matarrese, JCAP **0810**, 023 (2008).
387. O. F. Piattella, D. Bertacca, M. Bruni and D. Pietrobon, JCAP **1001**, 014 (2010).
388. Y. Urakawa and T. Kobayashi, arXiv:0907.1191 [astro-ph.CO].
389. O. Bertolami, C. G. Boehmer, T. Harko and F. S. N. Lobo, Phys. Rev. D **75**, 104016 (2007).
390. O. Bertolami and J. Paramos, Phys. Rev. D **77**, 084018 (2008).
391. V. Faraoni, Phys. Rev. D **80**, 124040 (2009).
392. M. E. Soussa and R. P. Woodard, Gen. Rel. Grav. **36**, 855 (2004).
393. G. Allemandi, A. Borowiec and M. Francaviglia, Phys. Rev. D **70**, 103503 (2004).
394. D. A. Easson, Int. J. Mod. Phys. A **19**, 5343 (2004).
395. R. Dick, Gen. Rel. Grav. **36**, 217 (2004).
396. G. Allemandi, A. Borowiec and M. Francaviglia, Phys. Rev. D **70**, 043524 (2004).
397. S. Carloni, P. K. S. Dunsby, S. Capozziello and A. Troisi, Class. Quant. Grav. **22**, 4839 (2005).
398. A. D. Dolgov and M. Kawasaki, Phys. Lett. B **573**, 1 (2003).
399. L. Amendola, D. Polarski and S. Tsujikawa, Phys. Rev. Lett. **98**, 131302 (2007).
400. L. Amendola, D. Polarski and S. Tsujikawa, Int. J. Mod. Phys. D **16**, 1555 (2007).
401. T. Chiba, Phys. Lett. B **575**, 1 (2003).
402. G. J. Olmo, Phys. Rev. Lett. **95**, 261102 (2005).
403. G. J. Olmo, Phys. Rev. D **72**, 083505 (2005).
404. I. Navarro and K. Van Acoleyen, JCAP **0702**, 022 (2007).
405. A. L. Erickcek, T. L. Smith and M. Kamionkowski, Phys. Rev. D **74**, 121501 (2006).
406. T. Chiba, T. L. Smith and A. L. Erickcek, Phys. Rev. D **75**, 124014 (2007).

407. S. M. Carroll, I. Sawicki, A. Silvestri and M. Trodden, New J. Phys. **8**, 323 (2006).
408. Y. S. Song, W. Hu and I. Sawicki, Phys. Rev. D **75**, 044004 (2007).
409. R. Bean, D. Bernat, L. Pogosian, A. Silvestri and M. Trodden, Phys. Rev. D **75**, 064020 (2007).
410. T. Faulkner, M. Tegmark, E. F. Bunn and Y. Mao, Phys. Rev. D **76**, 063505 (2007).
411. L. Pogosian and A. Silvestri, Phys. Rev. D **77**, 023503 (2008).
412. V. Muller, H. J. Schmidt and A. A. Starobinsky, Phys. Lett. B **202**, 198 (1988).
413. V. Faraoni, Phys. Rev. D **70**, 044037 (2004).
414. L. Amendola, R. Gannouji, D. Polarski and S. Tsujikawa, Phys. Rev. D **75**, 083504 (2007).
415. W. Hu and I. Sawicki, Phys. Rev. D **76**, 064004 (2007).
416. A. A. Starobinsky, JETP Lett. **86**, 157 (2007).
417. S. Tsujikawa, Phys. Rev. D **77**, 023507 (2008).
418. S. A. Appleby and R. A. Battye, Phys. Lett. B **654**, 7 (2007).
419. E. V. Linder, Phys. Rev. D **80**, 123528 (2009).
420. B. Li and J. D. Barrow, Phys. Rev. D **75**, 084010 (2007).
421. R. Gannouji, B. Moraes and D. Polarski, JCAP **0902**, 034 (2009).
422. D. F. Torres, Phys. Rev. D **66**, 043522 (2002).
423. B. Boisseau, G. Esposito-Farese, D. Polarski and A. A. Starobinsky, Phys. Rev. Lett. **85**, 2236 (2000).
424. H. Motohashi, A. A. Starobinsky and J. Yokoyama, arXiv:1002.1141 [astro-ph.CO].
425. K. Bamba, C. Q. Geng, S. Nojiri and S. D. Odintsov, Phys. Rev. D **79**, 083014 (2009).
426. K. Bamba, arXiv:0909.2991 [astro-ph.CO].
427. A. Dev, D. Jain, S. Jhingan, S. Nojiri, M. Sami and I. Thongkool, Phys. Rev. D **78**, 083515 (2008).
428. M. Martinelli, A. Melchiorri and L. Amendola, Phys. Rev. D **79**, 123516 (2009).
429. V. F. Cardone, A. Diaferio and S. Camera, arXiv:0907.4689 [astro-ph.CO].
430. A. Ali, R. Gannouji, M. Sami and A. A. Sen, arXiv:1001.5384 [astro-ph.CO].
431. J. c. Hwang and H. r. Noh, Phys. Rev. D **65**, 023512 (2002).
432. J. c. Hwang and H. Noh, Phys. Rev. D **71**, 063536 (2005).
433. A. De Felice and S. Tsujikawa, arXiv:1002.4928 [gr-qc].
434. S. Tsujikawa, Phys. Rev. D **76**, 023514 (2007).
435. S. Tsujikawa, K. Uddin and R. Tavakol, Phys. Rev. D **77**, 043007 (2008).
436. S. A. Appleby and R. A. Battye, JCAP **0805**, 019 (2008).
437. A. de la Cruz-Dombriz, A. Dobado and A. L. Maroto, Phys. Rev. Lett. **103**, 179001 (2009).
438. A. de la Cruz-Dombriz, A. Dobado and A. L. Maroto, Phys. Rev. D **77**, 123515 (2008).
439. H. Motohashi, A. A. Starobinsky and J. Yokoyama, Int. J. Mod. Phys. D **18**, 1731 (2009).
440. S. Tsujikawa, R. Gannouji, B. Moraes and D. Polarski, Phys. Rev. D **80**, 084044 (2009).
441. P. Zhang, Phys. Rev. D **73**, 123504 (2006).
442. Y. S. Song, W. Hu and I. Sawicki, Phys. Rev. D **75**, 044004 (2007).
443. Y. S. Song, H. Peiris and W. Hu, Phys. Rev. D **76**, 063517 (2007).
444. L. Amendola, M. Kunz and D. Sapone, JCAP **0804**, 013 (2008).
445. S. Tsujikawa and T. Tatekawa, Phys. Lett. B **665**, 325 (2008).
446. F. Schmidt, Phys. Rev. D **78**, 043002 (2008).
447. A. Borisov and B. Jain, Phys. Rev. D **79**, 103506 (2009).
448. T. Narikawa and K. Yamamoto, Phys. Rev. D **81**, 043528 (2010).
449. H. F. Stabenau and B. Jain, Phys. Rev. D **74**, 084007 (2006).
450. W. Hu and I. Sawicki, Phys. Rev. D **76**, 104043 (2007).
451. H. Oyaizu, Phys. Rev. D **78**, 123523 (2008).
452. H. Oyaizu, M. Lima and W. Hu, Phys. Rev. D **78**, 123524 (2008).
453. F. Schmidt, M. V. Lima, H. Oyaizu and W. Hu, Phys. Rev. D **79**, 083518 (2009).
454. I. Laszlo and R. Bean, Phys. Rev. D **77**, 024048 (2008).
455. T. Tatekawa and S. Tsujikawa, JCAP **0809**, 009 (2008).
456. K. Koyama, A. Taruya and T. Hiramatsu, Phys. Rev. D **79**, 123512 (2009).
457. F. Schmidt, A. Vikhlinin and W. Hu, Phys. Rev. D **80**, 083505 (2009).
458. K. i. Maeda, Phys. Rev. D **39**, 3159 (1989).

459. S. Capozziello and S. Tsujikawa, Phys. Rev. D **77**, 107501 (2008).
460. A. V. Frolov, Phys. Rev. Lett. **101**, 061103 (2008).
461. S. Appleby, R. Battye and A. Starobinsky, arXiv:0909.1737 [astro-ph.CO].
462. V. Miranda, S. E. Joras, I. Waga and M. Quartin, Phys. Rev. Lett. **102**, 221101 (2009).
463. I. Thongkool, M. Sami, R. Gannouji and S. Jhingan, Phys. Rev. D **80**, 043523 (2009).
464. A. de la Cruz-Dombriz, A. Dobado and A. L. Maroto, Phys. Rev. Lett. **103**, 179001 (2009).
465. T. Kobayashi and K. i. Maeda, Phys. Rev. D **78**, 064019 (2008).
466. T. Kobayashi and K. i. Maeda, Phys. Rev. D **79**, 024009 (2009).
467. S. Tsujikawa, T. Tamaki and R. Tavakol, JCAP **0905**, 020 (2009).
468. E. Babichev and D. Langlois, Phys. Rev. D **80**, 121501 (2009).
469. A. Upadhye and W. Hu, Phys. Rev. D **80**, 064002 (2009).
470. E. Babichev and D. Langlois, arXiv:0911.1297 [gr-qc].
471. A. Palatini, Rend. Circ. Mat. Palermo 43, 203 (1919).
472. M. Ferraris, M. Francaviglia and I. Volovich, Class. Quant. Grav. **11**, 1505 (1994).
473. D. N. Vollick, Phys. Rev. D **68**, 063510 (2003).
474. D. N. Vollick, Class. Quant. Grav. **21**, 3813 (2004).
475. E. E. Flanagan, Class. Quant. Grav. **21**, 417 (2003).
476. E. E. Flanagan, Phys. Rev. Lett. **92**, 071101 (2004).
477. E. E. Flanagan, Class. Quant. Grav. **21**, 3817 (2004).
478. F. W. Hehl and G. D. Kerling, Gen. Rel. Grav. **9**, 691 (1978).
479. T. P. Sotiriou and S. Liberati, Annals Phys. **322**, 935 (2007).
480. T. P. Sotiriou and S. Liberati, J. Phys. Conf. Ser. **68**, 012022 (2007).
481. S. Capozziello, R. Cianci, C. Stornaiolo and S. Vignolo, Class. Quant. Grav. **24**, 6417 (2007).
482. T. P. Sotiriou and V. Faraoni, arXiv:0805.1726 [gr-qc].
483. X. Meng and P. Wang, Class. Quant. Grav. 20, 4949 (2003).
484. X. Meng and P. Wang, Class. Quant. Grav. **21**, 951 (2004).
485. X. H. Meng and P. Wang, Phys. Lett. B **584**, 1 (2004).
486. T. P. Sotiriou, Class. Quant. Grav. **23**, 1253 (2006).
487. T. P. Sotiriou, Phys. Rev. D **73**, 063515 (2006).
488. M. Amarzguioui, O. Elgaroy, D. F. Mota and T. Multamaki, Astron. Astrophys. **454**, 707 (2006).
489. S. Fay, R. Tavakol and S. Tsujikawa, Phys. Rev. D **75**, 063509 (2007).
490. T. Koivisto, Phys. Rev. D **73**, 083517 (2006).
491. B. Li and M. C. Chu, Phys. Rev. D **74**, 104010 (2006).
492. B. Li, K. C. Chan and M. C. Chu, Phys. Rev. D **76**, 024002 (2007).
493. A. Iglesias, N. Kaloper, A. Padilla and M. Park, Phys. Rev. D **76**, 104001 (2007).
494. G. J. Olmo, Phys. Rev. Lett. **98**, 061101 (2007).
495. G. J. Olmo, Phys. Rev. D **77**, 084021 (2008).
496. G. J. Olmo, arXiv:0910.3734 [gr-qc].
497. E. Barausse, T. P. Sotiriou and J. C. Miller, Class. Quant. Grav. **25**, 105008 (2008).
498. E. Barausse, T. P. Sotiriou and J. C. Miller, Class. Quant. Grav. **25**, 062001 (2008).
499. G. J. Olmo and P. Singh, JCAP **0901**, 030 (2009).
500. C. Barragan, G. J. Olmo and H. Sanchis-Alepuz, Phys. Rev. D **80**, 024016 (2009).
501. V. Reijonen, arXiv:0912.0825 [gr-qc].
502. S. M. Carroll, A. De Felice, V. Duvvuri, D. A. Easson, M. Trodden and M. S. Turner, Phys. Rev. D **71**, 063513 (2005).
503. O. Mena, J. Santiago and J. Weller, Phys. Rev. Lett. **96**, 041103 (2006).
504. A. Nunez and S. Solganik, Phys. Lett. B **608**, 189 (2005).
505. G. Calcagni, S. Tsujikawa and M. Sami, Class. Quant. Grav. **22**, 3977 (2005).
506. A. De Felice, M. Hindmarsh and M. Trodden, JCAP **0608**, 005 (2006).
507. G. Calcagni, B. de Carlos and A. De Felice, Nucl. Phys. B **752**, 404 (2006).
508. S. Nojiri, S. D. Odintsov and M. Sasaki, Phys. Rev. D **71**, 123509 (2005).
509. T. Koivisto and D. F. Mota, Phys. Lett. B **644**, 104 (2007).
510. S. Tsujikawa and M. Sami, JCAP **0701**, 006 (2007).
511. T. Koivisto and D. F. Mota, Phys. Rev. D **75**, 023518 (2007).

512. B. M. Leith and I. P. Neupane, JCAP **0705**, 019 (2007).
513. Z. K. Guo, N. Ohta and S. Tsujikawa, Phys. Rev. D **75**, 023520 (2007).
514. L. Amendola, C. Charmousis and S. C. Davis, JCAP **0612**, 020 (2006).
515. S. Nojiri and S. D. Odintsov, Phys. Lett. B **631**, 1 (2005).
516. B. Li, J. D. Barrow and D. F. Mota, Phys. Rev. D **76**, 044027 (2007).
517. G. Cognola, E. Elizalde, S. Nojiri, S. Odintsov and S. Zerbini, Phys. Rev. D **75**, 086002 (2007).
518. A. De Felice and M. Hindmarsh, JCAP **0706**, 028 (2007).
519. S. C. Davis, arXiv:0709.4453 [hep-th].
520. A. De Felice and S. Tsujikawa, Phys. Lett. B **675**, 1 (2009).
521. S. Y. Zhou, E. J. Copeland and P. M. Saffin, JCAP **0907**, 009 (2009).
522. K. Uddin, J. E. Lidsey and R. Tavakol, Gen. Rel. Grav. **41**, 2725 (2009).
523. A. De Felice and S. Tsujikawa, Phys. Rev. D **80**, 063516 (2009).
524. A. De Felice, D. F. Mota and S. Tsujikawa, Phys. Rev. D **81**, 023532 (2010).
525. J. O'Hanlon, Phys. Rev. Lett. **29**, 137 (1972).
526. P. Teyssandier and Ph. Tourrenc, J. Math. Phys. **24** 2793 (1983).
527. J. P. Uzan, Phys. Rev. D **59**, 123510 (1999).
528. T. Chiba, Phys. Rev. D **60**, 083508 (1999).
529. N. Bartolo and M. Pietroni, Phys. Rev. D **61**, 023518 (2000).
530. F. Perrotta, C. Baccigalupi and S. Matarrese, Phys. Rev. D **61**, 023507 (2000).
531. C. Baccigalupi, S. Matarrese and F. Perrotta, Phys. Rev. D **62**, 123510 (2000).
532. A. Riazuelo and J. P. Uzan, Phys. Rev. D **66**, 023525 (2002).
533. L. Perivolaropoulos, JCAP **0510**, 001 (2005).
534. S. Nesseris and L. Perivolaropoulos, JCAP **0701**, 018 (2007).
535. J. Martin, C. Schimd and J. P. Uzan, Phys. Rev. Lett. **96**, 061303 (2006).
536. R. Gannouji, D. Polarski, A. Ranquet and A. A. Starobinsky, JCAP **0609**, 016 (2006).
537. V. Acquaviva and L. Verde, JCAP **0712**, 001 (2007).
538. Y. S. Song, L. Hollenstein, G. Caldera-Cabral and K. Koyama, arXiv:1001.0969 [astro-ph.CO].
539. P. Zhang, M. Liguori, R. Bean and S. Dodelson, Phys. Rev. Lett. **99**, 141302 (2007).
540. S. Wang, L. Hui, M. May and Z. Haiman, Phys. Rev. D **76**, 063503 (2007).
541. B. Jain and P. Zhang, Phys. Rev. D **78**, 063503 (2008).
542. S. F. Daniel, R. R. Caldwell, A. Cooray and A. Melchiorri, Phys. Rev. D **77**, 103513 (2008).
543. E. Bertschinger and P. Zukin, Phys. Rev. D **78**, 024015 (2008).
544. G. B. Zhao, L. Pogosian, A. Silvestri and J. Zylberberg, Phys. Rev. D **79**, 083513 (2009).
545. Y. S. Song and K. Koyama, JCAP **0901**, 048 (2009).
546. Y. S. Song and O. Dore, JCAP **0903**, 025 (2009).
547. J. Guzik, B. Jain and M. Takada, Phys. Rev. D **81**, 023503 (2010).
548. P. Brax, C. van de Bruck, A. C. Davis and D. Shaw, arXiv:0912.0462 [astro-ph.CO].
549. S. F. Daniel, E. V. Linder, T. L. Smith, R. R. Caldwell, A. Cooray, A. Leauthaud and L. Lombriser, arXiv:1002.1962 [astro-ph.CO].
550. R. Bean and M. Tangmatitham, arXiv:1002.4197 [astro-ph.CO].
551. G. B. Zhao et al., arXiv:1003.0001 [astro-ph.CO].
552. C. Deffayet, Phys. Lett. B **502** (2001), 199.
553. C. Deffayet, G. R. Dvali and G. Gabadadze, Phys. Rev. D **65** (2002), 044023.
554. V. Sahni and Y. Shtanov, JCAP **0311**, 014 (2003).
555. K. Hinterbichler, A. Nicolis and M. Porrati, JHEP **0909**, 089 (2009).
556. I. Sawicki and S. M. Carroll, arXiv:astro-ph/0510364.
557. M. Fairbairn and A. Goobar, Phys. Lett. B **642**, 432 (2006).
558. R. Maartens and E. Majerotto, Phys. Rev. D **74**, 023004 (2006).
559. U. Alam and V. Sahni, Phys. Rev. D **73**, 084024 (2006).
560. Y. S. Song, I. Sawicki and W. Hu, Phys. Rev. D **75**, 064003 (2007).
561. J. Q. Xia, Phys. Rev. D **79**, 103527 (2009).
562. G. Dvali and M. S. Turner, arXiv:astro-ph/0301510.
563. A. Lue, R. Scoccimarro and G. D. Starkman, Phys. Rev. D **69**, 124015 (2004).

564. K. Koyama and R. Maartens, JCAP **0601**, 016 (2006).
565. S. M. Carroll, I. Sawicki, A. Silvestri and M. Trodden, New J. Phys. **8**, 323 (2006).
566. K. Yamamoto, B. A. Bassett, R. C. Nichol and Y. Suto, Phys. Rev. D **74**, 063525 (2006).
567. K. Koyama and F. P. Silva, Phys. Rev. D **75**, 084040 (2007).
568. D. Huterer and E. V. Linder, Phys. Rev. D **75**, 023519 (2007).
569. H. Wei, Phys. Lett. B **664**, 1 (2008).
570. Y. Gong, M. Ishak and A. Wang, Phys. Rev. D **80**, 023002 (2009).
571. X. y. Fu, P. x. Wu and H. w. Yu, Phys. Lett. B **677**, 12 (2009).
572. A. I. Vainshtein, Phys. Lett. B **39**, 393 (1972).
573. C. Deffayet, G. R. Dvali, G. Gabadadze and A. I. Vainshtein, Phys. Rev. D **65**, 044026 (2002).
574. A. Gruzinov, New Astron. **10**, 311 (2005).
575. M. Porrati, Phys. Lett. B **534**, 209 (2002).
576. M. A. Luty, M. Porrati and R. Rattazzi, JHEP **0309**, 029 (2003).
577. A. Nicolis, R. Rattazzi and E. Trincherini, Phys. Rev. D **79**, 064036 (2009).
578. C. Deffayet, G. Esposito-Farese and A. Vikman, Phys. Rev. D **79**, 084003 (2009).
579. C. Deffayet, S. Deser and G. Esposito-Farese, Phys. Rev. D **80**, 064015 (2009).
580. F. P. Silva and K. Koyama, Phys. Rev. D **80**, 121301 (2009).
581. N. Chow and J. Khoury, Phys. Rev. D **80**, 024037 (2009).
582. T. Kobayashi, H. Tashiro and D. Suzuki, Phys. Rev. D **81**, 063513 (2010).
583. T. Kobayashi, arXiv:1003.3281 [astro-ph.CO].
584. T. Buchert, Gen. Rel. Grav. **40**, 467 (2008).
585. R. A. Vanderveld, E. E. Flanagan and I. Wasserman, Phys. Rev. D **74**, 023506 (2006).
586. A. Paranjape and T. P. Singh, Class. Quant. Grav. **23**, 6955 (2006).
587. J. W. Moffat, JCAP **0605**, 001 (2006).
588. D. Garfinkle, Class. Quant. Grav. **23**, 4811 (2006).
589. D. J. H. Chung and A. E. Romano, Phys. Rev. D **74**, 103507 (2006).
590. H. Alnes and M. Amarzguioui, Phys. Rev. D **74**, 103520 (2006).
591. H. Alnes, M. Amarzguioui and O. Gron, JCAP **0701**, 007 (2007).
592. H. Alnes and M. Amarzguioui, Phys. Rev. D **75**, 023506 (2007).
593. T. Kai, H. Kozaki, K. i. Nakao, Y. Nambu and C. M. Yoo, Prog. Theor. Phys. **117**, 229 (2007).
594. K. Enqvist and T. Mattsson, JCAP **0702**, 019 (2007).
595. T. Biswas, R. Mansouri and A. Notari, JCAP **0712**, 017 (2007).
596. N. Brouzakis, N. Tetradis and E. Tzavara, JCAP **0702**, 013 (2007).
597. M. Tanimoto and Y. Nambu, Class. Quant. Grav. **24**, 3843 (2007).
598. J. Garcia-Bellido and T. Haugboelle, JCAP **0804**, 003 (2008).
599. J. Garcia-Bellido and T. Haugboelle, JCAP **0809**, 016 (2008).
600. T. Clifton, P. G. Ferreira and K. Land, Phys. Rev. Lett. **101**, 131302 (2008).
601. S. Alexander, T. Biswas, A. Notari and D. Vaid, JCAP **0909**, 025 (2009).
602. C. Clarkson, T. Clifton and S. February, JCAP **0906**, 025 (2009).
603. K. Kainulainen and V. Marra, Phys. Rev. D **80**, 127301 (2009).
604. R. R. Caldwell and A. Stebbins, Phys. Rev. Lett. **100**, 191302 (2008).
605. C. Quercellini, M. Quartin and L. Amendola, Phys. Rev. Lett. **102**, 151302 (2009).
606. M. Quartin and L. Amendola, Phys. Rev. D **81**, 043522 (2010).
607. A. E. Romano and M. Sasaki, arXiv:0905.3342 [astro-ph.CO].
608. M. N. Celerier, K. Bolejko, A. Krasinski and C. Hellaby, arXiv:0906.0905 [astro-ph.CO].
609. P. Dunsby, N. Goheer, B. Osano and J. P. Uzan, arXiv:1002.2397 [astro-ph.CO].
610. K. Saito, A. Ishibashi and H. Kodama, arXiv:1002.3855 [astro-ph.CO].
611. F. Piazza, arXiv:0904.4299 [hep-th].
612. F. Piazza, New J. Phys. **11**, 113050 (2009).
613. S. Nesseris, F. Piazza and S. Tsujikawa, arXiv:0910.3949 [astro-ph.CO].
614. C. M. Hirata and U. Seljak, Phys. Rev. D **72**, 083501 (2005).
615. P. Martineau and R. H. Brandenberger, Phys. Rev. D **72**, 023507 (2005).
616. A. Ishibashi and R. M. Wald, Class. Quant. Grav. **23**, 235 (2006).
617. T. Buchert, Class. Quant. Grav. **23**, 817 (2006).
618. T. Buchert, J. Larena and J. M. Alimi, Class. Quant. Grav. **23**, 6379 (2006).

619. S. Rasanen, JCAP **0611**, 003 (2006).
620. V. Marra, E. W. Kolb, S. Matarrese and A. Riotto, Phys. Rev. D **76**, 123004 (2007).
621. M. Kasai, H. Asada and T. Futamase, Prog. Theor. Phys. **115**, 827 (2006).
622. M. Kasai, Prog. Theor. Phys. **117**, 1067 (2007).
623. D. L. Wiltshire, New J. Phys. **9**, 377 (2007).
624. J. Behrend, I. A. Brown and G. Robbers, JCAP **0801**, 013 (2008).
625. A. Paranjape, Phys. Rev. D **78**, 063522 (2008).
626. J. Larena, J. Alimi, T. Buchert, M. Kunz and P. S. Corasaniti, Phys. Rev. D **79**, 083011 (2009).
627. M. Gasperini, G. Marozzi and G. Veneziano, JCAP **0903**, 011 (2009).
628. E. W. Kolb, V. Marra and S. Matarrese, arXiv:0901.4566 [astro-ph.CO].

Index